Volume IV
Solvents 2

T0187472

Handbook of Environmental

FATE
and
EXPOSURE
DATA

For Organic Chemicals

Editor

Philip H. Howard

Associate Editors

Edward M. Michalenko   William M. Meylan

Dipak K. Basu   Julie A. Beauman

Gloria W. Sage   William F. Jarvis

D. Anthony Gray

CRC Press

Taylor & Francis Group

Boca Raton London New York

CRC Press is an imprint of the
Taylor & Francis Group, an informa business

First published 1993 by Lewis Publishers

Published 2019 by CRC Press
Taylor & Francis Group
6000 Broken Sound Parkway NW, Suite 300
Boca Raton, FL 33487-2742

© 1993 by Taylor & Francis Group, LLC
CRC Press is an imprint of Taylor & Francis Group, an Informa business

First issued in paperback 2019

No claim to original U.S. Government works

ISBN 13: 978-0-367-45000-7 (pbk)
ISBN 13: 978-0-87371-413-6 (hbk)

Visit the Taylor & Francis Web site at
http://www.taylorandfrancis.com

and the CRC Press Web site at
http://www.crcpress.com

Library of Congress Cataloging-in-Publication Data

Howard, Philip H. (Philip Hall), 1943-
    Handbook of environmental fate and exposure data
for organic chemicals.

    Includes bibliographical references and indexes.
    Contents: v. 1. Large production and priority
pollutants -- v. 2. Solvents -- v. 3. Pesticides. -- v. 4. Solvents and
Chemical Intermediates.
    1. Pollutants--Handbooks, manuals, etc.   2. Environmental
chemistry--Handbooks, manuals, etc.
I. Title.
TD176.4.H69 1989   363.7'38   89-2436

ISBN 13: 978-0-87371-151-7 (v. 1)
ISBN 13: 978-0-87371-204-0 (v. 2)
ISBN 13: 978-0-87371-328-3 (v. 3)
ISBN 13: 978-0-87371-413-6 (v. 4)

Associate Editors for Volume IV

The following individuals from the Syracuse Research Corporation's Chemical Hazard Assessment Division either were authors of the individual chemical records prepared for the Hazardous Substances Data Bank or edited the expanded and updated chemical chapters in this volume. The order of names, which will vary in each volume, is by the number of chemicals for which the individual was responsible.

Edward M. Michalenko, Ph.D.

Dipak K. Basu, Ph.D.

Gloria W. Sage, Ph.D.

William M. Meylan

Julie A. Beauman

William F. Jarvis, Ph.D.

D. Anthony Gray, Ph.D.

Philip H. Howard joined Syracuse Research Corporation in 1970 and has served as project director for numerous environmental fate and effects projects for federal agencies and industry. Dr. Howard's current research projects include development of structure/biodegradability correlations, development of estimation techniques for environmental fate physical properties and rate constants, and databases of information to support these efforts. He received a B.S. degree in chemistry from Norwich University in 1965 and a Ph.D. in organic chemistry from Syracuse University in 1970.

Edward M. Michalenko joined Syracuse Research Corporation in 1988 where he writes documents on the fate and transport of chemicals in the environment based on review of the available literature. Dr. Michalenko also co-authored a book of chemical degradation rates in environmental media. His current research includes assessing the aquatic fate of biocides used for controlling the zebra mussel in cooling waters. He received a B.S. in biology/chemistry education from SUNY College at Cortland in 1979, an M.S. in environmental communication/education and a Ph.D. in soil ecology from SUNY College of Environmental Science and Forestry in 1984 and 1991, respectively.

Dipak K. Basu joined Syracuse Research Corporation in 1975. As the principal investigator of projects funded by the U.S. EPA, Dr. Basu developed analytical methods for the determination of micropollutants in drinking water. He has a number of publications in analytical chemistry and toxicology. Dr. Basu has co-authored several documents for the Agency of Toxic Substances and Disease Registry, U.S. EPA and National Institute for Occupational Safety and Health. He received his M.S. degree from Calcutta University, India in 1959 and Ph.D. with specialization in physical photochemistry from Case Western Reserve University in 1970.

Gloria W. Sage joined Syracuse Research Corporation in 1980, at which time she was the codeveloper of the Environmental Fates Data Bases, DATALOG and CHEMFATE. Later on, she developed the environmental fate section of the Hazardous Substances Data Bank and contributed profiles to this section. Dr. Sage has been involved in many projects at Syracuse Research Corporation concerned with assessing the environmental fate of chemicals, making exposure assessments, and developing information tracking systems. She is also interested in using mosses to measure atmospheric deposition of metals and organics. Dr. Sage holds an A.B. with distinction and honors in chemistry from Cornell University, an A.M. in chemistry from Radcliffe College, and a Ph.D. in physical chemistry from Harvard University.

William M. Meylan joined Syracuse Research Corporation in 1975 and has participated as an environmental scientist in many environmental fate projects. He has contributed numerous profiles to the environmental fate sections of the Hazardous Substances Data Bank and co-authored many Health and Environmental Effects Documents for the U.S. EPA. Mr. Meylan is currently involved in the computer programming of estimation methods useful for predicting environmental reaction rates and physical properties. He received his B.S. degree in chemical engineering from Clarkson University in 1972.

Julie A. Beauman joined Syracuse Research Corporation in 1988 where she is currently assisting in the evaluation of chemical fate information and the development of information profiles for the TSCA Interagency Testing Committee. Ms. Beauman has interpreted data for the Environmental Fate Data Bases including the Fate/Exposure file of production, use and monitoring data and the Physprop file of structures and physical properties. She received her A.A.S in Chemical Engineering Technology from Broome Community College in 1986 and her B.S. in chemistry from SUNY Albany in 1988.

William F. Jarvis joined Syracuse Research Corporation in 1985 and has participated in the preparation of critical and comprehensive literature reviews on the environmental fate and effects associated with environmental pollutants. Dr. Jarvis' current research projects include risk evaluation projects funded by the U.S. EPA and Agency of Toxic Substances and Disease Registry. He received a B.S. degree in chemistry from SUNY College at Cortland in 1974 and a Ph.D. in organic chemistry from Syracuse University in 1988.

D. Anthony Gray joined Syracuse Research Corporation in 1979 where he currently is the Director of the Exposure Modeling Center. Since joining, Dr. Gray has served in a number of capacities including technical writer, project manager, and program manager. Dr. Gray's research interests include the assessment of exposure to multiple chemicals at hazardous waste sites by modeling their fate in soil, water, and air, and determining the most commonly found mixtures of chemicals occurring at hazardous waste sites. He received his B.S. degree in chemistry from Drexel University in 1973 and his Ph.D. in organic chemistry from Syracuse University in 1981.

Preface

Many articles and books have been written on how to review the environmental fate and exposure of organic chemicals (e.g., [11] and [19] - citations at end of Explanation of Data). Although these articles and books often give examples of the fate and exposure of several chemicals, rarely do they attempt to review large numbers of chemicals. These "how to" guides provide considerable insight into ways of estimating and using physical/chemical properties as well as mechanisms of environmental transport and transformation. However, when it comes to reviewing the fate and exposure of individual chemicals, there are discretionary factors that significantly affect the overall fate assessment. For example, is it reasonable to use regression equations for estimating soil or sediment adsorption for aromatic amine compounds? Is chemical oxidation likely to be important for phenols in surface waters? These discretionary factors are dependent upon the available data on the individual chemical or, when data are lacking, on chemicals of related structures.

This series of books outlines in detail how individual chemicals are released, transported, and degraded in the environment and how they are exposed to humans and environmental organisms. It is devoted to the review and evaluation of the available data on physical/chemical properties, commercial use and possible sources of environmental contamination, environmental fate, and monitoring data of individual chemicals. Each review of a chemical provides most of the data necessary for either a qualitative or quantitative exposure assessment.

Chemicals were selected from a large number of chemicals prepared by Syracuse Research Corporation (SRC) for inclusion in the National Library of Medicine's (NLM) Hazardous Substances Data Bank (HSDB). Chemicals selected for the first two volumes were picked from lists of high volume commercial chemicals, priority pollutants, and solvents. The chemicals in the first two volumes include most of the nonpesticidal priority pollutants and many of the chemicals on priority lists for a variety of environmental regulations (e.g., RCRA and CERCLA Reportable Quantities, Superfund, SARA). Pesticides selected for Volume III were identified from a list of significantly produced pesticides (Resources for the Future, National Pesticide Usage Data Base, 1986) and the priority pollutants. Chemicals selected for Volume IV were mostly solvents and chemical intermediates. More solvents, chemical intermediates, plasticizers, polycyclic aromatic hydrocarbons, pesticides, and other groups of chemicals will be included in later volumes.

The chemicals are listed in strict alphabetical order by the name considered to be the most easily recognized. Prefixes commonly used in organic chemistry which are not normally considered part of the name, such as ortho-, meta-, para-, α-, β-, γ-, n-, sec-, tert-, cis-, trans-, N-, as well as all numbers, have not been considered for alphabetical order. Other prefixes which normally are considered part of the name, such as iso-, di-, tri-, tetra-, and cyclo-, are used for alphabetical positioning. For example, 2,4-Dinitrotoluene is under D and tert-Butyl alcohol is under B. In addition, cumulative indices are provided at the end of each volume to allow the reader to find a given chemical by chemical name synonym, Chemical Abstracts Services (CAS) number, and chemical formula.

Acknowledgments

The following authors of the initial chemical records for the Hazardous Substances Data Bank were or are staff scientists with the Chemical Hazard Assessment Division of the Syracuse Research Corporation: Gloria W. Sage, William M. Meylan, William F. Jarvis, Erin K. Crosbie, Jeffery Jackson, Amy E. Hueber, Jeffrey Robinson, Edward M. Michalenko, Jay Tunkel, and Julie A. Beauman. We wish to thank several individuals at the National Library of Medicine (NLM) for their encouragement and support during the project. Special thanks go to our project officer, Vera Hudson, and to Bruno Vasta and Dalton Tidwell.

Explanation of Data

In the following outline, each field covered for the individual chemicals is reviewed with such information as the importance of the data, the type of data included in each field, how data are usually handled, and data sources. For each chemical, the physical properties as well as the environmental fate and monitoring data were identified by conducting searches of the Environmental Fate Data Bases of Syracuse Research Corporation (SRC) [13].

SUBSTANCE IDENTIFICATION

Synonyms: Only synonym names used fairly frequently were included.

Structure: Chemical structure.

CAS Registry Number: This number is assigned by the American Chemical Society's Chemical Abstracts Services as a unique identifier.

Molecular Formula: The formula is in Hill notation, which is given as the number of carbons followed by the number of hydrogens followed by any other elements in alphabetic order.

Wiswesser Line Notation: This is a chemical structure representation that can be used for substructure searching. It was designed back when computer notations had to fit into 80 characters and, therefore, is very abbreviated (e.g., Q is used for a benzene ring).

CHEMICAL AND PHYSICAL PROPERTIES

The Hazardous Substances Data Bank (HSDB) of the National Library of Medicine was used as a source of boiling points, melting points, and molecular weights. The dissociation constant, octanol/water partition coefficient, water solubility, vapor pressure, and Henry's Law constant were judiciously selected from the many values that were identified in SRC's DATALOG file. All values selected were referenced to the primary literature source when possible.

Boiling Point: The boiling point or boiling point range is given along with the pressure. When the pressure is not given it should be assumed that the value is at 760 mm Hg.

Melting Point: The melting point or melting point range is given.

Molecular Weight: The molecular weight to two decimal points is given.

Dissociation Constants: The acid dissociation constant as the negative log (pKa) is given for chemicals that are likely to dissociate at environmental pHs (between 5 and 9). Chemical classes where dissociation is important include, for example, phenols, carboxylic acids, and aliphatic and aromatic amines. Once the pKa is known, the percent in the dissociated and undissociated form can be determined. For example, for an acid with a pKa of 4.75, the following is true at different pHs:

> 1% dissociated at pH 2.75
> 10% dissociated at pH 3.75
> 50% dissociated at pH 4.75
> 90% dissociated at pH 5.75
> 99% dissociated at pH 6.75

The degree of dissociation affects such processes as photolysis (absorption spectra of chemicals that dissociate can be considerably affected by the pH), evaporation from water (ions do not evaporate), soil or sediment adsorption, and bioconcentration. Values from evaluated sources such as Perrin [22] and Serjeant and Dempsey [24] were used when available.

Log Octanol/Water Partition Coefficient: The octanol/water partition coefficient is the ratio of the chemical concentration in octanol divided by the concentration in water. The most reliable source of values is from the Medchem project at Pomona College [9]. When experimental values are unavailable, estimated values have been provided using a fragment constant estimation method, CLOGP3, from Medchem. Occasionally, chemical octanol/water partition coefficients were not calculated because a necessary fragment constant for the chemical was not available. The octanol/water partition coefficient has been shown to correlate well with bioconcentration factors in aquatic organisms [27] and adsorption to soil or sediment [14], and recommended regression equations have been reviewed [16].

Water Solubility: The water solubility of a chemical provides considerable insight into the fate and transport of a chemical in the environment. High water soluble chemicals, which have a tendency to remain dissolved in the water column and not partition to soil or sediment or bioconcentrate in aquatic organisms, are less likely to volatilize from water (depending upon the vapor pressure - see Henry's Law constant) and are generally more likely to biodegrade. Low water soluble chemicals are just the opposite; they partition to soil or sediments and bioconcentrate in aquatic organisms, volatilize more readily from water, and are less likely to be biodegradable. Other fate processes that are, or can be, affected by water solubility include photolysis, hydrolysis, oxidation, and washout from the atmosphere by rain or fog. Water solubility values were taken from either the ARIZONA dATABASE [28] or from SRC's DATALOG or CHEMFATE files. The values were reported in ppm at a temperature at or as close as possible to 25 °C. Occasionally, when no values were available, the value was estimated from the octanol/water partition coefficient using recommended regression equations [16].

Vapor Pressure: The vapor pressure of a chemical provides considerable insight into the transport of a chemical in the environment. The volatility of the pure chemical is dependent upon the vapor pressure, and volatilization from water is dependent upon the vapor pressure and water solubility (see Henry's Law constant). The form in which a chemical will be found in the atmosphere is dependent upon the vapor pressure; chemicals with a vapor pressure less than 10^{-6} mm Hg will be mostly found associated with particulate matter [8]. When available, sources such as Boublik et al [4], Riddick et al [23], and Daubert and Danner [6] were used, since the data in these sources were evaluated and some of them provided recommended values. Vapor pressure was reported in mm Hg at or as close as possible to 25 °C. In many cases, the vapor pressure was calculated from a vapor pressure/temperature equation.

Henry's Law Constant: The Henry's Law constant, H, is really the air/water partition coefficient, and therefore a nondimensional H relates the chemical concentration in the gas phase to its concentration in the water phase. The dimensional H can be determined by dividing the vapor pressure in atm by the water solubility in mole/m^3 to give H in atm-m^3/mole. H provides an indication of the partition between air and water at equilibrium and also is used to calculate the rate of evaporation from water (see discussion under Evaporation from Water/Soil). Henry's Law constants can be directly measured, calculated from the water solubility and vapor pressure, or estimated from structure using the method of Hine and Mookerjee [10], and this same order was used in selecting values. Some critical review data on Henry's Law constants are available (e.g., [17]).

ENVIRONMENTAL FATE/EXPOSURE POTENTIAL

Data for the following sections were identified with SRC's Environmental Fate Data Bases. Biodegradation data were selected from the DATALOG, BIOLOG, and BIODEG files. Abiotic degradation data were identified in the Hydrolysis, Photolysis, and Oxidation fields in DATALOG and CHEMFATE. Transport processes such as Bioconcentration, Soil Adsorption/Mobility, and Volatilization, as well as the monitoring data, were also identified in the DATALOG and CHEMFATE files.

Summary: This section is an abbreviated summary of all the data presented in the following sections and is not referenced; to find the citations the reader should refer to appropriate sections that follow. In general, this summary discusses how a chemical is used and released to the environment, how the chemical will behave in soil, water, and air, and how exposure to humans and environmental organisms is likely to occur.

Natural Sources: This section reviews any evidence that the chemical may have any natural sources of pollution, such as forest fires and volcanos, or may be a natural product that would lead to its detection in various media (e.g., methyl iodide is found in marine algae and is the major source of contamination in the ocean).

Artificial Sources: This section is a general review of any evidence that the chemical has anthropogenic sources of pollution. Quantitative data are reviewed in detail in Effluent Concentrations; this section provides a qualitative review of various sources based upon how the chemical is manufactured and used as well as the physical/chemical properties. For example, it is reasonable to assume that a highly volatile chemical which is used mostly as a solvent will be released to the atmosphere as well as the air of occupational settings, even if no monitoring data are available. Information on production volume and uses was obtained from a variety of chemical marketing sources including the Kirk-Othmer Encyclopedia of Chemical Technology, SRI International's Chemical Economics Handbook, and the Chemical Profiles of the Chemical Marketing Reporter.

Terrestrial Fate: This section reviews how a chemical will behave if released to soil or ground water. Field studies or terrestrial model ecosystems studies are used here when they provide insight into the overall behavior in soil. Studies which determine an individual process (e.g., biodegradation, hydrolysis, soil adsorption) in soil are reviewed in the appropriate sections that follow. Quite often, except with pesticides, field or terrestrial ecosystem studies either are not available or do not give enough data to make conclusions on the terrestrial fate of a chemical. In these cases, data from the sections on Biodegradation, Abiotic Degradation, Soil Adsorption/Mobility, Volatilization from Water/Soil, and any appropriate monitoring data will be used to synthesize how a chemical is likely to behave if released to soil.

Aquatic Fate: This section reviews how a chemical will behave if released to fresh, marine, or estuarine surface waters. Field studies or aquatic model ecosystems are used here when they provide insight into the overall behavior in water. Studies which determine an individual process (e.g., biodegradation, hydrolysis, photolysis, sediment adsorption, and bioconcentration in aquatic organisms) in water are reviewed in the appropriate sections that follow. When field or aquatic ecosystems studies are not available or do not give enough data to make conclusions on the aquatic fate of the chemical, data from the appropriate degradation, transport, or monitoring sections will be used to synthesize how a chemical is likely to behave if released to water.

Atmospheric Fate: This section reviews how a chemical will behave if released to the atmosphere. The vapor pressure will be used to determine if the chemical is likely to be in the vapor phase or adsorbed to particulate matter [8]. The water solubility will be used to assess the likelihood of washout with rain. Smog chamber studies or other studies where the mechanism of degradation is not determined will be reviewed in this section; studies of the rate of reaction with hydroxyl radical or ozone or direct photolysis will be reviewed in Abiotic Degradation and integrated into this section.

Biodegradation: The principles outlined by Howard and Banerjee [11] are used in this section to review the relevant biodegradation data pertinent to biodegradation in soil, water, or wastewater treatment. In general, the studies have been separated into screening studies (inoculum in defined nutrient media), biological treatment simulations, and grab samples (soil or water sample with chemical added and loss of concentration followed). Pure culture studies are only used to indicate potential metabolites, since the artificial nutrient conditions under which the pure cultures are isolated provide little assurance that these same organisms will be present in any quantity or that their enzymes will be functioning in various soil or water environments. Anaerobic biodegradation studies, which are pertinent to whether a chemical will biodegrade in biological treatment digestors, sediment, and some ground waters, are discussed separately.

Abiotic Degradation: Nonbiological degradation processes in air, water, or soil are reviewed in this section. For most chemicals in the vapor phase in the atmosphere, reaction with photochemically generated hydroxyl radicals is the most important degradation process. Occasionally, reaction in the atmosphere with ozone (for olefins), nitrate radicals at night, and direct photolysis (direct sunlight absorption resulting in photochemical alteration) are significant for some chemicals [3]. For many chemicals, experimental reaction rate constants for hydroxyl radical are available (e.g., [1]) and are used to calculate an estimated half-life by assuming an average hydroxyl radical concentration of $5 \times 10^{+5}$ molecules/cm^3 in nonsmog conditions (e.g., [3]). If experimental rate constants are not available, they have been estimated using the fragment constant method of Atkinson [1] and then a half-life estimated using the assumed radical concentration. The reaction rate for ozone reaction with olefins may be experimentally available or can be estimated using the Atmospheric Oxidation Program from SRC which is based upon the work of Atkinson and Carter [2]. Using either the experimental or estimated rate constant and an assumed concentration of $7.2 \times 10^{+11}$ molecules/m^3 [3], an estimated half-life for reaction with ozone can be calculated. Nitrate radicals are significant only with certain classes of chemicals such as higher alkenes, dimethyl sulfide and lower thiols, furan and pyrrole, and hydroxy-substituted aromatics [3].

The possibility of direct photolysis in air or water can be partially assessed by examining the ultraviolet spectrum of the chemical. If the chemical does not absorb light at wavelengths provided by sunlight (>290 nm), the chemical cannot directly photolyze. If it does absorb sunlight, it may or may not photodegrade depending upon the efficiency (quantum yield) of the photochemical process, and unfortunately, such data are rarely available. Indirect photolysis processes may be important for some chemicals in water [18]. For example, some chemicals can undergo sensitized photolysis by absorbing triplet state energy from the excited triplet state of chemicals commonly found in water, such as humic acids. Transient oxidants found in water, such as peroxy radicals, singlet oxygen, and hydroxyl radicals, may also contribute to abiotic degradation in water for some chemicals. For example, phenols and aromatic amines have half-lives of less than a day for reaction with peroxy radicals; substituted and unsubstituted olefins have half-lives of 7 to 8 days with singlet oxygen; and dialkyl sulfides have half-lives of 27 hours with singlet oxygen [18].

Chemical hydrolysis at pHs that are normally found in the environment (pHs 5 to 9) can be important for a variety of chemicals that have functional groups that are potentially hydrolyzable, such as alkyl halides, amides, carbamates, carboxylic acid esters, epoxides and lactones, phosphate esters, and sulfonic acid esters [19]. Half-lives at various pHs are usually reported in order to provide an indication of the influence of pH.

Bioconcentration: Certain chemicals, due to their hydrophobic nature, have a tendency to partition from the water column and bioconcentrate in aquatic organisms. This concentration of chemicals in aquatic organisms is of concern because it can lead to toxic concentrations being reached when the organism is consumed by higher organisms such as wildlife and humans. Such bioconcentrations are usually reported as the bioconcentration factor (BCF), which is the concentration of the chemical in the organism at equilibrium divided by the concentration of the chemical in water. This unitless BCF value can be determined experimentally by dosing water containing the organism and dividing the concentration in the organism by the concentration in the water once equilibrium is reached, or if equilibration is slow, the rate of uptake can be used to calculate the BCF at equilibrium. The BCF value can also be estimated by using recommended regression equations that have been shown to correlate well with physical properties such as the octanol/water partition coefficient and water solubility [16]; however, these estimation equations assume that little metabolism of the chemical occurs in the aquatic organism, which is not always correct. Therefore, when available, experimental values are preferred.

Soil Adsorption/Mobility: For many chemicals (especially pesticides), experimental soil or sediment partition coefficients are available. These values are measured by determining the concentration in both the solution (water) and solid (soil or sediment) phases after shaking for about 24 to 48 hours and using different initial concentrations. The data are then fit to a Freundlich equation to determine the adsorption coefficient, Kd. These Kd values for individual soils or sediments are normalized to the organic carbon content of the soil or sediment by dividing by the organic content (Koc), since of the numerous soil properties that affect sorption (organic carbon content, particle size, clay mineral composition, pH, cation-exchange capacity) [15], organic carbon is the most important for undissociated organic chemicals. Occasionally, the experimental adsorption coefficients are reported on a soil-organic matter basis (Kom) and these are converted to Koc by multiplying by 1.724 [16]. When experimental values are unavailable, estimated Koc values are calculated using either the water solubility or octanol/water partition coefficient and some recommended regression equations [16]. The measured or estimated adsorption values are used to determine the likelihood of leaching through soil or adsorbing to sediments using the criteria of Swann et al. [25]. Occasionally, experimental soil thin-layer chromatography studies are also available and can be used to assess the potential for leaching.

The above discussion applies generally to undissociated chemicals, but there are some exceptions. For example, aromatic amines have been shown to covalently bond to humic material [21] and this slow but nonreversible process can lead to aromatic amines being tightly bound to the humic material in soils. Methods to estimate the soil or sediment adsorption coefficient for dissociated chemicals which form anions are not yet available, so it is particularly important to know the pKa value for chemicals that can dissociate so that a determination of the relative amounts of the dissociated and undissociated forms can be determined at various pH conditions. Chemicals that form cations at ambient pH conditions are generally thought to sorb strongly to clay material, similar to what occurs with paraquat and diquat (pyridine cations).

Volatilization from Water/Soil: For many chemicals, volatilization can be an extremely important removal process, with half-lives as low as several hours. The Henry's Law constant can give qualitative indications of the importance of volatilization; for chemicals with values less than

10^{-7} atm-m^3/mole, the chemical is less volatile than water and as water evaporates the concentration will increase; for chemicals around 10^{-3} atm-m^3/mole, volatilization will be rapid. The volatilization process is dependent upon physical properties of the chemical (Henry's Law constant, diffusivity coefficient), the presence of modifying materials (adsorbents, organic films, electrolytes, emulsions), and the physical and chemical properties of the environment (water depth, flow rate, the presence of waves, sediment content, soil moisture, and organic content) [16]. Since the overall volatilization rate cannot be estimated for all the various environments to which a chemical may be released, common models have been used in order to give an indication of the relative importance of volatilization. For most chemicals that have a Henry's Law constant greater than 10^{-7} atm-m^3/mole, the simple volatilization model outlined in Lyman et al [16] was used; this model assumes a 20 °C river 1 meter deep flowing at 1 m/sec with a wind velocity of 3 m/sec and requires only the Henry's Law constant and the molecular weight of the chemical for input. This model gives relatively rapid volatilization rates for this model river and values for ponds, lakes, or deeper rivers will be considerably slower. Occasionally, a chemical's measured reaeration coefficient ratio relative to oxygen is available, and this can be used with typical oxygen reaeration rates in ponds, rivers, and streams to give volatilization rates for these types of bodies of water. For chemicals that have extremely high Koc values, the EXAMS-II model has been used to estimate volatilization both with and without sediment adsorption (extreme differences are noted for these high Koc chemicals). Soil volatilization models are less validated and only qualitative statements are given of the importance of volatilization from moist (about 2% or greater water content) or dry soil, based upon the Henry's Law constant or vapor pressure, respectively. This assumes that once the soil is saturated with a molecular layer of water, the volatilization rate will be mostly determined by the value of the Henry's Law constant, except for chemicals with high Koc values.

Water Concentrations: Ambient water concentrations of the chemical are reviewed in this section, with subcategories for surface water, drinking water, and ground water when data are available. In general, the number of samples, the percent positive, the range of concentrations, and the average concentration are reported when the data are available.

Effluents Concentration: Air emissions and wastewater effluents are reviewed in this section. In general, the number of samples, the percent positive, the range of concentrations, and the average concentration are reported when the data are available.

Sediment/Soil Concentrations: Sediment and soil concentrations are reviewed in this section. In general, the number of samples, the percent positive, the range of concentrations, and the average concentration are reported when the data are available.

Atmospheric Concentrations: Ambient atmospheric concentrations are reviewed in this section, with subcategories for rural/remote and urban/suburban when data are available in such sources as Brodzinsky and Singh [5]. In general, the number of samples, the percent positive, the range of concentrations, and the average concentration are reported when the data are available.

Food Survey Values: Market basket survey data such as found in Duggan et al [7] and individual studies of analysis of the chemical in processed food are reported in this section. In general, the number of samples, the percent positive, the range of concentrations, and the average concentration are reported when the data are available.

Plant Concentrations: Concentrations of the chemical in plants are reviewed in this section. If the plant has been processed for food, it is reported in Food Survey Values.

Fish/Seafood Concentrations: Concentrations in fish, seafood, shellfish, etc. are reviewed in this section. If the fish or seafood have been processed for food, the data are reported in Food Survey Values.

Animal Concentrations: Concentrations in animals are reviewed in this section. If the animals have been processed for food, the data are reported in Food Survey Values.

Milk Concentrations: Since dairy milk constitutes a high percentage of the human diet, concentrations of the chemical found in dairy milk are reviewed in this section and not in Food Survey Values.

Other Environmental Concentrations: Concentrations of the chemical found in other environmental media that may contribute to an understanding of how a chemical may be released to the environment or exposed to humans (e.g., detection in gasoline or cigarette smoke) are reviewed in this section.

Probable Routes of Human Exposure: The monitoring data and physical properties are used to provide conclusions on the routes (oral, dermal, inhalation) of exposure.

Average Daily Intake: The average daily intake is a calculated value of the amount of the chemical that is typically taken in daily by human adults. The value is determined by multiplying typical concentrations in drinking water, air, and food by average intake factors such as 2 liters of water, 20 m^3 of air, and 1600 grams of food [26].

Occupational Exposures: Monitoring data, usually air samples, from occupational sites are reviewed in this section. In addition, estimates of the number of workers exposed to the chemical from the two National Institute for Occupational Safety and Health (NIOSH) surveys are reviewed in this section. The National Occupational Hazard Survey (NOHS) conducted from 1972 to 1974 and the National Occupational Exposure Survey (NOES) conducted from 1981 to 1983 provided statistical estimates of worker exposures based upon limited walk-through industrial hygiene surveys.

Body Burdens: Any concentrations of the chemical found in human tissues or fluids is reviewed in this section. Included are blood, adipose tissue, urine, and human milk.

REFERENCES

1. Atkinson RA; Internat J Chem Kinet 19: 799-828 (1987)
2. Atkinson RA, Carter WP; Chem Rev 84: 437-70 (1984)
3. Atkinson RA; Chem Rev 85: 60-201 (1985)
4. Boublik T et al; The Vapor Pressures of Pure Substances. Amsterdam: Elsevier (1984)
5. Brodzinsky R, Singh HB; Volatile organic chemicals in the atmosphere: an assessment of available data. SRI Inter EPA contract 68-02-3452 Menlo Park, CA (1982)
6. Daubert TE, Danner RP; Data Compilation Tables of Properties of Pure Compounds. Amer Inst Chem Engr pp 450 (1985)
7. Duggan RE et al; Pesticide Residue Levels in Foods in the U.S. from July 1, 1969 to June 30, 1976. Washington, DC: Food Drug Administ. 240 pp (1983)
8. Eisenreich SJ et al; Environ Sci Technol 15: 30-8 (1981)

9. Hansch C, Leo AJ; Medchem Project Issue No 26. Claremont, CA: Pomona College (1985)
10. Hine J, Mookerjee PK; J Org Chem 40: 292-8 (1975)
11. Howard PH, Banerjee S; Environ Toxicol Chem 3: 551-562 (1984)
12. Howard PH et al; Environ Sci Technol 12: 398-407 (1978)
13. Howard PH et al; Environ Toxicol Chem 5: 977-88 (1986)
14. Karickhoff SW; Chemosphere 10: 833-46 (1981)
15. Karickhoff SW; In Environ Expos from Chemicals Vol I. ed Neely WB, Blau GE, Boca Raton, FL: CRC Press p 49-64 (1985)
16. Lyman WJ et al; Handbook of Chemical Property Estimation Methods. McGraw-Hill, NY (1982)
17. Mackay D, Shiu WY; J Phys Chem Ref Data 10: 1175-99 (1981)
18. Mill T, Mabey W; In Environ Expos from Chemicals Vol I. ed Neely WB, Blau GE, Boca Raton, FL: CRC Press p 175-216 (1985)
19. Neely WB; In Environ Expos from Chemicals Vol I. ed Neely WB, Blau GE, Boca Raton, FL: CRC Press p 157-73 (1985)
20. Neely WR, Blau GE; Environ Expos from Chemicals Vol I. Boca Raton, FL: CRC Press (1985)
21. Parris GE; Environ Sci Technol 14: 1099-1105 (1980)
22. Perrin DD; Dissociation Constants of Organic Bases in Aqueous Solution. IUPAC Chemical Data Series, London: Butterworth (1965)
23. Riddick JA et al; Organic Solvents: Physical Properties and Methods of Purification, 4th Edit. New York: J Wiley & Sons (1986)
24. Serjeant EP, Dempsey B; Ionisation Constants of Organic Acids in Aqueous Solution. IUPAC Chemical Data Series No 23, New York: Pergamon Press (1979)
25. Swann RL et al; Residue Reviews 85: 17-28 (1983)
26. U.S. EPA; Reference Values for Risk Assessment. Environ Criteria Assess Office, Off Health Environ Assess, Off Research Devel, ECAO-CIN-477, Cincinnati, OH: U.S. Environ Prot Agency (1986)
27. Veith GD et al; J Fish Res Board Can 36: 1-40-8 (1979)
28. Yalkowsky SH et al; ARIZONA dATABASE of Aqueous Solubility, U. Arizona, Tucson, AZ (1987)

Contents

Acetonitrile

Synonyms: Methyl cyanide

Structure:

$$H_3C\text{————}C\text{======}N$$

CAS Registry Number: 75-05-8

Molecular Formula: C_2H_3N

Wiswesser Line Notation:

CHEMICAL AND PHYSICAL PROPERTIES

Boiling Point: 81.6°C

Melting Point: -45.7°C

Molecular Weight: 41.05

Dissociation Constants: pKa= 29.1 [34]

Log Octanol/Water Partition Coefficient: -0.34 [17]

Water Solubility: miscible with water [34]

Vapor Pressure: 88.81 mm Hg at 25°C [34]

Henry's Law Constant: 2.93×10^{-5} atm-m^3/mole [38]

ENVIRONMENTAL FATE/EXPOSURE POTENTIAL

Summary: Acetonitrile is released to the environment during its manufacture and use, from shale oil retorting and coal gasification, incineration of polyacrylonitrile, from automobile exhaust and cigarette

1

smoke. If released to soil, aerobic biodegradation is likely to occur in subsoil. Acetonitrile is expected to be mobile in soil and may evaporate from soil surfaces. Biodegradation is expected to be a major loss process in water. Acclimatization increases the biodegradation rate substantially. Volatilization may become competitive with other loss processes, particularly at shallow water depths. Hydrolysis, photolysis, adsorption to suspended particles and sediments and bioconcentration in aquatic organisms are not likely to be important. Acetonitrile is likely to be unreactive towards direct photolysis in air and the half-lives for its reaction with OH radicals and ozone have been estimated to be 535 days and 860 days, respectively. Therefore, it will persist in the troposphere for a long time and may be transported long distance from its source of emission. Wet deposition may remove some of the atmospheric acetonitrile. Adequate data regarding its typical concentrations in air, water and total diet sample are not available to estimate intake from these exposure routes.

Natural Sources:

Artificial Sources: Acetonitrile is released in the atmosphere as a result of fugitive emission during its manufacture [14,44]. Emission of acetonitrile from industrial facilities during its use also occurs [14]. It is also released in the atmosphere from incineration of polyacrylonitrile polymers [14,44], automobile exhaust [14,35], tobacco smoke [20], synthetic rubber manufacture [14], off-gas of shale oil retorting [37], manufacture of acrylonitrile [43] and turbine engines [14]. Acetonitrile has been detected in shale oil wastewaters [19,32] and in wastewater from coal gasification process [30,32].

Terrestrial Fate: Although no conclusive study demonstrating the biodegradability of acetonitrile in grab soil samples is available, it can be inferred from the pure culture and biodegradability studies in water that the compound may biodegrade in soil. Photolysis studies in air [9,21] and hydrolysis study in water [11] suggest that acetonitrile would not undergo appreciable photolysis or hydrolysis in soil. Based on an estimated Koc value of 16 [17,29], acetonitrile would be weakly sorbed to most soils. The high water solubility, moderately high vapor pressure [8], and weak soil sorption of the compound suggest that volatilization from soil surfaces and leaching into ground water would be important.

Acetonitrile

Aquatic Fate: A number of biodegradation studies with sewage, activated sludge, and pure cultures serving as microbial organisms have shown that acetonitrile is biodegradable in water following acclimatization, as long as its original concentration is not too high (e.g., 500 mg/L). The decomposition of the compound (concn. 0.1 to 25 mg/L) in Ohio River water was 20% in 5 days and 40% in 12 days [26]. Biodegradation was faster in water following acclimatization. Photochemical studies in the vapor phase [9,21] suggest that photodegradation in water may not be important. Hydrolysis is unimportant at the pH range normally present in natural waters [11]. Based on the value of 2.93×10^{-5} atm-m^3/mole for Henry's Law constant (H) [38] and the relationship between H and volatility [41], volatilization of the compound from water may not be rapid, but may become competitive with other loss processes, particularly at shallow water depths. The high water solubility and low Koc of acetonitrile would suggest that adsorption of the compound to suspended solids and sediment in water and bioconcentration in aquatic organisms would be unimportant.

Atmospheric Fate: The rate constant for the reaction of acetonitrile with OH radicals in air has been determined to range from 1.9×10^{-14} to 4.94×10^{-14} cm^3/molecule-sec in the temperature range 20 to 27°C [3,15,16,18,45]. Based on a rate constant of 3×10^{-14} cm^3/molecule-sec and the average daily OH radicals concn of $5 \times 10^{+5}$ radicals/cm^3 in the atmosphere [3], the half-life of this reaction is 535 days. The rate constant for the reaction of acetonitrile in air with ozone is 1.3×10^{-20} cm^3/molecule-sec [4]. In a typical atmosphere where the average daily ozone concn is $7.2 \times 10^{+11}$ molecules/cm^3 [3], the half-life due to this reaction would be 860 days. The photochemical smog studies also show that this compound is unreactive towards photochemically generated free radicals [9]. Acetonitrile is also unreactive towards direct photolysis in the gas phase [13,21].

Biodegradation: The biodegradation studies of acetonitrile with mixed cultures of microorganisms from activated sludge and sewage show that degradation proceeds sluggishly without acclimatization of microorganisms, particularly at high concn [22,33]. Degradation is faster with acclimatization [5,25,40,42]. With activated sludge as microbial inoculum, the lag period of acetonitrile degradation was about 1 day after

3

which the compound degraded with a half-life of 1.2 days [42]. Acclimated mixed microbial cultures isolated by an enrichment culture technique degraded 58% acetonitrile in 5 days [5]. Nitrile degradation was postulated to proceed via an enzymatic hydrolysis leading to the formation of organic acids and ammonia, followed by nitrification [25]. The degradation was faster at higher temperature than at lower temperature (30 °C vs 4 °C) [25]. Results of anaerobic tests with acetonitrile suggest that it is not effective for removing the compound from wastewaters [28]. The biodegradability of acetonitrile was also observed with river water and the nitrile assimilation was found to be faster with the river water than with organisms derived from aged sewage. The 12-day ThOD (theoretical oxygen demand) with river water was 40%. Acclimation of the microorganisms was examined by redosing, and the degradation was 5 times faster after acclimation. The degradation was also 4 times faster at 20 °C than at 5 °C [26,27]. The biodegradation is expected to be much slower in seawater than in freshwater [39]. The pathway of acetonitrile degradation with a species of Pseudomonas has been postulated to proceed from acetonitrile to acetamide to acetate to tricarboxylic acid-cyclic intermediates (e.g., 5-pyrrolidone-2-carboxylic acid) [12].

Abiotic Degradation: The hydrolysis of acetonitrile in water is base-catalyzed. The rate constant for base catalyzed hydrolysis is 5.8 x 10^{-3}/M-hr. The estimated hydrolysis half-life at pH 7 is more than 150,000 yrs [11]. The reported rate constant for the reaction of acetonitrile with hydroxyl radicals in aqueous solution at pH 9 is in the range 2.1-3.5 x 10^{+6}/M-sec [1,10]. Assuming the concentration of hydroxyl radicals in natural surface waters as 10^{-17} M, a half-life of >600 yrs can be calculated. Acetonitrile absorbs light only in the far UV region and absorption of sunlight is not important [2]. Therefore, it can be concluded that photolysis of the compound in the atmosphere should not be an important process. The rate constant for the reaction of acetonitrile with OH radicals in air has been estimated to range from 1.9 x 10^{-14} to 4.94 x 10^{-14} cm^3/molecule-sec in the temperature range 20 to 27°C [3,15,16,18,45]. Based on a rate constant of 3 x 10^{-14} cm^3/molecule-sec and the average daily OH radicals concn of 5 x 10^{+5} radicals/cm^3 in the atmosphere [3], the half-life of this reaction is 535 days. The rate constant for the reaction of acetonitrile in air with ozone is 1.3 x 10^{-20} cm^3/molecule-sec [4]. In a typical atmosphere where the average daily ozone concn is $7.2 \times 10^{+11}$ molecules/cm^3 [3], the half-life due to this reaction would be 860 days. The photochemical smog studies also show that this compound is unreactive towards photochemically generated free radicals [9]. Due to nonreactivity of acetonitrile in the atmosphere, transport of the compound from troposphere to stratosphere is expected to occur and acetonitrile has been detected in the stratosphere [2].

Bioconcentration: Measured values for acetonitrile bioconcentration factor in aquatic organisms are not available. Based on a log Kow of -0.34 [17] and the regression equation [7], the estimated BCF of acetonitrile is 0.3. Therefore, bioconcentration in aquatic organisms should not be an important fate process for acetonitrile.

Soil Adsorption/Mobility: The sorption of acetonitrile on clayish soils is not due to any specific bonding, but rather due to distribution or partition between the interfacial phase of clay and the bulk solution phase [46]. Based on a log Kow value of -0.34 [17] and the regression equation [29], a Koc value of 16 can be estimated for the compound. Therefore, the sorption of the compound onto soil or sediments will not be important.

Volatilization from Water/Soil: The volatilization half-life of a chemical in water can be estimated from its Henry's Law constant and the overall liquid-phase mass transfer coefficient [41]. Based on a value of Henry's Law constant (H) of 2.93×10^{-5} atm-m^3/mole [38], the volatilization half-life of acetonitrile from a 1-m deep model river water flowing at a current speed of 1 m/sec and a wind speed of 3 m/sec would be 21 hr. Therefore, volatilization from water may become competitive with other loss processes, particularly at shallow water depths. The volatility of acetonitrile from soil has not been studied. The value of H and the moderately high vapor pressure of the compound indicate that volatilization from dry and moist soil surface would be important.

Water Concentrations:

Effluent Concentrations: Acetonitrile was qualitatively detected in shale oil wastewaters [19] and wastewater from coal gasification processes [30,32].

Acetonitrile

Sediment/Soil Concentrations:

Atmospheric Concentrations: Acetonitrile was detected at a mean concn of 0.024 ppb in two samples from a rural area but was not detected in 1 sample from an urban area and 1 sample from a source-dominated area [36]. It was detected in air near ground levels in both urban and rural areas at concn 2-7 ppb [6]. It was reported to be present in the upper stratosphere [2]. Acetonitrile was detected with a frequency of 20% in exhaled air of normal human (nonsmoking) at a mean concn 7.4 ng/L [23,24].

Food Survey Values: Acetonitrile was detected in some milk products, such as kefir culture [31].

Plant Concentrations:

Fish/Seafood Concentrations:

Animal Concentrations:

Milk Concentrations:

Other Environmental Concentrations:

Probable Routes of Human Exposure:

Average Daily Intake:

Occupational Exposures:

Body Burdens:

REFERENCES

1. Anbar M, Neta P; Inter J Appl Radiation Isotopes 18: 493-523 (1967)
2. Arijs E et al; Nature 303: 314-6 (1983)
3. Atkinson R; Chem Rev 85: 69-201 (1985)
4. Atkinson R, Carter WPL; Chem Rev 84: 437-70 (1984)
5. Babeu L, Vaishnav DD; J Ind Microbiol 2: 107-15 (1987)
6. Becker KM, Ionescu A; Geophys Res Lett 9: 1349-51 (1982)

Acetonitrile

7. Bysshe SE; Bioconcentration factor in aquatic organisms In: Handbook of Chemical Property Estimation Methods: Environmental Behavior of Organic Compounds Lyman WJ et al.(eds), NY:NY McGraw-Hill Book Co. pp 5-5 (1982)
8. Chiou CT et al; Environ International 3: 231-6 (1980)
9. Dimitriades B, Joshi SB; Application of reactivity criteria in oxidant-related emission control in the USA In: Internation Conf. on Photochemical Oxidant Pollution and Its Control, USEPA Report No. 600/3-77-001B NC: Research Triangle Park, US Environmental Protection Agency pp 705-11 (1977)
10. Dorfman LM, Adams GE; Reactivity of the hydroxyl radical in aqueous solution NDRD-NBS-46 Washington, DC: National Bureau of Standards, Available through NTIS order No. COM-73-50623 (1973)
11. Ellington JJ et al; Measurement of hydrolysis rate constants for evaluation of hazardous waste land disposal: vol 3. Data on 70 Chemical, USEPA Report No. 600/S3-88/028, Available through NTIS, Order No. PB88-234 042/AS (1988)
12. Firmin JL, Gray DO; Biochem J 158: 223-9 (1976)
13. Fujiki M et al; Simulation studies of degradation of chemicals in the environment: Simulation studies of degradation of chemicals in the atmosphere, Chem Res Report No. 1/1978, Japan: Tokyo, Environmental Agency Japan, Office of Health Studies (1978)
14. Graedel TE et al; Atmospheric Chemical Compounds NY, NY: Academic Press pp 389 (1986)
15. Gusten H et al; Atmos Environ 15: 1763-5 (1981)
16. Gusten H et al; J Atmos Chem 2: 83-93 (1984)
17. Hansch C, Leo AJ; Medchem Project Issue No. 26 CA: Claremont, Pomona College (1985)
18. Harris GW et al; Chem Phys Lett 80: 479-83 (1981)
19. Hawthorne SB, Sievers RE; Environ Sci Technol 18: 483-90 (1984)
20. Kadaba PK et al; Bull Environ Contam Toxicol 19: 104-11 (1978)
21. Kagiya T et al; Japan Chem Soc Spring Term Annual Meeting, Japan: Tokyo, April 1-4, Paper No. 1036 (1975)
22. Kalmykova GY, Rogovskaya TI; Biol Vnutr Vod 38: 79-83 (1978)
23. Krotoszyniski BK et al; J Anal Toxicol 3: 225-34 (1979)
24. Krotoszyniski BK et al; J Chromatog Sci 15: 239-44 (1977)
25. Ludzack FJ et al; Sewage Ind Wastes 31: 33-44 (1959)
26. Ludzack FJ et al; Proc Thirteenth Ind. Waste Conf., Purdue University Engineering Extension Service, pp 297-312 (1958)
27. Ludzack FJ, Ettinger MB; Water Pollut Control Assoc J 32: 1173-2000 (1960)
28. Ludzack FJ et al; J Water Pollut Control Fed 33: 492-505 (1961)
29. Lyman WJ; Adsorption coefficient for soils and sediments In: Handbook of Chemical Property Estimation Methods: Environmental Behavior of Organic Compounds Lyman WJ et al.(eds), NY:NY McGraw-Hill Book Co. pp 4-9 (1982)
30. Mohr DH, King CJ; Environ Sci Technol 19: 929-35 (1985)
31. Palo V, Ilkova H; J Chromatog 53: 363-7 (1970)
32. Pellizzari ED et al; Identification of organic components in aqueous effluents from energy-related processes, ASTM Spec. Tech. Publ. STP 686 p 256-74 (1979)
33. Placak OR, Ruchhoft CC; Sewage Works J 19: 423-40 (1946)

34. Riddick JA et al (eds); Organic Solvents: Physical Properties and Methods of Purification, 4th ed, John Wiley and Sons, NY pp 582-3 (1986)
35. Schuchmann HP, Laidler KJ; J Air Pollut Control Assoc 22: 52-3 (1972)
36. Shah JJ, Heyerdahl EK; National ambient volatile organic compounds (VOCs) data base update, Research Triangle Park: NC USEPA Report No. 600/3-88/010(a), Atmospheric Sciences Research Laboratory, Office of Research and Development, USEPA (1988)
37. Sklarew DS, Hayes DJ; Environ Sci Technol 18: 600-3 (1984)
38. Snider JR, Dawson GA; J Geophys Res 90(D2): 3797-3805 (1985)
39. Takemoto S et al; Suishitsu Odaku Kenkyu 4: 80-90 (1981)
40. Thom NS, Agg AR; Proc R Soc Lond B189: 347-57 (1975)
41. Thomas RG; Volatilization from Water In: Handbook of Chemical Property Estimation Methods: Environmental Behavior of Organic Compounds Lyman WJ et al.(eds), NY:NY McGraw-Hill Book Co. pp 15-16 (1982)
42. Urano K, Kato Z; J Hazardous Materials 13: 147-59 (1986)
43. USEPA; Locating and estimating air emissions from sources of acrylonitrile USEPA-450/4-84-007A Research Triangle Park, NC: USEPA (1984)
44. USEPA; Chemical Hazard Information Profiles, Report No. EPA-560/11-80-011 Washington, DC U.S. Environmental Protection Agency (1980)
45. Wallington TJ et al; J Phys Chem 92: 5024-8 (1988)
46. Zhang ZZ et al; Soil Sci Soc Am J 54: 351-6(1990)

Acetophenone

SUBSTANCE IDENTIFICATION

Synonyms:

Structure:

![Chemical structure of acetophenone]

CAS Registry Number: 98-86-2

Molecular Formula: C_8H_8O

Wiswesser Line Notation: 1VR

CHEMICAL AND PHYSICAL PROPERTIES

Boiling Point: 202 °C

Melting Point: 20.5 °C

Molecular Weight: 120.16

Dissociation Constants:

Log Octanol/Water Partition Coefficient: 1.58 [25]

Water Solubility: 6130 mg/L at 25 °C [57]

Vapor Pressure: 0.397 mm Hg at 25 °C [12]

Henry's Law Constant: 1.07×10^{-5} atm-m^3/mole at 25 °C [43]

ENVIRONMENTAL FATE/EXPOSURE POTENTIAL

Summary: Acetophenone is released to the environment from a variety of combustion processes and may be released during its manufacture and the manufacture of propylene oxide, kraft bleaching and its use in certain perfumes. If released to soil, microbial degradation is likely to be the

9

major degradation pathway. It is expected to be moderately to highly mobile in soil and may evaporate from dry soil surfaces. Biodegradation and volatilization are expected to be the major loss processes in water. The estimated biodegradation half-lives in ground water, river water and lake water samples were 32 days, 8 days and 4.5 days, respectively. The volatilization half-life from a river 1 m deep flowing at 1 m/sec with a wind speed of 3 m/sec is estimated to be 3.8 days. Hydrolysis, oxidation and adsorption to suspended particles and sediments and bioconcentration in aquatic organisms are not likely to be important fate processes. Oxidation by hydroxyl radicals in air has an estimated half-life of 2.2 days. Other oxidants (e.g., ozone) and photolysis do not appear to be important loss mechanism of this compound in air. Wet deposition may be important for the removal of atmospheric acetophenone.

Natural Sources:

Artificial Sources: Acetophenone is released into the atmosphere from vehicular exhausts [24], waste incineration [29], residential fuel oil [39], coal combustion [31], and vaporization of certain perfumes [1]. Acetophenone has been detected in waste waters from a petrochemical plant [34], a propylene oxide manufacturing plant [44], a shale oil processing plant [13], in a spent chlorination liquor from kraft bleaching [8] and in a surface water downstream from a tire fire location [50].

Terrestrial Fate: If released to soil, microbial degradation is likely to be the major loss process for acetophenone [28]. Based on various adsorption studies [6,7,22,27,35], it is expected to be moderately to highly mobile depending on the nature of soil [62]. It may also evaporate from dry soil surfaces.

Aquatic Fate: A number of biodegradation studies with sewage and natural water serving as inoculum for microbial organisms [26,41,54] have shown that acetophenone is readily biodegradable. The biodegradation half-lives in ground water, river water and lake water samples were 32 days, 8 days and 4.5 days, respectively [65,66]. Based on the value of its Henry's Law constant [43] and an estimation method [42], the volatilization half-life from a model river 1 m deep flowing at 1 m/sec with a wind speed of 3 m/sec is estimated to be 3.8 days. Therefore, microbial degradation and volatilization are expected to be the major loss processes for this compound water. Hydrolysis, oxidation, adsorption to suspended particles and sediments, and bioconcentration are generally not likely to be important.

Atmospheric Fate: Based on its vapor pressure, acetophenone is likely to exist in the vapor phase in the atmosphere [18]. The estimated half-life for the reaction of vapor phase acetophenone with photochemically produced hydroxyl radicals in the atmosphere (generally, one of the most important fate determining process for atmospheric pollutants) is 22 days [5]. Oxidation by other oxidants and photolysis do not appear to be important for the loss of this compound in the atmosphere. Because of its significant water solubility, wet deposition may be important for the removal of atmospheric acetophenone.

Biodegradation: BOD tests performed with sewage as microbial inoculum degraded 20-32% of the compound (based on theoretical oxygen demand) in 5 days [14,26]. The 5-day BOD test using acclimated mixed microbial cultures degraded 59% of acetophenone [67]. In a Japanese MITI test, the compound was confirmed to be significantly biodegradable (more than 30% degradation in 14 days) [54]. The lag time for biodegradation in a laboratory study was 15-20 hr [64]. On the other hand, the lag time for biodegradation in Ohio River water was about 3 days followed by rapid carbon dioxide production [41]. Half the theoretical carbon dioxide was recovered in three days after acclimation of the river water [41]. The biodegradation half-lives in ground water, river water and lake water samples were 32 days, 8 days and 4.5 days, respectively [65,66]. A laboratory rapid infiltration study with soil columns indicated biodegradation as an important fate process after a certain adaptation period [28].

Abiotic Degradation: Since ketones, in general, are resistant to hydrolysis [42], aquatic hydrolysis of acetophenone is not expected to be important. The rate constant for the reaction of acetophenone with hydroxyl radicals in water at room temperature is in the range $2.9 \times 10^{+9}$ to $4.8 \times 10^{+9}$/M-sec [3,15]. Based on an average hydroxyl radical concn of 3×10^{-17} M [47], the estimated minimum half-life of acetophenone due to this reaction is 56 days. Acetophenone absorbs light significantly in the environmentally important region of greater than 290 nm [16], which indicates its potential for direct photolysis. However, acetophenone is known to act as a photosensitizer whereby it transfers its excited energy from light absorption to a receptor molecule. The receptor molecule undergoes photoalteration and the excited acetophenone returns to its ground stated without degradation [38]. Based on the vapor pressure, acetophenone will exist in the vapor phase in the atmosphere [18]. The half-life for the vapor phase reaction of acetophenone with hydroxyl radicals is about 22 days based on an estimated rate constant [5] and a daily average hydroxyl radical concn of $5 \times 10^{+5}/cm^3$ [5]. Similarly, it can be concluded from an estimation method [4] that the atmospheric reaction of the compound with ozone is not important. The average rate constant for the vapor phase reaction of acetophenone with singlet molecular oxygen is $0.26 \times 10^{+4}/M\text{-sec}$ [11]. Based on an ambient atmospheric concn of singlet oxygen molecule of $3.32 \times 10^{-14} M$ [23], acetophenone reaction with this species is not important. Because of the high water solubility of acetophenone, the removal of this compound by wet deposition is possible.

Bioconcentration: Based on the log octanol/water partition coefficient and water solubility values and the recommended regression equations [42], the estimated bioconcentration factor for acetophenone is in the range of 5-9. Therefore, bioconcentration may not be important in aquatic organisms.

Soil Adsorption/Mobility: The experimental and estimated log Koc for acetophenone in several soils and sediments range 1.34-2.43 [6,7,22,27,35,52,57]. These values are indicative of medium to high mobility in soil and sediments [62].

Volatilization from Water/Soil: Based on the Henry's Law constant, the volatilization half-life of acetophenone from a model river 1 m deep flowing at 1 m/sec with a wind speed of 3 m/sec is estimated to be 3.8 days [42]. The vapor pressure of acetophenone would suggest that evaporation from dry soil surfaces may occur. Evaporation from moist soil may also be important based upon the value of Henry's Law constant.

Water Concentrations: DRINKING WATER: Acetophenone was qualitatively detected in drinking water from Philadelphia, PA [40,60,61]; Poplarville, MS [40]; Cincinnati, OH [40]; Miami, FL [40]; New Orleans,

Acetophenone

LA [40]; Ottumwa, IA [40]; Seattle, WA [40] and Bayonne-Elizabeth area, NJ [68]. The concn of the compound in Philadelphia, PA drinking water collected in 1974 was 1.0 ppb [33]. Its concn in a Japanese drinking water was 5.4 ppb [56]. SURFACE WATER: Acetophenone was qualitatively detected in the following river and sea waters in the U.S. and other parts of the world: Kanawha River, Nitro, WV [51]; unnamed river in England [21]; Waal River, Netherlands [45]; river in Kitakyushu area in Japan [2]; Hamilton Harbor, Bermuda [17]; and Sea water near Japan [2]. It was detected in 6 of 204 samples from 14 heavily industrialized river basins in the U.S. sampled during 1975-1976 [20]. Its concn range in these six samples was 1-2 ppb [20]. Acetophenone was detected at a concn 2 ug/L in water from Lake Michigan [37]. GROUND WATER: Acetophenone was qualitatively detected in an aquifer near an organic wastes dump site in Australia [59]. It was detected in the concentration range undetectable (less than 0.1 ppb) to 10 ppb in the ground water of a waste disposal site in Netherlands at depths of up to 39 m [70].

Effluent Concentrations: Acetophenone was qualitatively detected in waste waters from a petrochemical plant [29], a propylene oxide manufacturing plant [32] and in a surface water downstream from a tire fire location [50]. Acetophenone is released into the atmosphere from vehicular exhaust [24], waste incineration [34], residential fuel oil combustion [39], coal combustion [31], plant volatiles [44] and vaporization of certain perfumes [1]. Its concn in waste water from a shale oil processing plant in Australia was 10 mg/L [13] and in spent chlorination liquor from kraft bleaching was 0.2-0.5 g/ton of treated pulp [8]. It was also detected in secondary effluents from municipal treatment plants [19]. In one case, its concn in secondary effluent was 0.038-0.053 mg/L [28].

Sediment/Soil Concentrations: Acetophenone was tentatively identified in the sediments of a polluted lake in Saskatchewan, Canada [53]. It was also detected in Calcasieu River near an area of heavy industrial discharge in Lake Charles, LA [58].

Atmospheric Concentrations: SOURCE DOMINATED: In the US, acetophenone was detected at a median concn of 0.041 ppb in two urban air samples and at a median concn of 0.094 ppb in 32 samples near industrial areas [55].   URBAN/SUBURBAN: Acetophenone was qualitatively detected in the ambient air in Japan [69] and Belgium [9]. It was qualitatively detected in 4 of 8 indoor air samples from NJ and NC area homes and in 5 of 12 breath samples from residents of these homes [68]. Its frequency of occurrence in 36 area homes in Chicago was 17% in indoor air compared to 9% in outdoor air [30]. RURAL/REMOTE: In the US, acetophenone was not detected in the ambient air of a remote area [55].

Food Survey Values: Acetophenone has been detected in the volatile components of filbert nut [36], clove essential oil [47] and nectarines [63].

Plant Concentrations:

Fish/Seafood Concentrations:

Animal Concentrations:

Milk Concentrations:

Other Environmental Concentrations: Acetophenone was detected in the concn range 0-30 ug/m^3 in the extrusion area of an electric cables insulation plant [10].

Probable Routes of Human Exposure:

Average Daily Intake:

Occupational Exposure: NIOSH (NOES Survey 1981-1983) has statistically estimated that 39,880 workers are potentially exposed to acetophenone in the USA [48].

Body Burdens: Acetophenone was detected in 8 of 8 mother's milk collected from volunteers in Bridgeville, PA, Jersey City, NJ and Baton Rouge, LA [49].

Acetophenone

REFERENCES

1. Abrams EF et al; Identification of Organic Compounds in Effluents from Industrial Sources, Springfield, VA:Versar Inc USEPA-650/3-75-002 (1975)
2. Akiyama T et al; J UOEH 2: 285-300 (1980)
3. Anbar M, Neta P; Inter J Appl Rad Isotopes 18: 493-523 (1967)
4. Atkinson R, Carter WPL; Chem Rev 84: 437-70 (1984)
5. Atkinson R; Environ Toxicol Chem 7: 435-42 (1988)
6. Bahnick DA, Doucette WJ; Chemosphere 17: 1703-15 (1988)
7. Banwart WL et al; J Environ Sci Health B15: 165-79 (1980)
8. Carlberg GE et al; Sci Total Environ 48: 157-67 (1986)
9. Cautreels W, Cauwenberghe KV; Atmos Environ 12: 1133-41 (1978)
10. Cocheo V et al; Am Ind Hyg Assoc J 44: 521-7 (1983)
11. Datta RK, Rao KN; Indian J Chem 18A: 102-5 (1979)
12. Daubert TE, Danner RP; Data Compilation Tables of Properties of Pure Compounds, Am Inst Chem Engr pp 450 (1985)
13. Dobson KR et al; Water Res J 19: 849-56 (1985)
14. Dore M et al; Trib Cebedeau 28: 3-11 (1975)
15. Dorfman LM, Adams GE; Reactivity of the Hydroxyl radical in Aqueous Solution, NSRD-NBS-46, NTIS COM-73-50623, Natl Bur Stand Washington, DC pp 51 (1973)
16. Draper WM, Crosby DG; J Agric Food Chem 31: 734-7 (1983)
17. Ehrhardt M; Marine Chem 22: 85-94 (1987)
18. Eisenreich SJ et al; Environ Sci Technol 15: 30-8 (1981)
19. Ellis DD et al; Arch Environ Contam Toxicol 11: 373-82 (1982)
20. Ewing BB et al; Monitoring to Detect Previously Unrecognized Pollutants in Surface Waters. Appendix: Organic Analysis Data. USEPA-560/6-77-015A Washington, DC (1977)
21. Fielding M et al; Organic Micropollutants in Drinking Water, Report No TR-159, Water Research Centre, Medmenham, England pp 49 (1981)
22. Gerstl Z, Mingelgrin U; J Environ Sci Health B19: 297-312 (1984)
23. Graedel TE; Chemical Compounds in the Atmosphere, Academic Press, NY pp 7 (1978)
24. Hampton CF et al; Environ Sci Technol 16: 287-98 (1982)
25. Hansch C, Leo AJ; Medchem Project, Issue No. 26, Pomona College, Claremont CA (1985)
26. Heukelekian H, Rand MC; J Water Pollut Control Assoc 29: 1040-53 (1955)
27. Hodson J, Williams NA; Chemosphere 17: 67-77 (1988)
28. Hutchins SR et al; Environ Toxicol Chem 2: 195-216 (1983)
29. James RH et al; J Proc Air Pollut Control Assoc, 77th Ann Meeting, June 24-29, San Francisco, CA (1984)
30. Jarke FH et al; ASHRAE Trans 87: 153-66 (1981)
31. Junk GA et al; pp 109-23 in ACS Symp Ser 319 Fossil Fuels Utilization (1986)
32. Juttner F; Chemosphere 15: 985-92 (1986)
33. Keith LR et al; pp 329-73 in Identification and Analysis of Organic Pollutants in Water, Keith LH, eds, Ann Arbor,MI:Ann Arbor Science Publ (1976)

Acetophenone

34. Keith LH; Sci Total Environ 3: 87-102 (1974)
35. Khan A et al; Soil Sci 128: 297-302 (1979)
36. Kinlin TE et al; J Agric Food Chem 20: 1021 (1972)
37. Konasewich D et al; Status Report on Organic and Heavy Metal Contaminants in the Lake Erie, Michigan, Huron and Superior Basins, Great Lakes Water Quality Board (1978)
38. Lande SS et al; Investigation of Selected Potential Environmental Contaminants USEPA-560/2-76/003, Off Tox Sub Washington DC (1976)
39. Leary JA et al; Environ Health Perspectives 73: 223-34 (1987)
40. Lucas SV; GC/MS Analysis of Organics in Drinking Water Concentrates and Advanced Waste Treatment Concentrates: Vol I USEPA-600/1-84-020A, NTIS PB 85-128 221, Health Effects Research Laboratory, Columbus, OH (1984)
41. Ludzack FJ, Ettinger MB; pp 278-82 in Proc Eighteen Ind Waste Conf, April 30-May 2, Purdue Univ Lafayette, IN (1963)
42. Lyman WJ et al; Handbook of Chemical Property Estimation Methods NY: McGraw-Hill pp 5-5 (1982)
43. Mackay D et al; Volatilization of Organic Pollutants from Water USEPA-600/53-82-019, NTIS PB 82-230 939, Athens, GA (1982)
44. Mamedova VM et al; Azerb Khim Zh 2: 121-32 (1973)
45. Meijers AP, Vanderleer RC; Water Res 10: 597-604 (1976)
46. Muchalal M, Crouzet J; Agric Biol Chem 49: 1583-9 (1985)
47. Neely WB, Blau GE; Environmental Exposure from Chemicals Vol I Boca Raton, FL: CRC Press pp. 207 (1985)
48. NIOSH; National Occupational Exposure Survey (NOES) (1983)
49. Pellizzari ED et al; Bull Environ Contam Toxicol 28: 322-8 (1982)
50. Peterson JC et al; Anal Chem 58: 70A-74A (1986)
51. Rosen AA et al; J Water Pollut Control Fed 35: 777-82 (1963)
52. Sabljic A; Environ Sci Technol 21: 358-66 (1987)
53. Samolloff MR et al; Environ Sci Technol 17: 329-34 (1983)
54. Sasaki S; pp 283-98 in Aquatic Pollutants: Transformation and Biological Effects, Hutzinger O et al, eds, Pergamon Press, Oxford (1978)
55. Shah JJ, Heyerdahl EK; National Organic Volatile Organic Chemical Compounds (VOCS) Database Update USEPA-600/3-88/010a Atmos Sci Res Lab Research Triangle Park, NC (1988)
56. Shinohara R et al; Water Res 15: 535-42 (1981)
57. Southworth GR, Keller JL; Water Air Soil Poll 28: 239-48 (1986)
58. Steinheimer TR et al; Anal Chim Acta 129: 57-67 (1981)
59. Stepan S et al; Movement and Chemical Change of Organic Pollutants in an Aquifer, Australian Water Resources Council Conf. Ser 1: 415-24 (1981)
60. Suffet IH et al; Water Res 14: 853-67 (1980)
61. Suffet IH et al; pp 375-97 in Identification and Analysis of Organic Pollutants in Water, Keith LH, eds, Ann Arbor, MI:Ann Arbor Science Publ (1976)
62. Swann RL et al; Res Rev 85: 17-28 (1983)
63. Takeoka GR et al; J Agric Food Chem 36: 553-60 (1988)
64. Urano K, Kato Z; J Hazard Mater 13: 147-59 (1986)
65. Vaishnav DD, Babeu L; J Great Lakes Res 12: 184-91 (1986)

Acetophenone

66. Vaishnav DD, Babeu L; Bull Environ Contam Toxicol 39: 237-44 (1987)
67. Vaishnav DD et al; Chemosphere 16: 695-703 (1987)
68. Wallace LA et al; Environ Res 35: 293-319 (1984)
69. Yokouchi Y et al; Chemosphere 16: 1143-7 (1987)
70. Zoeteman BCJ et al; Sci Total Environ 21: 187-202 (1981)

1-Amino-2-propanol

SUBSTANCE IDENTIFICATION

Synonyms:

Structure:

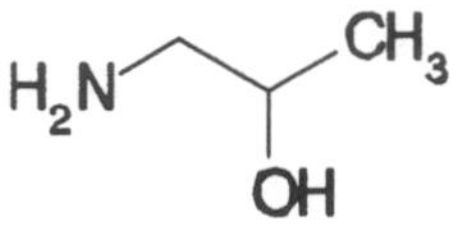

H₂N—CH₂—CH(OH)—CH₃

CAS Registry Number: 78-96-6

Molecular Formula: C_3H_9NO

Wiswesser Line Notation: Z1YQ1

CHEMICAL AND PHYSICAL PROPERTIES

Boiling Point: 159.46 °C at 760 mm Hg [15]

Melting Point: 1.74 °C [15]

Molecular Weight: 75.11

Dissociation Constants:

Log Octanol/Water Partition Coefficient: -0.96 [8]

Water Solubility: Miscible at 25 °C [11]

Vapor Pressure: 0.47 mm Hg at 25 °C [4]

Henry's Law Constant: 4.9 x 10^{-10} atm-m³/mole at 25 °C (estimated using the bond method) [10]

ENVIRONMENTAL FATE/EXPOSURE POTENTIAL

Summary: 1-Amino-2-propanol can be released to the environment in waste streams generated at sites of its commercial production and use. If released to the atmosphere, it will degrade relatively rapidly by reaction with photochemically produced hydroxyl radicals (estimated half-life of 9.9 hr). If released to soil or water, 1-amino-2-propanol is expected to degrade via biodegradation. Several biodegradation studies have demonstrated that 1-amino-2-propanol is biodegradable. It may leach readily in soils; however, concurrent biodegradation may lessen the importance of leaching. Occupational exposure to 1-amino-2-propanol occurs through dermal contact and inhalation of vapor.

Natural Sources:

Artificial Sources: Since 1-amino-2-propanol is primarily used as a chemical intermediate in the production of derivatives for many industries [7], it may be released to the environment in waste streams generated at sites of its commercial production and use.

Terrestrial Fate: The dominant degradation process for 1-amino-2-propanol in soil is expected to be biodegradation. Several biodegradation studies have demonstrated that 1-amino-2-propanol is biodegradable [2,3,6]. Based upon an estimated Koc of 7.1, 1-amino-2-propanol is expected to leach readily in soil [9]. The importance of leaching may be lessened by concurrent biodegradation. Based upon the vapor pressure, 1-amino-2-propanol should evaporate slowly from dry surfaces.

Aquatic Fate: The dominant removal process for 1-amino-2-propanol in water is expected to be biodegradation. Several biodegradation studies have demonstrated that 1-amino-2-propanol is readily biodegradable [2,3,6]. Aquatic volatilization, bioconcentration, and adsorption to sediment are not expected to be important.

Atmospheric Fate: Based upon the measured vapor pressure, 1-amino-2-propanol is expected to exist almost entirely in the vapor-phase in the ambient atmosphere [5]. It is expected to degrade relatively rapidly in an average ambient atmosphere (estimated half-life of about 9.9 hr) by reaction with photochemically produced hydroxyl radicals [1]. Since 1-amino-2-propanol is miscible in water [11], physical removal via wet deposition is likely to occur.

Biodegradation: A 5-day theoretical BOD of 4% was observed for 1-amino-2-propanol using a nonacclimated sewage inocula and a standard

BOD dilution method [2]; adaptation of the sewage inocula resulted in a 5-day theoretical BOD of 43% [2]. Using a sewage inocula and a BOD dilution method, 5-day, 10-day, 15-day and 20-day theoretical BODs of 5.1, 34.0, 43.4 and 46.0% were measured respectively for 1-amino-2-propanol [6]. In anaerobic serum bottle degradation studies, 1-amino-2-propanol exhibited a lag period of 9 days followed by a removal rate of 22 mg/L/day [3]; during the observation period, 65% of initial 1-amino-2-propanol was removed compared to 100% removal for 1-propanol [3]. 1-Amino-2-propanol is considered to be amenable to anaerobic biotechnology for industrial wastewater treatment [13].

Abiotic Degradation: The rate constant for the vapor-phase reaction of 1-amino-2-propanol with photochemically produced hydroxyl radicals has been estimated to be 39.1×10^{-12} cm^3/molecule-sec at room temperature which corresponds to an atmospheric half-life of about 9.9 hr at an atmospheric concn of $5 \times 10^{+5}$ hydroxyl radicals per cm^3 [1].

Bioconcentration: Based upon the experimental log Kow, the BCF for 1-amino-2-propanol can be estimated to be 0.11 from a regression-derived equation [9]. This BCF value suggests that 1-amino-2-propanol will not bioconcentrate significantly in aquatic organisms.

Soil Adsorption/Mobility: Based upon the experimental log Kow, the Koc for 1-amino-2-propanol can be estimated to be 7.1 from a regression-derived equation [9]. This Koc value suggests that 1-amino-2-propanol has very high soil mobility [14].

Volatilization from Water/Soil: The estimated Henry's Law constant indicates that 1-amino-2-propanol is essentially nonvolatile from water [9]. Based upon the vapor pressure, 1-amino-2-propanol should evaporate slowly from dry surfaces.

Water Concentrations:

Effluent Concentrations:

Sediment/Soil Concentrations:

Atmospheric Concentrations:

1-Amino-2-propanol

Food Survey Values:

Plant Concentrations:

Fish/Seafood Concentrations:

Animal Concentrations:

Milk Concentrations:

Other Environmental Concentrations:

Probable Routes of Human Exposure: Occupational exposure to 1-amino-2-propanol occurs through dermal contact and inhalation of vapor [7]; because 1-amino-2-propanol has a relatively low vapor pressure, the acute inhalation hazard is low [7].

Average Daily Intake:

Occupational Exposure: NIOSH (NOES Survey 1981-1983) has statistically estimated that 99,320 workers are potentially exposed to 1-amino-2-propanol in the USA [12].

Body Burdens:

REFERENCES

1. Atkinson R; J Inter Chem Kinet 19: 799-828 (1987)
2. Bridie AL et al; Water Res 13: 627-30 (1979)
3. Chou WL et al; Biotechnol Bioeng Symp 8: 391-414 (1979)
4. Daubert TE; Danner RP; Physical and Thermodynamic Properties of Pure Chemicals: Data Compilation, NY: Hemisphere Pub Corp (1989)
5. Eisenreich SJ et al; Environ Sci Technol 15: 30-8 (1981)
6. Ettinger MB; Ind Eng Chem 48: 256-9 (1956)
7. Hammer H et al; Ullmann's Encycl of Indust Chem, NY: VCH Publishers A10: 13 (1987)
8. Hansch C, Leo AJ; Medchem Project Issue No 26. Claremont CA: Pomona College (1985)

1-Amino-2-propanol

9. Lyman WJ, Reehl WF, and Rosenblatt DH; Handbook of Chemical Property Estimation Methods. Environmental Behavior of Organic Compounds. Washington DC: American Chemical Society pp. 4-9, 54, 5-10, 7-4, 7-5, 15-15 to 15-32 (1990)
10. Meylan WM, Howard PH. Environmental Toxicology and Chemistry. In press Volume 10:1283-1293 (1991)
11. Mullins RM; Kirk-Othmer Encycl Chem Technol 3rd ed NY: John Wiley & Sons 1: 944-60 (1978)
12. NIOSH; National Occupational Exposure Survey (NOES) (1983)
13. Speece RE; Environ Sci Technol 17: 416A-27A (1983)
14. Swann RL et al; Res Rev 85: 23 (1983)
15. Weast RC et al. CRC Handbook of Chemistry and Physics. CRC Press, Inc: Boca Raton, FLA p C-447 (1985)

Benzyl Alcohol

SUBSTANCE IDENTIFICATION

Synonyms: (Hydroxymethyl)benzene

Structure:

CAS Registry Number: 100-51-6

Molecular Formula: C_7H_8O

Wiswesser Line Notation: Q1R

CHEMICAL AND PHYSICAL PROPERTIES

Boiling Point: 204.7 °C

Melting Point: -15.19 °C

Molecular Weight: 108.13

Dissociation Constants:

Log Octanol/Water Partition Coefficient: 1.10 [18]

Water Solubility: 40 g/L [39]

Vapor Pressure: 0.11 mm Hg at 25 °C [38]

Henry's Law Constant: 3.91×10^{-7} atm cu-m/mole at 25 °C, calculated from vapor pressure and water solubility

ENVIRONMENTAL FATE/EXPOSURE POTENTIAL

Summary: Benzyl alcohol may enter the environment through fugitive emissions during its production, and during its formulation and use in commercial products. It may also enter the environment from the exhaust of motor vehicles. If released to soil, benzyl alcohol is expected to display high mobility and readily leach through soil. Volatilization from dry soil to the atmosphere may be an important fate process; however, it is not expected to be an important process in moist soils. Microbial degradation in soil may occur, based on limited data. If released to water, benzyl alcohol is expected to undergo microbial degradation under aerobic and anaerobic conditions. Neither volatilization to the atmosphere, hydrolysis, direct photolytic degradation, chemical oxidation, bioconcentration in fish and aquatic organisms, nor adsorption to sediment and suspended organic matter are expected to be significant processes in environmental waters. In the atmosphere, benzyl alcohol is expected to exist almost entirely in the vapor phase. The estimated half-life for the vapor phase reaction of benzyl alcohol with photochemically produced hydroxyl radicals is 2 days. Its water solubility suggests that benzyl alcohol may undergo deposition to the surface by rain washout and other wet deposition processes. Occupational exposure to benzyl alcohol may occur by dermal contact and inhalation during its production or formulation. Benzyl alcohol finds use in numerous consumer products which may result in exposure by dermal contact, inhalation, or ingestion during their use.

Natural Sources: Benzyl alcohol occurs either free or as an ester in oils of jasmine, castoreum, gardenia, and ylang-ylang, and in balsams of Peru and Tolu [39].

Artificial Sources: Data from the USEPA TSCA production file [47] indicates that during 1977, 24 USA manufacturing plants produced between 115 and 1,150 thousand pounds of benzyl alcohol. More recent production data could not be located. Benzyl alcohol finds use in photographic developers for color films, and as a degreasing agent in rug cleaners. It is used as a solvent for dyestuffs, ball point inks, cellulose esters, casein, waxes, etc. It is also used for the preservation of aqueous and oily parenteral drugs, in cough syrups, ointments, ophthalmic, burn, and dental solutions, insect ointments and repellents, and dermatological aerosol sprays. Benzyl alcohol is used in cosmetics, such as nail lacquers and hair dyes. The aliphatic esters of benzyl alcohol are used in flavors, fragrances, soaps, and perfumes [39,40]. It may be released to the environment during the formulation, use or disposal of these products. Benzyl alcohol can also be released to the environment in the exhaust from gasoline and diesel engines [17,36,41], in waste water emissions during its manufacture, formulation, and use in commercial products [10,26]. Benzyl alcohol may also be released to the environment in the effluent from sewage treatment plants [15,16] and in the leachate from landfills [1,19]. Benzyl alcohol may be released to the atmosphere from incinerator emissions [25].

Terrestrial Fate: If released to soil, benzyl alcohol is expected to display high mobility and readily leach through soil [6,43,44]. Microbial degradation in soil may occur, based on limited data [8,30]. Volatilization from dry soil to the atmosphere may be an important fate process [39]; however, it is not expected to be a significant process in moist soils [22].

Aquatic Fate: If released to water, benzyl alcohol is expected to undergo microbial degradation under aerobic [3,5,21,31,33,34,46] and anaerobic [4,24,42] conditions.

Atmospheric Fate: Based on the vapor pressure, benzyl alcohol is expected to exist almost entirely in the vapor phase in the ambient atmosphere [14]. The estimated half-life for the vapor phase reaction of benzyl alcohol with photochemically produced hydroxyl radicals in the atmosphere is 2 days [2]. Based on its water solubility, it may undergo dissolution into clouds and subsequently be removed from the atmosphere in precipitation.

Biodegradation: Benzyl alcohol underwent 70% biological oxygen demand in 5 days under aerobic conditions using an acclimated mixed microbial culture [3]. At an initial concentration of 250 ppm, benzyl alcohol underwent 29% theoretical oxidation after 12 hours in a sewage sludge acclimated to this compound, and 31% oxidation in a sludge acclimated to mandelic acid [34]. At an initial concentration of 500 ppm, it underwent 52%, 42%, and 43% theoretical oxidation in 12 hours using a settled sewage sludge acclimated to phenol, benzoic acid, and catechol, respectively, under aerobic conditions [34]. It is listed as a synthetic organic chemical easily biodegradable by biological sewage treatment [46]. Benzyl alcohol at an initial concentration of 500 mg/L was shown to undergo rapid oxygen uptake under aerobic conditions when inoculated with municipal sewage sludge [31,33]. It underwent 48% biological oxygen demand in 5 days using a sewage sludge seed [21]. Benzyl alcohol underwent 60.8% degradation using an industrial sludge inoculum under aerobic conditions in 5 days [5]. Under anaerobic conditions, benzyl alcohol underwent 100% mineralization within 2 weeks when inoculated with a municipal digester sludge [24]. Benzyl alcohol at an initial concentration of 50 ppm underwent greater than 75% mineralization to carbon dioxide and methane within 8 weeks using a municipal sewage sludge inocula under anaerobic conditions [42]. Using sediment from an anoxic salt marsh, 10 mM benzyl alcohol underwent degradation to carbon dioxide and methane after a 2-month incubation period [4].

Abiotic Degradation: Based on an experimentally determined rate constant for the reaction of benzyl alcohol with alkylperoxy radicals, 2.4 L/mole-s [20] and an estimated alkylperoxy concentration in water of 1 x 10^{-9} mole/L [35], the half-life for this reaction is 9 years. The half-life for the reaction of benzyl alcohol with photochemically produced hydroxyl radicals in water can be estimated at approximately 100 days using an experimentally determined rate constant of 8.4 x 10^{+9} L/mole-s [11] and an optimal hydroxyl radical concentration of 1 x 10^{-17} mole/L in natural waters [35]. Exposure of benzyl alcohol to sunlight for 4 hours in natural water did not produce any detectable oxidizing species (detection limit 1.5 uM), demonstrating that photochemical induced oxidation did not occur within that time frame [12].

Bioconcentration: Based on the log octanol/water partition coefficient (Kow), a bioconcentration factor of 4.0 can be calculated from an appropriate regression equation [32]. This value implies that bioconcentration in fish and aquatic organisms will not occur to any significant extent.

Soil Adsorption/Mobility: Experimental Koc values for benzyl alcohol are <5 for three different soils; Apison (0.11% organic carbon), Fullerton (0.06% organic carbon), and Dormont (1.2% organic carbon) [43]. An experimental Koc of 15.6 was determined for benzyl alcohol on a red-brown Australian soil (1.09 % organic carbon) [6,7]. These values suggest that benzyl alcohol will display very high mobility in soil [7].

Benzyl Alcohol

Volatilization from Water/Soil: The calculated Henry's Law constant is in agreement with an estimated value of 2.28×10^{-7} atm cu-m/mol at 25 °C [22]. These values suggest that volatilization of benzyl alcohol from water should not be an important fate process. The estimated volatilization half-life from a model river 1 m deep, flowing at 1 m/sec, and a wind velocity of 3 m/sec is 97 days [32]. The Henry's Law constant suggests that volatilization from moist soil will not be an important process; however, the vapor pressure suggests that volatilization from dry soil may occur.

Water Concentrations: GROUND WATER: Benzyl alcohol was found in a ground water sample taken at an unauthorized waste site, at a concentration of 170 ug/L [48].

Effluent Concentrations: Benzyl alcohol has been identified in the waste water effluent from the photographic processing industry [10] and in the effluent from kraft paper mills in 6 out of 8 samples at a concentration up to 0.025 mg/L [28]. Benzyl alcohol has been qualitatively determined in the secondary effluent from wastewater treatment plants in Illinois [15]. The effluent from a Los Angeles county waste water treatment plant contained 500 ug/L of benzyl alcohol [16]. Benzyl alcohol has been detected in the leachate from a Barcelona, Spain, sanitary landfill [1] and from a municipal refuse disposal site in the Netherlands [19]. Benzyl alcohol was qualitatively detected in the wastewater of a petrochemical company producing olefins and oxygenated hydrocarbons [27]. Benzyl alcohol was found in the effluent of 1 out of 4 test waste incinerators [25].

Sediment/Soil Concentrations:

Atmospheric Concentrations:

Food Survey Values: Benzyl alcohol has been identified as a volatile flavor component of baked potatoes [9], Beaufort (Gruyere) cheese [13], bacon [23], and roasted filberts (nuts) [29]. It has been identified as a volatile component of blended nectarines, but not in a headspace analysis of the intact fruit [45].

Plant Concentrations:

Benzyl Alcohol

Fish/Seafood Concentrations:

Animal Concentrations:

Milk Concentrations:

Other Environmental Concentrations:

Probable Routes of Human Exposure: Exposure to benzyl alcohol may occur by dermal contact and inhalation during its production or formulation into commercial products of during the use of commercial products in which it is contained.

Average Daily Intake:

Occupational Exposure: NIOSH (NOES Survey 1981-1984) has statistically estimated that 273,896 workers are exposed to benzyl alcohol in the USA [37]. Benzyl alcohol find use in perfumes, the preservation of aqueous and oily parenteral drugs, ointments, ophthalmic and burn solutions, dermatological aerosol sprays, and cosmetics [39,40]. Dermal exposure to the general population is thus a necessary outcome during the use of these products. Benzyl alcohol also finds use in flavors, in cough syrups, and dental solutions [39,40], thus, ingestion of benzyl alcohol will occur during the use of these common products.

Body Burdens:

REFERENCES

1. Albaiges J et al; Wat Res 20: 1153-9 (1986)
2. Atkinson R et al; Chem Rev 85: 69-201 (1985)
3. Babeu L, Vaishnav DD; J Ind Microbiol 2: 107-15 (1987)
4. Balba MTM et al; Biochem Soc Trans 9: 230-1 (1981)
5. Belly RT, Goodhue CT; Proc Int Biodegrad Symp 3: 1103-7 (1976)
6. Briggs GG; Aust J Soil Res 19: 61-8 (1981)
7. Briggs GG; J Agric Food Chem 29: 1050-9 (1981)
8. Chambers CW et al; J Water Pollut Contr Fed 35: 1517-28 (1963)
9. Coleman EC et al; J Agric Food Chem 29: 42-8 (1981)
10. Dagon TJ; J Water Pollut Control Fed 45: 2123-35 (1973)
11. Dorfman LM, Adams GE; Reactivity of the Hydroxyl Radical in Aqueous Solution NSRD-NBS-46 Washington DC: National Bureau of Standards pp 51 (1973)

Benzyl Alcohol

12. Draper WM, Crosby DG; Arch Environ Contam Toxicol 12: 121-126 (1983)
13. Dumont JP, Adda J; J Agric Food Chem 26: 364-7 (1978)
14. Eisenreich SJ et al; Environ Sci Tech 15: 30-38 (1981)
15. Ellis DD et al; Arch Environ Contam Toxicol 11: 373-82 (1982)
16. Gossett RW et al; Mar Pollut Bull 14: 387-92 (1983)
17. Hampton CV et al; Environ Sci Technol 16: 287-98 (1982)
18. Hansch C, Leo AJ; Medchem project Issue No. 26 Claremont, CA Pomona College (1985)
19. Harmsen J; Water Res 17: 699-705 (1983)
20. Hendry DG et al; J Phys Chem Ref Data 3: 937-78 (1974)
21. Heukelekian H, Rand MC; J Water Pollut Contr Assoc 29: 1040-53 (1955)
22. Hine J, Mookerjee PK; J Org Chem 40: 292-8 (1975)
23. Ho CT et al; J Agric Food Chem 31: 336-42 (1983)
24. Horowitz A et al; Dev Ind Microbiol 23: 435-44 (1982)
25. James RH et al; J Proc APCA 77th Ann Meeting: Paper 84-18.5 pp 1-25 (1984)
26. Keith LH et al; pp 327-73 in Ident Anal Org Pollut Water, Keith LH Ed Ann Arbor MI: Ann Arbor Press (1976)
27. Keith LH; Sci Total Environ 3: 87-102 (1974)
28. Keith LH; Environ Sci Technol 10: 555-64 (1976)
29. Kinlin TE et al; J Agr Food Chem 20: 1021-8 (1972)
30. Kramer N, Doetsch RN; Arch Biochem Biophys 26: 401-5 (1950)
31. Lutin PA et al; Purdue Univ Eng Bull Ext Series 118: 131-45 (1965)
32. Lyman WJ et al; Handbook of Chemical Property Estimation Methods. McGraw Hill NY pp 5-1 to 5-30, 15- 1 to 15-29 (1982)
33. Marion CV, Malaney, GW; Proc Ind Waste Cong 18: 297-308 (1964)
34. McKinney RE et al; Sew Indust Wastes 28: 547-57 (1956)
35. Mill T et al; Science 207; 886-7 (1980)
36. Mulawa PA, Cadle SH; Anal Lett 14: 671-88 (1981)
37. NIOSH; National Occupational Exposure Survey (NOES) (1984)
38. Riddick JA et al; Organic Solvents: Physical Properties and Methods of Purification 4th ed Wiley Interscience NY (1986)
39. Ringk W, Theimer ET; Kirk-Othmer Encycl Chem Tech 3rd Ed. John-Wiley NY 3: 793-802 (1978)
40. Sax NI, Lewis RJSR; Hawley's Condensed Chemical Dictionary 11th ed NY: Van Nostrand Reinhold Co p 134-5 (1987)
41. Seizinger DE, Dimitriades B; J Air Pollut Control Assoc 22: 47-51 (1972)
42. Shelton DR, Tiedje JM; App Env Microbiol 47: 850-7 (1984)
43. Southworth GR, Keller JL; Water Air Soil Poll 28: 239-48 (1986)
44. Swann RL et al; Res Rev 85: 17-28 (1983)
45. Takeoka GR et al; J Agric Food Chem 36: 553-60 (1988)
46. Thom NS and Agg AR; Proc Royal Soc Lond B 189: 347-57 (1975)
47. USEPA; Nonconfidential Initial TSCA Inventory. Washington DC Off Toxic Sub (1977)
48. USEPA; Superfund Record of Decision USEPA/ROD/RO2-86/031 (1987)

Bis(2-chloroethoxy)methane

SUBSTANCE IDENTIFICATION

Synonyms:

Structure:

CAS Registry Number: 111-91-1

Molecular Formula: $C_5H_{10}Cl_2O_2$

Wiswesser Line Notation: G20102G

CHEMICAL AND PHYSICAL PROPERTIES

Boiling Point: 218.1 °C

Freezing Point: -32.8 °C

Molecular Weight: 173.05

Dissociation Constants:

Log Octanol/Water Partition Coefficient: 0.750 [21] (estimated)

Water Solubility: 1.2 x 10^5 mg/L at 25 °C [21] (estimated from the estimated octanol/water partition coefficient)

Vapor Pressure: 1.4 x 10^{-4} mm Hg at 25 °C [14]

Henry's Law Constant: 1.7 x 10^{-7} atm m^3/mol at 25 °C [7]

ENVIRONMENTAL FATE/EXPOSURE POTENTIAL

Summary: Bis(2-chloroethoxy)methane is a synthetic organic compound, chiefly used on site in the production of polysulfide polymers. If released to soil, bis(2-chloroethoxy)methane would be expected to display high to very high mobility. Biodegradation of bis(2-chloroethoxy)methane is not expected to be an important fate process. Volatilization from water should be a slow process; the estimated half-life for volatilization from a model pond is 11 years. Hydrolysis of bis(2-chloroethoxy)methane can be expected by comparison to other chlorine containing compounds; the half-life for this pH-independent process has been estimated at 0.5-2 years. Direct photochemical degradation of bis(2-chloroethoxy)methane in the atmosphere or in the upper layers of surface waters should not be an important fate process. The estimated half-life for the atmospheric reaction with photochemically produced hydroxyl radicals is 10 hours. The probable route of exposure to bis(2-chloroethoxy)methane is through dermal contact and inhalation during its manufacture and formulation in polymers, and during its use as a solvent.

Natural Sources: Bis(2-chloroethoxy)methane is an anthropogenic compound, and is not believed to occur in nature.

Artificial Sources: The 1977 TSCA inventory listed one company which produced bis(2-chloroethoxy)methane. Production volume was given as ten to fifty million pounds, which was used on site [20]. The vast majority of bis(2-chloroethoxy)methane is used in the synthesis of polysulfides, and it also finds use as a solvent [5,15]. Release of bis(2-chloroethoxy)methane to the environment could occur by volatilization during its manufacture and formulation in polysulfides, and during its use as a solvent.

Terrestrial Fate: If released to soil, Koc values for bis(2-chloroethoxy)methane lying in the range 7-115 can be calculated [2,8,21,22], suggesting high to very high mobility for bis(2-chloroethoxy)methane [17]. Biodegradation should not be an important fate process [18,19]. The vapor pressure and the estimated Henry's Law constant suggests that volatilization from dry soil to the atmosphere might occur, but that it should be a slow process; volatilization from moist soil, however, should not be an important fate process.

Aquatic Fate: If released to water, bis(2-chloroethoxy)methane would not be expected to biodegrade [18,19]. The Henry's Law constant suggests that volatilization to the atmosphere will not be an important fate process. The estimated volatilization half-life from a model pond is

Bis(2-chloroethoxy)methane

11 years [22]. The hydrolysis half-life for similar β-chloro ethers has been estimated at 0.5-2 years, a process which is pH independent [2].

Atmospheric Fate: If released to the atmosphere, bis(2-chloroethoxy)methane should react with photochemically produced hydroxyl radicals; the half life for this process can be estimated at 10 hours [1]. Its high water solubility suggests that rain washout may be an important fate process. Direct photochemical degradation would not be expected for bis(2-chloroethoxy)methane [2].

Biodegradation: Bis(2-chloroethoxy)methane was reported as undergoing 0% biodegradation using settled domestic wastewater as an inoculum; biodegradation remained at 0% through three subcultures. Bis(2-chloroethoxy)methane was thus listed as being not significantly degraded under the conditions of the method, the lowest of a three-tier rating system [18,19]. When the influent to a full scale activated sludge treatment system was spiked with 0.24 ug/L bis(2-chloroethoxy)methane, 60% of the pollutant was removed [13]; however, it was not reported if removal was due to biodegradation or some other removal process.

Abiotic Degradation: Bis(2-chloroethoxy)methane can be decomposed by mineral acids [15], and thus decomposition in highly acidic waters might be a method for degradation, although this situation should not commonly occur in the environment. The estimated half-life for hydrolysis was reported as 0.5-2 years (pH independent), based on analogy to similar β-chloro ethers [2]. The estimated half-life for the atmospheric reaction with photochemically produced hydroxyl radicals is 10.2 hours [1]. Direct degradation by photolysis should not be an important fate process since bis(2-chlorothoxy)methane does not have any chromophores that absorb radiation in the visible or near ultraviolet regions of the spectrum [2].

Bioconcentration: Based on the estimated log octanol/water partition coefficient (Kow) and the estimated water solubility, the bioconcentration factors are estimated to lie in the range 0.84-2.2, using several recommended regression equations [8]. Therefore, bioconcentration in aquatic organisms should not be an important fate process. Bis(2-chloroethoxy)methane was not found in a wide variety of fish collected from fourteen Lake Michigan tributaries and embayments (detection limits not given) [3].

Soil Adsorption/Mobility: Based on estimated log octanol/water partition coefficients (Kow) and an estimated water solubility, Koc values which lie in the range 7 to 61 can be estimated using several recommended regression equations [8]. Therefore bis(2-chloroethoxy)methane should have high to very high mobility in soil [17].

Volatilization from Water/Soil: The low Henry's Law constant for bis(2-chloroethoxy)methane suggests that its volatilization from water might be slower than for water itself, and thus should not be an important fate process [9]. The estimated volatilization half-life from a model pond is 10.9 years [22]. The vapor pressure suggests that volatilization from dry soil might occur, but that it should not be a rapid process. Based upon the Henry's Law constant, volatilization from moist soil will also be very slow.

Water Concentrations: SURFACE WATER: USEPA STORET Data Base, for bis(2-chloroethoxy)methane 834 samples, 0.1% positive, median concn less than 10 ug/L [16]. Not detected in eighty-six samples from fifty-one rainwater runoff catchments located throughout the USA (detection limits not given) [4]. Identified in one of eleven samples and three of ten samples along the Ohio River, Wheeling, WV, (mile point 86.8) and Evansville, KT, (mile point 791.5), respectively, in concn less than or equal to 0.1 ug/L, 1977-1978. It was not found in eighty-two other stations along the river [11].

Effluent Concentrations: Bis(2-chloroethoxy)methane was qualitatively detected in disposal trench leachate samples collected from commercially operated low-level radioactive disposal sites at Maxey Flats, KY, and at West Valley, NY [6]. USEPA STORET Data Base, for bis(2-chloroethoxy)methane 1,243 samples, 1.0% positive, median concn less than 10 ug/L [16]. Detected in eleven of one hundred twenty-nine raw water extracts at eleven water utilities in the Ohio River Basin, 1977-78, at concn less than 0.1 ug/L [11].

Sediment/Soil Concentrations: USEPA STORET Data Base, for bis(2-chloroethoxy)methane 344 samples, 0% positive, median concn less than

Bis(2-chloroethoxy)methane

500 ug/kg [16]. Bis(2-chloroethoxy)methane was not detected in sediment samples in Lake Pontchartrain, LA, but was found in benthic organisms of the lake [10].

Atmospheric Concentrations:

Food Survey Values:

Plant Concentrations:

Fish/Seafood Concentrations: Bis(2-chloroethoxy)methane was not found in a wide variety of fish collected from fourteen Lake Michigan tributaries and embayments (detection limits not given) [3]. Detected in Chef Menteur Clams in Lake Pontchartrain, LA, at a concn of 12 ng/g wet weight (composite sample) [10].

Animal Concentrations:

Milk Concentrations:

Other Environmental Concentrations:

Probable Routes of Human Exposure: The probable route of exposure for bis(2-chloroethoxy)methane is by inhalation and dermal contact during its manufacture and formulation in polymers, and during its use as a solvents.

Average Daily Intake:

Occupational Exposure:

Body Burdens:

REFERENCES

1. Atkinson R; Int J Chem Kinet 19: 799-828 (1987)
2. Callahan MA et al; Water Related Fate of 129 Priority Pollutants Vol II USEPA-440/4-79-029B (1979)
3. Camanzo J et al; J Great Lakes Res 13: 296-309 (1987)
4. Cole RH et al; J Water Pollut Contr Fed 56: 898-908 (1984)

Bis(2-chloroethoxy)methane

5. Ellerstein SM, Bertuzzi ER; Kirk-Othmer Encycl Chem Tech 3rd Ed. John-Wiley NY 18: 814-31 (1981)
6. Francis AJ et al; Nuc Tech 50: 158-63 (1980)
7. Hine J, Mookerjee PK; J Org Chem 40: 292-8 (1975)
8. Lyman WJ et al; Handbook of Chemical Property Estimation Methods NY: McGraw-Hill pp 4-1 to 4-33 (1982)
9. Lyman WJ et al; Handbook of Chemical Property Estimation Methods NY: McGraw-Hill NY pp 15-15 to 15-29 (1982)
10. McFall JA et al; Chemosphere 14: 1561-9 (1985)
11. Ohio River Valley Water Sanit Comm; Water Treatment Process Modification for Trihalomethane Control and Organic Substances in the Ohio River. EPA Grant R-804615 (1979)
12. Ohio River Valley Water Sanit Comm; Assessment of Water Quality Conditions. Ohio River Mainstream 1978-9 Ohio River Valley Water Sanit Comm Cincinnati OH (1980)
13. Patterson JW, Kodukala PS; Chem Eng Prog 77: 48-55 (1981)
14. Perry RH et al; Perry's Chemical Engineers' Handbook McGraw Hill (1984)
15. Sax NI, Lewis RJ SR; Hawley's Condensed Chemical Dictionary 11th ed Van Nostrand Reinhold Co. NY (1987)
16. Staples CA et al: Environ Toxicol Contam 4: 1314-42 (1985)
17. Swann RL et al; Res Rev 85: 17-28 (1983)
18. Tabak HH et al; pp 267-328 in Test Protocols for Environmental Fate and Movement of Toxicants. Proc Symp Assoc Off Anal Chem, 94th Ann Mtg Washington DC (1981)
19. Tabak HH et al; J Water Pollut Contr Fed 53: 1503-18 (1981)
20. USEPA; Nonconfidential Initial TSCA Inventory. Washington DC Office of Toxic Substances (1977)
21. USEPA; GEMS Graphic Exposure Modeling System CLOGP (1987)
22. USEPA; Exams II Computer Simulation (1987)

1-Bromo-2-chloroethane

SUBSTANCE IDENTIFICATION

Synonyms:

Structure:

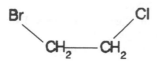

CAS Registry Number: 107-04-0

Molecular Formula: C_2H_4BrCl

Wiswesser Line Notation:

CHEMICAL AND PHYSICAL PROPERTIES

Boiling Point: 107 °C at 760 mm Hg

Melting Point: -16.7 °C

Molecular Weight: 143.41

Dissociation Constants:

Log Octanol/Water Partition Coefficient: 1.598 (estimated) [4]

Water Solubility: 6,830 mg/L at 30 °C [11]

Vapor Pressure: 40 mm Hg at 29.7 °C [11]

Henry's Law Constant: 9.08×10^{-4} atm-m^3/mole (estimated) [6]

ENVIRONMENTAL FATE/EXPOSURE POTENTIAL

Summary: 1-Bromo-2-chloroethane may be released into the environment via air and wastewater during its production and use as a solvent, chemical intermediate, and fumigant. If released on soil as a result of accidental spill or disposal, 1-bromo-2-chloroethane would evaporate from the soil surface or leach into the soil. If released into water, it would primarily be lost by volatilization (half-life 4.7 hr from a model river). Hydrolysis is possible but hydrolysis rates are unknown. Bioconcentration in aquatic organisms and adsorption to sediment would not be significant. It is not known whether 1-bromo-2-chloroethane will undergo photolytic and microbial reactions in water and soil. In the atmosphere, 1-bromo-2-chloroethane will degrade very slowly by reacting with photochemically produced hydroxyl radicals, disperse and be scavenged by rain. Human exposure will be primarily occupational.

Natural Sources:

Artificial Sources: 1-Bromo-2-chloroethane may be released into air and wastewater during its production and use as a solvent for cellulose esters and ethers, chemical intermediate, and fumigant for fruits and vegetables [5]. It can be released into soil by accidental spill or as result of disposal in waste sites.

Terrestrial Fate: If released on soil, 1-bromo-2-chloroethane may leach into the ground and will evaporate from the soil surface. Pertinent data on microbial and photochemical fate of the compound in soil is not available.

Aquatic Fate: If released into water, 1-bromo-2-chloroethane would primarily be lost by volatilization (half-life 4.7 hr from a model river). Degradation due to hydrolysis is possible but no data regarding rates could be found. Adsorption to sediment would not be significant. The photolytic and biodegradative fate of the compound in water is unknown.

Atmospheric Fate: In the atmosphere, 1-bromo-2-chloroethane will degrade very slowly by reacting with photochemically produced hydroxyl radicals (half-life 49.4 days). It is soluble in water and as a result should be scavenged by rain. Due to its persistence towards chemical reactions in the troposphere, considerable dispersion from source areas would be expected.

Biodegradation:

1-Bromo-2-chloroethane

Abiotic Degradation: Based on an estimation method [1], the rate constant for the reaction of 1-bromo-2-chloroethane with OH radicals is 3.248×10^{-13} cm^3/molecule-sec. Using the estimated rate constant and the concentration of OH radicals in unpolluted air as 5×10^5/cm^3, 1-bromo-2-chloroethane will degrade in the atmosphere by reacting with photochemically produced hydroxyl radicals by H-atom abstraction with an estimated half-life of 49.4 days. Alkyl halides are susceptible to hydrolysis by neutral- and base-catalyzed reactions to give alcohols. The hydrolysis half-lives of ethyl chloride and ethyl bromide are 38 and 30 days, respectively, at pH 7 [9]. However, the rate of hydrolysis for polyhalides is usually slower than simple halides [9]. No experimental hydrolysis data could be found for 1-bromo-2-chloroethane. Since it is a polyhalide, the hydrolysis half-life may be longer than both ethyl chloride and ethyl bromide.

Bioconcentration: Using the estimated log octanol/water partition coefficient of 1.598 [4] and a recommended regression equation [8], the BCF for 1-bromo-2-chloroethane can be estimated as 9.6. Therefore 1-bromo-2-chloroethane would not be expected to bioconcentrate in aquatic organisms.

Soil Adsorption/Mobility: Using the water solubility of 6,830 mg/L at 30 °C [11] and a recommended regression equation [8], the Koc for 1-bromo-2-chloroethane can be estimated as 34. Therefore 1-bromo-2-chloroethane would not adsorb appreciably to soil and sediment.

Volatilization from Water/Soil: Using the Henry's Law constant of 9.08×10^{-4} atm-m^3/mol [6] and an estimation method for determining the diffusion-controlled volatilization rate of a compound in water [8], the volatilization half-life of 1-bromo-2-chloroethane from a model river 1 m deep with a 1 m/sec current and a 3 m/sec overhead wind speed can be estimated to be 4.7 hr. 1-Bromo-2-chloroethane has a high vapor pressure (40 mm Hg at 29.7 °C [11]) and will evaporate rapidly from dry soil surfaces as well.

Water Concentrations: DRINKING WATER: 1-Bromo-2-chloroethane has been identified, but not quantified in unspecified drinking waters [7]. SURFACE WATER: In a sampling of 204 surface waters near major industrial areas across the continental USA, 1-bromo-2-chloroethane was detected, but not quantified, at one site [3].

Effluent Concentrations: 1-Bromo-2-chloroethane was found in water samples collected from a geographical area associated with the bromine industry [10].

Sediment/Soil Concentrations: 1-Bromo-2-chloroethane was found in sediment samples collected from a geographical area associated with the bromine industry [10].

Atmospheric Concentrations: RURAL/REMOTE: Not detected at the Grand Canyon, AZ [2]. SOURCE AREAS: Three source areas in the USA: Edison, NJ, El Dorado, AK, and Magnolia, AK (74 samples) - 6.1 ppt median, 1100 ppt maximum [2]. It was not found in three routine EPA field monitoring samples [12].

Food Survey Values:

Plant Concentrations:

Fish/Seafood Concentrations:

Animal Concentrations:

Milk Concentrations:

Other Environmental Concentrations:

Probable Routes of Human Exposure: Exposure to 1-bromo-2-chloroethane is primarily occupational by inhalation and dermal contact.

Average Daily Intake:

Occupational Exposure:

Body Burdens:

1-Bromo-2-chloroethane

REFERENCES

1. Atkinson R; Environ Toxicol Chem 7:435-62 (1988)
2. Brodzinsky R, Singh HB; Volatile Organic Chemicals In The Atmosphere: An Assessment Of Available Data. Contract 68-02-3452 Menlo Park CA: SRI International (1982)
3. Ewing BB et al; Monitoring To Detect Previously Unrecognized Pollutants In Surface Waters EPA-560/6-77-015 (1977)
4. GEMS; Graphical Exposure Modeling System. CLOGP. USEPA (1987)
5. Hawley GG; Condensed Chem Dictionary 10th ed Van Nostrand Reinhold NY p 151 (1981)
6. Hine J, Mookerjee PK; J Org Chem 40: 292-8 (1975)
7. Kool HJ et al; Crit Rev Environ Control 12: 307-57 (1982)
8. Lyman WJ et al; Handbook of Chem Property Estimation Methods. NY: McGraw-Hill pp 4-1 to 4-33, 5-1 to 5-34 (1982)
9. Mabey W, Mills T; J Phy Chem Ref Data 7: 383-415 (1978)
10. Pellizzari ED et al; Environmental Monitoring near Industrial Sites: Brominated Chemicals USEPA Contract 68-01-1978 (1978)
11. Riddick JA et al; Organic Solvents. Physical Properties and Methods of Purification. 4th ed. New York, NY: John Wiley & Sons pp 572-73 (1986)
12. Scott DR et al; Environ Sci Technol 21: 891-7 (1987)

Bromochloromethane

SUBSTANCE IDENTIFICATION

Synonyms:

Structure:

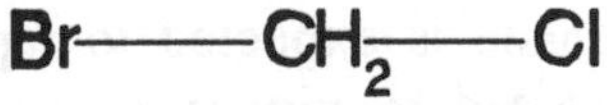

$$Br\!-\!\!-\!CH_2\!-\!\!-\!Cl$$

CAS Registry Number: 74-97-5

Molecular Formula: CH_2BrCl

Wiswesser Line Notation: G1E

CHEMICAL AND PHYSICAL PROPERTIES

Boiling Point: 68.1 °C

Melting Point: -86.5 °C

Molecular Weight: 129.38

Dissociation Constants:

Log Octanol/Water Partition Coefficient: 1.41 [6]

Water Solubility: 16,700 mg/L [25]

Vapor Pressure: 147.2 mm Hg at 25 °C [15]

Henry's Law Constant: 0.0015 atm-m^3/mol (estimated from vapor pressure and water solubility data)

ENVIRONMENTAL FATE/EXPOSURE POTENTIAL

Summary: Bromochloromethane, which finds use in fire extinguishers, may be released to the environment as a fugitive emission during its manufacture, and during the use of fire extinguishers containing the

41

compound. If released to the soil, bromochloromethane is expected to display high mobility and it has the potential to leach into ground water. Volatilization from the soil surface to the atmosphere is expected to be a significant process. Limited data suggests that the microbial degradation of this compound may occur in soil under anoxic conditions. If released to water, bromochloromethane is expected to rapidly volatilize to the atmosphere. It is not expected to bioconcentrate in fish and aquatic organisms, nor is it expected to adsorb to sediment and suspended organic matter. Limited data indicate that microbial degradation of bromochloromethane under aerobic conditions may occur. Hydrolysis and direct photochemical degradation are not expected to be significant processes. In the atmosphere, bromochloromethane is expected to exist predominately in the vapor phase. The vapor phase reaction with photochemically produced hydroxyl radicals and direct photochemical degradation are not expected to be significant processes in the atmosphere. The relatively high water solubility of this compound suggests that wet deposition may occur; however, bromochloromethane deposited by this process would be expected to revolatilize to the atmosphere. Because of its expected long lifetime in the troposphere, some of tropospheric compound may be transported to the stratosphere. Occupational exposure to bromochloromethane may occur by inhalation and dermal contact during its manufacture and formulation into fire extinguishers. Low levels of exposure to the general population may occur by ingestion of contaminated water of low bromochloromethane concentration. The level of exposure may be higher as a result of inhalation or dermal contact after the discharge of fire extinguishers containing this compound.

Natural Sources: Bromochloromethane along with other bromo- or chloromethanes may be produced by macro algae using haloperoxidase enzymes [3]. However, when brown algae from the Sargasso Sea (Fucales sargassum) was incubated in water, dichlorobromomethane and dibromochloromethane were detected in the trapped volatile compounds, but no bromochloromethane was detected [3].

Artificial Sources: Bromochloromethane finds use as a fire extinguisher fluid, especially in aircraft and in portable fire extinguishing units [16,17]. It may enter the environment as a fugitive emission during its production and formulation, and during the discharge of fire extinguishers containing this compounds.

Terrestrial Fate: If released to soil, bromochloromethane is expected to display high mobility and it has the potential to leach into ground water. Volatilization from the soil surface to the atmosphere is expected to be an important fate process. Based on limited data, microbial degradation of bromochloromethane may occur in soil under anoxic conditions.

Aquatic Fate: If released to water, bromochloromethane is expected to rapidly volatilize to the atmosphere. The estimated half-life for volatilization from a model river is approximately 4 hours [12,15,23,25]. Limited data suggest that this compound may undergo microbial degradation under aerobic conditions. Based on estimation methods [12] and the known water solubility of bromochloromethane of 16,700 mg/L at 25 °C [23,25], the bioconcentration of the compound in fish and aquatic organisms as well as adsorption to sediment and suspended organic matter may not be important. Hydrolysis and direct photochemical degradation are not expected to occur in natural waters.

Atmospheric Fate: In the atmosphere, bromochloromethane is expected to exist predominantly in the vapor phase. The vapor phase reaction with photochemically produced hydroxyl radicals is not expected to be a significant fate process as the estimated half-life for this process is 160 days [1]. Direct photochemical degradation is not expected to be important since chlorobromoethane does not absorb light available in the troposphere [2]. The water solubility of bromochloromethane of 16,700 mg/L at 25 °C [23,25] suggests that physical removal by wet deposition may occur; however bromochloromethane deposited by this process is expected to re-volatilize to the atmosphere.

Biodegradation: In a screening test, bromochloromethane at an initial concn of 5 or 10 mg/L underwent 100% degradation within seven days using a settled domestic wastewater inoculum under aerobic conditions [21,22]. Complete degradation ensued with 3 successive subcultures [21,22]. Bromochloromethane has been reported to undergo microbial degradation under anoxic conditions when cultured with soil bacteria, although no details were provided [9].

Bromochloromethane

Abiotic Degradation: Bromochloromethane does not absorb UV light at wavelengths >290 nm [2]. Therefore, it is not expected to undergo direct photochemical degradation in the atmosphere or in water. Based on an estimation method [1], the estimated half-life for the vapor phase reaction of bromochloromethane with photochemically produced hydroxyl radicals in the atmosphere is 160 days. Therefore, this compound may persist in the atmosphere for a long period allowing its long distance transport in the troposphere and some transfer into the stratosphere. Hydrolysis of bromochloromethane in natural waters is not expected to be a significant process as the half-life for this process under natural conditions at 25 °C has been estimated at 44 years [13].

Bioconcentration: Based on the regression equations [12] and the water solubility of 16,700 mg/L at 25 °C [23,25] and log octanol/water partition coefficient of 1.41 [23,25] for bromochloromethane, the bioconcentration factor (BCF) can be estimated to be 3-7. This range of values suggests that bioconcentration of bromochloromethane in fish and aquatic organisms will not be important.

Soil Adsorption/Mobility: Based on regression equations and the water solubility of 16,700 mg/L at 25 °C [23,25], and log octanol/water partition coefficient of 1.41 [23,25] for bromochloromethane, soil adsorption coefficient (Koc) can be calculated to be in the range 21-139. This range of values suggest that bromochloromethane will display high to very high mobility in soil [20].

Volatilization from Water/Soil: Based on the water solubility of bromochloromethane of 16,700 mg/L at 25 °C [23,25] and its vapor pressure of 147.2 mm Hg at 25 °C [15], a Henry's Law constant of 0.0015 atm cu-m/mol at 25 °C can be calculated [12]. Using this value and an estimation method [12], the volatilization half-life from a model river 1 m deep, flowing at 1 m/sec, and a wind velocity of 3 m/sec is approximately 4 hours. The Henry's Law constant and the vapor pressure suggest that volatilization from both moist and dry soil will be important processes.

Water Concentrations: DRINKING WATER: Bromochloromethane was qualitatively detected in Philadelphia's drinking water supply during 1975-1977 [19]. It was identified in unspecified drinking waters [4,10].

Bromochloromethane

Bromochloromethane was qualitatively detected in treated, but not raw, drinking water in the UK [5]. SURFACE WATER: Bromochloromethane was detected in 15 out of 91 stations in Lake Ontario at a concn of trace to 10 ng/L [8]. It was detected in 1% of 83 water samples taken from Lake Ontario, 1981 [18]. The baseline concn of bromochloromethane in the open Atlantic Ocean is 0.02 ng/L [3]. It was qualitatively detected in Narragansett Bay, RI during 1979-81 [24]. Bromochloromethane was qualitatively detected in rivers in the UK [5]. GROUND WATER: Detected in ground water samples in the Netherlands, at a maximum concn of 8 ug/L [26].

Effluent Concentrations:

Sediment/Soil Concentrations:

Atmospheric Concentrations: RURAL/REMOTE: In 1983, bromochloromethane was detected in the concentration range of 2.3 to 3.1 ppt (v/v) (ground level) in the arctic at Point Barrow, AL [14]. The concentration of the compound was maximum during the winter and spring seasons [14]. The average concn of bromochloromethane also depended on the so-called arctic haze with a value 2.3 ppt (v/v) in the haze and 2.0 ppt (v/v) outside the haze [14]. The mean concn of bromochloromethane in air over the open Atlantic (30 deg S to 40 deg N) is 0.4 ppt (v/v) in samples taken below the tropospheric boundary level, and 0.3 ppt (v/v) above that boundary [3]. The baseline concn of this compound in air over the north Atlantic was 0.002 ng/L [3]. SOURCE DOMINATED: Bromochloromethane was qualitatively identified in the air at a hazardous waste site in NJ [11].

Food Survey Values:

Plant Concentrations:

Fish/Seafood Concentrations: Bromochloromethane was detected in tissue from rainbow trout collected from the Colorado River. The estimated concentration in the whole fish sample was 8 ppb (µg/kg) [7].

Animal Concentrations:

Bromochloromethane

Milk Concentrations:

Other Environmental Concentrations:

Probable Routes of Human Exposure: Exposure of the general population to low levels of bromochloromethane may occur by ingestion of contaminated drinking water, or to higher levels by inhalation and dermal contact during the use of fire extinguishers containing the compound.

Average Daily Intake:

Occupational Exposure: Occupational exposure to bromochloromethane may occur by inhalation or dermal exposure during its manufacture, and to fire fighters during the use of fire extinguishers containing this compound.

Body Burdens:

REFERENCES

1. Atkinson R et al; Chem Rev 85: 69-70 (1985)
2. Cadman P, Simons JP; Trans Faraday Soc 62: 631-41 (1966)
3. Class TH, Ballachmiter K; J Atmos Chem 6: 35-46 (1988)
4. Fawell JK, Fielding M; Sci Total Environ 47: 317-41 (1985)
5. Fielding M et al; Organic Micropollutants in Drinking Water TR-159 Medmenham, Eng. Water Res Cent 49 pp (1981)
6. Hansch C, Leo AJ; Medchem Project Issue No. 26. Claremont CA: Pomona College (1985)
7. Hiatt MH; Anal Chem 55: 506-16 (1983)
8. Kaiser KLE et al; J Great Lakes Res 9: 212-23 (1983)
9. Kobayashi H, Rittman BE; Environ Sci Tech 16: 170A-83A (1982)
10. Kool HJ et al; CRC Crit Rev Env Control 12: 307-57 (1982)
11. Laregina J et al; Environ Prog 5: 18-27 (1986)
12. Lyman WJ et al; Handbook of Chemical Property Estimation Methods. McGraw Hill NY pp 4-1 to 4- 33, 5-1 to 5-33, 15-1 to 15-29 (1982)
13. Mill T et al; Validation of Estimation Techniques for Predicting Environmental Transformation of Chemicals USEPA 68- 01-6269: Washington DC (1982)
14. Rasmussen RA, Khalil MA; Geophys Res Lett 11: 433-36 (1984)
15. Riddick JA et al; Organic Solvents: Physical Properties and Methods of Purification 4th ed Wiley Interscience NY pp 562-3 (1986)

Bromochloromethane

16. Sax NI, Lewis RJ; Hawley's Condensed Chemical Dictionary 11th ed NY: Van Nostrand Reinhold Co. p 171 (1987)
17. Stenger VA; Kirk-Othmer Encycl Chem Tech 3rd Ed John-Wiley NY 4: 243-263 (1978)
18. Strachan WMJ, Edwards CJ; Organic Pollutants in Lake Ontario. Adv Environ Sci Technol 14: 239-64 (1984)
19. Suffet IH et al; Water Res 14: 853-67 (1980)
20. Swann RL et al; Res Rev 85: 17-28 (1983)
21. Tabak HH e al; pp. 267-328 in Test Protocols for Environmental Fate and Movement of Toxicants. Proc of Sym Assoc Off Anal Chem, 94th Annual Mtg Washington DC (1981)
22. Tabak HH et al; J Water Pollut Contr Fed 53: 1503-18 (1981)
23. Tewari YB et al; J Chem Eng Data 27: 451-4 (1982)
24. Wakeham SG et al; Can J Fish Aq Sci 40: 304-21 (1983)
25. Wasik SP et al; Octanol Water Partition Coefficients and Aqueous Solubilities of Organic Compounds NBSIR81-2406. Washington DC: US Dept Comm Natl Bur Std pp 66 (1981)
26. Zoeteman BCJ et al; Sci Total Environ 21: 187-202 (1981)

2-Butoxyethyl Acetate

SUBSTANCE IDENTIFICATION

Synonyms: Ethylene glycol monobutyl ether acetate

Structure:

CAS Registry Number: 112-07-2

Molecular Formula: $C_8H_{16}O_3$

Wiswesser Line Notation: 4O2OV1

CHEMICAL AND PHYSICAL PROPERTIES

Boiling Point: 192.3 °C

Melting Point: -64.5 °C

Molecular Weight:: 160.24

Dissociation Constants:

Log Octanol/Water Partition Coefficient:

Water Solubility: 11,000 mg/L [7]

Vapor Pressure: 0.375 mm Hg at 20 °C [11]

Henry's Law Constant: 7.19 x 10^{-6} atm-m^3/mole (calculated from water solubility and vapor pressure)

2-Butoxyethyl Acetate

ENVIRONMENTAL FATE/EXPOSURE POTENTIAL

Summary: Ethylene glycol monobutyl ether acetate is directly released to the atmosphere by evaporation in its use as a solvent in paints, lacquers, thinners, inks, and resins. If released to the atmosphere, it will degrade primarily by reaction with photochemically produced hydroxyl radicals (estimated half-life of 11.8 hr). If release to soil or water, ethylene glycol monobutyl ether acetate is expected to degrade via biodegradation. One biodegradation screening study has demonstrated that ethylene glycol monobutyl ether acetate is readily biodegradable. It may leach readily in soils based upon an estimated Koc of 26. The importance of leaching may be lessened if rapid biodegradation occurs. The major routes of occupational exposure to ethylene glycol monobutyl ether acetate are inhalation and dermal contact. Exposure to the general population can also occur through inhalation and dermal contact.

Natural Sources:

Artificial Sources: The emission rate of ethylene glycol monobutyl ether acetate into the atmosphere from painting operations at an automotive assembly plant in Janesville, WI was estimated to be 37.9 gallons/hr [8]. Ethylene glycol monobutyl ether acetate is directly released to the atmosphere through its use as a solvent in paints, lacquers, thinners, inks, stains, and varnishes.

Terrestrial Fate: The dominant degradation process for ethylene glycol monobutyl ether acetate in soil is expected to be biodegradation. One biodegradation screening study has demonstrated that ethylene glycol monobutyl ether acetate is readily biodegradable [12]. Based upon an estimated Koc of 26, ethylene glycol monobutyl ether acetate is expected to leach readily in soil [4]. If rapid biodegradation occurs, the importance of leaching will be lessened.

Aquatic Fate: The dominant removal process for ethylene glycol monobutyl ether acetate in water is expected to be biodegradation. One biodegradation screening study has demonstrated that ethylene glycol monobutyl ether acetate is readily biodegradable [12]. Volatilization from water is relatively slow. The volatilization half-lives from a model environmental river (1 meter deep) and model pond have been estimated to be 6.6 and 74 days, respectively [4,10]. Aquatic bioconcentration and adsorption to sediment are not expected to be important.

Atmospheric Fate: Based upon its vapor pressure, ethylene glycol monobutyl ether acetate is expected to exist almost entirely in the vapor-phase in the ambient atmosphere [2]. It is expected to degrade by reaction with photochemically produced hydroxyl radicals with an estimated half-life of about 18.4 hr [1]. Physical removal via wet deposition is likely since it is relatively soluble in water.

Biodegradation: Ethylene glycol monobutyl ether acetate was determined to be "completely" biodegradable using the Zahn-Wellens screening method [12]; total degradation exceeded 90% with a measured rate of 12%/day under the test conditions [12]; no observable lag period was required before onset of degradation [12]).

Abiotic Degradation: The rate constant for the vapor-phase reaction of ethylene glycol monobutyl ether acetate with photochemically produced hydroxyl radicals has been estimated to be 20.9×10^{-12} cm^3/molecule-sec at room temperature which corresponds to an atmospheric half-life of about 18.4 hr at an atmospheric concn of 5×10^5 hydroxyl radicals per cm^3 [1].

Bioconcentration: Based upon its water solubility, the BCF for ethylene glycol monobutyl ether acetate can be estimated to be 3.2 from a regression-derived equation [4]. This BCF value suggests that ethylene glycol monobutyl ether acetate will not bioconcentrate significantly in aquatic organisms.

Soil Adsorption/Mobility: Based upon its water solubility, the Koc for ethylene glycol monobutyl ether acetate can be estimated to be 26 from a regression-derived equation [4]. This Koc value suggests that ethylene glycol monobutyl ether acetate has very high mobility in soil [9].

Volatilization from Water/Soil: Based upon a calculated Henry's Law constant for ethylene glycol monobutyl ether acetate, volatilization from environmental waters is expected to be slow, with the possible exception of very shallow rivers [4]. Based on the Henry's Law constant, the

2-Butoxyethyl Acetate

volatilization half-life from a model river (1 m deep flowing 1 m/sec with a wind velocity of 3 m/sec) can be estimated to be about 74 days [10].

Water Concentrations:

Effluent Concentrations: The emission rate of ethylene glycol monobutyl ether acetate into the atmosphere from painting operations at an automotive assembly plant in Janesville, WI was estimated to be 3.5 gallons/hr [8].

Sediment/Soil Concentrations:

Atmospheric Concentrations:

Food Survey Values:

Plant Concentrations:

Fish/Seafood Concentrations:

Animal Concentrations:

Milk Concentrations:

Other Environmental Concentrations: Ethylene glycol monobutyl ether acetate was detected in 0.4% of 275 solvent products that were sampled in various industry workplaces and analyzed between 1978 and 1982 [3]; the solvents were commonly used for inks, thinners, degreasers, and paints [3].

Probable Routes of Human Exposure: Ethylene glycol monobutyl ether acetate's use as a solvent in paints, lacquers, thinners, inks, stains, and varnishes can result in occupational and general population exposure through inhalation of vapors and dermal contact.

Average Daily Intake:

2-Butoxyethyl Acetate

Occupational Exposure: NIOSH (NOES Survey 1981-1983) has statistically estimated that 123,911 workers are potentially exposed to ethylene glycol monobutyl ether acetate in the USA [6]. NIOSH (NOHS Survey 1972-1974) has statistically estimated that 36,525 workers are potentially exposed to ethylene glycol monobutyl ether acetate in the USA [5].

Body Burdens:

REFERENCES

1. Atkinson R; J Inter Chem Kinet 19: 799-828 (1987)
2. Eisenreich SJ et al; Environ Sci Technol 15: 30-8 (1981)
3. Lehman E et al; pp 31-41 in Safety and Health Aspects of Organic Solvents. Proc Int Course Safety Health Asp Org Solv Espoo Finland April 1985. Riihimaki, V & Ulfvarson, U eds NY: Alan R Liss Inc pp 31-41 (1986)
4. Lyman WJ et al; Handbook of Chemical Property Estimation Methods NY: McGraw-Hill p 5-10 (1982)
5. NIOSH; National Occupational Hazard Survey (NOHS) (1974)
6. NIOSH; National Occupational Exposure Survey (NOES) (1983)
7. Parrish CF; Kirk-Othmer Encycl Chem Technol 3rd ed. NY: John Wiley & Sons 21: 384 (1983)
8. Sexton K; Westberg H; Environ Sci Technol 14: 329-32 (1980)
9. Swann RL et al; Res Rev 85: 23 (1983)
10. USEPA; EXAMS II Computer Simulation (1987)
11. Weber RC et al; Vapor Pressure Distribution of Selected Organic Chemicals. USEPA-600/2-81-021 Cincinnati, OH: USEPA p 21 (1981)
12. Zahn R, Wellens H; Z Wasser Abwasser Forsch 13: 1-7 (1980)

sec-Butyl Alcohol

SUBSTANCE IDENTIFICATION

Synonyms: 1-Methyl-1-propanol

Structure:

$$H_3C \quad \overset{\displaystyle OH}{\underset{\displaystyle CH_2}{CH}} \quad CH_3$$

CAS Registry Number: 78-92-2

Molecular Formula: $C_4H_{10}O$

Wiswesser Line Notation: QY2&1

CHEMICAL AND PHYSICAL PROPERTIES

Boiling Point: 98 °C

Melting Point: -115 °C

Molecular Weight: 74.12

Dissociation Constants:

Log Octanol/Water Partition Coefficient: 0.81 [17]

Water Solubility: 181 g/L at 25 °C [20]

Vapor Pressure: 18.3 mm Hg at 25 °C [5]

Henry's Law Constant: 9.1 x 10^{-6} atm-m^3/mole at 25 °C [34]

53

sec-Butyl Alcohol

ENVIRONMENTAL FATE/EXPOSURE POTENTIAL

Summary: sec-Butyl alcohol enters the environment from natural sources as well as during its production and use in the manufacture of methyl ethyl ketone, its use as a solvent, and as an ingredient in paint remover and industrial cleaning agents. Based on limited information, it also appears to be released to the atmosphere in some combustion processes. If released on soil, sec-butyl alcohol will leach into the ground. It should also volatilize from dry soil. Based on the results of biodegradability screening tests and a few other studies, biodegradation will probably be the key process affecting sec-butyl alcohol's fate in soil. If released in water, biodegradation will probably also be the primary factor affecting its loss. In one study, sec-butyl alcohol's half-life was 5 days in river water. Volatilization will only be significant at a fairly high temperature (half-life 3.2 days in a model river at 25 °C). Adsorption to sediment and bioconcentration in fish will not be significant transport processes. In the atmosphere, sec-butyl alcohol will be lost by reaction with photochemically produced hydroxyl radicals. Its estimated half-life is 2 days. Occupational exposure to sec-butyl alcohol will occur by inhalation and dermal contact. The general population will ingest sec-butyl alcohol in food.

Natural Sources: sec-Butyl alcohol is produced naturally. For example, it is an aroma component of apples and pears [8] and also found in poultry manure [37].

Artificial Sources: sec-Butyl alcohol may be released to the environment during its production, transport, storage, and use. Practically the entire production of sec-butyl alcohol goes into the manufacture of methyl ethyl ketone (MEK) [33] and sec-butyl alcohol is released into the atmosphere from MEK manufacture [28]. Other uses for sec-butyl alcohol in other organic syntheses, as a solvent, paint remover, and industrial cleaner could result in consumer and work exposure [19]. One investigator found sec-butyl alcohol in exhaust gas from gasoline-powered motor vehicles [16], but it was not found in air in a tunnel on the Pennsylvania Turnpike [16].

Terrestrial Fate: If released on land, sec-butyl alcohol will be expected to leach into the ground. Volatilization should also occur from dry soil.

sec-Butyl Alcohol

Based on the results of screening studies, sec-butyl alcohol will probably biodegrade. No sec-butyl alcohol was found in ground water under a paint factory where sec-butyl alcohol and other solvents were stored in leaking underground storage tanks [1]. It is presumed to have degraded in the subsoil and ground water.

Aquatic Fate: The fate of sec-butyl alcohol in water is not well known. sec-Butyl alcohol is readily biodegradable in screening tests which suggest that biodegradation may be a key factor in its removal from natural waters. In one river die-away test, the half-life of sec-butyl alcohol was about 5 days [15]. The volatilization half-life in a model river is estimated to be 3.5 days at 25 °C [26,34]. Little sec-butyl alcohol should adsorb to sediment or particulate matter in the water column.

Atmospheric Fate: In the atmosphere, sec- butyl alcohol will be lost by reaction with photochemically produced hydroxyl radicals. The estimated half-life for this reaction is approximately 2 days [10]. sec-Butyl alcohol is very soluble in water and therefore, it should be readily scavenged by rain and snow.

Biodegradation: sec-Butyl alcohol biodegrades rapidly in screening tests using a sewage seed or activated sludge. The percentage of theoretical BOD reported include: 83% in 5 days [2]; 33% in 5 days [7]; 44.2 and 72.3% after 10 and 20 days, respectively [11], 81.7% after 5 days [36]; and 9.3% after 1 day [13]; and 98.5% in 5 days [29]. In a semi-automatic activated sludge system (SCAS) simulating a treatment plant, 98% BODT reduction was obtained in 4 hours [18]. In a river die-away test, 55% BODT reduction after 5 days [15]. sec-Butyl alcohol is also amenable to anaerobic biodegradation [30]. In one study, 100% degradation was obtained after a 14 day lag by acetate-acclimated methane cultures [39]. In a long-term study using anaerobic upflow filters and acetate-enriched cultures, a 93% utilization rate was obtained after 52 days of operation [3].

Abiotic Degradation: sec-Butyl alcohol reacts with photochemically produced hydroxyl radicals in the atmosphere by H-atom abstraction. The average rate constants at 24.0 °C obtained for this reaction by two investigators were 9.4×10^{-12} and 7.37×10^{-12} cm^3/molecule-sec [10]. Assuming an average concentration of hydroxyl radicals of $5 \times 10^{+5}$ per

cm^3, the resulting half-life of sec-butyl alcohol is 1.7 and 2.2 days, respectively. Under simulated atmospheric conditions and in the presence of 5 ppm of NO, the half-life of sec-butyl alcohol (10 ppm) was 4 hr [6]. In this experiment the light intensity was 2.6 times that of summer, noon sunlight in Freeport, TX. sec-Butyl alcohol does not contain chromophores that adsorb light >290 nm and therefore direct photolysis will not occur. Alcohols are resistant to hydrolysis and therefore sec-butyl alcohol will not be lost by this process.

Bioconcentration: Using the recommended value of the log octanol/water partition coefficient for sec-butyl alcohol, one can estimate a BCF of 1.7 [26], indicating that the bioconcentration of sec-butyl in fish should be negligible.

Soil Adsorption/Mobility: Using the value for the water solubility and a recommended regression equation, one can estimate a Koc of 5.6 [26]. Therefore, sec-butyl alcohol should not adsorb significantly to soil or sediment.

Volatilization from Water/Soil: The measured Henry's Law constant for sec-butyl alcohol is 9.1×10^{-6} atm-m^3/mole at 25 °C and 9.6×10^{-7} atm-m^3/mole at 0 °C [34]. Using these Henry's Law constants, one estimates that the volatilization half-life of sec-butyl alcohol in a model river 1 m deep flowing at 1 m/sec with a 3 m/sec wind is 3.5 days and 30 days at 25 and 0 °C, respectively [26]. sec-Butyl alcohol has a high vapor pressure, and low adsorption to soil. Therefore, it should volatilize from dry soil surfaces.

Water Concentrations: DRINKING WATER: sec-Butyl alcohol was identified in drinking water from Ottumwa, IA [25]. It was found in drinking water wells in China underneath land that was spread with sewage [35]. SURFACE WATER: sec-Butyl alcohol has been identified in the Niagara River (Lake Ontario Basin), but not in the western basin of Lake Ontario [14].

Effluent Concentrations: In a comprehensive survey of wastewater from 4000 industrial and publicly owned treatment works (POTWs) sponsored by the Effluent Guidelines Division of the U.S. EPA, sec-Butyl alcohol was identified in discharges of the following industrial category (positive occurrences, median concentration in ppb): leather tanning (1; 46.7), petroleum refining (1; 149.3), paint and ink (1; 324.7), organics and plastics (2; 35.4), pesticides manufacture (1; 36.2), pharmaceuticals (1; 4.6), foundries (1; 41.0), electronics (1; 19.9), mechanical products (3; 90.6), publicly owned treatment works (3; 12.4) [32]. The highest effluent concentration was 920.1 ppb in the mechanical products industry [32]. sec-Butyl alcohol was found in landfill leachate from 1 of 5 sites in Connecticut [31]. The concentration of sec-butyl alcohol in this leachate ranged from 6.2 to 14.9 ppm [31]. Trench leachate from the Maxey Flats, KY and West Valley, NY disposal sites also contained sec-butyl alcohol [12].

Sediment/Soil Concentrations:

Atmospheric Concentrations: sec-Butyl alcohol was detected, but not quantified, in forest air in the Southern Black Forest of Germany [22]. Sampling at Tucson, AZ and two rural sites 40 km away for light weight alcohols failed to detect any sec-butyl alcohol either in air or precipitation [34].

Food Survey Values: sec-Butyl alcohol is found in many diverse food items. It has been identified as a volatile component of baked potatoes [4], apples and pears [8], a gruyere-type cheese, fried bacon, roasted filberts [23], and dry legumes [24].

Plant Concentrations:

Fish/Seafood Concentrations: sec-Butyl alcohol was found in a sample of mussel at a concentration of 0.27 ppm [38]. The mussel sample was collected off the coast of Japan [38].

Animal Concentrations:

Milk Concentrations:

Other Environmental Concentrations: The concentration of sec-butyl alcohol in 3 samples of poultry manure was 0.150, 4.63, and 6.58 ppm [37].

sec-Butyl Alcohol

Probable Routes of Human Exposure: Ingestion, inhalation, skin and eye contact.

Average Daily Intake:

Occupational Exposure: NIOSH (NOES Survey, 1981-83) has statistically estimated that 47,211 workers, including 25,058 women, are exposed to sec-butyl alcohol in the USA [27]. 89% of these exposures are to sec-butyl alcohol contained in formulations that are known by trade names [27]. The NOES was based on field surveys of 4490 facilities and was designed as a nationwide survey based on a statistical sample of virtually all workplace environments, except mining and agriculture, in the United States where eight or more persons are employed. Workers using sec-butyl alcohol as a solvent, cleaner, or paint remover would be most likely to be exposed to high levels of the chemical.

Body Burdens:

REFERENCES

1. Botta D et al; Comm Eur Comm Eur 8518 (Anal Org Micropollut Water): 261-75 (1984)
2. Bridie AL et al; Water Res 13: 627-30 (1979)
3. Chou WL et al; Bioeng Symp 8: 391-414 (1979)
4. Coleman EC et al; J Agric Food Chem 29: 42-8 (1981)
5. Daubert TE, Danner RP; Data compilation tables of properties of pure compounds. Amer Institute of Chem Eng (1985)
6. Dilling WL et al; Environ Sci Technol 10: 351-6 (1976)
7. Dore M et al; Trib Cebedeau 28: 3-11 (1975)
8. Drawert F et al; Intern Fruchtsalt-union Ber Wiss - Tech Komm 4: 235-42 (1962)
9. Dumont JP, Adda J; J Agric Food Chem 26: 364-7 (1978)
10. Edney EO, Corse EW; Hydroxyl radical rate constant intercomparison study. USEPA, Washington, DC NTIS PB87-111 142/AS (1986)
11. Ettinger MB; Ind Eng Chem 48: 256-9 (1956)
12. Francis AJ et al; Nuclear Tech 50: 158-63 (1980)
13. Gerhold RM, Malaney GW; J Water Pollut Control Fed 38: 562-79 (1966)
14. Great Lakes Water Quality Board; Inventory Chem Subst Identified in the Great Lakes Ecosystem. Report to the Great Lakes Water Quality Board, Windsor, Canada (1983)
15. Hammerton C; J Appl Chem 5: 517-24 (1955)
16. Hampton CV et al; Environ Sci Technol 16: 287-98 (1982)

sec-Butyl Alcohol

17. Hansch C, Leo AJ; MEDCHEM Project Claremont, CA: Pomona College (1985)
18. Hatfield R; Ind Eng Chem 49: 192-6 (1957)
19. Hawley GG; Condensed Chem Dictionary 10th ed NY, NY: Von Nostrand Reinhold p.161 (1981)
20. Hefter GT; pp 94-119 in Solubility data Series 15 Perth, Australia Murdoch Univ (1984)
21. Ho CT et al; J Agric Food Chem 31: 336-42 (1983)
22. Juttner F; Chemosphere 15: 985-92 (1986)
23. Kinlin TE et al; J Agric Food Chem 20: 1021 (1972)
24. Lovegren NV et al; J Agric Food Chem 27: 851-3 (1979)
25. Lucas SV; GC/MS Anal of Org in Drinking Water Concentrates and Advanced Treatment Concentrates Vol 1 USEPA-600/1-84-020a (NTIS PB85-128239) (1984)
26. Lyman WJ et al; Handbook of Chem Property Estimation Methods NY: McGraw-Hill pp 15-1 to 15-34 1982)
27. NIOSH; National Occupational Exposure Survey computer printout 3/29/89
28. PES Inc; Toxic Air Pollutant/Source Crosswalk USEPA-450/4-87-023a (1987)
29. Pitter P; Water Res 10: 231-5 (1976)
30. Sasaki S; pp 283-98 in Aquat Pollutants: Transform and Biolog Effects. Hutzinger O et al. (ed) Oxf: Pergamon Press (1978)
31. Sawhney BL, Kozloski RP; J Environ Qual 13: 349-52 (1984)
32. Shackelford WM et al; Analyt Chim Acta 146: 15-27 (supplemental data) (1983)
33. Sherman PD Jr; Kirk-Othmer Encycl Chem Tech 3rd ed NY,NY: Wiley 4: 338-45 (1978)
34. Snider JR, Dawson GA; J Geophys Res 90: 3797-805 (1985)
35. Tu J et al; Huanjing Huaxue 5: 60-74 (1986)
36. Wagner R; Vom Wasser 42: 271-305 (1974)
37. Yashuhara A; J Chromatogr 387: 371-8 (1987)
38. Yashuhara A, Morita M; Chemosphere 16: 2559-65 (1987)
39. Yonezawa Y, Urushigawa Y; Chemosphere 8: 139-42 (1979)

sec-Butyl Acetate

SUBSTANCE IDENTIFICATION

Synonyms: Acetic acid, 1-methylpropyl ester

Structure:

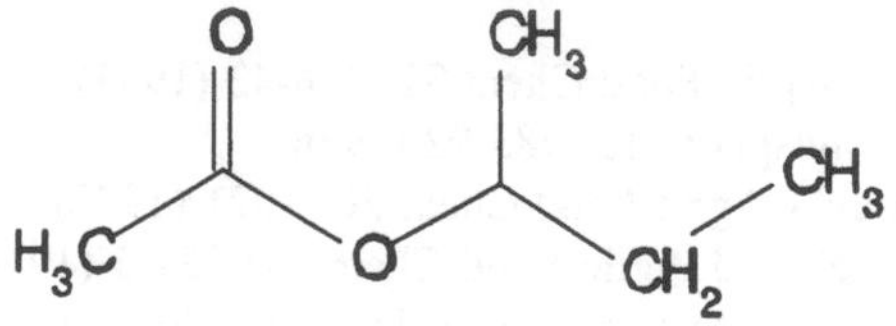

CAS Registry Number: 105-46-4

Molecular Formula: $C_6H_{12}O_2$

Wiswesser Line Notation: 2Y1&OV1

CHEMICAL AND PHYSICAL PROPERTIES

Boiling Point: 112 °C

Melting Point: -73.5 °C

Molecular Weight: 116.16

Dissociation Constants:

Log Octanol/Water Partition Coefficient: 1.51 [6]

Water Solubility: 8,700 g/L at 25 °C [14]

Vapor Pressure: 22.2 mm Hg at 25 °C [3]

Henry's Law Constant: 4.2 x 10^{-4} atm-m^3/mole at 25 °C (calculated from the vapor pressure and water solubility)

ENVIRONMENTAL FATE/EXPOSURE POTENTIAL

Summary: Evaporation of sec-butyl acetate solvent from lacquers and enamels is the dominant anthropogenic emission source of sec-butyl acetate into the environment. If released to soil, sec-butyl acetate may be susceptible to biodegradation based on the demonstrated biodegradability of the similarily structured compounds n-butyl acetate and isobutyl acetate in standard BOD tests. Chemical hydrolysis in moist alkaline soils (pH approaching 9 or higher) may be important, but not in neutral or acidic soils. sec-Butyl acetate may be subject to moderate-to-high leaching based on estimated Koc values of 30 and 158. Volatilization from dry soil surfaces is likely to be rapid. If released to water, volatilization is expected to be an important removal mechanism. The volatilization half-life from a river one meter deep flowing 1 m/sec with a wind velocity of 3 m/sec has been estimated to be 5.4 hours. The hydrolysis half-lives of sec-butyl acetate at pHs 7.0, 8.0, and 9.0 are about 12.6 years, 1.26 years and 46 days, respectively, at 25 °C, indicating that hydrolysis might only be important in very alkaline environmental waters. Aquatic adsorption and bioconcentration are not expected to be significant. Biodegradation in natural water may occur based on the demonstrated biodegradability of n-butyl acetate and isobutyl acetate in standard BOD tests. If released to air, sec-butyl acetate will exist almost entirely in the vapor phase in the ambient atmosphere. The dominant degradation mechanism in the atmosphere will be the vapor phase reaction with photochemically produced hydroxyl radicals which has an estimated half-life of about 1.86 days in an average atmosphere. General population exposure to sec-butyl acetate may occur through consumption of food (in which it may occur naturally) and by inhalation of contaminated air, especially where lacquers or enamels containing sec-butyl acetate solvents are used. Occupational exposure by inhalation and dermal routes may be significant.

Natural Sources: sec-Butyl acetate is produced during fermentation in the manufacture of malt vinegar [8]. It is also a volatile component of some foods [2,5].

Artificial Sources: sec-Butyl acetate has been predominantly used as a solvent for nitrocellulose lacquers and for thinners and nail enamels [7]; from these uses, sec-butyl acetate evaporates directly into the surrounding air, thereby becoming a source of sec-butyl acetate to the environment.

Terrestrial Fate: Chemical hydrolysis of sec-butyl acetate in moist alkaline soils (pH approaching 9 or higher) may be important, but hydrolysis in neutral or acidic soils is not expected to be important. Based on estimated Koc values of 30 and 158, sec-butyl acetate may be subject to moderate-to-high leaching. Volatilization from dry soil surfaces is likely to be rapid. Since the similarily structured compounds n-butyl acetate and isobutyl acetate have been found to be significantly biodegradable in standard BOD dilution water tests, sec-butyl acetate may be susceptible to biodegradation.

Aquatic Fate: The volatilization half-life of sec-butyl acetate from a river one meter deep flowing 1 m/sec with a wind velocity of 3 m/sec has been estimated to be 5.4 hours; the volatilization half-life from a similar river 10-m deep has been estimated to be 7.1 days. The hydrolysis half-lives of sec-butyl acetate at pHs 7.0, 8.0, and 9.0 are about 12.6 years, 1.26 years and 46 days, respectively, at 25 °C, indicating that hydrolysis will be important only in very alkaline environmental waters. Aquatic adsorption and bioconcentration are not expected to be significant. Since the similarily structured compounds n-butyl acetate and isobutyl acetate have been found to be significantly biodegradable in standard BOD dilution water tests, sec-butyl acetate may be susceptible to biodegradation.

Atmospheric Fate: Sec-butyl acetate will exist almost entirely in the vapor-phase in the ambient atmosphere due to its relatively high vapor pressure. The half-life for the vapor-phase reaction of sec-butyl acetate with photochemically produced hydroxyl radicals had been estimated to be about 1.86 days in an average atmosphere indicating that this reaction will be the dominant degradation mechanism. Physical removal via washout may be possible.

Biodegradation: The similarly structured compounds n-butyl acetate and isobutyl acetate have been found to be significantly biodegradable in standard BOD dilution water tests with 5-day theoretical BOD of 20.9-60% and 20-day theoretical BOD's of 55.4%-83% [9,15,17] suggesting that sec-butyl acetate may be susceptible to biodegradation.

Abiotic Degradation: Based on an experimentally derived base-catalyzed hydrolysis rate constant of 0.01738/M-sec at 25 °C [11], the hydrolysis half-lives of sec-butyl acetate at pHs 7.0, 8.0, and 9.0 are estimated to be 12.6 years, 1.26 years and 46 days, respectively. Based on an experimentally determined rate constant of 5.4×10^{-12} cm^3/molecule-sec at 32 °C [1], the half-life for the vapor-phase reaction of sec-butyl acetate with photochemically produced hydroxyl radicals in the atmosphere has been estimated to be 1.86 days assuming an average atmospheric hydroxyl radical concn of 8×10^5 molecules/cm^3.

Bioconcentration: Based on the log Kow and water solubility, the BCF value for sec-butyl acetate is estimated to be 8 and 4, respectively, by regression-derived equations [10]. These BCF values suggest that bioconcentration is not significant.

Soil Adsorption/Mobility: Based on the log Kow and water solubility, the Koc value for sec-butyl acetate is estimated to be 158 and 30, respectively, by regression-derived equations [10]. These Koc values indicate a high to medium soil mobility [16].

Volatilization from Water/Soil: The value of Henry's Law Constant suggests that volatilization of sec-butyl acetate is probably significant from environmental bodies of water [10]. The volatilization half-life from a river one meter deep flowing 1 m/sec with a wind velocity of 3 m/sec is estimated to be 5.4 hours [10]; the volatilization half-life from a similar river 10 meters deep is estimated to be 7.1 days [10]. sec-Butyl acetate evaporates relatively rapidly from solid surfaces with a half-life of about 30 min as measured by a standard solvent-coating test method at 19 °C [13]; the evaporative half-life of sec-butyl acetate from a solid surface at 25 °C, as measured by a different standard test method, was 2.25 min [4].

Water Concentrations:

Effluent Concentrations:

Sediment/Soil Concentrations:

Atmospheric Concentrations:

Food Survey Values: sec-Butyl acetate was detected as a volatile flavor component of baked potatoes [2]. It is also a volatile component of Beaufort cheese, a gruyere-type cheese made in the French Alps [5].

sec-Butyl Acetate

Plant Concentrations:

Fish/Seafood Concentrations:

Animal Concentrations:

Milk Concentrations:

Other Environmental Concentrations:

Probable Routes of Human Exposure: General population exposure to sec-butyl acetate can occur through consumption of food and inhalation of contaminated air. sec-Butyl acetate has been identified as a flavor component of baked potatoes. Pulmonary and dermal exposure from air in the vicinity of where lacquers, thinners, or enamels containing sec-butyl acetate solvent are used may be significant.

Average Daily Intake:

Occupational Exposure: NIOSH (NOHS Survey 1972-1974) statistically estimates that 696,165 workers are potentially exposed to sec-butyl acetate in the USA [12].

Body Burdens:

REFERENCES

1. Atkinson R; Chem Rev 85: 69-290 (1985)
2. Coleman EC et al; J Agric Food Chem 29: 42 (1981)
3. Daubert TE, Danner RP; Data Compilation, Tables of Properties of Pure Cmpds, Design Inst for Phys Prop Data, Am Inst for Phys Prop Data, NY (1989)
4. Davis DS; Am Perfumer Cosmet 81: 32 (1966)
5. Dumont JP, Adda J; J Agric Food Chem 26: 364-7 (1978)
6. GEMS; Graphical Exposure Modeling System. CLOGP3 (1986)
7. Hawley GG; Condensed Chem Dictionary 10th ed pp 159-60 Van Nostrand Reinhold NY (1981)
8. Jones DD, Greenshield RN; J Institute of Brewing 77: 160-63 (1971)
9. Lamb CB, Jenkins GF; pp 326-9 in Proc 8th Indust Waste Conf: Purdue Univ (1952)
10. Lyman WJ et al; Handbook of Chemical Property Estimation Methods. Environmental Behavior of Organic Compounds. McGraw-Hill NY p 4-9 (1982)

sec-Butyl Acetate

11. Mabey W, Mill T; J Phys Chem Ref Data 7: 383-415 (1978)
12. NIOSH; National Occupational Health Survey (1975)
13. Park JG, Hofmann HE; Ind Eng Chem 24: 132 (1932)
14. Park JG, Hopkins MB; Ind Eng Chem 22: 826 (1930)
15. Price KS et al; J Water Pollut Contr Fed 46: 63 (1974)
16. Swann RL et al; Res Rev 85: 16-28 (1983)
17. Takemoto J et al; Suishitsu Odaku Kenkyu 4: 80 (1981)

tert-Butyl Acetate

SUBSTANCE IDENTIFICATION

Synonyms: Acetic acid, 1,1-dimethylethyl ester

Structure:

CAS Registry Number: 540-88-5

Molecular Formula: $C_6H_{12}O_2$

Wiswesser Line Notation: 1X1&1&OV1

CHEMICAL AND PHYSICAL PROPERTIES

Boiling Point: 97.8 °C

Melting Point:

Molecular Weight: 116.16

Dissociation Constants:

Log Octanol/Water Partition Coefficient: 1.38 [2]

Water Solubility: Practically insoluble in water [7]

Vapor Pressure:

Henry's Law Constant: 3.3 x 10^{-4} atm m^3/mole at 25 °C [3]

tert-Butyl Acetate

ENVIRONMENTAL FATE/EXPOSURE POTENTIAL

Summary: If released to soil, chemical hydrolysis of tert-butyl acetate in moist alkaline soils (pH approaching 10 or higher) may be important, but hydrolysis in soils of pH 9 or lower is not expected to be important. tert-Butyl acetate may be susceptible to significant leaching in soil based on an estimated Koc value of 134. Volatilization from dry soil surfaces may be rapid. If released to water, volatilization is expected to be an important removal mechanism. The volatilization half-life from a river one meter deep flowing 1 m/sec with a wind velocity of 3 m/sec has been estimated to be 6.0 hours. The hydrolysis half-lives of butyl acetate at pHs 7.0, 8.0 and 9.0 are about 135 years, 14.6 years and 1.46 years, respectively, at 25 °C indicating that hydrolysis will be important only in extremely alkaline environmental waters with pHs approaching 10 or higher. Aquatic adsorption and bioconcentration are not expected to be significant. If released to air, tert-butyl acetate will exist almost entirely in the vapor-phase in the ambient atmosphere. The dominant degradation mechanism in the atmosphere may be the vapor-phase reaction with photochemically produced hydroxyl radicals which has an estimated half-life of about 26 days in an average atmosphere. Physical removal via washout may be possible. General population exposure to tert-butyl acetate does not currently appear to be important since the chemical is produced only in small quantities.

Natural Sources: Occurs naturally in bananas and related fruit [1].

Artificial Sources: tert-Butyl acetate can be used as a solvent and its use as an antiknock agent in motor fuels has been suggested [4]. Relatively minor amounts of tert-butyl acetate may be released to the environment at its manufacturing site via wastestreams. Solvent uses, if similar in nature to other butyl acetate isomers, would release tert-butyl acetate directly into the air by evaporation.

Terrestrial Fate: Chemical hydrolysis of tert-butyl acetate in moist, very alkaline soils (pH approaching 10 or higher) may be important, but hydrolysis in soils of pH 9 or lower is not expected to be important. Based on an estimated Koc value of 134, tert-butyl acetate may be subject to significant leaching in soil. Volatilization from dry soil surfaces may be rapid.

tert-Butyl Acetate

Aquatic Fate: The volatilization half-life of tert-butyl acetate from a river one meter deep flowing 1 m/sec with a wind velocity of 3 m/sec has been estimated to be 6.0 hours; the volatilization half-life from a similar river 10 m deep has been estimated to be 7.4 days. The hydrolysis half-lives of tert-butyl acetate at pHs 7.0, 8.0, and 9.0 are about 135 years, 14.6 years, and 1.46 years, respectively, at 25 °C indicating that hydrolysis will be important only in extremely alkaline environmental waters with pHs approaching 10 or higher. Aquatic adsorption and bioconcentration are not expected to be significant.

Atmospheric Fate: tert-Butyl acetate will exist almost entirely in the vapor phase in the ambient atmosphere due to its expected high vapor pressure. The half-life for the vapor-phase reaction of tert-butyl acetate with photochemically produced hydroxyl radicals has been estimated to be about 26 days in an average atmosphere indicating that this reaction may be the dominant atmospheric degradation mechanism. Physical removal via washout may be possible.

Biodegradation:

Abiotic Degradation: Based on experimentally derived acid- and base-catalyzed hydrolysis rate constants at 25 °C, the hydrolysis half-lives of tert-butyl acetate at pHs 7.0, 8.0 and 9.0 are estimated to be 135 years, 14.6 years and 1.46 years, respectively [6]. The half-life for the vapor-phase reaction of tert-butyl acetate with photochemically produced hydroxyl radicals in the atmosphere has been estimated to be about 26 days assuming an average atmospheric hydroxyl radical concn of $8 \times 10^{+5}$ molecules/cm^3 [2].

Bioconcentration: Based on the log Kow, the BCF value for tert-butyl acetate can be estimated to be 6.6 by a recommended regression-derived equation [5]. This BCF value suggests that bioconcentration is not significant.

Soil Adsorption/Mobility: Based on the log Kow, the Koc value for tert-butyl acetate can be estimated to be 134 by a regression-derived equation [5]. This Koc value indicates a high soil mobility [8].

tert-Butyl Acetate

Volatilization from Water/Soil: The estimated Henry's Law Constant for tert-butyl acetate suggests that volatilization is probably significant from environmental bodies of water [5]. The volatilization half-life from a river one meter deep flowing 1 m/sec with a wind velocity of 3 m/sec is estimated to be 6.0 hours [5]; the volatilization half-life from a similar river 10 meters deep is estimated to be 7.4 days [5]. Based on an expectedly high vapor pressure, tert-butyl acetate will probably evaporate relatively rapidly from solid surfaces.

Water Concentrations:

Effluent Concentrations:

Sediment/Soil Concentrations:

Atmospheric Concentrations: SOURCE DOMINATED: Air samples were collected in 336 different industrial plants in Northern Belgium [9]; 262 facilities contained one or more glycol esters; of these 262 facilities iso- or t-butyl acetate was identified in 5% of operations using printing pastes and inks, 17% of those using paints and varnishes, 45% of car repair shops, and 30% of other industries [9].

Food Survey Values:

Plant Concentrations:

Fish/Seafood Concentrations:

Animal Concentrations:

Milk Concentrations:

Other Environmental Concentrations:

Probable Routes of Human Exposure: Inhalation, ingestion, skin, and eye contact.

Average Daily Intake:

tert-Butyl Acetate

Occupational Exposure: Occupational exposure to tert-butyl acetate may occur at sites of manufacture and repackaging and at sites of use such as laboratories. General population exposure to tert-butyl acetate does not currently appear to be likely since the chemical is produced only in small quantities.

Body Burdens:

REFERENCES

1. Della Porta G et al; Tumori 56:325-334 (1970)
2. GEMS; Graphical Exposure Modeling System. USEPA (1986)
3. Hine J, Mookerjee PK; J Org Chem 40: 292 (1975)
4. Lowenheim FA, Moran MK; Faith, Keyes, and Clark's Industrial Chemicals, 4th ed. NY, John Wiley & Sons (1975)
5. Lyman WJ et al; Handbook of Chemical Property Estimation Methods. Environmental Behavior of Organic Compounds. McGraw-Hill NY p 5-4 (1982)
6. Mabey W, Mill T; J Phys Chem Ref Data 7: 383-415 (1978)
7. Merck; The Merck Index An Encyclopedia of Chemicals, Drugs, and Biologicals 10th ed Rahway, NJ: Merck & Co p 214 (1983)
8. Swann RL et al; Res Rev 85: 17 (1983)
9. Veulemans H et al; Am Ind Hyg Assoc H 48: 67-6 (1987)

t-Butyl Methyl Ether

SUBSTANCE IDENTIFICATION

Synonyms: MTBE

Structure:

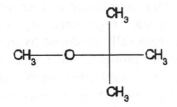

CAS Registry Number: 1634-04-4

Molecular Formula: $C_5H_{12}O$

Wiswesser Line Notation:

CHEMICAL AND PHYSICAL PROPERTIES

Boiling Point: 55.2 °C

Melting Point: -109 °C

Molecular Weight: 88.15

Dissociation Constants:

Log Octanol/Water Partition Coefficient: 1.24 [10]

Water Solubility: 51,000 mg/L at 25 °C [4]

Vapor Pressure: 249 mm Hg at 25 °C [8]

Henry's Law Constant: 5.87×10^{-4} atm-m^3/mole [13]

ENVIRONMENTAL FATE/EXPOSURE POTENTIAL

Summary: Methyl t-butyl ether (MTBE) is released to the environment from its storage and use as an octane booster in unleaded gasoline.

71

MTBE will volatilize from the surface of soils, but it will be highly mobile in soil and may leach to ground water. It will not be expected to hydrolyze in soil. Using estimated physical-chemical properties of MTBE or by analogies to other structurally related aliphatic ethers, MTBE in water will not adsorb significantly to sediment or suspended particulate matter, bioconcentrate in aquatic organisms, hydrolyze, directly photolyze, or react with photochemically produced hydroxyl radicals in the water. MTBE in surface water probably will rapidly volatilize with an estimated half-life of 4.1 hr for volatilization from a river one meter deep flowing 1 m/sec with a wind velocity of 3 m/sec and 2.0 days for a model pond. It may be resistent to biodegradation in environmental media based on a screening test using activated sludge inocula. Many ethers are known to be resistant to biodegradation. If MTBE is released to the atmosphere, it will exist almost entirely in the vapor phase based on its vapor pressure. It will be susceptible to photoxidation by vapor-phase reaction with photochemically produced hydroxyl radicals with an estimated half-life of 5-6 days for this process. Direct photolysis will not be an important removal process since aliphatic ethers do not adsorb light at wavelengths >290 nm. The most probable route of general population exposure to MTBE probably is via inhalation of contaminated air. Exposures through dermal contact may occur in occupational settings, although general population dermal contact may occur at self-service gasoline stations.

Natural Sources:

Artificial Sources: MTBE will be released as a result of its use as an octane booster in unleaded gasoline (up to 7% by volume) [7,12].

Terrestrial Fate: If MTBE is released to soil, it will volatilize using a reported Henry's Law constant of 5.87×10^{-4} atm-m^3/mole [13] and vapor pressure of 249 mm Hg at 25 °C [8]. It also will be expected to exhibit very high mobility [17] in soil (based on an estimated Koc of 11.2) and may leach to ground water [4,14]. It will not hydrolyze in soil [14]. MTBE may be resistent to biodegradation in soil based on screening test data from a study using activated sludge inocula [10]. Many ethers are known to be resistant to biodegradation [1].

t-Butyl Methyl Ether

Aquatic Fate: Using estimated physical-chemical properties and/or analogies to other structurally related aliphatic ethers [2,4,6], MTBE in water will not adsorb significantly to sediment or suspended particulate matter [2,4], bioconcentrate in aquatic organisms [2,4], hydrolyze [2], directly photolyze [6], nor react with photochemically produced hydroxyl radicals in the water [3]. MTBE in surface water will rapidly volatilize [2,13]. Using a reported Henry's Law constant of 5.87×10^{-4} atm-m^3/mole [13], a half-life for volatilization of 4.1 hours at 25 °C for MTBE from a river one meter deep flowing 1 m/sec with a wind velocity of 3 m/sec has been estimated [2]. The volatilization half-life from a model pond, which considers the effect of adsorption, has been estimated to be 2.0 days [18]. MTBE may be resistant to biodegradation in environmental media based on screening test data from a study using activated sludge inocula [10]. Many ethers are known to be resistant to biodegradation [1].

Atmospheric Fate: In air, MTBE is expected to exist almost entirely in the vapor phase [9] given its vapor pressure of 249 mm Hg at 25 °C [3]. It will be susceptible to photooxidation via vapor-phase reaction with photochemically produced hydroxyl radicals. An atmospheric half-life of 5-6 days at an atmospheric concentration of $5 \times 10^{+5}$ hydroxyl radicals per cm^{-3} has been calculated using measured rate constants [5,20]. Direct photolysis will not be an important removal process since aliphatic ethers do not absorb light at wavelengths >290 nm [6].

Biodegradation: No data concerning the biodegradation of MTBE in environmental media were located. An activated sludge aqueous screening study found that MTBE was degraded very slowly; 1% of the theoretical biochemical oxygen demand was measured after a 21-day incubation [1]. This screening test result suggests that MTBE may be resistant to biodegradation in the environment. Studies of three biological treatment processes indicated that most of the compound could be removed from wastewater by treatment, but it was not determined whether the removal was due to biological activity or to some other processes such as volatilization [19]. The percentages of the compound removed by a conventional activated sludge process, an activated sludge process with powder activated carbon treatment (PACT), and the PACT-process in combination with wet-air regeneration of activated carbon containing surplus sludge were: 85%, 94%, and 95%, respectively [19]. Many ethers are known to be resistant to biodegradation [3].

73

Abiotic Degradation: The rate constant for the vapor phase reaction of MTBE with photochemically produced hydroxyl radicals has been measured to be 2.84 x 10^{-12} to 3.2 x 10^{-12} cm^3 molecule^{-1} sec^{-1} at 25 °C [5,20,21], which corresponds to an atmospheric half-life of 5-6 days at an atmospheric concentration of 5 x 10^{+5} hydroxyl radicals cm^{-3}. Hydrolysis is not expected to be significant under normal environmental conditions (pH 5-9) [2]. Direct photolysis will not be an important removal process since aliphatic ethers do not absorb light at wavelengths >290 nm [6].

Bioconcentration: A bioconcentration factor of 1.5 has been reported in Japanese carp [10]. This value suggests that MTBE will not bioconcentrate in aquatic organisms.

Soil Adsorption/Mobility: A soil sorption factor (Koc) of 11.2 has been estimated using the water solubility and a regression equation [14]. This low Koc value suggests that MTBE will exhibit very high mobility in soil [17] and may leach through soil to ground water if volatilization or biodegradation do not occur first.

Volatilization from Water/Soil: From the Henry's law constant, the estimated half-life for volatilization of MTBE from a river one meter deep flowing 1 m/sec with a wind velocity of 3 m/sec is 4.1 hr at 25 °C [14]. The volatilization half-life from a model pond, which considers the effect of adsorption, has been estimated to be 2.0 days [18]. The vapor pressure and Henry's Law constant of MTBE suggest that it will volatilize from surfaces and near-surface soil.

Water Concentrations: GROUND WATER: MTBE has been detected in ground water in Maine [11]. Monitoring data indicate that MTBE reached ground water from a leaky underground gasoline storage tank. The leak was detected by the odor of the water. Concentrations ranged from 0.197-236 ppm. MTBE has been detected at concentrations up to 50 ppb in the Old Bridge aquifer under an industrial plant in South Brunswick Township, NJ (no sampling dates specified) [2]. A contamination abatement system installed at this aquifer, including 7 extraction wells and a water treatment facility, reduced the MTBE concentration by an estimated 26% [2]. A well in Rockaway Township, NJ was reported to have 25-40 ppm MTBE [15].

t-Butyl Methyl Ether

Effluent Concentrations:

Sediment/Soil Concentrations:

Atmospheric Concentrations:

Food Survey Values:

Plant Concentrations:

Fish/Seafood Concentrations:

Animal Concentrations:

Milk Concentrations:

Other Environmental Concentrations:

Probable Routes of Human Exposure: General population exposure to MTBE is probably via inhalation of contaminated air and dermal exposure. Exposure through inhalation and dermal contact may occur in occupational settings.

Average Daily Intake:

Occupational Exposure: NIOSH (NOES Survey 1981-1983) has statistically estimated that 3,522 workers are exposed to MTBE in the USA [16].

Body Burdens:

REFERENCES

1. Alexander M; Biotechnol Bioeng 15: 611-47 (1973)
2. Althoff WF et al; Groundwater 19: 495-504 (1981)
3. Anbar M, Neta P; Int J Appl Radiation Isotopes 18: 493-523 (1967)
4. Bennett GM, Philip WG; J Chem Soc pp 1930-7 (1989)
5. Bennett PJ, Kerr JA; J Atmos Chem 10: 29-38 (1990)
6. Calvert JG, Pitts JN Jr; Photochemistry. John Wiley & Sons: NY p 441-2 (1966)
7. Chemical Marketing Reporter; Chemical Profile MTBE February 20, p 42 (1990)

t-Butyl Methyl Ether

8. Daubert TE, Danner RP; Data Compilation Tables of Properties of Pure Compounds. Am Inst Chem Eng. (1989)
9. Eisenreich SJ et al; Environ Sci Technol 15: 30-8 (1981)
10. Fujiwara Y et al; Yukagaku 33: 111-14 (1984)
11. Garrett et al; Proc. Petroleum Hydrocarbons and Organic Chemicals in Ground Water Conference sponsored by the National Water Well Association and the American Petroleum Institute Nov. (1986)
12. Hawley GG; Condensed Chemical Dictionary 10th ed Van Nostrand Reinhold NY p 671-2 (1981)
13. Hine J, Mookerjee PK; J Org Chem 40: 292-8 (1975)
14. Lyman WJ et al; Handbook of Chem Property Estimation Methods Environ Behavior of Org Compounds NY: McGraw-Hill pp 4-9, 7-4 4-9, 5-5, 7-4, 15-16 to 15-29 15-16 to 15-29 (1982)
15. McKinnon and Dyksen; J. Amer. Water Works Assoc. 76: 42-47 (1984)
16. NIOSH; The National Occupational Exposure Survey (NOES) (1983)
17. Swann RL et al; Res Rev 85: 17-28 (1983)
18. USEPA; EXAMS II Computer Simulation (1987)
19. Van Luin AB, Teurlinckx LVM; Manage Haz Toxic Wastes Proc Int Congr p 476-85 (1987)
20. Wallington et al; Environ Sci Technol 22: 842-844 (1988)
21. Wallington et al; Int J Chem Kinet 21: 993-1001 (1989)

Butyl Vinyl Ether

SUBSTANCE IDENTIFICATION

Synonyms: n-Butyl vinyl ether

Structure:

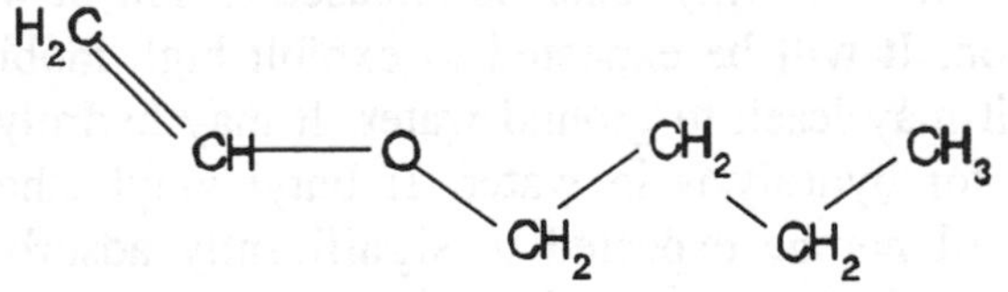

$$H_2C=CH-O-CH_2-CH_2-CH_2-CH_3$$

CAS Registry Number: 111-34-2

Molecular Formula: $C_6H_{12}O$

Wiswesser Line Notation:

CHEMICAL AND PHYSICAL PROPERTIES

Boiling Point: 93.8 °C

Melting Point: -92 °C

Molecular Weight: 100.16

Dissociation Constants:

Log Octanol/Water Partition Coefficient:

Water Solubility: 3,000 mg/L [11]

Vapor Pressure: 51 mm Hg at 25 °C [11]

Henry's Law Constant: 2.24×10^{-3} atm-m^3/mole (calculated from vapor pressure and water solubility)

Butyl Vinyl Ether

ENVIRONMENTAL FATE/EXPOSURE POTENTIAL

Summary: Butyl vinyl ether may be released to the environment as a result of its manufacture and use including use in the synthesis of copolymers. If butyl vinyl ether is released to soil, it will be subject to volatilization. It will be expected to exhibit high mobility in soil and, therefore, it may leach to ground water. It may hydrolyze in soil based upon data for hydrolysis in water. If butyl vinyl ether is released to water, it will not be expected to significantly adsorb to sediment or suspended particulate matter, bioconcentrate in aquatic organisms, or directly photolyze, based upon estimated physical-chemical properties or analogies to other structurally related aliphatic ethers. Butyl vinyl ether will be susceptible to appreciable hydrolysis in certain environmental waters, especially at acidic pH, with calculated half-lives for hydrolysis of 9.5 hr, 40 days, and 10.9 yr at pH 5, 7, and 9, respectively. In surface water, it will be subject to rapid volatilization with estimated half-lives for volatilization of 3.3 hr and 42 hr for volatilization from a river one meter deep flowing 1 m/sec with a wind velocity of 3 m/sec and a model pond, respectively. It is unknown whether butyl vinyl ether biodegrades in environmental media. If butyl vinyl ether is released to the atmosphere, it will be expected to exist almost entirely in the vapor phase based upon the vapor pressure. It will be susceptible to photooxidation via vapor-phase reaction with photochemically produced hydroxyl radicals and ozone with an overall atmospheric half-life of 9 hours for these processes based upon estimated rate constants. It will not be expected to photolyze in the atmosphere. The most probable routes of general population exposure to butyl vinyl ether are via inhalation of contaminated air and ingestion of contaminated drinking water. Exposure through dermal exposure will be expected to be highest in workplaces where butyl vinyl ether is made and used.

Natural Sources:

Artificial Sources: Butyl vinyl ether may be released to the environment as a result of its manufacture and use, including use in the synthesis of copolymers [6].

Terrestrial Fate: If butyl vinyl ether is released to soil, it will be subject to volatilization based upon an estimated Henry's Law constant [8] and

78

its vapor pressure. It will be expected to exhibit high mobility [13] in soil and, therefore, it may leach to ground water, based upon an estimated Koc of 53 [8,11]. It may hydrolyze in soil depending upon the pH based upon data for hydrolysis in water [12]. It is unknown whether butyl vinyl ether will biodegrade in soil.

Aquatic Fate: If butyl vinyl ether is released to water, it is not expected to adsorb significantly to sediment or suspended particulate matter [8,11], bioconcentrate in aquatic organisms [8,11], or directly photolyze [3]. This assessment is based upon estimated physical-chemical properties or analogies to other structurally related aliphatic ethers [3,8,11]. Butyl vinyl ether will be susceptible to appreciable hydrolysis in certain environmental waters, especially at acidic pH; half-lives for hydrolysis of 9.5 hr, 40 days, and 10.9 yr at pH 5, 7, and 9, respectively, are calculated from a reported acid-catalyzed hydrolysis rate constant [12]. Butyl vinyl ether in surface water will be subject to rapid volatilization [8,11,14]. Using the estimated Henry's Law constant, a half-life for volatilization of butyl vinyl ether from a river one meter deep flowing 1 m/sec with a wind velocity of 3 m/sec has been estimated to be 3.3 hr at 25 °C [8]. The volatilization half-life from a model pond, which considers the effect of adsorption, has been estimated to be 42 hours [14]. It is not known whether butyl vinyl ether will biodegrade in environmental waters.

Atmospheric Fate: If butyl vinyl ether is released to the atmosphere, it is expected to exist almost entirely in the vapor phase [4], based upon its vapor pressure. It will be susceptible to photooxidation via vapor-phase reaction with photochemically produced hydroxyl radicals. An overall atmospheric half-life of 9 hours at an atmospheric concentration of 5×10^5 hydroxyl radicals per cm^3 and 7×10^{11} ozone molecules per cm^3 has been calculated for these processes based upon estimated rate constants [1,2]. It is not expected to photolyze in the atmosphere [3].

Biodegradation: No data were located concerning the biodegradation of butyl vinyl ether in environmental media or in screening tests.

Abiotic Degradation: The rate constants for the vapor-phase reactions of butyl vinyl ether with photochemically produced hydroxyl radicals and ozone have been estimated to be 43×10^{-12} cm^3/molecule-sec [1] and 1.75×10^{-18} cm^3/molecule-sec [2] at 25 °C, respectively, which correspond to an atmospheric half-life of 9 hours at an atmospheric concentration of 5 x 10^5 hydroxyl radicals per cm^3 and 7 x 10^{11} ozone molecules per cm^3. Direct photolysis will not be an important removal process since aliphatic ethers do not absorb light at wavelengths >290 nm [3]. Butyl vinyl ether is susceptible to appreciable hydrolysis in certain environmental waters, especially at acidic pH, based upon a measured acid-catalyzed hydrolysis rate constant of 2.02 M-1 s-1 at 25 °C [12]. This hydrolysis rate constant corresponds to half-lives for hydrolysis of 9.5 hr, 40 days, and 10.9 yr at pH 5, 7, and 9, respectively [12].

Bioconcentration: Based upon its reported water solubility, a BCF of 6.8 has been estimated using a recommended regression equation [8]. Based upon this estimated BCF, butyl vinyl ether will not be expected to bioconcentrate in aquatic organisms.

Soil Adsorption/Mobility: Based upon its reported water solubility, a Koc of 53 has been estimated using a recommended regression equation [8]. Based upon this estimated Koc, butyl vinyl ether will be expected to exhibit high mobility in soil [13]. Butyl vinyl ether, therefore, may leach through soil to ground water if it does not volatilize or biodegrade first.

Volatilization from Water/Soil: Based upon an estimated Henry's Law constant, the volatilization half-life from a model river (1 meter deep flowing 1 m/sec with a wind speed of 3 m/sec) has been estimated to be 3.3 hr [8]. The volatilization half-life from a model pond, which considers the effect of adsorption, has been estimated to be 42 hr [14]. Based upon the Henry's Law constant and vapor pressure, butyl vinyl ether will be subject to volatilization from surfaces and near-surface soil.

Water Concentrations: DRINKING WATER: Butyl vinyl ether was tentatively identified, but not quantified, in a drinking water concentrate from Seattle, WA sampled in Nov 1976 [7]. SURFACE WATER: Butyl vinyl ether has been detected, not quantified, in samples of water from the southern basin of Lake Michigan [5].

Effluent Concentrations:

Sediment/Soil Concentrations:

Butyl Vinyl Ether

Atmospheric Concentrations: Butyl vinyl ether was detected, but not quantified, at 1 of 4 sites in the Houston, TX, area sampled in Nov 1974 [10]. It was not found in any of the samples from the 5 sites in the Los Angeles Basin sampled in March and April 1975 or the 2 sites in the Kanawha Valley, WV sampled in Sept 1974 and it was not detected in any of the night ambient air samples in the study [10].

Food Survey Values:

Plant Concentrations:

Fish/Seafood Concentrations:

Animal Concentrations:

Milk Concentrations:

Other Environmental Concentrations:

Probable Routes of Human Exposure: The most probable routes of general population exposure to butyl vinyl ether are via inhalation of contaminated air [10] and ingestion of contaminated drinking water [3,7]. Exposure through dermal contact may occur in occupational settings. Inhalation and dermal exposure will be expected to be highest in workplaces where butyl vinyl ether is made and used.

Average Daily Intake:

Occupational Exposure: NIOSH (NOES Survey 1981-1983) has statistically estimated that 851 workers are exposed to butyl vinyl ether in the USA [9].

Body Burdens:

REFERENCES

1. Atkinson R; Internat J Chem Kin 19: 799-828 (1987)
2. Atkinson R, Carter WP; Chem Rev 84: 437-70 (1984)
3. Calvert JG, Pitts JNJr; Photochemistry John Wiley & Sons: NY pp 441-2 (1966)
4. Eisenreich SJ et al; Environ Sci Technol 15: 30-8 (1981)

5. Great Lakes Water Quality Board; Inventory Chem Subst Id Great Lakes Ecos p 68 (1983)
6. Hawley GG; Condensed Chemical Dictionary 10th ed Van Nostrand Reinhold NY p 1084 (1981)
7. Lucas SV; GC/MS Anal of Org in Drinking Water Concentrates and Advanced Treatment Concentrates Vol 2 USEPA-600/1-84-020B (NTIS PB85-128239) p 41 (1984)
8. Lyman WJ et al; Handbook of Chem Property Estimation Methods NY: McGraw-Hill p 5-5 (1982)
9. NIOSH; The National Occupational Exposure Survey (NOES) (1983)
10. Pellizzari ED et al; Development of Analytical Techniques for Measuring Ambient Atmospheric Carcinogenic Vapors USEPA-600/2-75-076. pp 75, 115 (1975)
11. Riddick JA et al; Organic Solvents John Wiley & Sons Inc. NY (1984)
12. Salomaa P et al; Acta Chem Scand 20: 1790-801 (1966)
13. Swann RL et al; Res Rev 85: 17-28 (1983)
14. USEPA; EXAMS II Computer Simulation (1987)

Chlorodifluoromethane

SUBSTANCE IDENTIFICATION

Synonyms: Freon 22

Structure:

$$
\begin{array}{c}
F \\
| \\
H - C - Cl \\
| \\
F
\end{array}
$$

CAS Registry Number: 75-45-6

Molecular Formula: $CHClF_2$

Wiswesser Line Notation: GYFF

CHEMICAL AND PHYSICAL PROPERTIES

Boiling Point: -40.8 °C at 760 mm Hg

Melting Point: -146.5 °C

Molecular Weight: 86.47

Dissociation Constants:

Log Octanol/Water Partition Coefficient: 1.08 [11]

Water Solubility: 2,899 mg/L at 25 °C and 1 atm [14]

Vapor Pressure: 4279 mm Hg at 4.5 °C [23]

Henry's Law Constant: 0.0294 atm-m³/mole [13]

Chlorodifluoromethane

ENVIRONMENTAL FATE/EXPOSURE POTENTIAL

Summary: Chlorodifluoromethane, may be released to the environment during its production, storage, transport, and use as a refrigerant, low-temperature solvent, aerosol propellent, and in the production of fluorocarbon resins, especially tetrafluoroethylene monomer. When used as a refrigerant, solvent, or aerosol propellant, chlorodifluoromethane will be lost to the atmosphere unless it is captured and recycled. In the case of refrigerant use, there is a delay of several years before it is vented into the air. Chlorodifluoromethane is an extremely unreactive gas and losses due to photolysis, photooxidation and hydrolysis in air, water and soil will not be significant. Biodegradation in water and soil should not be important. If released on land, most of the chlorodifluoromethane will volatilize into the air. It is highly mobile in soil and therefore will have a potential for leaching into ground water. If released in water, it will be removed by volatilization. Its half-life in a model river of depth 1 m flowing at 1 m/sec and a wind speed of 3 m/sec is estimated to be 2.7 hr. In the atmosphere, chlorodifluoromethane is mainly removed by reaction with hydroxyl radicals. It is estimated that the half-life for chlorodifluoromethane in the troposphere is 11.1-17.3 yrs. As a result of its long half-life, the chlorodifluoromethane will accumulate and disperse all over the globe. The concentration of chlorodifluoromethane in the atmosphere will continue to increase as more chlorodifluoromethane is produced and released. About 44% of the chlorodifluoromethane released will slowly diffuse to the stratosphere where it will react with UV light. In the process, free halogens are released which catalyze the destruction of the ozone layer. Chlorodifluoromethane will also be removed from the atmosphere by wet deposition. However, the chlorodifluoromethane removed in this manner will volatilize back into the atmosphere. Environmental exposure occurs to low levels of chlorodifluoromethane in ambient air. Occupation exposure may occur via inhalation and dermal contact with the vapor or liquified gas.

Natural Sources:

Artificial Sources: Chlorodifluoromethane may be released to the environment during its production, storage, transport, and use as a refrigerant, low-temperature solvent, aerosol propellant, and in the production of fluorocarbon resins, especially tetrafluoroethylene monomer

Chlorodifluoromethane

[1-3]. The use of chlorodifluoromethane has been increasing markedly since it is one of the few fluorocarbons not restricted by the Montreal Protocol for protecting the ozone layer. When used as a refrigerant, solvent, or propellant, all chlorodifluoromethane will be released to the air unless captured and recycled. The principal use of chlorodifluoromethane has been in refrigeration. There is a delay of six to seven years between production of chlorodifluoromethane and its release into the atmosphere from the refrigerated units [1].

Terrestrial Fate: Since chlorodifluoromethane is an inert gas with a low adsorption to soil (estimated Koc of 57.5), most of the chemical released on land will be lost by volatilization. Its low Koc also indicates that it is highly mobile in soil and therefore will have a high potential for leaching into ground water [24]. However, chlorodifluoromethane's high volatility should effectively reduce this potential. Because of its very high volatility and unreactivity, it is unlikely that photooxidation, hydrolysis, or biodegradation will be significant in soil.

Aquatic Fate: Since chlorodifluoromethane has a very high Henry's Law constant, is extremely stable in water, and does not adsorb appreciably to sediment, it will be removed from water predominantly by volatilization. Its half-life in a model river of depth 1 m flowing at 1 m/sec and a wind speed of 3 m/sec is estimated to be 2.7 hr.

Atmospheric Fate: In the atmosphere, chlorodifluoromethane is mainly removed by reaction with hydroxyl radicals. It is estimated that the half-life for chlorodifluoromethane in the troposphere is 11.1 to 17.3 yr [5]. As a result of its long half-life, the chlorodifluoromethane released to the atmosphere will accumulate and disperse all over the globe; its background concentration is relatively uniform. A substantial fraction of the chlorodifluoromethane will slowly diffuse to the stratosphere where it will be destroyed as a result of photodissociation and reaction with hydroxyl radicals and O(1D) atoms [2-4]. In the process, radicals are released which catalyze the destruction of the ozone layer. Chlorodifluoromethane will also be removed from the atmosphere by wet deposition. However, the chlorodifluoromethane removed in this manner will volatilize back into the atmosphere.

Biodegradation:

85

Chlorodifluoromethane

Abiotic Degradation: Chlorodifluoromethane does not adsorb UV radiation >290 nm [15] and therefore will not photolyze in the troposphere, waters or in soil surfaces. Chlorodifluoromethane is mainly removed from the troposphere by reaction with hydroxyl radicals by H-atom abstraction [2]. Assuming a rate constant of 3.7×10^{-15} cm^3/molecule sec at 25 °C, an average hydroxyl radical concentration of $1 \times 10^{+6}$ radicals/cc, and utilizing a two-dimensional model which takes account of the temperature dependence of the reaction rate, the half-life for chlorodifluoromethane in the troposphere is estimated to be 11.1 to 17.3 yr [5]. Therefore, chlorodifluoromethane will disperse all over the world and a substantial fraction, about 44% [8], of the chlorodifluoromethane will slowly diffuse to the stratosphere [5,10] where it will be destroyed as a result of photodissociation and reaction with hydroxyl radicals and O(1D) atoms [3,7,15]. In the process, free radicals are released which catalyze the destruction of the ozone layer. Fluorocarbons are chemically inert under tropospheric environmental conditions [6,9]. The rate of hydrolysis of chlorodifluoromethane is very low, <0.01 g/L-yr) at 30 °C [9].

Bioconcentration: Based on its log octanol/water partition coefficient, 1.08 [11], a bioconcentration factor (BCF) of 3.9 may be estimated from the equation log BCF = 0.76 log Kow - 0.23 [19]. This indicates that chlorodifluoromethane should not bioconcentrate in aquatic organisms.

Soil Adsorption/Mobility: Based on its water solubility, 2,899 mg/L at 25 °C [14], a Koc of 57.5 was estimated using a recommended regression equation [19]. Therefore, chlorodifluoromethane would not adsorb appreciably to sediment and suspended solids in the water column.

Volatilization from Water/Soil: Chlorodifluoromethane is a gas with a moderate water solubility, 2,899 mg/L [14], and high volatility, and would, therefore, be expected to volatilize rapidly from water. The experimental Henry's Law constant for chlorodifluoromethane is 0.0294 atm-m^3/mole [13]. Hence, its volatilization from water will be very rapid; the volatilization rate will be limited by chlorodifluoromethane's diffusion through water [19]. The half-life of chlorodifluoromethane in a model river 1 m deep, flowing at 1 m/sec, with a wind of 3 m/sec is estimated to be 2.7 hr [19].

Chlorodifluoromethane

Water Concentrations:

Effluent Concentrations:

Sediment/Soil Concentrations:

Atmospheric Concentrations: Chlorodifluoromethane is the seventh most common halocarbon in the atmosphere, having an average concentration of 60 ppt [10]. Because of its long atmospheric half-life, chlorodifluoromethane will disperse over the globe and accumulate in the atmosphere. Its concentration levels, removed from local sources, will be similar at different locations and will increase with time. In January 1980, the concentration of chlorodifluoromethane in the remote Pacific Northwest and the South Pole was 63 and 45 ppt, respectively [22]. At Point Barrows, Alaska, the concentration was highest in January, 62.5 ppt and lowest in July, 55.3 ppt [17]. The median concentration of chlorodifluoromethane at two rural/remote sites and four urban/suburban sites in the United States from the 1970s was 14 and 25 ppt, respectively [4]. The maximum concentration reported at the urban/suburban sites was 150 ppt [4].

Food Survey Values:

Plant Concentrations:

Fish/Seafood Concentrations:

Animal Concentrations:

Milk Concentrations: A pilot study of volatile organic chemicals in mother's milk found that two of eight samples of milk obtained from women in 4 urban areas in the United States contained chlorodifluoromethane; the levels were not quantified [21].

Other Environmental Concentrations:

Probable Routes of Human Exposure: The general population will be exposed to chlorodifluoromethane by inhalation of ambient air.

Chlorodifluoromethane

Occupational exposure may occur via inhalation and dermal contact with the vapor or liquified gas.

Average Daily Intake: AIR INTAKE: Assuming air concn of 60 ppt and an inhalation rate of 20 m³/day, the inhalation intake of the general population has been estimated to be 4.25 ug/day. WATER INTAKE: insufficient data. FOOD INTAKE: insufficient data.

Occupational Exposure: NIOSH (NOES Survey 1981-1983) has statistically estimated that 149,149 workers, including 8054 women, are potentially exposed to chlorodifluoromethane in the USA [20]. In one study, workers in a fluorocarbon shipping and packaging plant were exposed to 4.7-13.5 ppm of chlorodifluoromethane [16]. Workers involved in the manufacture, servicing and disposal of refrigeration units, food processing, plastic foam blowing and fire extinguishing are likely of higher exposure to the compound.

Body Burdens: A pilot study of volatile organic chemicals in mother's milk found that two of eight samples of milk obtained from women in 4 urban areas in the United States contained chlorodifluoromethane; the levels were not quantified [21].

REFERENCES

1. Altshuller AP; Adv Environ Sci Technol 10: 181-219 (1979)
2. Atkinson R et al; Adv Photochem 11: 375-488 (1979)
3. Atkinson R et al; J Chem Phys 63: 1703-6 (1975)
4. Brodzinsky R, Singh HB; Volatile organic chemicals in the atmosphere Menlo Park: SRI Inter Atmos Sci Ctr Contract 68-02-3452 (1982)
5. Chang JS, Kaufman F; J Chem Phys 66: 4989-94 (1977)
6. Council on Environmental Quality; Fluorocarbons and the environment. June 1975 (1975)
7. Davidson JA et al; J Chem Phys 69: 4277-9 (1978)
8. Derwent RG, Eggleton AEJ; Atmos Environ 12: 126-69 (1978)
9. Du Pont de Nemours Co; Freon Product Information B-2. Wilmington, DE: E.I. Du Pont de Nemours and Co (1980)
10. Fabian P; p 23-51 in Handbook of Environmental Chemistry. Vol 4 Part A. Berlin: Springer-Verlag, Hutzinger O, ed. (1986)
11. Hansch C, Leo AJ; MedChem Project, Issue No. 26, Claremont, CA: Pomona College (1985)
12. Hawley GG; Condensed Chem Dictionary 10th ed NY: Von Nostrand Reinhold (1981)

Chlorodifluoromethane

13. Hine J, Mookerjee PK; J Org Chem 40: 292-8 (1975)
14. Horvath, AL; Halogenated Hydrocarbons NY: Marcel Dekker Inc p 482 (1982)
15. Hubrich C, Stahl F; J Photochem 12: 93-107 (1980)
16. IARC; Monograph on the Evaluation of the Carcinogenic Risk of Chemicals to Humans 41: 238-52 (1986)
17. Khalil MAK, Rasmussen RA; Environ Sci Technol 17: 157-64 (1983)
18. Kuney JH; Chemicyclopedia Vol 7 Washington, DC: Amer Chem Soc p 56 (1989)
19. Lyman WJ et al; Handbook of Chem Property Estimation Methods NY: McGraw-Hill pp 4-1 to 4-33, 5-1 to 5-30, 15-1 to 15-34 (1982)
20. NIOSH; National Occupational Exposure Survey. Cincinnati, OH: NIOSH computer printout 5/10/88 (1988)
21. Pellizzari ED et al; Bull Environ Contam Toxicol 28: 322-8 (1982)
22. Rasmussen RA et al; Science 211: 285-7 (1981)
23. Riddick JA et al; Techniques of Chemistry Vol II, Organic Solvents - Physical Properties and Methods of Purification 4th ed, NY: John Wiley and Sons p 559 (1986)
24. Roy WR, Griffin RA; Environ Geol Water Sci 7: 241-7 (1985)
25. Smart, BE; Kirk-Othmer Encycl Chem Tech 3rd ed NY: John Wiley 10: 856-70 (1980)

2-Chloroethyl Vinyl Ether

Synonyms:

Structure:

$$H_2C = CH - O - CH_2 - CH_2 - Cl$$

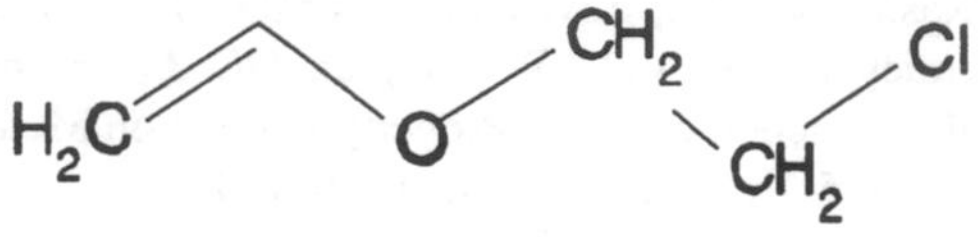

CAS Registry Number: 110-75-8

Molecular Formula: C_4H_7ClO

Wiswesser Line Notation: G2O1U1

CHEMICAL AND PHYSICAL PROPERTIES

Boiling Point: 109 °C at 740 mm Hg

Melting Point: -70 °C

Molecular Weight: 106.55

Dissociation Constants:

Log Octanol/Water Partition Coefficient: 0.99 [6]

Water Solubility: 6000 mg/L at 20 °C [16]

Vapor Pressure: 26.75 mm Hg at 20 °C (extrapolated) [2]

Henry's Law Constant: 6.25×10^{-4} atm-m^3/mol (calculated from the vapor pressure and water solubility)

ENVIRONMENTAL FATE/EXPOSURE POTENTIAL

Summary: 2-Chloroethyl vinyl ether is a synthetic organic chemical, designated as a priority pollutant by the USEPA, that is no longer produced commercially in the United States. If released to soil, 2-

chloroethyl vinyl ether is expected to leach readily. Volatilization from the soil surfaces to the atmosphere may be an important fate process. Biodegradation in soil may occur, and hydrolysis may be an important fate process in acidic soils or soils possessing acidic sites. If released to water, hydrolysis at neutral pH will occur slowly with a half-life of about 50 years; however, at pH 5 the hydrolysis half-life is about 6.9 days. Volatilization from water should be an important fate process; the volatilization half-life for a model river and pond can be estimated at 4.4 and 52 hours, respectively. Direct photochemical degradation in water should not occur, nor should 2-chloroethyl vinyl ether be expected to bioaccumulate in aquatic organisms. The half-life for the reaction with photochemically produced hydroxyl radicals in the atmosphere can be estimated to be about 12 hours. The high reactivity of the double bond on this molecule makes it a candidate for reaction with other radicals and oxidants which may be present in the atmosphere. The half-life for the reaction with ozone in the atmosphere can be estimated at 1.3 days.

Natural Sources: 2-Chloroethyl vinyl ether is an anthropogenic compound, and is not known to exist as a natural product.

Biodegradation: 2-Chloroethyl vinyl ether gave a 76% and 52% (initial concentration 5 and 10 mg/L, respectively) theoretical biological oxygen demand in seven days using a settled domestic wastewater as a microbial inoculum. Complete biodegradation was obtained in seven days using the third subculture. 2-Chloroethyl vinyl ether was listed as showing significant degradation with rapid adaptation (the fastest of a three-tier rating system) [21,22].

Abiotic Degradation: The experimental rate constant for acid catalyzed cleavage of 2-chloroethyl vinyl ether at 25 °C is 0.168 L/mol-sec [15] which translates to a half-life of 69 days at a pH of 6, and to 6.9 days at pH of 5. The rate constant for hydrolysis of 2-chloroethyl vinyl ether at a pH of 7 is 4.4×10^{-10} sec^{-1} [2] which corresponds to a half-life of 50 years in neutral waters. In the laboratory, 2-chloroethyl vinyl ether was shown to disappear in autoclaved samples of a model aquifer system, suggesting that abiotic processes were involved [24]. A laboratory rapid-infiltration microcosm study showed that 2-chloroethyl vinyl ether was removed from the influent stream [4,14]. The half-life for the reaction of 2-chloroethyl vinyl ether with photochemically produced hydroxyl radicals in the atmosphere can be estimated at 6.6 hours [1]. The half-life for the reaction with ozone in the atmosphere can be estimated at 1.3 days [6].

Bioconcentration: Based on the log octanol/water partition coefficient and the water solubility, BCF values in the range 3-5 can be calculated for 2-chloroethyl vinyl ether [11], suggesting that bioconcentration in aquatic organisms should not be an important fate process.

Soil Adsorption/Mobility: Based upon the log octanol/water partition and the water solubility, Koc values can be calculated to fall in the range 22-118 [11], suggesting high to very high mobility in soil [20].

Volatilization from Water/Soil: The Henry's Law constant for 2-chloroethyl vinyl ether indicates that volatilization from environmental waters is probably significant, but not rapid [11]. The estimated volatilization half-life for a model river 1 m deep, flowing at 1 m/sec, and a wind velocity of 3 m/sec is 4.4 hours [11]. The estimated volatilization half-life from a model environmental pond is 52 hours [23]. 2-Chloroethyl vinyl ether has a relatively high vapor that suggests important evaporation from soil surfaces.

Water Concentrations: SURFACE WATER: The US EPA STORET Data Base contained the following data for 2-chloroethyl vinyl ether: 929 samples, 0.8% positive, median concentration less than 10 ug/L [19]. 2-Chloroethyl vinyl ether was not detected in eighty-six samples from fifty-one rainwater runoff catchments located throughout the USA (detection limits not given) [3]. GROUND WATER: 2-Chloroethyl vinyl ether was detected in three of three wells on site at Amphenol Products Division, Allied Corporation, Broadville IL, at concentrations less than 1 ug/L [8,9]. It was detected in eleven of eleven off-site wells in concentrations ranging less than 1 ug/L to less than 10 ug/L [8,9]. Not found in 1174 community and 617 private wells throughout WI, early 1980s, detection limits ca 5 ug/L [10]. Not detected in NJ coastal Plain Aquifer System [5].

Effluent Concentrations: The US EPA Storet Data Base contained the following effluent data for 2-chloroethyl vinyl ether: 1,291 samples, 1% positive, median concentration less than 5 ug/L [19]. 2-Chloroethyl vinyl ether was not detected in Oak Ridge Gaseous Diffusion Plant wastewater (detection limit 10 ppb) [12].

Sediment/Soil Concentrations: The US EPA Storet Data Base contained the following sediment data for 2-chloroethyl vinyl ether: 339 samples, 0% positive [19]. 2-Chloroethyl vinyl ether was found in six of six off-site sediment samples near Amphenol Products Division Broadview Il plant in concentrations less than 100 ug/kg [9].

Atmospheric Concentrations:

Food Survey Values:

Plant Concentrations:

Fish/Seafood Concentrations:

Animal Concentrations:

Milk Concentrations:

Other Environmental Concentrations:

Probable Routes of Human Exposure: 2-Chloroethyl vinyl ether is no longer produced industrially in the USA [17], and there are no known current routes of exposure to the general population.

Average Daily Intake:

Occupational Exposure: NIOSH (NOHS Survey 1972-1974) has statistically estimated that 23,221 workers are exposed to 2-chloroethyl vinyl ether in the USA [13].

Body Burdens:

REFERENCES

1. Atkinson R; Int J Chem Kinet 19: 799-828 (1987)
2. Callahan MA et al; Water Related Fate of 129 Priority Pollutants Vol II USEPA-440/4-79-029B (1979)

2-Chloroethyl Vinyl Ether

3. Cole RH et al; J Water Pollut Contr Fed 56: 898-908 (1984)
4. Enfield CG et al; Haz Was Haz Mat 3: 57-76 (1986)
5. Fusillo TV et al; Ground Water 23: 354-60 (1985)
6. GEMS; Graphic Exposure Modeling System. CLOGP, PCFAP USEPA (1987)
7. Hort EV, Gasman RC; Kirk-Othmer Encycl Chem Tech 3rd Ed. John-Wiley NY 23: 937-960 (1981)
8. IT Corporation; Preliminary Site Assessment, Broadview, IL Plant, Amphenol Products Div (1985)
9. IT Corporation; Phase II Site Assessment, Broadview, IL Plant, Amphenol Products Div (1985)
10. Krill RM, Sonzogni WC; J Am Water Works Assoc 78: 70-5 (1986)
11. Lyman WJ et al; Handbook of Chemical Property Estimation Methods Washington, DC: Amer Chem Soc pp 4-9, 5-4, 5-10, 15-15 to 15-31 (1990)
12. McMahon LW; Organic Priority Pollutants in Wastewater NTIS DE83-010817 CONF 820418 (1983)
13. NIOSH; National Occupational Hazard Survey (NOHS) (1974)
14. Piwoni MD et al; Haz Was Haz Mat 3: 43-55 (1986)
15. Salomaa P et al; Acta Chem Scand 20: 1790-801 (1966)
16. Schildknecht CE; Kirk-Othmer Encycl Chem Technol (2nd ed) NY: John Wiley & Sons 21: 414 (1970)
17. SRC; The Verification of the Production of 56 Chemicals; Prepared for Test Rules Development Branch, Office of Toxic Substances USEPA (1986)
18. SRC; Statement of Research Needs for 2-Chloroethyl vinyl Ether, Prepared for Environmental Criteria and Assessment Office, Contract# 68-03-3521, USEPA. Cincinnati OH (1988)
19. Staples CA et al: Environ Toxicol Contam 4: 1313-42 (1985)
20. Swann RL et al; Res Rev 85: 17-28 (1983)
21. Tabak HH et al; J Water Pollut Contr Fed 53: 1503-18 (1981)
22. Tabak HH et al; pp 267-38 in Test Protocols for Environmental Fate and Movement of Toxicants. Proc Symp Assoc Official Anal Chem, 94th Annual Mtg Washington DC (1981)
23. USEPA; Exams II Computer Simulation (1987)
24. Wilson J, Noonan MJ; Microbial Activity in Model AquiferSystems. USEPA-600/D-84-136 (1984)

4-Chlorotoluene

SUBSTANCE IDENTIFICATION

Synonyms: 4-Chloro-1-methylbenzene

Structure:

CAS Registry Number: 106-43-4

Molecular Formula: C_7H_7Cl

Wiswesser Line Notation: GR D

CHEMICAL AND PHYSICAL PROPERTIES

Boiling Point: 161.99 °C

Melting Point: 7.5 °C

Molecular Weight: 126.59

Dissociation Constants:

Log Octanol/Water Partition Coefficient: 3.33 [9]

Water Solubility: 106 mg/L at 20 °C [29]

Vapor Pressure: 2.59 mm Hg at 20 °C [22]

Henry's Law Constant: 0.00407 atm-m³/mol (calculated from the vapor pressure and water solubility)

4-Chlorotoluene

ENVIRONMENTAL FATE/EXPOSURE POTENTIAL

Summary: p-Chlorotoluene may be released to the environment in emissions and effluents from sites of its manufacture or industrial use, from venting during storage and transport, and from disposal of industrial waste products which contain this compound [such as spent solvent]. p-Chlorotoluene may be formed in the environment as a photodegradation product of p-chlorobenzyl chloride which is used as a chemical intermediate. If released to soil, p-chlorotoluene is expected to have low mobility and should volatilize fairly rapidly from soil surfaces. If released to water, volatilization (half-life in a model river of 3.5 hours), sensitized photolysis, and adsorption to suspended solids and sediments may be important fate processes. The relative importance of these fate processes and the rate of compound loss are expected to vary depending upon ambient conditions and characteristics of the water body. Based on monitoring data, the half-life of p-chlorotoluene in a river 4-5 m deep during midsummer was estimated to be 1.2 days. This compound is not expected to undergo chemical hydrolysis, react with oxidants found in natural waters or bioaccumulate significantly in aquatic organisms. Due to a lack of data, the significance of biodegradation in soil or water is not known. One screening test (Japanese MITI) has found that p-chlorotoluene is resistant to biodegradation. If released to the atmosphere, the dominant removal mechanism for p-chlorotoluene is expected to be reaction with photochemically generated hydroxyl radicals (estimated half-life of 8.4 days). A slight potential also exists for direct photolysis in the atmosphere. The most probable route of human exposure to p-chlorotoluene is inhalation of contaminated air. Segments of the population may also be exposed to ingestion of contaminated drinking water.

Natural Sources:

Artificial Sources: p-Chlorotoluene may be released to the environment in emissions and effluents from sites of its manufacture or industrial use, from venting during storage and transport, and from disposal of industrial waste products which contain this compound [5,19]. This compound is widely used as a solvent and intermediate in the synthesis of other organic chemicals, dyes, pharmaceuticals, and synthetic rubber chemicals [1]. p-Chlorotoluene may be formed in the environment as a photodegradation product of p-chlorobenzyl chloride, an intermediate used in the manufacture of pharmaceuticals, dyes and other organic chemicals [3,12].

Terrestrial Fate: If released to soil, p-chlorotoluene is expected to leach slowly with a relatively low soil mobility. Volatilization may occur fairly rapidly from dry soil surfaces. Chemical hydrolysis should not be environmentally important. Due to a lack of data, the significance of biodegradation in soil is not known. One screening test (Japanese MITI) has found that p-chlorotoluene is resistant to biodegradation [16,23].

Aquatic Fate: If released to water, volatilization (half-life in a model river of 3.5 hours), sensitized photolysis, and adsorption to suspended solids and sediments are potentially important fate processes. The relative importance of these fate processes and the rate of compound loss are expected to vary depending on ambient conditions and characteristics of the water body. This compound is not expected to undergo chemical hydrolysis, react with oxidants found in natural waters or bioaccumulate significantly in aquatic organisms. Due to a lack of data, the significance of biodegradation is not known. One screening test (Japanese MITI) has found that p-chlorotoluene is resistant to biodegradation [16,23]. Based on monitoring data from the River Rhine, the half-life of p-chlorotoluene in a river 4-5 m deep during midsummer was estimated to be 1.2 days [30].

Atmospheric Fate: Based on the vapor pressure, p-chlorotoluene is expected to exist almost entirely in the vapor phase in the ambient atmosphere [4]. The dominant removal mechanism is expected to be reaction with photochemically generated hydroxyl radicals which has an estimated half-life of 8.4 days. A slight potential exists for direct photolysis in the atmosphere. Physical removal from air via washout may occur; however, volatilization will probably return the p-chlorotoluene to the atmosphere.

Biodegradation: Using the Japanese MITI protocol (initial concentration 100 ppm, 14 days incubation and an activated sludge inoculum), p-chlorotoluene was found to be resistant to biodegradation because the theoretical biological oxygen demand was less than 30 percent [16,23]. An isolated strain of Pseudomonas putida 39/D, oxidized p-chlorotoluene to (+)-cis-4-chloro-2,3-dihydroxy-1-methylcyclohexa-4,6-diene [6]. p-Chlorotoluene is metabolized via cis-dihydrodiol to its respective catechol which is resistent to further degradation [7].

Abiotic Degradation: p-Chlorotoluene is inert to chemical hydrolysis under environmental conditions [14]. Reaction of p-chlorotoluene with free radical oxidants found in natural waters is not an environmentally important fate process [13,14]. When exposed to UV light at wavelengths about 300 nm, p-chlorotoluene, in deaerated methanol solution, underwent 2.9% degradation in 48 hours [3]. Under the same conditions with acetone added as a photosensitizer, p-chlorotoluene underwent 54.5% degradation in 9 hours [3]. Of the p-chlorotoluene degraded in this latter study, 52 percent was converted into toluene [3]. The half-life for p-chlorotoluene vapor reacting with photochemically generated hydroxyl radicals in the atmosphere has been estimated to be 8.4 days based on a reaction rate constant of 1.9×10^{-12} cm^3/molecule-sec at 25 °C and an average hydroxyl radical concentration of $5.0 \times 10^{+5}$ molecules/cm^3 [2].

Bioconcentration: Based on the water solubility and log Kow, bioconcentration factors (BCF) of 45-200 can be estimated for p-chlorotoluene from recommended regression equations [20]. These BCF values suggest that slight bioaccumulation in aquatic organisms may occur.

Soil Adsorption/Mobility: Soil adsorption coefficients (Koc) of 446-1544 can be estimated for p-chlorotoluene using linear regression equations based on the water solubility and log Kow [20]. These Koc value suggests that p-chlorotoluene would have low mobility in soil [24] and that adsorption to suspended solids and sediments in water may have some environmental significance.

Volatilization from Water/Soil: The value of the Henry's Law Constant for p-chlorotoluene suggests that volatilization is probably significant from all bodies of water [20]. Based on the Henry's Law Constant, the volatilization half-life from a model river 1 m deep, flowing 1 m/sec with a wind speed of 3 m/sec was estimated to be 3.5 hours [20]. The estimated volatilization half-life from a model environmental pond is 6.6 days [25]. Due to its relatively high vapor pressure, p-chlorotoluene is expected to volatilize fairly rapidly from dry soil surfaces.

4-Chlorotoluene

Water Concentrations: DRINKING WATER: 1975 National Organics Reconnaissance Survey (NORS) 10-city survey identified p-chlorotoluene in finished drinking water from 1 out of 10 cities [17]; Miami, FL had a concentration of 1.5 ug/L [17]. GROUND WATER: As of June 1984, p-chlorotoluene not found (although analyzed for at a detection limit of 1.0-5.0 ug/L) in 1174 community wells and 617 private wells in Wisconsin [18]. During 1981-1982, p-chlorotoluene was analyzed for but not found in 945 wells scattered throughout the USA (detection limit 0.2-0.5 ug/L) [28]. SURFACE WATER: In 1976, the River Maas at Eysden (The Netherlands) had a median concentration of 0.1 ug/L and a maximum concentration of 0.3 ug/L [27]. In 1976, the River Maas at Keizersveer (The Netherlands) had a median concentration 0.1 ug/L and a maximum concentration of 0.2 ug/L [27]. In July 1979, a concentration of 0.03 ug/L p-chlorotoluene was detected in the River Rhine at Lobith (The Netherlands) [30].

Effluent Concentrations: p-Chlorotoluene was identified in chlorinated leachate from a simulated landfill lysimeter used to study codisposal of metal plating sludge with municipal solid waste [8]. It was identified as a principal organic hazardous constituent (POHC) in the emissions from an incinerator test burn [15].

Sediment/Soil Concentrations:

Atmospheric Concentrations: Ambient monitoring in New Jersey during July-Aug 1981 had the following results: Newark, NJ, 38 samples, 95% positive, mean concentration 0.21 ppb; Elizabeth, NJ, 37 samples, 97% positive, mean concentration 0.25 ppb; Camden, NJ, 35 samples, 97% positive, mean concentration 0.22 ppb [10]. Ambient monitoring in New Jersey during Jan-Feb 1982 had the following results: Newark, NJ, 28 samples, 96% positive, mean concentration 0.17 ppb; Elizabeth, NJ, 38 samples, 92% positive, mean concentration 0.14 ppb; Camden, NJ, 37 samples, 78% positive, mean concentration 0.07 ppb [10]. During 1983-84, p-chlorotoluene was detected in air samples collected at 6 hazardous waste sites in NJ at a mean concentration range of <0.03-0.70 ppb [11]; and one sanitary landfill in NJ had a mean concentration of 0.04 ppb [11]. Combined chlorotoluene isomers (o-, m-, and p-) were identified at the following sites [21,26]: Love Canal in Niagara Falls, NY, 15 samples, 80% positive, concentration range trace-12,274 ng/m^3 [21]; Love Canal,

4-Chlorotoluene

10 samples, 50% positive, concentration range <0.010-1.64 ng/m^3 [26]; Baton Rouge, LA, 11 samples, 9% positive, 35 ng/m^3 [21].

Food Survey Values:

Plant Concentrations:

Fish/Seafood Concentrations:

Animal Concentrations:

Milk Concentrations:

Other Environmental Concentrations:

Probable Routes of Human Exposure: The most probable route of human exposure to p-chlorotoluene is inhalation of contaminated air [10,21]. Segments of the general population may also be exposed to p-chlorotoluene by ingestion of contaminated drinking water [17,21]. Workers involved in the manufacture, use, packaging, or transport of this compound may be exposed by inhalation and/or dermal contact [1].

Average Daily Intake:

Occupational Exposure:

Body Burdens:

REFERENCES

1. ACGIH; Documentation of the Threshold Limit Value and Biological Exposure Indices 5th ed Cincinnati, OH: ACGIH pp 95-6 (1986)
2. Atkinson R; Inter J Chem Kinetics 19: 799-828 (1987)
3. Choudhry GG et al; Toxicol Environ Chem 13: 27-84 (1986)
4. Eisenreich SJ et al; Environ Sci Tech 15: 30-8 (1981)
5. Gelfand S; Kirk-Othmer Encycl Chem Tech 3rd ed NY: Wiley 5: 819-27 (1979)
6. Gibson DT et al; Biochemistry 7: 3795-802 (1968)
7. Gibson DT; pp 187-204 in Aquatic Pollutants: Transformation and Biological Effects; Hutzinger O et al eds Oxford: Pergamon Press (1978)
8. Gould JP et al; Water Chlorination: Environ Impact Health Eff 4: 525-39 (1983)

4-Chlorotoluene

9. Hansch C, Leo AJ; Medchem Project Issue No. 26 Claremont, CA: Pomona College (1985)
10. Harkov R et al; Sci Tot Environ 38: 259-74 (1984)
11. Harkov R et al; J Environ Sci H A20: 491-501 (1985)
12. Hawley GG; The Condensed Chemical Dictionary 10th ed NY: Van Nostrand Reinhold p 235 (1981)
13. Howard JA, Chenier JHB; J Am Chem Soc 95: 3054-9 (1973)
14. Jaber HM et al; Data Acquisition for Environmental Transport and Fate Screening p USEPA-600/6-84/009 NTIS PB84-243906 (1984)
15. James RH et al; J Proc - APCA 77th Annual Mtg June 24-9, 1984 Paper 84-18 (1984)
16. Kawasaki M; Ecotox Environ Safety 4: 444-54 (1980)
17. Keith LH et al; pp 329-73 in Ident Anal Organ Pollut Water; Keith LH ed Ann Arbor, MI: Ann Arbor Press (1976)
18. Krill RM, Sonzogni WC; J Am Water Works Assoc 78: 70-5 (1986)
19. Liepins R et al; p 6-795 in Industrial Process Profiles for Environmental Use: The Industrial Organic Chemicals Industry USEPA-600/2-77-023f NTIS PB-281 478 (1977)
20. Lyman WJ et al; Handbook of Chemical Property Estimation Methods Washington, DC: Amer Chem Soc pp 4-9, 5-4, 5-10, 15-12 to 5-32 (1990)
21. Pellizzari ED et al; Formulation of Preliminary Assessment of Halogenated Organic Compounds in Man and Environmental Media pp 55,57,72 USEPA-560/13-79-006 (1979)
22. Riddick JA et al; Organic Solvents: Physical Properties and Methods of Purification 4th ed NY: Wiley-Interscience p 485 (1986)
23. Sasaki S; pp 283-98 in Aquatic Pollutants: Transformation and Biological Effects; Hutzinger O et al eds Oxford: Pergamon Press (1978)
24. Swann RL et al; Res Rev 85: 17-28 (1983)
25. USEPA; Exams II Computer Simulation (1987)
26. VanTassel S et al; Anal Chem 53: 2130-5 (1981)
27. Verschueren K; Handbook of Environmental Data on Organic Chemicals 2nd ed NY: Van Nostrand Reinhold p 387 (1983)
28. Westrick JJ et al; J Am Water Works Assoc 76: 52-9 (1984)
29. Yalkowsky SH et al; Arizona Data Base of Aqueous Solubilities, Univ of Ariz, College of Pharmacy (1987)
30. Zoeteman BCJ et al; Chemosphere 9: 231-49 (1980)

Cyclohexanol

SUBSTANCE IDENTIFICATION

Synonyms:

Structure:

CAS Registry Number: 108-93-0

Molecular Formula: $C_6H_{12}O$

Wiswesser Line Notation: L6TJ AQ

CHEMICAL AND PHYSICAL PROPERTIES

Boiling Point: 161.1 °C

Melting Point: 25.1 °C

Molecular Weight: 100.16

Dissociation Constants:

Log Octanol/Water Partition Coefficient: 1.23 [12]

Water Solubility: 38,000 mg/L at 25 °C [3]

Vapor Pressure: 0.80 mm Hg at 25 °C [7]

Henry's Law Constant: 2.77 x 10^{-6} atm-m^3/mol (calculated from the vapor pressure and water solubility)

ENVIRONMENTAL FATE/EXPOSURE POTENTIAL

Summary: Cyclohexanol may be released to the environment as a result of its manufacture and use as a solvent in various processes and its use in various product formulations. If released to soil, it is expected to exhibit high to very high mobility in soil based upon estimated Koc values. In the absence of biodegradation, it may leach through soil to ground water. It will not chemically hydrolyze in moist soil, but it may volatilize slowly from soil surfaces. If released to water, it is not expected to adsorb to sediment or suspended particulate matter or to bioconcentrate significantly in aquatic organisms. It will not chemically hydrolyze or directly photolyze in environmental waters. The results of several laboratory biodegradation screening tests, using sewage and activated sludge inocula, indicate that cyclohexanol may biodegrade readily. Therefore, the dominant transformation process in soil and water is probably biodegradation. Cyclohexanol will volatilize slowly from surface waters. If released to the atmosphere, it is expected to exist mainly in the vapor phase where it will degrade by reaction with photochemically produced hydroxyl radicals. The estimated half-life for this vapor-phase reaction is 22 hr at an atmospheric concentration of 5 x 10^{+5} hydroxyl radicals/cm^3. Cyclohexanol will not directly photolyze in the atmosphere. Exposure of cyclohexanol to the general population may occur via the ingestion of contaminated drinking water and food, and the inhalation of contaminated ambient air. Minor exposure may occur via dermal contact with contaminated water.

Natural Sources: Cyclohexanol was qualitatively detected in the mixture of flavor volatiles from fried chicken [31] and baked potatoes [5] which indicates that the compound may be formed naturally.

Artificial Sources: Cyclohexanol may be released to the environment as a result of its manufacture and use as a solvent for alkyd and phenolic resins and cellulosics [13]; its use as a source of adipic acid for the manufacture of nylon [13]; its use in the production of soap, plasticizers, and plastics [13]; its use in insecticide formulations, lacquers, paints, varnishes, finish removers, polishes, emulsified products, and germicides [13]; and its use in textile finishing and leather degreasing [13].

Terrestrial Fate: If cyclohexanol is released to soil, it is expected to exhibit high to very high mobility in soil [30] based upon Koc values of 111 and 13 estimated from linear regression equations [18]. It may, therefore, leach through soil to ground water if it does not biodegrade first. It will not chemically hydrolyze in moist soil [18], although it may volatilizate slowly from soil surfaces. It may biodegrade readily in soil based upon the results observed in laboratory aqueous aerobic screening tests using sewage and activated sludge inocula [21,24]. Biodegradation is probably the dominant transformation process in soil.

Aquatic Fate: If cyclohexanol is released to water, it is not expected to adsorb to sediment or suspended particulate matter or to bioconcentrate significantly in aquatic organisms based upon its estimated Koc (13-111) and BCF (1.5-5.1) values [18]. It is not expected to hydrolyze [18] or directly photolyze in environmental waters [28]. It may biodegrade readily in natural waters based upon results observed in laboratory biodegradation aqueous aerobic screening tests using sewage and activated sludge inocula [21,24]. Biodegradation is probably the dominant transformation process in natural water. Volatilization from surface waters is slow based upon estimated half-lives of 13.3 days from a model river one meter deep flowing 1 m/sec with a wind velocity of 3 m/sec [18] and 145 days for volatilization from a model environmental pond [32].

Atmospheric Fate: If cyclohexanol is released to the atmosphere, it can be expected to exist almost entirely in the vapor phase in the ambient atmosphere [9] based upon the vapor pressure [9]. The estimated rate constant for vapor-phase reaction with photochemically produced hydroxyl radicals of 17.4×10^{-12} cm^3/molecule-sec at 25 °C [1] corresponds to an atmospheric half-life of 22 hr at an atmospheric concentration of $5 \times 10^{+5}$ hydroxyl radicals/cm^3. Cyclohexanol is not expected to directly photolyze in the atmosphere since it does not absorb ultraviolet light above 290 nm [28].

Biodegradation: A theoretical BOD of 74% was observed for cyclohexanol after 5 days in aqueous aerobic screening tests using the standard dilution technique and acclimated sewage as inoculum [21]. A 96% removal was observed in 5 days in aerobic screening tests using a vigorous activated sludge system which was acclimated for 20 days prior to the experiments [26]. Tests using acclimated mixed microbial cultures as inoculum gave a theoretical BOD of 57 percent after 5 days under aerobic conditions [2]. In an aqueous aerobic sewage die-away screening test, a percent theoretical BOD of 78% was observed after 5 days; initial cyclohexanol concentration was 6,000 ppm [33]. Although cyclohexanol was readily biodegraded in most aerobic screening tests, only 3% theoretical BOD was observed in 5 days in one aqueous aerobic screening test using sewage as inoculum [15]. Aqueous aerobic screening tests with an inoculum of trench leachate from a low-level radioactive waste disposal site added to a medium consisting of filtered, sterilized leachate gave 49% and 17% removal of cyclohexanol (initial concentration 1.2 ppm) after 21 days with and without supplemental addition of ammonium nitrate, respectively [10]. Cyclohexanol appears to be removable by biological treatment based upon the 98-99% reduction in BODT observed in tests of an acclimated activated sludge biological simulator treating caprolactam production waste that contains cyclohexanol, cyclohexanone, cyclohexanone oxime, and caprolactam [24]. No information was found regarding biodegradation obtained under anaerobic conditions nor in natural water or soil.

Abiotic Degradation: The rate constant for the vapor phase reaction of cyclohexanol with photochemically produced hydroxyl radicals has been estimated to be 17.4×10^{-12} cm^3/molecule-sec at 25 °C [1] which corresponds to an atmospheric half-life of 22 hr at an atmospheric concentration of $5 \times 10^{+5}$ hydroxyl radicals/cm^3. Oxidation by peroxyl radicals will not be an important removal process in water based upon a measured rate constant for the reaction of 0.17 L/mol-sec at 27 °C [14], which corresponds to a half-life of 129 years at 1×10^{-9} moles of peroxyl radicals per liter [20]. Because alcohols do not absorb light at wavelengths above 290 nm [28], cyclohexanol will not be expected to directly photolyze in the environment. Aqueous hydrolysis is not expected to be significant under normal environmental conditions (pH 5-9) [18].

Bioconcentration: Based upon the experimental log Kow, a BCF of 5.1 has been estimated using a recommended regression equation [18]. Based upon the experimental water solubility, a BCF of 1.5 has been estimated using a recommended regression equation [18]. These estimated BCF values suggest that cyclohexanol will not bioconcentrate significantly in aquatic organisms.

Soil Adsorption/Mobility: Based upon the experimental log Kow, a Koc of 111 has been estimated using a linear regression equation [18]. Based upon the experimental water solubility, a Koc of 13 has been estimated using a linear regression equation [18]. Based upon these estimated Koc values, cyclohexanol will be expected to exhibit high to very high mobility in soil [30]. Cyclohexanol, therefore, may leach through soil to ground water if it does not volatilize or biodegrade first.

Volatilization from Water/Soil: Based upon the Henry's Law constant, the half-life for volatilization of cyclohexanol from a model river one meter deep flowing 1 m/sec with a wind velocity of 3 m/sec is estimated to be 13.3 days at 25 °C [18]. The volatilization half-life from a model environmental pond, which considers the effect of adsorption, has been estimated to be 145 days [32].

Water Concentrations: DRINKING WATER: Cyclohexanol was qualitatively detected in 8 of 16 samples drinking water concentrate derived from large volume (>400 gallons) samples from 6 of 7 cities (Cincinnati, OH, Oct 1978; Miami, FL, Feb 1976; New Orleans, LA, Jan 1976; Philadelphia, PA, Feb 1976; Ottumwa, IA, Sept 1976; Seattle, WA, Nov 1976) [17]. GROUND WATER: Cyclohexanol was qualitatively detected in samples of ground water taken between 1975 and 1978 from an aquifer polluted by organic waste from western suburbs of Melbourne, Australia [29]. It was detected at a maximum concentration of 1 ppb in samples of ground water taken between 1976 and 1978 at contaminated dump sites in the Netherlands [35]. It was detected at a concentration of 1 ppb in ground water taken from a landfill site near Norman, OK [8].

Effluent Concentrations: Cyclohexanol was qualitatively detected in 5 of 16 samples of concentrate derived from large volume (>400 gallons) samples of advanced treatment concentrate from 4 of 6 cities (Lake Tahoe, CA, Oct 1974; Orange County, CA, Feb 1976; Dallas, TX, Nov 1974; Blue Plains, Washington, DC, May 1975) [17]. Cyclohexanol was found at concentrations ranging from 0.24 to 1.2 mg/L in trench leachate from low level radioactive waste sites at Maxey Flats, KY and West Valley, NY [10,11]. It has been found at significant concentrations (>0.5 ppm) in wastewaters from unspecified industries in the organic chemicals, and plastics and synthetic fibers industrial categories [34]. A concentration of 1 ppb was found in samples of leachate or samples of ground water plume taken at an unspecified municipal landfill at an unspecified time [4]. Concentration found in wastewater from a Russian caprolactam manufacturing facility was reported to be 3.5 mg/L [27].

Cyclohexanol

Sediment/Soil Concentrations:

Atmospheric Concentrations: Cyclohexanol was qualitatively detected in ambient air samples taken from an unspecified site in the Los Angeles Basin between 1974 and 1975 [25].

Food Survey Values: Cyclohexanol was qualitatively detected in mixtures of volatile flavor components from fried chicken [31] and baked potatoes (Idaho Russet Burbank potatoes) [5].

Plant Concentrations:

Fish/Seafood Concentrations:

Animal Concentrations:

Milk Concentrations:

Other Environmental Concentrations:

Probable Routes of Human Exposure: General population exposure to cyclohexanol will occur mainly via the ingestion of contaminated drinking water [17] and food [5,31], and the inhalation of contaminated ambient air [25]. Minor exposure may occur via dermal contact with contaminated water. Occupational exposure may occur through inhalation of contaminated air and dermal contact with solutions containing the compound.

Average Daily Intake:

Occupational Exposure: NIOSH (NOES Survey 1981-1983) has statistically estimated that 68,715 workers are potentially exposed to cyclohexanol in the USA [22]. NIOSH (NOHS Survey 1972-1974) has statistically estimated that 898,132 workers are potentially exposed to cyclohexanol in the USA [23]. Cyclohexanol was detected at a Russian caprolactam fiber production plant at concentrations higher than the maximum allowable limit (maximum allowable limit not specified); highest levels were found in the spinning area and in the summer months [19].

107

Cyclohexanol

Body Burdens: Cyclohexanol was detected in the expired air of 5 of 8 volunteer employees from the U.S. Air Force School of Aerospace Medicine, Brooks Air Force Base, Texas; concentrations averaged 0.78 ug/hr (for positive samples) and ranged between 0.28 and 1.6 ug/hr [6]. It was detected in 65.6% of 387 samples taken from 54 study subjects who were carefully selected from urban areas and were normal, healthy, and nonsmoking individuals; the geometric average was 2.6 ng/L of expired air [16].

REFERENCES

1. Atkinson R; Intern J Chem Kinetics 19: 799-828 (1987)
2. Babeu L, Vaishnav DD; J Indust Microbiol 2: 107-15 (1987)
3. Barton AFM; Solubility Data Series. Vol 15. Alcohols with Water. NY: Pergamon Press Inc (1984)
4. Brown KW, Donnelly KC; Haz Waste Haz Mater 5: 1-30 (1988)
5. Coleman EC et al; J Agric Food Chem 29: 42-8 (1981)
6. Conkle JP et al; Arch Environ Health 30: 290-5 (1975)
7. Daubert TE, Danner RP; Physical & Thermodynamic Properties of Pure Compounds. Data Compilation. Am Inst Chem Eng. (1989)
8. Dunlap WJ et al; Identification Anal Org Pollut Water pp 453-77 (1976)
9. Eisenreich SJ et al; Environ Sci Technol 15: 30-8 (1981)
10. Francis AJ; Environmental Migration of Long-Lived Radionuclides. Internat Atomic Energy Agency. Vienna, Austria. IAEA-SM-257/72. pp 415-29 (1982)
11. Francis AJ et al; Nuclear Technol 50: 153-63 (1980)
12. Hansch C, Leo AJ; Medchem Project Issue No.26 Claremont, CA: Pomona College (1985)
13. Hawley GG; Condensed Chemical Dictionary 10th ed Van Nostrand Reinhold NY p 297 (1981)
14. Hendry DG et al; J Phys Chem Ref Data 3: 944-78 (1974)
15. Heukelekian H, Rand MC; J Water Pollut Control Assoc 27: 1040-53 (1955)
16. Krotoszynski BK et al; J Anal Toxicol 3: 225-34 (1979)
17. Lucas SV; GC/MS Anal of Org in Drinking Water Concentrates and Advanced Treatment Concentrates Vol 1 USEPA-600/1-84-020a (NTIS PB85-128221) p 133, 169 (1984)
18. Lyman WJ et al; Handbook of Chem Property Estimation Methods Washington DC: Amer Chem Soc pp 4-9, 5-4, 5-10, 7-14, 15-16 to 15-29 (1990)
19. Mandzhgaladze RN, Kuchukhidze GE; Soobshch Akad Nauk Gruz SSR 3: 733-76 (1972)
20. Mill T; Environ Toxicol Chem 1: 135-41 (1982)
21. Niemi GJ et al; Environ Toxicol Chem 6: 515-27 (1987)
22. NIOSH; The National Occupational Exposure Survey (NOES) (1983)
23. NIOSH; The National Occupational Hazard Survey (NOHS) (1974)
24. Patel MD, Patel DR; Indian J Environ Health 19: 310-8 (1977)

Cyclohexanol

25. Pellizzari ED; Development of Analytical Techniques for Measuring Ambient Atmospheric Carcinogenic Vapors. USEPA Environ Res Lab, Research Triangle Park, NC. USEPA-600/2-75-076 p 75 (1975)
26. Pitter P; Water Res 10: 231-5 (1976)
27. Savelova VA, Sergeev AN; Gig Sanit 35: 21-25 (1970)
28. Silverstein RM et al; Spectrometric Id of Org Cmpd, J Wiley and Sons Inc 3rd ed p 238-55 (1974)
29. Stepan S et al; Australian Water Resources Council Conf Ser 1: 415-24 (1981)
30. Swann RL et al; Res Rev 85: 17-28 (1983)
31. Tang J et al; J Agric Food Chem 31: 1287-92 (1983)
32. USEPA; EXAMS II Computer Simulation (1987)
33. Wagner R; Vom Wasser 47: 241-65 (1976)
34. Wise HE, Fahrenthold PD; in 181st Amer Chem Soc Natl Mtg, Div Indust Chem Symp Atlanta, GA, March 29-April 3 (1981)
35. Zoeteman BCJ et al, Sci Total Environ 21: 187-202 (1981)

Cyclopentane

SUBSTANCE IDENTIFICATION

Synonyms:

Structure:

CAS Registry Number: 287-92-3

Molecular Formula: C_5H_{10}

Wiswesser Line Notation: L5TJ

CHEMICAL AND PHYSICAL PROPERTIES

Boiling Point: 49.3 °C

Melting Point: -94.4 °C

Molecular Weight: 70.13

Dissociation Constants:

Log Octanol/Water Partition Coefficient: 3.00 [12]

Water Solubility: 156 mg/L at 25 °C [43]

Vapor Pressure: 317.8 mm Hg at 25 °C [5]

Henry's Law Constant: 1.88 x 10^{-1} atm-m^3/mole at 25 °C, calculated from VP/Wsol

ENVIRONMENTAL FATE/EXPOSURE POTENTIAL

Summary: Cyclopentane is a highly volatile constituent in the cycloparaffin fraction of crude oil. Cyclopentane is released to the environment via the manufacture, use and disposal of many products

Cyclopentane

associated with the petroleum and gasoline industries. Extensive data show release of cyclopentane into the environment from waste incinerators and the combustion of gasoline and diesel fueled engines. Photolysis or hydrolysis of cyclopentane are not expected to be important fate processes. Limited data suggests that cyclopentane is recalcitrant to biodegradation. Varying estimates of Koc indicate a wide range of adsorption characteristics for cyclopentane and the mobility class in soil may range from low to medium. In aquatic systems, it may partition from the water column to organic matter in sediments and suspended solids. The potential for bioconcentration of cyclopentane in aquatic organisms is low. The Henry's Law constant suggests rapid volatilization of cyclopentane from natural waters and moist soils. The volatilization half-lives from a model river and a model pond (the latter considers the effect of adsorption) have been estimated to be 2.5 hr and 5.2 days, respectively. Based on the vapor pressure, cyclopentane should rapidly evaporate from dry surfaces, especially when present in high concentrations such as in spill situations. Cyclopentane is expected to exist almost entirely in the vapor phase in ambient air. Reactions with photochemically produced hydroxyl radicals in the atmosphere have been shown to be important (average half-life of 3.3 days). Physical removal from air by precipitation and dissolution in clouds may occur; however, the short atmospheric residence time of cyclopentane suggests that wet deposition is of limited importance. The most probable human exposure to cyclopentane would be occupational exposure, which may occur through dermal contact or inhalation at places where it is produced or used. Workplace exposures have been documented. Common nonoccupational exposure would include inhalation; cyclopentane is a widely occurring atmospheric pollutant.

Natural Sources: Cyclopentane is a constituent in the cycloparaffin fraction of crude oil [42].

Artificial Sources: Cyclopentane is released to the environment via the manufacture, use and disposal of many products associated with the petroleum and gasoline industries [1,26,29]. Cyclopentane is a component of gasoline [10], and has been identified as a product of combustion engines [11,21,22,38,44]. Municipal waste incinerators also release cyclopentane as a stack emission [14].

111

Cyclopentane

Terrestrial Fate: Photolysis [35] or hydrolysis [18] of cyclopentane are not expected to be important in terrestrial environments. Limited data suggests that cyclopentane is recalcitrant to biodegradation [13]. An estimated range for Koc from 272 to 1020 [18] indicates a wide range of adsorption characteristics for cyclopentane and the mobility class in soil may range from low to medium [39]. Based upon the calculated Henry's Law constant, cyclopentane is expected to rapidly volatilize from surface soils [18]. Based on the vapor pressure, cyclopentane should rapidly evaporate from dry surfaces, especially when present in high concentrations such as in spill situations.

Aquatic Fate: Photolysis [18] or hydrolysis [18] of cyclopentane in aquatic systems are not expected to be important fate processes. Limited data suggests that cyclopentane is recalcitrant to biodegradation [13]. The log bioconcentration factor (log BCF) has been estimated to range from 0.08 to 2.05 [18] indicating the potential for cyclopentane to bioconcentrate in aquatic organisms is low. An estimated range for Koc from 272 to 1020 [35] indicates cyclopentane may partition from the water column to organic matter contained in sediments and suspended solids. The calculated Henry's Law constant suggests rapid volatilization of cyclopentane from environmental waters [18]. Based on the Henry's Law constant, the volatilization half-life from a model river has been estimated to be 2.5 hr [18]. The volatilization half-life from a model pond, which considers the effect of adsorption, can be estimated to be about 5.2 days [40].

Atmospheric Fate: Cyclopentane does not absorb UV light in the environmentally significant range, >290 nm [35], and will not undergo direct photolysis in the atmosphere. Based on the vapor pressure, cyclopentane is expected to exist almost entirely in the vapor phase in ambient air [7]. Vapor phase reactions with photochemically produced hydroxyl radicals in the atmosphere have been shown to be important. Rate constants for cyclopentane were measured to be 5.02×10^{-12}, 5.26×10^{-12} and 4.43×10^{-12} cm^3/molecule-sec at 22 [6], 26 [2], and 27 [4] °C, respectively, which correspond to atmospheric half-lives of about 3.2 [6], 3.1 [4] and 3.6 [6] days at an atmospheric concentration of $5 \times 10^{+5}$ hydroxyl radicals per cm^3 [2]. The water solubility for cyclopentane indicates that physical removal from air by precipitation and dissolution in clouds may occur; however, the short atmospheric residence time of cyclopentane suggests that wet deposition is of limited importance.

Biodegradation: Pure culture studies showed various isolated species of bacteria were unable to utilize cyclopentane as a single carbon source in soil [3,27]. Mixed populations from ground water contaminated with gasoline did not biodegrade cyclopentane at an initial concentration of 0.17 ppm in a gasoline mixture [13]. After 192 hr, the concentration was 0.04 ppm; however, the concentration in a sterilized control was 0.05 ppm [13].

Abiotic Degradation: Alkanes are generally resistant to hydrolysis [18]. Based on data for cyclohexane, cyclopentane is not expected to absorb UV light in the environmentally significant range, >290 nm [35]. Therefore, cyclopentane will not undergo hydrolysis or direct photolysis in the environment. The rate constants for the vapor-phase reaction of cyclopentane with photochemically produced hydroxyl radicals were measured to be 5.02×10^{-12}, 5.26×10^{-12} and 4.43×10^{-12} cm^3/molecule-sec at 22 [6], 26 [2], and 27 [4] °C, respectively, which correspond to atmospheric half-lives of about 3.2 [6], 3.1 [4] and 3.6 [6] days at an atmospheric concentration of $5 \times 10^{+5}$ hydroxyl radicals per cm^3 [2].

Bioconcentration: Based upon the water solubility and log Kow, bioconcentration factors (log BCF) for cyclopentane have been calculated to be 0.08 and 2.05, respectively, from recommended regression-derived equations [18]. These BCF values indicate the potential for cyclopentane to bioconcentrate in aquatic organisms is low.

Soil Adsorption/Mobility: Based on the water solubility and a log Kow, respective Koc values of 272 and 1020 for cyclopentane have been calculated from various regression-derived equations [18]. These Koc values indicate a wide range of adsorption characteristics for cyclopentane and the mobility class in soil may range from low to medium [39].

Volatilization from Water/Soil: The Henry's Law constant indicates cyclopentane should volatilize rapidly from natural waters [18]. The volatilization half-life from a model river (1 meter deep flowing 1 m/sec with a wind speed of 3 m/sec) has been estimated to be 2.5 hr [18]. The

Cyclopentane

volatilization half-life from a model pond, which considers the effect of adsorption, has been estimated to be 5.2 days [40].

Water Concentrations: DRINKING WATER: Cyclopentane was listed as one of the many organic chemicals identified in drinking water as of 1974 [15].

Effluent Concentrations: Cyclopentane was identified as a stack emission from waste incinerators [14]. A refinery located in Tulsa OK was attributed with emissions to the surrounding atmosphere where the cyclopentane concentration was measured to be 5.2 and 6.4 ppbC for two min before and after 1:33 PM [1]. An oil fire emitted cyclopentane to surrounding air that contained a concentration of 0.21 mg/m^3 [26]. Underwater hydrocarbon vent discharges from offshore oil production platforms were found to contain cyclopentane concentrations in the liquid and gas phase at 40 ng/L and 1 umol/L of gas, respectively [29]. Cyclopentane has been identified as a product of combustion engines [38,44]. Data from Sept 2, 1979 identified cyclopentane as a gaseous emission of the vehicle traffic through the Allegheny Mountain Tunnel of the Pennsylvania Turnpike [11]. The average exhaust from 67 gasoline fueled vehicles was found to contain cyclopentane at a concentration of 0.4% by weight [22]. Cyclopentane from car exhaust ranged in concentration from 0.005 to 0.02 ppmV with an average for 8 samples of 0.012 ppmV [21].

Sediment/Soil Concentrations: Cyclopentane was detected in 4 of 4 sediment samples of 4-8, 28-32, 52-56, and 76-80 cm deep from Walvis Bay of the Namibian shelf of SW Africa at respective concentrations of 2.1, 1.8, 0.61 and 1.1 ng/g [41]. At two other deepwater sampling stations, cyclopentane was not detected [41].

Atmospheric Concentrations: URBAN: Cyclopentane was listed as one of 64 most abundant air pollutants in US cities [31]. For 823 air samples, collected from 1984-6, in 39 US cities, the median cyclopentane concentration was 2.1 ppbC with a minimum and maximum concentration of 0.1 and 104 ppbC [31]. According to the National Ambient Volatile Organic Compounds (VOCs) Database, the median urban atmospheric concentration of cyclopentane is 0.354 ppbV for 457 samples [34]. The cyclopentane concentration ranged from 1 to 8 ppbV at a downtown Los

Cyclopentane

Angeles location where it was detected in 16 of 17 samples in the Fall of 1981 [9]. The cyclopentane concentration for 6 sites in Rio Blanco, CO ranged from 0.5 to 9.3 ppbC with an average of 2.8 ppbC [1]. The cyclopentane concentrations for 2 air samples in downtown Tulsa, OK were 5.8 and 4.6 ppbC [1]. The median cyclopentane concentrations for 5 can and 10 bag air samples from Houston, TX were 0.6 and 1.4 ppbC [32]. The 1977 maximum and average concentration of 682 points for cyclopentane at a site in Houston, Texas was 2 and 29 ppbC, respectively [20]. The average cyclopentane concentration from 6:00 to 9:00 AM in Houston, Texas was 4 ppbC for 73 points in 1977 [20]. Cyclopentane was detected in 7 of 16 air samples from Houston, TX ranging in concentration from 19.9 to 81.5 ppm with an average of 43.8 ppm [16]. The ground level atmospheric concentration of cyclopentane at 1:25 PM was 0.6 ppb and 7.2 ppb at 8:00 AM for Huntington Park, CA [30]. At 1500 ft the cyclopentane concentration was 0.2 ppb at 7:43 AM and at 8:07 AM at a height of 2,200 ft the cyclopentane concentration was 0.1 ppb [30]. From Aug 12, 1960 to Nov 18, 1960 the air of downtown Los Angeles ranged in cyclopentane concentration from 0.02 to 0.03 ppm with an average for 16 samples of 0.03 ppm [21]. The ambient air of Riverside, CA was found to contain a cyclopentane concentration of 3.2 ppb at 7:30 AM Sept 24, 1968 [37]. At 8:05-8:25 AM on Mar 3, 1966 the cyclopentane concentrations for a 2nd floor and roof top sample at the County Health and Finance Building in Riverside, CA were 1.4 and 2.0 ppb, respectively [36,37]. The cyclopentane concentration for a 2nd floor sample at the County Health and Finance Building in Riverside, CA was 6.0 ppb at 7:40-8:00 AM Dec 22, 1965 [36]. Cyclopentane was listed as an atmospheric contaminant at Durban and Johannesburg, Pretoria [17]. Air of Sidney, Australia contained cyclopentane at an average concentration of 0.8 ppbV [23]. SUBURBAN: According to the National Ambient Volatile Organic Compounds (VOCs) Database, the median suburban atmospheric concentration of cyclopentane is 0.440 ppbV for 218 samples [34]. The cyclopentane concentrations for 2 air samples in suburban Tulsa, OK were 2.5 and 3.2 ppbC [1]. RURAL/REMOTE: The cyclopentane concentrations for 2 air samples in the Liberty Mts outside Tulsa, OK were 5.3 and 8.5 ppbC [1]. Smoky Mt. air contained cyclopentane at concentrations ranging from 0.4 to 1.7 ppbC, with an average concentration of 0.9 ppbC [1]. The atmospheric concentration of cyclopentane at the Jones State Forest in TX ranged from 0.3 to 6.1 ppbC [32]. The respective median, minimum and maximum atmospheric concentrations of cyclopentane for 5 rural locations in NC ranged from 0.0 to 1.6, 0.0 to 1.0 and 0.2 to 2.4 ppb [33]. SOURCE DOMINATED: According to the National Ambient Volatile Organic Compounds (VOCs) Database, the median concentration of cyclopentane for source dominated air is 0.349 ppbV for 29 samples [34]. Air downwind of a natural gas facility in Rio Blanco, CO contained cyclopentane at an average concentration of 9.3 ppbC [1]. The average cyclopentane concentration for air samples in urban plumes taken at altitudes of 2000, 2500 and 300 ft over Lake Michigan on Aug 27, 1976 was 2.4 ppbV [19]. The average cyclopentane concentration for samples taken at altitudes of 1000 and 1500 ft over Lake Michigan on Aug 28, 1976 was 1.2 ppbV [19].

Food Survey Values:

Plant Concentrations:

Fish/Seafood Concentrations:

Animal Concentrations:

Milk Concentrations: Cyclopentane was detected in 6 of 12 samples of mothers breast milk from the cities of Bayonne, NJ, Jersey City, NJ, Bridgeville, PA and Baton Rouge, LA [25].

Other Environmental Concentrations: Cyclopentane was emitted from 8 adhesives used in building materials [8]. An air sample taken near an oil fire was found to contain cyclopentane at a concentration of 0.21 mg/m^3 [26].

Probable Routes of Human Exposure: The most probable route of human exposure to cyclopentane is by inhalation. Atmospheric workplace exposures have been documented [10,28]. Cyclopentane is a highly volatile compound and monitoring data indicates that it is a widely occurring atmospheric pollutant.

Average Daily Intake:

Cyclopentane

Occupational Exposure: The most probable human exposure to cyclopentane would be occupational exposure, which may occur through dermal contact or inhalation at places where it is produced or used. A 1984 study showed cyclopentane was emitted from gasoline exposing transport drivers to an average atmospheric concentration of 0.102 mg/m^3; it was detected in 30 of 49 samples [28]. Gas station attendants were exposed to cyclopentane at an average atmospheric concentration of 0.030 mg/m^3; it was detected in 17 of 49 samples [28]. Workers at gasoline bulk handling facilities were exposed to vapors that contained cyclopentane at a concentration of 0.7% by volume of total hydrocarbons [10]. Exposures to total hydrocarbons at one of the facilities exceeded 240 ppm for 5% of the sampling time [10].

Body Burdens: Cyclopentane was detected in 6 of 12 samples of mothers breast milk from the cities of Bayonne NJ, Jersey City NJ, Bridgeville PA and Baton Rouge LA [25].

REFERENCES

1. Arnts RR, Meeks SA; Atmos Environ 15: 1643-51 (1981)
2. Atkinson R et al; Internat J Chem Kin 14: 781-8 (1982)
3. Beam, HW, Perry JJ; J Gen Microb 82: 163-9 (1974)
4. Darnall KR et al; J Phys Chem 82: 1581-4 (1978)
5. Daubert TE, Danner RP; Data Compilation, Tables of Properties of Pure Cmpds, Design Inst for Phys Prop Data, Am Inst for Phys Prop Data, NY,NY (1989)
6. Droege AT, Tully FP; J Phys Chem 91: 1222-5 (1987)
7. Eisenreich SJ et al; Environ Sci Technol 15: 30-8 (1981)
8. Girman JR et al; Environ Int 12: 317-21 (1986)
9. Grosjean D, Fung K; J Air Pollut Control Assoc 34: 537-43 (1984)
10. Halder CA et al; Am Ind Hyg Assoc J 47: 164-72 (1986)
11. Hampton CV et al; Environ Sci Technol 16: 287-98 (1982)
12. Hansch C, Leo AJ; Medchem Project Issue No 26. Claremont CA: Pomona College (1985)
13. Jamison VW et al; pp 187-96 in Proc Int Biodeg Symp 3rd Sharpley JM, Kaplan AM eds Essex Eng (1976)
14. Junk GA, Ford CS; Chemosphere 9: 187-230 (1980)
15. Kool HJ et al; Crit Rev Env Control 12: 307-57 (1982)
16. Lonneman WA et al; Hydrocarbons in Houston Air USEPA-600/3-79/018 p 44 (1979)
17. Louw CW et al; Atmos Environ 11: 703-17 (1977)
18. Lyman WJ et al; Handbook of Chemical Property Estimation Methods NY: McGraw-Hill pp 4-9, 5-4, 10, 7-4 (1982)
19. Miller MM, Alkezweeny AJ; Ann NY Acad Sci 338: 219-32 (1980)

Cyclopentane

20. Monson PR et al; Houston Oxidant Field Study, 71st Ann Meeting Air Pollut Contr Assoc Paper 78-50.4 (1978)
21. Neligan RE; Arch Environ Health 5: 581-91 (1962)
22. Nelson PF, Quigley SM; Atmos Environ 18: 79-87 (1984)
23. Nelson PF, Quigley SM; Environ Sci Technol 16: 650-5 (1982)
24. Novak et al; J Chromatography 76: 45-50 (1973)
25. Pellizzari ED et al; Bull Environ Contam Toxicol 28: 322-8 (1982)
26. Perry R; pp 130-7 in Int Symp Ident Meas Environ Pollut (1971)
27. Perry, JJ; The Role of Co-oxidation and Cometabolism in the biodegradation of Recalcitrant Molecules NTIS PB-80-28-034 US Army Res Off p 21 (1980)
28. Rappaport SM et al; Appl Ind Hyg 2: 148-54 (1987)
29. Sauer TC Jr; Org Geochem 7: 1-16 (1981)
30. Scott Research Labs Inc; Atmospheric Reaction Studies in the Los Angeles Basin, NTIS PB-194-058 p 86 (1969)
31. Seila RL et al; Determination of C2 to C12 Ambient Air Hydrocarbons in 39 US Cities, from 1984 through 1986 USEPA/600/S3-89/058 (1989)
32. Seila RL; Non-urban Hydrocarbons Concentration in Ambient Air No of Houston, TX USEPA-500/3-79-010 p 38 (1979)
33. Seila RL et al; Atmospheric Volatile Hydrocarbon Composition at Five Remote Sites in NW NC USEPA-600/D-84-092 NTIS PB84-177930 (1984)
34. Shah JJ, Heyerdahl EK; National Ambient VOC Database Update USEPA-600/3-88/010 (1988)
35. Silverstein RM, Bassler GC; Spectrometric Id Org Cmpd J Wiley & Sons Inc pp 148-69 (1963)
36. Stephens ER, Burleson FR; J Air Pollut Control Assoc 17: 147-53 (1967)
37. Stephens ER; Hydrocarbons in Polluted Air NTIS PB-230 993-/8 p 86 (1973)
38. Stump FD, Dropkin DL; Anal Chem 57: 2629-34 (1985)
39. Swann RL et al; Res Rev 85: 16-28 (1983)
40. USEPA; EXAMS II Computer Simulation (1987)
41. Whelan JK et al; Geochim Cosmochim Acta 44: 1767-85 (1980)
42. Windholz M et al; Merck Index 10th ed. Rahway, NJ (1983)
43. Yalkowsky SH et al; Arizona Data Base of Water Solubility (1989)
44. Zweidinger RB et al; Environ Sci Tech 22: 956-62 (1988)

Cyclopentanone

Synonyms:

Structure:

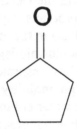

CAS Registry Number: 120-92-3

Molecular Formula: C_5H_8O

Wiswesser Line Notation: L5VTJ

CHEMICAL AND PHYSICAL PROPERTIES

Boiling Point: 130.6 °C

Melting Point: -58.2 °C

Molecular Weight: 84.12

Dissociation Constants:

Log Octanol/Water Partition Coefficient: 0.24 estimate [2]

Water Solubility: 9,177 mg/L estimate [13]

Vapor Pressure: 11.4 mm Hg at 25 °C [5]

Henry's Law Constant: 1.0×10^{-5} atm-m³/mole [11]

ENVIRONMENTAL FATE/EXPOSURE POTENTIAL

Summary: Cyclopentanone may be released to the environment via effluents at sites where it is produced or used as a chemical intermediate for pharmaceuticals, biologicals, insecticides and rubber chemicals. Cyclopentanone is also released to the environment via effluents from the manufacture and use of coal-derived liquid fuels and the disposal of coal liquefication and gasification waste by-products. Cyclopentanone is not expected to undergo hydrolysis or photolysis in the environment. Limited data suggests that cyclopentanone should biodegrade rapidly upon acclimation in soil and water. A low estimated Koc indicates cyclopentanone should have a very high mobility in soil. In aquatic systems, it should not partition from the water column to organic matter in sediments and suspended solids, nor should it bioconcentrate in aquatic organisms. The Henry's Law constant suggests that the volatilization of cyclopentanone from natural waters will be an important fate process. Volatilization half-lives from a model river and a model pond (the latter considers the effect of adsorption) have been estimated to be about 3.5 and 40 days, respectively. Based on its vapor pressure, cyclopentanone should evaporate from dry surfaces, especially when present in high concentrations such as in spill situations. Cyclopentanone is expected to exist entirely in the vapor phase in ambient air. Vapor phase reactions with photochemically produced hydroxyl radicals in the atmosphere have been shown to be important (half-life of 5.5 days). The most probable human exposures to cyclopentanone would be occupational exposure, which may occur through dermal contact or inhalation at places where it is produced or used. Atmospheric workplace exposures have been documented. Nonoccupational exposures are likely to occur among populations with contaminated drinking water supplies or from the ingestion of certain foods.

Natural Sources:

Artificial Sources: Cyclopentanone may be released to the environment via effluents at sites where it is produced or used as an intermediate for pharmaceuticals, biologicals, insecticides and rubber chemicals [10]. Cyclopentanone is also released to the environment via effluents from the manufacture and use of coal-derived liquid fuels and the disposal of coal liquefaction and gasification waste by-products [6,8,14].

Terrestrial Fate: Cyclopentanone is not expected to hydrolyze in soils [13] or undergo photolysis on sunlit soil surfaces due to lack of >290 nm UV light adsorption [17]. A single 5-day BOD screening test, which utilized activated sludge for inocula, suggests that cyclopentanone will biodegrade rapidly upon acclimation in terrestrial environments [16]. An estimated Koc of 30 [13] indicates cyclopentanone should have a very high mobility in soil [19]. The Henry's Law constant suggests volatilization of cyclopentanone from moist soils may be an important fate process [13]. Based on the vapor pressure, cyclopentanone should evaporate from dry surfaces, especially when present in high concentrations such as in spill situations.

Aquatic Fate: Cyclopentanone is not expected to hydrolyze [13] or undergo direct photolysis in aquatic systems [17]. A single 5-day BOD screening test, which utilized activated sludge for inocula, suggests that cyclopentanone will biodegrade rapidly upon acclimation in natural waters [16]. An estimated log bioconcentration factor (log BCF) of -0.05 [13] indicates cyclopentanone should not bioconcentrate in aquatic organisms. An estimated Koc of 30 [13] indicates cyclopentanone should not partition from the water column to organic matter contained in sediments and suspended solids. The Henry's Law constant indicates that volatilization of cyclopentanone from natural bodies of water should be an important fate process [13]. Based on this Henry's Law constant, the volatilization half-life from a model river has been estimated to be 3.5 days [13]. The volatilization half-life from an model pond, which considers the effect of adsorption, has been estimated to be about 40 days [20].

Atmospheric Fate: Based on the vapor pressure, cyclopentanone is expected to exist almost entirely in the vapor phase in ambient air [7]. Vapor-phase reactions with photochemically produced hydroxyl radicals in the atmosphere have been shown to be important. A rate constant for cyclopentanone of 2.94×10^{-12} cm^3/molecule-sec at 25 °C corresponds to an atmospheric half-life of about 5.5 days at an atmospheric concentration of $5 \times 10^{+5}$ hydroxyl radicals per cm^3 [4]. Direct photolysis in air is not expected to be an important environmental fate process [17].

Biodegradation: Pure culture studies showed various isolated species of bacteria [1,15] and yeast [9] were unable to utilize cyclopentanone as a single carbon source. A single 5-day BOD screening test, which utilized activated sludge for inocula, indicates cyclopentanone biodegraded rapidly with acclimation [16]. After a 20-day acclimation period, 95.4%

COD of an initial concentration of 100 mg/L was removed in a closed bottle maintained at 20 °C and a pH of 7.2 [16].

Abiotic Degradation: Ketones are generally resistant to hydrolysis [17]. The photolysis of cyclopentanone in the environment should not be an important fate process [17]. The rate constant for the vapor-phase reaction of cyclopentanone with photochemically produced hydroxyl radicals has been measured to be 2.94 x 10^{-12} cm^3/molecule-sec at 25 °C, which corresponds to an atmospheric half-life of about 5.5 days at an atmospheric concentration of 5 x 10^{+5} hydroxyl radicals per cm^3 [4].

Bioconcentration: Based upon the estimated log Kow, the bioconcentration factor (log BCF) for cyclopentanone has been calculated to be -0.05, from a recommended regression-derived equation [13]. This BCF value indicates cyclopentanone should not bioconcentrate in aquatic organisms.

Soil Adsorption/Mobility: Based on the estimated water solubility, a Koc value of 30 for cyclopentanone has been calculated from a regression-derived equation [13]. This Koc value indicates cyclopentanone should be very highly mobile in soil [19].

Volatilization from Water/Soil: The Henry's Law constant indicates that volatilization of cyclopentanone from natural bodies of water should be an important fate process [13]. Based upon the Henry's Law constant, the volatilization half-life from a model river (1 meter deep flowing 1 m/sec with a wind speed of 3 m/sec) has been estimated to be 3.5 days [13]. The volatilization half-life from a model pond, which considers the effect of adsorption, has been estimated to be about 40 days [20]. Based on the vapor pressure, cyclopentanone should evaporate from dry surfaces, especially when present in high concentrations such as in spill situations.

Water Concentrations: DRINKING WATER: Cyclopentanone was listed as a contaminant found in drinking water for a survey of US cities including Pomona, Escondido, Lake Tahoe and Orange Co, CA, and Dallas, Washington, DC, Cincinnati, Philadelphia, Miami, New Orleans, Ottumwa, IA, and Seattle [12]. GROUND WATER: Cyclopentanone was detected in 3 of 3 ground water samples near a coal gasification site near

Cyclopentanone

Hoe Creek in northeastern WY at concentrations of 13, 37 and 180 ppb [18].

Effluent Concentrations: Cyclopentanone was detected in 4 of 7 wastewater effluents from energy related processes [14]. A ground water water sample from a coal gasification facility in Hanna, WY contained cyclopentanone at an average concentration of 32 ppb; the process water at a coal gasification facility in Gillette, WY contained cyclopentanone at an average concentration of 110 ppb; effluent from the gasification of Rosebud coal in Morgantown, WV contained cyclopentanone at an average concentration of 56 ppb; and retort water from an in situ shale oil processing facility in Rock Springs, WY contained cyclopentanone at an average concentration of 5 ppb [14]. Wastewater from coal gasification at the Grand Fork's Energy Technology Center, ND was also reported to contain cyclopentanone at a concentration of 0.8 mg/L [8]. In addition, wastewater effluent from a shale oil facility in Queensland, Australia was shown to contain cyclopentanone at a concentration of 53 mg/L [6].

Sediment/Soil Concentrations:

Atmospheric Concentrations: SOURCE DOMINATED: In Nov. 1982, cyclopentanone was detected in the air outside an oil shale wastewater facility of Occidental Oil Shale Inc. at Logan Wash, CO [11].

Food Survey Values: Cyclopentanone was identified as a volatile component of baked potatoes produced by the baking process [3].

Plant Concentrations:

Fish/Seafood Concentrations:

Animal Concentrations:

Milk Concentrations:

Other Environmental Concentrations:

Cyclopentanone

Probable Routes of Human Exposure: The most probable route of human exposure to cyclopentanone is by inhalation, dermal contact and ingestion. Atmospheric workplace exposures have been documented [11]. Drinking water supplies [12] and baked potatoes [3] have been shown to contain cyclopentanone.

Average Daily Intake:

Occupational Exposure: The most probable human exposure to cyclopentanone would be occupational exposure, which may occur through dermal contact or inhalation at places where it is produced or used. A 1982 study showed cyclopentanone was emitted to the air from wastewaters at a shale oil facility exposing inside workers [11]. Nonoccupational exposures are likely to occur among populations with contaminated drinking water supplies [12] or from the ingestion of certain foods [3].

Body Burdens:

REFERENCES

1. Beam HW, Perry JJ; J Gen Microb 82: 163-9 (1974)
2. CLOGP; PCGEMS Graphical Exposure Modeling System USEPA (1986)
3. Coleman EC et al; J Agric Food Chem 29: 42-8 (1981)
4. Dagaut et al; J Phys Chem 92: 4375-7 (1988)
5. Daubert TE, Danner RP; Data Compilation, Tables of Properties of Pure Cmpds, Design Inst for Phys Prop Data, Am Inst for Phys Prop Data, NY, NY (1989)
6. Dobson KR et al; Water Res 19: 849-56 (1985)
7. Eisenreich SJ et al; Environ Sci Technol 15: 30-8 (1981)
8. Giabbai, MF et al; Intern J Environ Anal Chem 20: 113-29 (1985)
9. Hasegawa Y et al; Can J Microbiol 28: 942-4 (1982)
10. Hawley GG; Condensed Chemical Dictionary 10th ed Van Nostrand Reinhold NY p 375 (1981)
11. Hawthorne SB et al; Environ Sci Technol 19: 992-7 (1985)
12. Lucas SV; GC/MS Anal of Org in Drinking Water Concentrates and Advanced Treatment Concentrates Vol 1 USEPA-600/1-84-020A (NTIS PB85-128239) p 397 (1984)
13. Lyman WJ et al; Handbook of Chemical Property Estimation Methods NY: McGraw-Hill pp 4-9, 5-4, 10, 7-4 (1982)
14. Pelizzari E et al; ASTM Spec Tech Publ STP 686: 256-74 (1979)
15. Perry, JJ; The Role of Co-oxidation and Cometabolism in the biodegradation of Recalcitrant Molecules NTIS PB-80-28-034 US Army Res Off pp 21 (1980)

Cyclopentanone

16. Pitter P; Water Res 10: 231-5 (1976)
17. Silverstein RM, Bassler GC; Spectrometric Id of Org Cmpd, J Wiley & Sons Inc p 148-169 (1963)
18. Stuermer DH et al; Environ Toxicol Chem 16: 582-7 (1982)
19. Swann RL et al; Res Rev 85: 16-28 (1983)
20. USEPA; EXAMS II Computer Simulation (1987)

Decahydronaphthalene

SUBSTANCE IDENTIFICATION

Synonyms:

Structure:

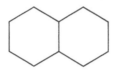

CAS Registry Number: 91-17-8

Molecular Formula: $C_{10}H_{18}$

Wiswesser Line Notation: L66TJ

CHEMICAL AND PHYSICAL PROPERTIES

Boiling Point: 189-191 °C

Melting Point: 57 °C

Molecular Weight: 138.25

Dissociation Constants:

Log Octanol/Water Partition Coefficient: 4.79 estimate [5]

Water Solubility: 0.889 mg/L at 25 °C [20]

Vapor Pressure: 2.3 mm Hg at 25 °C [4]

Henry's Law Constant: 4.70×10^{-1} atm-m^3/mole at 25 °C, calculated from VP/Wsol

ENVIRONMENTAL FATE/EXPOSURE POTENTIAL

Summary: Decahydronaphthalene, a component of crude oil and a product of combustion, is produced and released to the environment during natural fires. Emissions from petroleum refining, coal tar distillation, and gasoline and diesel fueled engines are major contributors of decahydronaphthalene to the environment. Decahydronaphthalene is also used as a chemical intermediate and a general solvent. Consequently, decahydronaphthalene is released to the environment via manufacturing effluents where decahydronaphthalene is produced and used. Because of the widespread use of decahydronaphthalene in a variety of products, decahydronaphthalene is also released to the environment through the disposal of these products and municipal waste water treatment facilities. Decahydronaphthalene should biodegrade in acclimated environments under the proper conditions. Decahydronaphthalene is not expected to undergo hydrolysis or photolysis in the environment. A calculated Koc range of 4700 to 9600 indicates that decahydronaphthalene will be slightly mobile to immobile in soil. In aquatic systems, decahydronaphthalene may partition from the water column to organic matter contained in sediments and suspended solids. Decahydronaphthalene has the potential to bioconcentrate in aquatic systems. The Henry's Law constant suggests volatilization of decahydronaphthalene from environmental waters should be rapid. The volatilization half-lives from a model river and model pond (the latter considers the effect of adsorption) have been estimated to be 3.4 hr and 28.1 days, respectively. Decahydronaphthalene is expected to exist entirely in the vapor phase in ambient air. Reaction with photochemically produced hydroxyl radicals (half-life of 20.3 hr) is likely to be an important fate process in the atmosphere. The most probable human exposure would be occupational exposure, which may occur through dermal contact or inhalation at places where decahydronaphthalene is produced or used. Nonoccupational exposures would most likely occur via urban atmospheres, contaminated drinking water supplies and recreational activities at contaminated waterways.

Natural Sources:

Natural Sources: Decahydronaphthalene is a natural component of crude oil [8]. Decahydronaphthalene is also a product of combustion and can be released to the environment via natural fires associated with lightening, volcanic activity, and spontaneous combustion.

Decahydronaphthalene

Artificial Sources: Decahydronaphthalene is emitted to the environment by effluents from petroleum refining and coal tar distillation [8]. The combustion of gasoline and diesel fuels releases decahydronaphthalene to the atmosphere [9]. Decahydronaphthalene is used as a chemical intermediate and in the manufacture of paints, lacquers, varnishes, shoe creams and floor waxes and as a powerful solvent for oils, resins, waxes, rubber, asphalt, aromatic hydrocarbons and printing ink [8]. Consequently, decahydronaphthalene is released to the environment via manufacturing effluents where it is produced or used. Because of the widespread use of decahydronaphthalene in a variety products [8], decahydronaphthalene is also released to the environment through the disposal of these products and municipal waste water treatment facilities [2].

Terrestrial Fate: Data regarding the biodegradation of decahydronaphthalene in soil were not available. However, limited marine water and sediment grab sample data suggest that decahydronaphthalene will biodegrade in acclimated soils under the proper conditions. Decahydronaphthalene is not expected to undergo hydrolysis or photolysis in soil. A calculated Koc range of 4700 to 9600 [10] indicates decahydronaphthalene will be slightly mobile to immobile in soil [17]. The Henry's Law constant suggests volatilization of decahydronaphthalene from moist soils with a low organic matter content may be rapid.

Aquatic Fate: Limited data suggest that decahydronaphthalene will biodegrade in acclimated aquatic systems under the proper conditions. Decahydronaphthalene is not expected to undergo hydrolysis or photolysis in environmental waters. Decahydronaphthalene has the potential to bioconcentrate in aquatic systems. Decahydronaphthalene may also partition from the water column to organic matter contained in sediments and suspended solids [17]. The calculated Henry's Law constant suggests volatilization of decahydronaphthalene from environmental waters should be rapid [10]. Based on the Henry's Law constant, the volatilization half-life from a model river has been estimated to be 3.4 hr [10]. The volatilization half-life from a model pond, which considers the effect of adsorption, has been estimated to be about 28.1 days [18].

Decahydronaphthalene

Atmospheric Fate: Based upon the vapor pressure, decahydronaphthalene is expected to exist entirely in the vapor phase in ambient air [7]. In the atmosphere, direct photolysis or hydrolysis of decahydronaphthalene is unlikely to occur. Reactions of decahydronaphthalene with photochemically produced hydroxyl radicals is likely to be an important fate process in ambient air. An estimated rate constant at 25 °C of 1.90×10^{-11} cm^3/molecule-sec for the vapor-phase reaction with hydroxyl radicals corresponds to a half-life of 20.3 hours at an atmospheric concentration of $5 \times 10^{+5}$ hydroxyl radicals [1].

Biodegradation: Both marine water and sediment grab samples from oiled and pristine beach areas and mud from an intertidal zone were unable to degrade decahydronaphthalene [11]. However, decahydronaphthalene was degraded in water from a stagnant pond that had been acclimated to oil [11]. Acclimated mixed cultures in mineral salt media were able to degrade 50% of a crude oil containing decahydronaphthalene within 48 hr [16].

Abiotic Degradation: Alkanes are generally resistant to hydrolysis [10] and cyclic alkanes do not absorb UV at wavelengths greater than 290 nm [15]. Therefore, decahydronaphthalene probably will not undergo hydrolysis or photolysis in the environment. The rate constant for the vapor-phase reaction of decahydronaphthalene with photochemically produced hydroxyl radicals was estimated to be 1.90×10^{-11} cm^3/molecule-sec at 25 °C, which corresponds to an atmospheric half-life of about 20.3 hours at an atmospheric concentration of $5 \times 10^{+5}$ hydroxyl radicals per cm^3 [1].

Bioconcentration: Based on the water solubility and the estimated log Kow, the log BCF of decahydronaphthalene has been calculated to range from 2.82 to 3.27 from various regression-derived equations [10]. These log BCF values suggest decahydronaphthalene has the potential to bioconcentrate in aquatic systems.

Soil Adsorption/Mobility: Based on the water solubility and the estimated log Kow, the Koc of decahydronaphthalene has been calculated to range from 4700 to 9600 from various regression-derived equations [10]. These Koc values indicate decahydronaphthalene will be slightly mobile to immobile in soil [17].

Decahydronaphthalene

Volatilization from Water/Soil: The Henry's Law constant indicates volatilization of decahydronaphthalene from environmental waters should be rapid [10]. The volatilization half-life from a model river (1 meter deep flowing 1 m/sec with a wind speed of 3 m/sec) has been estimated to be 3.4 hr [10]. The volatilization half-life from a model pond, which considers the effect of adsorption, has been estimated to be 28.1 days [18].

Water Concentrations: DRINKING WATER: Decahydronaphthalene was listed as a contaminant found in drinking water at Tuscaloosa, Al [3]. SURFACE WATER: Decahydronaphthalene is also listed as a contaminant of coastal waters off Narragansett Bay, RI [19] and Los Angeles River stormwaters [6]. GROUND WATER: Decahydronaphthalene was identified in ground water contaminated by sewage treatment facility at Falmouth, MA [2].

Effluent Concentrations: Data from Sept 2, 1979 identified decahydronaphthalene as a gaseous emission of the vehicle traffic through the Allegheny Mountain Tunnel of the Pennsylvania Turnpike [9]. Decahydronaphthalene was detected in the municipal wastewaters from sewage treatment plants in Falmouth, MA [2].

Sediment/Soil Concentrations:

Atmospheric Concentrations: Data from Sept 2, 1979 identified decahydronaphthalene as air pollution contaminant in the Allegheny Mountain Tunnel of the Pennsylvania Turnpike [9]. Decahydronaphthalene was detected in the ambient air of Paris, France in 1972 [14].

Food Survey Values:

Plant Concentrations:

Fish/Seafood Concentrations:

Animal Concentrations:

Milk Concentrations:

Decahydronaphthalene

Other Environmental Concentrations:

Probable Routes of Human Exposure:

Average Daily Intake:

Occupational Exposure: NIOSH (NOHS Survey 1972-1974) has statistically estimated that 935 workers are potentially exposed to decahydronaphthalene in the USA [13]. NIOSH (NOES Survey 1981-1983) has statistically estimated that 28 workers are potentially exposed to decahydronaphthalene in the USA [12]. The most probable human exposure would be occupational exposure, which may occur through dermal contact or inhalation at places where decahydronaphthalene is produced or used. Nonoccupational exposures would most likely occur via urban atmospheres, contaminated drinking water supplies and recreational activities at contaminated waterways.

Body Burdens:

REFERENCES

1. Atkinson R; Intern J Chem Kin 19: 799-828 (1987)
2. Barber LB et al; Environ Sci Technol 22: 205-11 (1988)
3. Bertsch W et al; J Chromat 112: 701-18 (1975)
4. Boublik T et al; Vapor Pressures of Pure Substances. Elsevier NY p 607 (1984)
5. CLOGP; PCGEMS Graphical Exposure Modeling System USEPA (1986)
6. Eaganhouse RP et al; Environ Sci Technol 15: 315-26 (1981)
7. Eisenreich SJ et al; Environ Sci Technol 15: 30-8 (1981)
8. Gaydos RM; Kirk-Othmer Encycl Chem Tech 3rd NY, NY: Wiley 15: 698-719 (1981)
9. Hampton CV et al; Environ Sci Technol 16: 287-98 (1982)
10. Lyman WJ et al; Handbook of Chemical Property Estimation Methods NY: McGraw-Hill (1982)
11. Mulkins-Phillips GJ, Stewart JE; Appl Microbiol 28: 915-922 (1974)
12. NIOSH; National Occupational Exposure Survey (NOES) (1983)
13. NIOSH; National Occupational Hazard Survey (NOHS) (1974)
14. Raymond A, Guiochon G; Environ Sci Technol 8: 143-8 (1974)
15. Silverstein RM, Bassler GC; Spectrometric Id of Org Cmpd, J Wiley and Sons Inc pp 148-169 (1963)
16. Soli G, Bens EM; Biotech Bioeng 24: 319-30 (1972)
17. Swann RL et al; Res Rev 85: 16-28 (1983)
18. USEPA; EXAMS II Computer Simulation (1987)

Decahydronaphthalene

19. Wakeham SG et al; Can J Fish Aquat Sci 40: 304-21 (1983)
20. Yalkowsky SH et al; Arizona Data Base of Water Solubility (1987)

Decane

Synonyms:

Structure:

$$H_3C \diagdown \diagup \diagdown \diagup \diagdown \diagup \diagdown CH_3$$

CAS Registry Number: 124-18-5

Molecular Formula: $C_{10}H_{22}$

Wiswesser Line Notation:

CHEMICAL AND PHYSICAL PROPERTIES

Boiling Point: 174.3 °C

Melting Point: -29.5 °C

Molecular Weight: 142.28

Dissociation Constants:

Log Octanol/Water Partition Coefficient: 5.98 estimate [4]

Water Solubility: 0.052 mg/L at 25 °C [66]

Vapor Pressure: 1.43 mm Hg at 25 °C [7]

Henry's Law Constant: 5.15 atm-m^3/mole at 25 °C, calculated from VP/Wsol

ENVIRONMENTAL FATE/EXPOSURE POTENTIAL

Summary: n-Decane is a constituent in the paraffin fraction of crude oil and natural gas. n-Decane is released to the environment via the manufacture, use, and disposal of many products associated with the

133

petroleum and gasoline industries. Extensive data show release of n-decane into the environment from solvent based building materials, printing pastes, paints, varnishes, adhesives and other coatings; landfills and waste incinerators; vulcanization and extrusion operations during rubber and synthetic production; and the combustion of gasoline, diesel fuels, and plastics. Photolysis or hydrolysis of n-decane is not expected to be environmentally important. Biodegradation of n-decane may occur in soil and water; however, volatilization and adsorption are expected to be far more important environmental fate processes. A high Koc indicates n-decane will be immobile in soil and may partition from the water column to organic matter contained in sediments and suspended solids. The bioconcentration of n-decane may be important in aquatic systems. The Henry's Law constant suggests rapid volatilization of n-decane from environmental waters. The volatilization half-lives from a model river and a model pond (the latter considers the effect of adsorption) have been estimated to be 3.5 hr and 130 days, respectively. n-Decane is expected to exist almost entirely in the vapor phase in ambient air. Reactions with photochemically produced hydroxyl radicals in the atmosphere have been shown to be important (estimated half-life of 1.4 days). The most probable route of human exposure to n-decane is by inhalation. Extensive monitoring data indicates n-decane is a widely occurring atmospheric pollutant.

Natural Sources: n-Decane is a constituent in the paraffin fraction of crude oil and natural gas [61].

Artificial Sources: n-Decane is released to the environment via the manufacture, use, and disposal of many products associated with the petroleum [41,43,46] and gasoline industries [43,53]. The combustion of plastics [30], gasoline and diesel fuels have been shown to release n-decane into the atmosphere [17,18,36]. Vulcanization and extrusion operations during rubber and synthetic production as with shoes, tires and electrical insulation also emit n-decane to the air [5]. Other well-documented materials that are responsible for the release of n-decane to the environment include solvent-based building materials, printing pastes, paints, varnishes, adhesives and other coatings [10,12,30,60]. Landfills [14,67,68] and waste incinerators [23] also release n-decane into the environment.

Decane

Terrestrial Fate: Photolysis or hydrolysis [32] of n-decane is not expected to be important in terrestrial environments. The biodegradation of n-decane may occur in soils; however, volatilization and adsorption are expected to be far more important fate processes. An estimated range for Koc from 22,200 to 42,700 [32] indicates n-decane will be immobile in most soils [3]. Based upon the estimated Henry's Law constant, n-decane is also expected to rapidly volatilize from moist surface soils [59].

Aquatic Fate: Photolysis and hydrolysis [32] of n-decane in aquatic systems are not expected to be important. The log bioconcentration factor (log BCF) for n-decane has been estimated to range from 3.52 to 4.31 [32], suggesting bioconcentration may be an important fate process in aquatic systems. Biodegradation of n-decane may occur in aquatic environments; however, volatilization and adsorption are expected to be far more important fate processes. An estimated range for Koc from 22,200 to 42,700 [32] indicates n-decane may partition from the water column to organic matter [59] contained in sediments and suspended solids. The estimated Henry's Law constant suggests rapid volatilization of n-decane from environmental waters [32]. Based on the Henry's Law constant, the volatilization half-life from a model river has been estimated to be 3.5 hr [32]. The volatilization half-life from an model pond, which considers the effect of adsorption, has been estimated to be about 130 days [62].

Atmospheric Fate: Based on the vapor pressure, n-decane is expected to exist almost entirely in the vapor phase in ambient air [10]. n-Decane is not expected to undergo direct photolysis in ambient air. However, vapor-phase reactions with photochemically produced hydroxyl radicals in the atmosphere have been shown to be important. The rate constant for n-decane was measured to be 11.4×10^{-12} cm^3/molecule-sec at 26 °C, which corresponds to an atmospheric half-life of about 1.4 days at an atmospheric concentration of $5 \times 10^{+5}$ hydroxyl radicals per cm^3 [3].

Biodegradation: At intervals of 6, 12 and 24 hr endogenous respiration was 1.3, 2.6, and 4.7 %, respectively, of the TOD for n-decane at a concentration of 500 mg/L in 20 mL volume soln of supernatant liquid with a suspended solids concentration adjusted to 2500 mg/L from differing aeration units of sewage treatment facilities [13]. A 25% loss of n-decane loss occurred within 5 days and completely disappeared within 15 days from 1 mL of crude oil added to a 100 mL simulated seawater soln inoculated with sediment samples from Fukae of Kobe harbor, Japan and incubated at 20 °C [34]. Loss of 13 and 52% of n-decane was observed within 5 and 15 days, respectively, from 1 mL of crude oil added to a 100 mL seawater soln collected at Fukae of Kobe harbor, Japan and incubated at 20 °C [34]. Complete recovery was reported for all the control samples [34]. Bacteria from Colgate Creek sediment cultured in water from both Colgate Creek and Eastern Bay in Chesapeake Bay, MD were able to utilize 30 and 34%, respectively, of the n-decane from a petroleum hydrocarbon mixture after 28 days incubation at 20 °C [65]. Bacteria from Eastern Bay sediment cultured in water from both Colgate Creek and Eastern Bay in Chesapeake Bay, MD were able to utilize 14 and 10%, respectively, of the n-decane from a petroleum hydrocarbon mixture after 28 days incubation at 20 °C [65].

Abiotic Degradation: Alkanes are generally resistant to hydrolysis [32]. Based on data for n-hexane and iso-octane, n-decane is not expected to absorb UV light in the environmentally significant range, >290 nm [55]. Therefore, n-decane probably will not undergo hydrolysis or direct photolysis in the environment. The rate constant for the vapor-phase reaction of n-decane with photochemically produced hydroxyl radicals was measured to be 11.4×10^{-12} cm^3/molecule-sec at 26 °C, which corresponds to an atmospheric half-life of about 1.4 days at an atmospheric concentration of $5 \times 10^{+5}$ hydroxyl radicals per cm^3 [3]. At 39 °C the rate constant for the vapor phase reaction of n-decane with photochemically produced hydroxyl radicals was determined to be 11.7×10^{-12} cm^3/molecule-sec, which also corresponds to an atmospheric half-life of 1.4 days [39].

Bioconcentration: Based upon the water solubility and the estimated log Kow, the bioconcentration factor (log BCF) for n-decane has been calculated, using recommended regression derived equations, to be 3.52 and 4.31, respectively [32]. These bioconcentration factor values indicate bioconcentration may be important.

Soil Adsorption/Mobility: Based on the water solubility and the estimated log Kow, the Koc of n-decane has been calculated, using recommended regression-derived equations, to range from 22,200 to 42,700 [32]. These Koc values indicate n-decane will be immobile in soil [59].

Volatilization from Water/Soil: Based upon the Henry's Law constant, n-decane should undergo extremely rapid volatilization from natural waters [32]. The volatilization half-life from a model river (1 meter deep flowing 1 m/sec with a wind speed of 3 m/sec) has been estimated to be 3.5 hr [32]. The volatilization half-life from an model pond, which considers the effect of adsorption, has been estimated to be 130 days [62].

Water Concentrations: DRINKING WATER: n-Decane was identified in 14 of 14 treated water supplies in England [12]. n-Decane was listed as one of the many organic chemicals identified in drinking water in the USA as of 1974 [1] and as of 1982 [27]. The drinking water supply for the District of Columbia was found to contain n-decane; the concentration was estimated to be 0.03 ppm [49]. n-Decane was detected in New York City and New Orleans drinking water [25]. The Torresdale water supply of the City of Philadelphia, PA also contained n-decane [58]. SURFACE WATER: The average n-decane concentration of 6 water samples from both Little Britain Lake and Welsh Harp Lake, England were 1.8 and 12.2 ppb, respectively [6]. The n-decane average concentration of weekly samples taken over an approximate period of 1 year for Luton Brook, England was 7.9 ppb [6]. The average n-decane concentration for water samples taken from the River Pinn at Brunel University, England was 1.6 ppb [6]. SEAWATER WATER: Only trace quantities to 2 ng/L of n-decane were detected in open surface waters of the north central Gulf of Mexico [47]. The n-decane concentration of 3 surface water samples from an unpolluted coastal area of the north central Gulf of Mexico also ranged from trace levels to 2 ng/L [47]. n-Decane was detected in 6 of 8 surface water samples in the Gulf of Mexico ranging in concentration from 0.9 to 1.6 ng/L with an average concentration of 1.3 ng/L [48]. RAIN WATER: The n-decane concentration of rain water collected at Brunel University, England was 27.4 ppb [6].

Effluent Concentrations: n-Decane was identified as a stack emission from waste incinerators [23]. n-Decane was also identified as a vapor emitted from landfills [64,68]. A clay pit landfill in England that received municipal, industrial and liquid wastes emitted n-decane gas at a concentration of 51 mg/L [67]. The combustion of plastics also emits n-decane [30]. Building materials such as petroleum-based solvents such as floor adhesives and waxes, wood stains, polyurethane finish and air fresheners emit n-decane to indoor air [60]. n-Decane was detected in 1 of 63 industrial wastewater effluents at a concentration less than 10 ug/L [41]. Underwater hydrocarbon vents and formation water discharges from offshore oil production platforms were found to contain n-decane concentrations in the vapor phase at trace quantities and in the liquid state at 20 ng/L, respectively [46]. Formation water contained n-decane at a concentration of 410 ug/L [46]. Data from Sept 2, 1979 identified n-decane as a gaseous emission of the vehicle traffic through the Allegheny Mountain Tunnel of the Pennsylvania Turnpike [18]. Data from Aug 25 to Sept 7, 1979 showed for a speed of 80 km/hr on straight and level highway, gasoline-powered vehicles emitted n-decane at an average rate of 1.7 mg/km and diesel trucks emitted n-decane at an average of 7.2 mg/km [17]. The average exhaust from 67 gasoline-fueled vehicles was found to contain n-decane at a concentration 0.4% by weight of the fuel [36]. Motorboats emitted n-decane to canal water with resultant concentrations ranging from 4 to 19 ng/L with an average of 13 ng/L for 7 samples [24].

Sediment/Soil Concentrations:

Atmospheric Concentrations: URBAN: n-Decane occurred in 29% of the indoor air samples and 18% of the outdoor air samples taken in Chicago Illinois [22]. The average n-decane concentration for 2 samples per 4 sites in Tulsa, OK was 2.7 ppb carbon with a range of 0.6 to 6.9 ppb carbon [2]. The n-decane concentration for 6 sites in Rio Blanco, CO averaged 2.6 ppb carbon with a range from 1.3 to 5.4 [2]. The average concentration of n-decane in 3 samples indoor and outdoor air from Neenah, WI were 0.44 and 0.16 ug/m^3, respectively [54]. The average concentration of n-decane in 3 samples indoor and outdoor air from Newark, NJ were 6.21 and 2.23 ug/m^3, respectively [54]. The ground level atmospheric concentration of n-decane at 1325 ft was 0.8 ppb and 4.5 ppb at 0800 for Huntington Park, CA [50]. At 1500 ft the n-decane concentration was 0.4 ppb at 0743 and at 0807 at a height of 2,200 ft the n-decane concentration was 0.3 ppb [50]. The n-decane concentration

Decane

ranged from 1 to 9 ppbV at a downtown Los Angeles location for the fall of 1981 [16]. The average atmospheric gas phase concentration of n-decane was 284 ng/m^3 for 7 rain events in Portland, Oregon from Feb to Apr 1984 [29]. According to the National Ambient Volatile Organic Compounds (VOCs) Database, the median urban atmospheric concentration of n-decane is 0.356 ppbV for 790 samples [52]. The average n-decane concentration of air samples taken at Brunel University, England was 12.2 ppb [6]. n-Decane was also identified in the ambient air of Paris, France at concentrations ranging from 4.3 to 11.2 ug/m^3 [44]. In 1983-4, the respective minimum, maximum and average outdoor air concentration of n-decane in northern Italy were less than 1, 10 and 3.1 ug/m^3 [8]. The same study determined the respective minimum, maximum and average indoor air concentration of n-decane for 14 homes and an office building were less than 2, 1100 and 92 ug/m^3 [8]. The concentration of n-decane in the downtown air of Zurich, Switzerland was 1.6 ppb [15]. At Deuselbach, Hunsruck in Germany, the atmospheric n-decane concentration was less than 0.01 ppb for October 23, 1983 [45]. n-Decane was detected in the atmospheres of 6 industrialized cities of the USSR ranging in size of population from 0.4 to 4.5 million people [19,20,21]. The average n-decane concentration in the atmospheres of Pretoria, Johannesburg and Durban, South Africa were 0.6, 0.7, and 0.5 ppb, respectively [31]. n-Decane was identified in the ambient air of Sydney, Australia [33] ranging in concentrations from 0.1 to 3.3 ppbV with an average concentration of 0.7 ppbV [35]. n-Decane was detected in the atmosphere over the British Columbia Research Council Laboratory at the University of British Columbia [57]. In the Netherlands, the average n-decane concentrations in ambient air were 7.6 and 1.3 ppb for 20 and 24 samples, respectively [56]. SUBURBAN: According to the National Ambient Volatile Organic Compounds (VOCs) Database, the median suburban atmospheric concentration of n-decane is 0.198 ppb by volume for 336 samples [52]. RURAL: At a rural site near Duren, Germany, the atmospheric n-decane concentration was less than 0.23 ppb for March 1984 [45]. The atmospheric concentration of n-decane for Jones State Forest, TX ranged from 0.4 to 27.7 ppb with an average of 5.6 ppb for 10 samples [51]. According to the National Ambient Volatile Organic Compounds (VOCs) Database, the median rural atmospheric concentration of n-decane is 0.000 ppb by volume for 80 samples [52]. REMOTE: For 9 samples collected over a 30-hour period, the average n-decane concentration in the Smokey Mountains, NC was 0.6 ppbC with a range from 0.4 to 0.9 ppb carbon [2]. According to the National Ambient Volatile Organic Compounds (VOCs) Database, the median remote atmospheric concentration of n-decane is 0.185 ppb by volume for 6 samples [52]. SOURCE DOMINATED: Construction activities accounted for an indoor air concentration of 88 ug/m^3 for n-decane; whereas, 42 days after construction was completed, the indoor air concentration of n-decane was 3.33 ug/m^3 [54]. According to the National Ambient Volatile Organic Compounds (VOCs) Database, the median source dominated and indoor atmospheric concentrations of n-decane are 1.120 and 0.900 ppbV for 26 and 96 samples, respectively [52]. According to the National Ambient Volatile Organic Compounds (VOCs) Database, the median workplace atmospheric concentration of n-decane is 1.160 ppb by volume for 18 samples [52].

Food Survey Values: n-Decane was identified as a volatile component of roasted filberts [26], and beaufort cheese [9].

Plant Concentrations:

Fish/Seafood Concentrations: The average n-decane concentrations for the oyster and Rigolets clam population of Lake Pontchartrain, a shallow oligohaline estuary located in the deltaic plain of the Mississippi River near New Orleans, was 21 and 2.5 ng/g of wet weight [11].

Animal Concentrations:

Milk Concentrations: n-Decane was detected in 7 of 12 samples of mothers breast milk from the cities of Bayonne, NJ, Jersey City, NJ, Bridgeville, PA and Baton Rouge, LA [40].

Other Environmental Concentrations: n-Decane was emitted from 8 adhesives used in building materials for a two-week drying period [14]. An air sample taken near an oil fire was found to contain n-decane at a concentration of 0.51 mg/m^3 [42]. Two sets of 6 air samples both resulted in average n-decane concentrations of 14.8 and 13.7 mg/m^3 for a room applied with alkyl resin paint diluted with white spirits [63].

Probable Routes of Human Exposure: The most probable route of human exposure to n-decane is by inhalation. Atmospheric workplace exposures have been documented [5,14,43,53,63]. Extensive monitoring data indicates n-decane is a widely occurring atmospheric pollutant and breath samples have demonstrated n-decane exposure among urban residents [28].

Average Daily Intake:

Occupational Exposure: NIOSH (NOHS Survey 1972-1974) has estimated that 10,988 workers are exposed to n-decane in the USA [38]. NIOSH (NOES Survey 1981-1983) has estimated that 1,667 workers are exposed to n-decane in the USA [37]. The atmospheric concentration of n-decane ranged from 1 to 370 ug/m^3 for the vulcanization area of a shoe sole manufacturing plant; from 0 to 20 ug/m^3 for the vulcanization area and 0 to 2 ug/m^3 for the extrusion area of a tire retreading factory; and from 0 to 20 ug/m^3 for the extrusion area of electrical insulation manufacturing plant [5]. A 1984 study showed that outside operators at oil refineries, truck drivers during transport, and gas station attendants were exposed to n-decane emitted to air from gasoline [43]. For outside refinery operators, n-decane was detected in 24 of 56 air samples with an average concentration of 0.144 mg/m^3 [43]. n-Decane was detected in 44 of 49 air samples from the truck cabs, exposing drivers to an average n-decane concentration of 0.075 mg/m^3 [43]. n-Decane was detected in 28 of 49 air samples exposing gas station attendants to an average concentration of 0.073 mg/m^3 [43]. Painters were exposed to an average atmospheric concentration of n-decane of 6.8 mg/m^3 when applying an alkyl resin paint diluted with white spirits [63]. According to the National Ambient Volatile Organic Compounds (VOCs) Database, the median workplace atmospheric concentration of n-decane is 1.160 ppb by volume for 18 samples [52].

Body Burdens: n-Decane was detected in 7 of 12 samples of mothers breast milk from the cities of Bayonne, NJ, Jersey City, NJ, Bridgeville, PA and Baton Rouge, LA [40]. The air expired from humans contained n-decane in 8.5% of the 387 samples collected from 54 subjects [28]. The average n-decane concentration of 0.198 ng/L was expressed as the geometric mean with upper and lower limits of 0.893 and 0.044 ng/L, respectively [28].

Decane

REFERENCES

1. Abrams EF et al; Identification of Organic Compounds in Effluents from Industrial Sources USEPA-560/3-75-002 (1975)
2. Arnts RR, Meeks SA; Atmos Environ 15: 1643-51 (1981)
3. Atkinson R et al; Inter J Chem Kin 14: 781-8 (1982)
4. CLOGP; PCGEMS Graphical Exposure Modeling System USEPA (1986)
5. Cocheo V et al; Am Ind Hyg Assoc J 44: 521-7 (1983)
6. Colenutt BA, Thornburn S; Int J Environ Stud 15: 25-32 (1980)
7. Daubert TE, Danner RP; Data Compilation, Tables of Properties of Pure Cmpds, Design Inst for Phys Prop Data, Am Inst for Phys Prop Data, NY NY (1989)
8. DeBortoli M et al; Environ Int 12: 343-50 (1986)
9. Dumont JP, Adda J; J Agric Food Chem 26: 364-7 (1978)
10. Eisenreich SJ et al; Environ Sci Technol 15: 30-8 (1981)
11. Ferrario JB et al; Bull Environ Contam Toxicol 34: 246-55 (1985)
12. Fielding M et al; Organic Pollutants in Drinking Water, TR-159 Water Res Cent p 49 (1981)
13. Gerhold RM, Malaney GW; J Water Pollut Contr Fed 38: 562-79 (1966)
14. Girman JR et al; Environ Int 12: 317-21 (1986)
15. Grob K, Grob G; J Chromatogr 62: 1-13 (1971)
16. Grosjean D, Fung K; J Air Pollut Control Assoc 34: 537-43 (1984)
17. Hampton CV et al; Environ Sci Technol 17: 699-708 (1983)
18. Hampton CV et al; Environ Sci Technol 16: 287-98 (1982)
19. Ioffe BV et al; Dokl Akad Nauk Sssr 243: 1186-9 (1978)
20. Ioffe BV et al; Environ Sci Technol 13: 864-8 (1979)
21. Ioffe BV et al; J Chromatogr 142: 787-95 (1977)
22. Jarke FH et al; Ashrae Trans 87: 153-66 (1981)
23. Junk GA, Ford CS; Chemosphere 9: 187-230 (1980)
24. Juttner F; Z Wasser-Abwasser-Forrsch 21: 36-9 (1988)
25. Keith LH et al; pp 329-73 in Ident Anal Org Pollut Water, Keith LH (ed) Ann Arbor MI (1976)
26. Kinlin TE et al; J Agric Food Chem 20: 1021 (1972)
27. Kool HJ et al; Crit Rev Env Control 12: 307-57 (1982)
28. Krotoszynski BK et al; J Anal Toxicol 3: 225-34 (1979)
29. Ligocki MP et al; Atmos Environ 19: 1607-17 (1985)
30. Linak, WP et al; JAPCA 39: 836-46 (1989)
31. Louw CW et al; Atmos Environ 11: 703-17 (1977)
32. Lyman WJ et al; Handbook of Chemical Property Estimation Methods NY: McGraw-Hill p 4-9, 5-4 to 5-10, 7-4, 15-15 to 15-29 (1982)
33. Mulcahy MFR et al; Paper IV p 17 in Occurrence Contr Photochem Pollut, Proc Symp Workshop Sess (1976)
34. Nagata S, Kondo G; pp 617-20 in Photooxidation of Crude Oils, Proc 1977 Oil Spill Conf, Am Petrol Inst (1977)
35. Nelson PF, Quigley SM; Environ Sci Technol 16: 650-5 (1982)
36. Nelson PF, Quigley SM; Atmos Environ 18: 79-87 (1984)

Decane

37. NIOSH; National Occupational Exposure Survey (NOES) (1989)
38. NIOSH; National Occupational Hazard Survey (NOHS) (1974)
39. Nolting F et al; J Atmos Chem 6: 47-59 (1988)
40. Pellizzari LD et al; Bull Environ Contam Toxicol 28: 322-8 (1982)
41. Perry DL et al; Ident of Org Compounds in Ind Effluent discharges USEPA-600/4-79-016 p 230 (1979)
42. Perry R; Mass Spectroscopy in the Detection and Identification of Air Pollutants, Int Symp Ident Meas Environ Pollut p 130-7 (1971)
43. Rappaport SM et al; Appl Ind Hyg 2: 148-54 (1987)
44. Raymond A, Guiochon G; Environ Sci Technol 8: 143-8 (1974)
45. Rudolph J, Khedim A; Int J Environ Anal Chem 290: 265-82 (1985)
46. Sauer TC Jr; Org Geochem 7: 1-16 (1981)
47. Sauer TC Jr; Org Geochem 3: 91-101 (1981)
48. Sauer TC Jr et al; Mar Chem 7: 1-16 (1978)
49. Scheiman MA et al; Biomed Mass Spectrom 4: 209-11 (1974)
50. Scott Research Labs Inc; Atmospheric Reaction Studies in the Los Angeles Basin, NTIS PB-194-058 p 86 (1969)
51. Seila RL; Non-urban Hydrocarbons Concentrations in Ambient Air No of Houston TX USEPA-500/3- 79-010 p 38 (1979)
52. Shah JJ, Heyerdahl EK; National Ambient VOC Database Update USEPA-600/3-88/010 (1988)
53. Shamsky S, Samimi B; App Ind Hyg 2: 242-5 (1987)
54. Shields HC, Weschler CJ; J Air Pollut Control Fed 37: 1039-45 (1987)
55. Silverstein RM, Bassler GC; Spectrometric Id of Org Cmpd, J Wiley and Sons Inc pp 148-169 (1963)
56. Smeyers-Verbeke J et al; Atmos Environ 18: 2471-8 (1984)
57. Stump FD, Dropkin DL; Anal Chem 57: 2629-34 (1985)
58. Suffet IH, Radziul JV; Anal Org Pollut in Drinking Water, Inter Conf Environ Sensing Assess 1975, pp 30-1 to 30-7 (1976)
59. Swann RL et al; Res Rev 85: 16-28 (1983)
60. Tichenor BA, Mason MA; JAPCA 38: 264-8 (1988)
61. USEPA; Drinking water Criteria Document for Gasoline ECAO-CIN-D006, 8006-61-9 (1986)
62. USEPA; EXAMS II Computer Simulation (1987)
63. Van der Wal JF, Moerkerken A; Ann Occup Hyg 28: 39-47 (1984)
64. Vogt WG, Walsh JJ; Volatile Organic Compounds in Gases from Landfill Simulators Proc APCA Annu Meet 78th 6: 2-17 (1985)
65. Walker JD, Colwell RA; Can J Microbiol 21: 305-13 (1975)
66. Yalkowsky SH et al; Arizona Data Base of Water Solubility (1987)
67. Young P, Parker A; Vapors Odors and Toxic Gases from Landfills ASTM Spec Tech Publ 851: 24-41 (1984)
68. Zimmerman RE et al; Landfill Methane Trace Volatile Org Constituent, Proc Int Gas Res Conf p 230-9 (1983)

1,2-Dibromoethylene

SUBSTANCE IDENTIFICATION

Synonyms:

Structure:

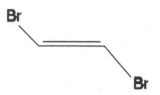

CAS Registry Number: 540-49-8

Molecular Formula: $C_2H_2Br_2$

Wiswesser Line Notation:

CHEMICAL AND PHYSICAL PROPERTIES

Boiling Point: 112.5 °C at 760 mm Hg (cis-isomer)

Melting Point: - 53 °C (cis-isomer)

Molecular Weight: 185.87

Dissociation Constants:

Log Octanol/Water Partition Coefficient: 1.89 [13]

Water Solubility : 4,100 mg/L at 25 °C [8]

Vapor Pressure:

Henry's Law Constant: 0.674×10^{-2} atm-m^3/mol [7]

ENVIRONMENTAL FATE/EXPOSURE POTENTIAL

Summary: There is no evidence that 1,2-dibromoethylene is used commercially and therefore little of the chemical may be released to the environment. If released to soil, 1,2-dibromoethylene would be expected to rapidly volatilize from the soil surface or leach into the soil. Hydrolysis in moist soil or water is not expected to be important. If released into water, 1,2-dibromoethylene would rapidly volatilize (half-life 4.2 hr in a model river). Bioconcentration in aquatic organism or adsorption to sediment would not be significant. In the atmosphere 1,2-dibromoethylene should degrade by reacting with photochemically produced hydroxyl radicals with a half-life of about 4 days. Wet deposition may partially remove the compound from the atmosphere.

Natural Sources:

Artificial Sources: No evidence could be found that 1,2-dibromoethylene is used commercially. Therefore, significant releases to the environment would not be expected. However, smaller amounts of the compound are manufactured [1], at least for laboratory use.

Terrestrial Fate: If released to soil, 1,2-dibromoethylene would be expected to rapidly volatilize from the soil surface. Based on its low soil adsorption coefficient, leaching of the compound from subsurface soil may occur. Its fate in soil with respect to abiotic and biotic degradative processes is unknown; however, based on hydrolysis characteristics of 1,2-dichloroethylene, hydrolysis of 1,2-dibromoethylene should be a slow process. Based on the biodegradation characteristics of the chlorinated analog, anaerobic degradation of 1,2-dibromoethylene is likely in anaerobic zone of the soil. However, the loss of the compound due to volatilization is expected to be far more important than anaerobic biodegradation.

Aquatic Fate: If released into water, 1,2-dibromoethylene would rapidly volatilize. It has been estimated that the half-life for the volatilization of the compound from a model river 1 m deep flowing at 1 m/sec and a wind velocity of 3 m/sec is 4.2 hr. Based on the hydrolysis characteristics of 1,2-dichloroethylene, 1,2-dibromoethylene may slowly hydrolyze in water but rate data are lacking. The biodegradation of similar compounds in water indicates that although anaerobic biodegradation may occur, the importance of the process should be far less important than volatilization from water. Based on Koc value, adsorption to sediments and suspended solids in water would not be significant. The estimated bioconcentration factor suggests bioconcentration of 1,2-dibromoethylene in aquatic organisms should not be important.

Atmospheric Fate: Based on an estimation method and the average concentration of hydroxyl radicals in the atmosphere [2,3], 1,2-dibromoethylene will degrade by reacting with photochemically produced hydroxyl radicals with a half-life of about 4 days. Its moderate solubility in water [8] suggests that 1,2-dibromoethylene may be scavenged by rain.

Biodegradation: Experimental data regarding the biodegradability of 1,2-dibromoethylene in water or soil are lacking. 1,2-Dichloroethylene, a structurally similar compound, was found to undergo slow microbial degradation especially under anaerobic conditions by reductive dechlorination [5,6]. Therefore, slow anaerobic biodegradation is likely for 1,2-dibromoethylene.

Abiotic Degradation: Since 1,2-dichloroethylene, a compound structurally similar to 1,2-dibromoethylene, does not absorb sunlight available in the troposphere [4], the direct photolysis of 1,2-dibromoethylene should also be unimportant in air, water or soil surfaces. Data on the rates of hydrolysis of 1,2-dibromoethylene at different pHs are lacking. The hydrolysis of 1,2-dichloroethylene in water is not important [9]. Although the bromide salts generally undergo hydrolysis at a faster rate than the chloride salts [11], it is likely that the hydrolysis of 1,2-dibromoethylene may still be noncompetitive compared to volatilization loss. Based on an estimation method [3], the rate constants for the reaction of 1,2-dibromoethylene with hydroxyl radicals have been estimated to be 3.79 x 10^{-12} cm^3/molecule-sec for the cis-isomer and 4.31 x 10^{-12} cm^3/molecule-sec for the trans-isomer. If the daily average concentration of hydroxyl radicals in the atmosphere is assumed to be 10^5/cm^3 [2], the half-life for these reactions can be estimated to be 4.2 days for the cis-isomer and 3.7 days for the trans-isomer. Based on an estimation method [3], the rate constants for the reactions of cis- and trans-1,2-dibromoethylene with ozone have been estimated to be 1.72 x 10^{-20} and 3.43 x 10^{-20} cm^3/molecule-sec, respectively. If the daily average atmospheric concentration of ozone is assumed to be 7 x 10^{11} molecules/cm^3 [2], the half-lives for these reactions can be estimated to 668.2 days for the cis-isomer and 334.1 days for the trans-isomer.

1,2-Dibromoethylene

Therefore, the atmospheric reaction with ozone should not be important.

Bioconcentration: From the estimated log octanol/water partition coefficient of 1.89 [13] and a recommended regression equation [10], the bioconcentration factor for 1,2-dibromoethylene has been estimated to be 16. This indicates that bioconcentration of 1,2-dibromoethylene in aquatic organisms will not be important.

Soil Adsorption/Mobility: Using the water solubility data of 4,100 mg/L [8] and a recommended regression equation [10], one can estimate a KOC of 45 for 1,2-dibromoethylene. One measured value of KOC is 77 [12]. Therefore, 1,2-dibromoethylene should not adsorb significantly to soil or sediment.

Volatilization from Water/Soil: The Henry's Law constant for 1,2-dibromoethylene is 0.674×10^{-2} atm-m^3/mol [7]. Using this value and an estimation method [10], the half-life for volatilization of 1,2-dibromoethylene from a model river 1 m deep with a 1 m/sec current and 3 m/sec wind speed can be estimated to be 4.2. Volatilization is controlled by diffusion in the liquid phase [10].

Water Concentrations:

Effluent Concentrations:

Sediment/Soil Concentrations:

Atmospheric Concentrations:

Food Survey Values:

Plant Concentrations:

Fish/Seafood Concentrations:

Animal Concentrations:

Milk Concentrations:

1,2-Dibromoethylene

Other Environmental Concentrations:

Probable Routes of Human Exposure:

Average Daily Intake:

Occupational Exposure:

Body Burdens:

REFERENCES

1. Aldrich; Catalog Handbook of Fine Chemicals, Milwaukee, WI: Aldrich Chemical Co. p 410 (1991)
2. Atkinson R; Chem Rev 85: 69-201 (1985)
3. Atkinson R; Environ Toxicol Chem 7: 435-42 (1988)
4. Ausbel R, Wijnen MHJ; Int J Chem Kinetics 75: 739-51 (1975)
5. Barrio-Lage G et al; Environ Sci Technol 20: 96-98 (1986)
6. Fogel MM et al; Appl Environ Microbiol 51: 720-24 (1986)
7. Hine J, Mookerjee PK; J Org Chem 40: 292-8 (1975)
8. Horvath AL; Halogenated Hydrocarbons. Solubility-Miscibility with Water. NY: Marcel Dekker p 685 (1982)
9. Jaber HM et al; Data Acquisition for Environmental Transport and Fate Screening, NTIS PB 84-243955, Springfield, VA (1984)
10. Lyman WJ et al; Handbook of Chemical Property Estimation Methods New York: McGraw-Hill pp 4-1 to 4-33, 5-1 to 5-30, 15-1 to 15-34 (1982)
11. Mabey W, Mills T; J Phys Chem Ref Data 7: 383-415 (1978)
12. Sabljic A; J Agric Food Chem 32: 243-6 (1984)
13. USEPA; Graphical Exposure Modeling System. PCGEMS, USEPA, Washington, DC: Office of Toxic Substances (1987)

Dibromomethane

SUBSTANCE IDENTIFICATION

Synonyms: Methylene bromide

Structure:

CAS Registry Number: 74-95-3

Molecular Formula: CH_2Br_2

Wiswesser Line Notation: E1E

CHEMICAL AND PHYSICAL PROPERTIES

Boiling Point: 96.95 °C

Melting Point: -52.5 °C

Molecular Weight: 173.83

Dissociation Constants:

Log Octanol/Water Partition Coefficient: 1.22 [8]

Water Solubility: 11,442 mg/L at 25 °C [9]

Vapor Pressure: 340 mm Hg at 20 °C [22]

Henry's Law Constant: 8.88 x 10^{-4} atm-m^3/mol [15]

ENVIRONMENTAL FATE/EXPOSURE POTENTIAL

Summary: Dibromomethane finds limited use in chemical synthesis, as a solvent, and as a gage fluid. It may be released to the environment during these uses as well as in its production and transport. Natural production by marine algae also adds to its environmental input. If released on soil, dibromomethane should volatilize from the soil surface and leach into the ground. If released in water, dibromomethane would be primarily lost by volatilization with an estimated half-life of 5.2 hr from a model river. Adsorption to sediment and bioconcentration in aquatic organisms should not be significant. No significant biotic or abiotic degradative processes have been reported in natural waters or soil. However, catalyzed photolysis may occur in surface layers of some natural waters or soil. In the atmosphere, dibromomethane will be lost by reaction with photochemically produced hydroxyl radicals. The estimated half-life for this reaction is 213 days. Dibromomethane should also be readily scavenged by rain and snow. However, this dibromomethane will reenter the atmosphere by volatilization. The general population will be exposed to low levels of dibromomethane in the atmosphere from both natural and anthropogenic sources. Limited occupational exposure via inhalation and dermal contact will also occur.

Natural Sources: Dibromomethane is a primary emission product of macroalgae (e.g., Fucales sargassum, Laminariales lamanaria) [5]. Macroalgae are often concentrated along beaches and coastlines and releases of dibromomethane occur though dissolution into seawater followed by volatilization into air or directly by the algae [5]. Four of six species of intertidal macroalgae collected from three sites around Cape Cod produced and released dibromomethane into seawater at release rates that ranged up to 2100 ng/g algae (dry wt) [7]. Representative species of brown and green algae released dibromomethane while the red algae did not [7].

Artificial Sources: Dibromomethane may be released to the environment during its production and use in chemical synthesis, as a solvent, or as a gage fluid [21]. It finds limited use in these applications [21]. A suggested secondary source of bromomethanes is from the chlorination of seawater in which bromine is relatively abundant [23]. However, this process predominantly generates bromoform.

Terrestrial Fate: Dibromomethane has a high vapor pressure and low adsorptivity to soil. Consequently, if released on soil, it will be expected to volatilize from the soil surface and leach into the ground. Catalyzed photolysis may take place on the soil surface. Its biotic and abiotic fate in soil and ground water is unknown.

Aquatic Fate: If released in water, dibromomethane would be primarily lost by volatilization with a half-life of 5.1 hr from a model river of 1 m depth, flowing at 1 m/sec with a wind speed of 3 m/sec. Catalyzed photolysis may occur in surface layers of some natural waters.

Atmospheric Fate: In the atmosphere, dibromomethane will be lost by reaction with photochemically produced hydroxyl radicals. The estimated half-life for this reaction is 213 days [1]. Because of its high solubility in water (11,442 mg/L at 25 °C) [9], dibromomethane should be readily scavenged by rain and snow.

Biodegradation:

Abiotic Degradation: Dibromomethane solution at a concentration of 35 mg/L completely mineralized to carbon dioxide and HBr in about 8.3 hr when illuminated with black-light fluorescent lamps at wavelengths less than 360 nm while heterogeneously photocatalyzed by titanium dioxide slurry in water [16]. In the presence of a 0.1% slurry of TiO2 in a photoreactor (300 to <400 nm), the rate of conversion of dibromomethane was 4.1 ppm/min-g(catalyst) [16]. This reaction has been observed to occur in several rivers near populated areas [16]. Dibromomethane is resistant to hydrolysis; the hydrolysis half-life at pH 7 and 25 °C is 183 yr [13]. The major atmospheric reaction affecting the fate of dibromomethane is its reaction with photochemically produced hydroxyl radicals; the estimated half-life is 213 days [19].

Bioconcentration: The log Kow for dibromomethane is 1.22 [8], from which one can estimate a BCF of 5 [12]. Therefore, bioconcentration of dibromomethane in aquatic organisms should not be significant.

Soil Adsorption/Mobility: From the water solubility of dibromomethane of 11,442 mg/L at 25 °C [9] and a recommended regression equation [12], a Koc value of 39 can be estimated. This indicates that dibromomethane will not adsorb significantly to soil or sediment.

Dibromomethane

Volatilization from Water/Soil: The measured value of the Henry's Law constant for dibromomethane is 8.88×10^{-4} atm-m^3/mol [15]. Using this Henry's Law constant, one can estimate a volatilization half-life of 5.1 hr for dibromomethane from a model river 1 m deep flowing at a speed of 1 m/sec with a 3 m/sec [12].

Water Concentrations: DRINKING WATER: In a survey of 14 treated drinking water supplies of varied sources in England, dibromomethane was detected in seven supplies [6]. These supplies were derived from ground water and surface water sources. Dibromomethane was detected in treated drinking water from the Niagara River at a concn range of 0.2-0.8 ppb [11]. GROUND WATER: No detectible dibromomethane was found in samples from a study of ground water contamination at 19 municipal and 6 industrial landfill sites in Wisconsin [2]. Of the 377 and 282 representative samples of ground water and surface water in New Jersey that were analyzed for dibromomethane, 12% and 28%, respectively, contained dibromomethane [17]. Ninety percent of the samples of both types contained equal or less than 0.1 ppb of dibromomethane [17]. The maximum dibromomethane concentration in ground water was 44.9 and that in surface water was 358.6 ppb [17]. Dibromomethane was found at 9 of 17 stations in the Lower Niagara River; levels up to 5 ppt were found [10]. SURFACE WATER: Dibromomethane is a major volatile organic hydrocarbon in Narraganset Bay [23]. However, levels and distribution of the chemical were not reported. Surface seawater concentrations of dibromomethane at a site in the South Atlantic, and two sites in the North Atlantic (south of the Canary Islands and west of the Strait of Gibraltar) were 0.26, >1, and 0.3 ng/L [5]. In Lake Ontario, dibromomethane was detected in 66% of the 82 stations that were sampled [10]. Only 8 samples contained more than trace amounts of dibromomethane and the highest level was 7 ppt. The detection limit was 0.7 ppt. RAIN/SNOW: In a fast moving front coming from the North Atlantic, the concentration of dibromomethane in rain water collected in Ulm, southern Germany was 1.4 ng/L [5].

Effluent Concentrations: In a comprehensive survey of wastewater from 4000 industrial and publicly owned treatment works (POTWs) sponsored by the Effluent Guidelines Division of the U.S. EPA, dibromomethane was identified in discharges of the following industrial categories (positive occurrences, median concn in ppb): nonferrous metals (8; 2.2),

152

organics and plastics (2; 32.9), inorganic chemicals (2; 1.9), pesticides manufacture (2; 104.6), publicly owned treatment works (9; 0.3) [20]. Maximum effluent concn >100 ppb were found in the nonferrous metals industry (286 ppb) and in pesticide manufacturing (151 ppb) [20]. In a previous survey of 63 wastewaters from a wide range of chemical manufacturers across the U.S., 1 effluent contained dibromomethane [18]. The level of dibromomethane in that sample was >100 ppb.

Sediment/Soil Concentrations:

Atmospheric Concentrations: RURAL/REMOTE: The baseline concentration of dibromomethane in marine air far removed from coastal areas and large concentrations of macroalgae is 2.4 ppt/volume [5]. The level of dibromomethane in air decreases with altitude to 1.2 ppt/volume above the marine boundary layer and tradewind inversion [5]. The dibromomethane concentrations in air samples collected in the South and North Atlantic on a cruise from Capetown to Bremerhaven, on the Azores, Madeira, Bermuda, and Tenerife was reported to range from 0.8 to 4 ppt (v/v) with the exception of the beach at Sao Miguel, the Azores where the concentration was 50 ppt (v/v) [5]. The concentrations of dibromomethane measured across the Arctic from Anchorage, AK to Norway and the North Pole during March and April 1983 ranged from 3-60 ppt/volume with a mean (SD) of 15 [12] ppt/volume [3]. The monthly average concentration of dibromomethane at Point Barrows, AK during 1983 ranged from 4.7 to 5.6 ppt/volume [19]. The concentration within the arctic haze that seasonally occurs in Point Barrows is estimated to be 4.2 ppt/volume, while that outside the haze is 2.5 ppt (v/v) [19]. The concentration is highest in winter and spring at which times meteorological and atmospheric conditions favor the formation of arctic haze [3,19]. This suggests that dibromomethane may be transported from industrial sources in the mid-latitudes [19]. SOURCE DOMINATED: The median concentration of dibromomethane from six source-related areas in the U.S. (22 measurement), namely, Edison, NJ, Magnolia,AR, Phoenix, AZ, Seattle, WA, Sugas Creek, MO, and Westwood Village, CA was 980 ppt; the range was 190-13,000 ppt [4].

Food Survey Values:

Plant Concentrations:

153

Dibromomethane

Fish/Seafood Concentrations:

Animal Concentrations:

Milk Concentrations:

Other Environmental Concentrations:

Probable Routes of Human Exposure: The general population will be exposed to low levels of dibromomethane in the atmosphere. Limited occupational exposure via inhalation and dermal contact will also be expected.

Average Daily Intake: AIR INTAKE: Remote areas (assume air concentration of 2.4-15 ppt and inhalation rate of 20 m^3/day) - 0.3-2.2 ug/day. Source areas (assume air concentration of 900 ppt and inhalation rate of 20 m^3/day) - 130 ug/day. WATER INTAKE: insufficient data. FOOD INTAKE: insufficient data.

Occupational Exposure: NIOSH (NOES Survey, 1981-83) has statistically estimated that 1065 workers are exposed to dibromomethane in the USA [14]. The National Occupational Exposure Survey was based on field surveys of 4490 facilities and was designed as a nationwide survey based on a statistical sample of virtually all workplace environments, except mining and agriculture, in the United States where eight or more persons are employed.

Body Burdens:

REFERENCES

1. Atkinson R; Int J Chem Kinet 19: 799-828 (1987)
2. Battista J, Connelly JP; Wisconsin Dept of Natural Resources, Madison, WI PUBL SW-094 89 (1989)
3. Berg WW et al; Geophys Res Lett 11: 429-32 (1984)
4. Brodzinsky R, Singh HB; Volatile Organic Chemicals in the Atmosphere. Menlo Park, CA: Atmos Sci Ctr. SRI Inter. Contract 68-02-3452 (1982)
5. Class T et al; Chemosphere 15: 429-36 (1986)
6. Fielding M et al; Organic micropollutants in drinking water. TR-159. Medmenham, England: Water Res Ctr (1981)
7. Gschwend PM et al; Science 227: 1033-5 (1985)

Dibromomethane

8. Hansch C, Leo AJ; MedChem Project Claremont CA: Pomona College (1985)
9. Horvath AL; Halogenated Hydrocarbons: Solubility- Miscibility with Water, NY: Marcel Dekker, Inc p 479 (1982)
10. Kaiser KLE et al; J Great Lakes Res 9: 212-23 (1983)
11. Komsta E et al; Bull Environ Contam Toxicol 41: 515-22 (1988)
12. Lyman WJ et al; Handbook of Chem Property Estimation Methods NY: McGraw-Hill p 3-1 to 4-33, 5-1 to 5-30, 15-1 to 15-34 (1982)
13. Mabey W, Mill T; J Phys Chem Ref Data 7: 383-403 (1978)
14. NIOSH; National Occupational Exposure Survey (NOES) Computer printout 3/29/89 (1985)
15. Nirmalakhandan NN, Speece RE; Environ Sci Technol 22: 1349-57 (1988)
16. Ollis DF; Environ Sci Technol 19: 480-84 (1985)
17. Page GW; Environ Sci Technol 15: 1475-81 (1981)
18. Perry DL et al; Identification of organic compounds in industrial effluent discharges. USEPA-600/4-79-016 (1979)
19. Rasmussen RA, Khalil MAK; Geophys Res Lett 11: 433-6 (1984)
20. Shackelford WM et al; Analyt Chim Acta 146: 15-27 (supplemental data) (1983)
21. Stenger VA; Kirk-Othmer Encycl Chem Tech 3rd ed NY: Wiley 4: 243-63 (1978)
22. Verschueren K; Handbook of Environmental Data on Organic Chemicals, NY: Van Nostrand Reinhold Co. p 450-51 (1977)
23. Wakeham SG et al; Can J Fish Aquat Sci 40: 304-21 (1983)

Dibutyl Ether

SUBSTANCE IDENTIFICATION

Synonyms:

Structure:

H₃C$\diagup\diagdown\diagup$O$\diagdown\diagup\diagdown$CH₃

CAS Registry Number: 142-96-1

Molecular Formula: $C_8H_{18}O$

Wiswesser Line Notation: 4O4

CHEMICAL AND PHYSICAL PROPERTIES

Boiling Point: 142 °C at 760 mm Hg

Melting Point: -95.3 °C

Molecular Weight: 130.26

Dissociation Constants:

Log Octanol/Water Partition Coefficient: 3.08.

Water Solubility: 300 mg/L at 25 °C [18]

Vapor Pressure: 6.01 mm Hg at 25 °C [7]

Henry's Law Constant: 5.89 x 10^{-3} atm-m³/mole [10]

ENVIRONMENTAL FATE/EXPOSURE POTENTIAL

Summary: Dibutyl ether may be released to the environment as a result of the manufacture and use of various consumer products which have been shown to contain the compound, such as latex paint. It may be released as a result of its manufacture and use as a solvent and extraction agent. If dibutyl ether is released to soil, it will be subject to volatilization. It will be expected to exhibit low mobility in soil and, therefore, leaching to ground water is unlikely. It will not be expected to hydrolyze in soil. If dibutyl ether is released to water, it will not be expected to significantly adsorb to sediment or suspended particulate matter, bioconcentrate in aquatic organisms, hydrolyze, directly photolyze, or photooxidize via reaction with photochemically produced hydroxyl radicals in the water, based upon estimated physical-chemical properties or analogies to other structurally related aliphatic ethers. Dibutyl ether in surface water will be subject to rapid volatilization with estimated half-lives of 3.5 hr and 4.4 days for volatilization from a river one meter deep flowing 1 m/sec with a wind velocity of 3 m/sec and a model pond, respectively. It may be resistent to biodegradation in environmental media based upon screening test data which utilized acclimated mixed microbial cultures. Many ethers are known to be resistant to biodegradation. If dibutyl ether is released to the atmosphere, it will be expected to exist almost entirely in the vapor phase based upon its vapor pressure. It will be susceptible to photooxidation via vapor-phase reaction with photochemically produced hydroxyl radicals with an estimated half-life of 23 hours calculated for this process. Direct photolysis will not be an important removal process since aliphatic ethers do not absorb light at wavelengths >290 nm. The most probable route of general population exposure to dibutyl ether is via inhalation of contaminated air. Inhalation and dermal exposure may occur where dibutyl ether is made and used, including occupational settings as well as where various consumer products which contain the product, such as latex paint, are applied.

Natural Sources:

Artificial Sources: Dibutyl ether may be released to the environment as a result of the manufacture and use of various consumer products which have been shown to contain the compound, such as latex paint [21]. It may be released as a result of its manufacture and use as a solvent and extraction agent [9].

Terrestrial Fate: If dibutyl ether is released to soil, it will be subject to volatilization based upon its Henry's Law constant and vapor pressure. It will be expected to exhibit low mobility [20] in soil and, therefore, leaching to ground water is unlikely based upon a Koc of 1130 estimated from its octanol/water partition coefficient [14]. It will not be expected to hydrolyze in soil [14]. Dibutyl ether may be resistent to biodegradation in environmental media based upon aqueous screening test data from studies which utilized acclimated mixed microbial cultures [4]. Many ethers are known to be resistant to biodegradation [2].

Aquatic Fate: If dibutyl ether is released to water, it will not be expected to significantly adsorb to sediment or suspended particulate matter [1,14], bioconcentrate in aquatic organisms [1,14], hydrolyze [14], directly photolyze [6], or photooxidize via reaction with photochemically produced hydroxyl radicals in water [3], based upon estimated physical-chemical properties or analogies to other structurally related aliphatic ethers [1,6,14]. Dibutyl ether in surface water will be subject to rapid volatilization [5,14]. Using its reported Henry's Law constant, a half-life for volatilization of dibutyl ether from a river one meter deep flowing 1 m/sec with a wind velocity of 3 m/sec has been estimated to be 3.5 hr at 25 °C [14]. The volatilization half-life from a model pond, which considers the effect of adsorption, has been estimated to be 4.4 days [23]. Dibutyl ether may be resistent to biodegradation in environmental media based upon screening test data which utilized acclimated mixed microbial cultures [4]. Many ethers are known to be resistant to biodegradation [2].

Atmospheric Fate: If dibutyl ether is released to the atmosphere, it will be expected to exist almost entirely in the vapor phase [8] in view of its high vapor pressure. It will be susceptible to photooxidation via vapor phase reaction with photochemically produced hydroxyl radicals. An atmospheric half-life of 23 hours at an atmospheric concentration of 5 x 10^5 hydroxyl radicals per cm^3 has been calculated for this process based upon a measured rate constant [5]. Direct photolysis will not be an important removal process since aliphatic ethers do not absorb light at wavelengths >290 nm [6].

Biodegradation: No data concerning the biodegradation of dibutyl ether in environmental media were located. The 5-day biological oxygen demand was 16% of theoretical in screening tests which utilized acclimated mixed microbial cultures [4]. These data suggest that dibutyl ether may slowly be biodegraded in environmental media. Many ethers are known to be resistant to biodegradation [2].

Dibutyl Ether

Abiotic Degradation: The rate constant for the vapor phase reaction of dibutyl ether with photochemically produced hydroxyl radicals has been measured to be 17.1×10^{-12} cm^3/molecule-sec at 25 °C [5] which corresponds to an atmospheric half-life of 23 hours at an atmospheric concentration of 5×10^5 hydroxyl radicals per cubic cm. Hydrolysis is not expected to be significant under normal environmental conditions (pH 5-9) [14]. Direct photolysis will not be an important removal process since aliphatic ethers as a class do not absorb light at wavelengths >290 nm [6].

Bioconcentration: Based upon its log Kow, a BCF of 129 has been estimated using a recommended regression equation [14]. Based upon this estimated BCF, dibutyl ether will not be expected to bioconcentrate in aquatic organisms.

Soil Adsorption/Mobility: Based upon its log Kow, a Koc of 1130 has been estimated using recommended regression equation [14]. Based upon this estimated Koc, dibutyl ether will be expected to exhibit low mobility in soil [20].

Volatilization from Water/Soil: The half-life for volatilization of dibutyl ether from a river one meter deep flowing 1 m/sec with a wind velocity of 3 m/sec is estimated to be 3.5 hr at 25 °C [14] based on its Henry's Law constant. The volatilization half-life from a model pond, which considers the effect of adsorption, has been estimated to be 4.4 days [23]. Based upon its Henry's Law constant and vapor pressure, dibutyl ether will be subject to volatilization from surfaces and near-surface soil.

Water Concentrations: DRINKING WATER: Dibutyl ether has been detected, but not quantified, in drinking water from unidentified sources [11]. GROUND WATER: Dibutyl ether has been found in contaminated ground water in The Netherlands at a max concn of 1 ug/L [24]. SURFACE WATER: Dibutyl ether was detected at 1 ug/L in water from 3 sites on the River Rhine in The Netherlands sampled in 1979 [12].

Effluent Concentrations: Dibutyl ether has been detected in 1 out of 63 samples of industrial effluents collected from a wide variety of industries across the U.S. at a concn of <10 ug/L (dates not reported) [17]. Dibutyl ether was tentatively identified, but not quantified, in an advanced waste treatment concentrate from Lake Tahoe, CA, sampled in Oct 1974 [13]. Dibutyl ether was detected at concn of 5.6 and 0.2 ug/L in 2 of 7 sites sampled in Feb 1979 for standing water and impoundments which were discharged into Wilson Creek from a hazardous waste dump site called "Valley of the Drums" in Bullitt County, KY [19].

Sediment/Soil Concentrations:

Atmospheric Concentrations: Dibutyl ether was qualitatively detected in the atmosphere of a chamber containing latex paint which suggests that the compound may be present in indoor air in buildings that have been painted with interior latex paint [22].

Food Survey Values:

Plant Concentrations:

Fish/Seafood Concentrations:

Animal Concentrations:

Milk Concentrations:

Other Environmental Concentrations:

Probable Routes of Human Exposure: The most probable route of general population exposure to ether is via inhalation of contaminated air [24]. Inhalation and dermal exposure may occur where dibutyl ether is made and used, including occupational settings as well as where various consumer products which contain the compound, such as latex paint, are used [21].

Average Daily Intake:

Occupational Exposure: NIOSH (NOES Survey 1981-1983) has statistically estimated that 4,570 workers are exposed to dibutyl ether in the USA [16]. NIOSH (NOHS Survey 1972-1974) has statistically estimated that 1,003 workers are exposed to dibutyl ether in the USA [15].

Dibutyl Ether

Body Burdens:

REFERENCES

1.   Abernethy SG et al; Environ Toxicol Chem 7: 469-81 (1988)
2.   Alexander M; Biotechnol Bioeng 15: 611-47 (1973)
3.   Anbar M, Neta P; Int J Appl Radiation Isotopes 18: 493-523 (1967)
4.   Babeu L, Vaishnav DD; J Indust Microbiol 2: 107-15 (1987)
5.   Bennett PJ, Kerr JA; J Atmos Chem 8: 87-94 (1989)
6.   Calvert JG, Pitts JNJr; Photochemistry John Wiley & Sons NY p 441-2 (1966)
7.   Daubert TE, Danner RP; Data Compilation Tables of Properties of Pure Compounds. Am Inst Chem Eng. (1989)
8.   Eisenreich SJ et al; Environ Sci Technol 15: 30-8 (1981)
9.   Hawley GG; Condensed Chemical Dictionary 10th ed Van Nostrand Reinhold NY p 165 (1981)
10. Hine J, Mookerjee PK; J Org Chem 40: 292-8 (1975)
11. Kool HJ et al; Crit Rev Env Control 12: 307-57 (1982)
12. Linders JBHJ; Inventory of organic substances in the River Rhine in 1979 (NTIS PB83-200543) p 14 (1981)
13. Lucas SV; GC/MS Anal of Org in Drinking Water Concentrates and Advanced Treatment Concentrates Vol 2 USEPA-600/1-84-020B (NTIS PB85-128239) p 149 (1984)
14. Lyman WJ et al; Handbook of Chem Property Estimation Methods NY: McGraw-Hill p 4-9 (1982)
15. NIOSH; The National Occupational Hazard Survey (NOHS) (1974)
16. NIOSH; The National Occupational Exposure Survey (NOES) (1983)
17. Perry DL et al; Identification of Organic Compounds in Industrial Effluent Discharges. USEPA-600/4-79-016. NTIS PB-294794. p 45 (1979)
18. Riddick JA et al; Organic Solvents: Physical properties and methods of purification 4th ed John Wiley and Sons Inc. NY (1986)
19. Stonebreaker RD, Smith AJ; Containment and Treatment of a Mixed Chemical Discharge "Valley of Drums" Louisville KY, Contr Haz Mater Spills, Proc Natl Conf pp 1-10 (1980)
20. Swann RL et al; Res Rev 85: 17-28 (1983)
21. Tichenor BA; Organic emission measurements via small chamber testing USEPA/600/D-87/187 (NTIS PB87-199154) p 8 (1987)
22. Tichenor BA, Mason MA; J Air Pollut Control Assoc 38: 264-8 (1988)
23. USEPA; EXAMS II Computer Simulation (1987)
24. Zoeteman BCJ et al, Sci Total Environ 21: 187-202 (1981)

161

Dibutyl Ketone

Synonyms: 5-Nonanone

Structure:

H₃C~~~~C(=O)~~~~CH₃

$$H_3C \diagdown \diagup \diagdown \diagup^{\displaystyle O}_{\parallel} \diagdown \diagup \diagdown CH_3$$

CAS Registry Number: 502-56-7

Molecular Formula: $C_9H_{18}O$

Wiswesser Line Notation: 4V4

CHEMICAL AND PHYSICAL PROPERTIES

Boiling Point: 188.4 °C [15]

Melting Point: -4.8 °C [15]

Molecular Weight: 142.24

Dissociation Constants:

Log Octanol/Water Partition Coefficient: 2.9 (estimated) [4]

Water Solubility: 379 ppm at 25 °C [8]

Vapor Pressure: 0.59 mm Hg at 25 °C [1]

Henry's Law Constant: 2.91×10^{-4} atm-m³/mole at 25 °C (estimated using the experimental water solubility and experimental vapor pressure values)

ENVIRONMENTAL FATE/EXPOSURE POTENTIAL

Summary: Dibutyl ketone can be released to the environment by wastewater effluents from chemical manufacturing, domestic sewage and

162

by leachates from low-level radioactive waste disposal sites. It can also be released to the atmosphere by evaporation from solvents in which it occurs as a contaminant. If released to the atmosphere, dibutyl ketone is degraded relatively rapidly by reaction with photochemically produced hydroxyl radicals (estimated half-life of 28 hr in air). If released to water, volatilization is expected to be important. Volatilization half-lives of 7.1 and 80 hr have been estimated for a model river (one meter deep) and a model environmental pond, respectively. If released to soil, dibutyl ketone may leach based upon an estimated Koc value of 167. Dibutyl ketone can be expected to evaporate from dry surfaces. Insufficient experimental data are available to predict the relative importance of biodegradation in soil or water. Occupational exposure to dibutyl ketone may occur by inhalation and by skin or eye contact.

Natural Sources:

Artificial Sources: Dibutyl ketone has been detected in leachates from low-level radioactive waste disposal sites, in wastewater effluents from domestic sewage, and in wastewater effluents from chemical manufacturing [6,11,12]. Dibutyl ketone may be found as a contaminant of other solvents [10] and can therefore be released to the atmosphere by evaporation.

Terrestrial Fate: An estimated Koc value of 167 suggests that dibutyl ketone will have medium mobility in soil and a potential for leaching [9,13]. Insufficient experimental data are available to predict the relative importance of biodegradation in soil. Based upon the experimental vapor pressure, dibutyl ketone can be expected to evaporate from dry surfaces.

Aquatic Fate: Volatilization is expected to be an important environmental fate process for dibutyl ketone in water. Volatilization half-lives of 7.1 and 80 hr have been estimated for a model river (one meter deep) and a model environmental pond, respectively [9,14]. Insufficient experimental data are available to predict the relative importance of biodegradation. The chemical structure of dibutyl ketone suggests that aquatic hydrolysis is not important [9]. An estimated BCF of 22 indicates that bioconcentration in aquatic organisms may not be a significant fate process.

Dibutyl Ketone

Atmospheric Fate: Based on the experimental vapor pressure, dibutyl ketone can be expected to exist almost entirely in the vapor phase in the ambient atmosphere [5]. Vapor phase dibutyl ketone is degraded relatively rapidly in the ambient atmosphere by reaction with photochemically formed hydroxyl radicals; the half-life for this reaction in typical air can be estimated to be about 28 hr [2].

Biodegradation: Using a Warburg respirometer, a municipal activated sludge inocula, and a 24-hr incubation period, dibutyl ketone was observed to have a theoretical BOD of 7.9 percent at a concn of 500 ppm [7].

Abiotic Degradation: The rate constant for the vapor phase reaction of dibutyl ketone with photochemically produced hydroxyl radicals has been estimated to be 13.7×10^{-12} cm^3/molecule-sec at 25 °C which corresponds to an atmospheric half-life of about 28 hr at an atmospheric concn of $5 \times 10^{+5}$ hydroxyl radicals per cm^3 [3]. Ketones are generally resistant to aqueous environmental hydrolysis [9]; therefore, dibutyl ketone is not expected to hydrolyze in water.

Bioconcentration: Based upon the measured water solubility, the BCF for dibutyl ketone can be estimated to be about 22 from a recommended regression-derived equation [9]. This BCF value suggests that bioconcentration in aquatic organisms is not significant.

Soil Adsorption/Mobility: Based upon the measured water solubility, the Koc for dibutyl ketone can be estimated to be 167 from a regression-derived equation [9]. This Koc value suggests that dibutyl ketone has medium soil mobility [13].

Volatilization from Water/Soil: The estimated Henry's Law constant indicates that volatilization from environmental waters is probably significant, but may not be rapid [9]. Based on this Henry's Law constant, the volatilization half-life from a model river (1 m deep flowing 1 m/sec with a wind velocity of 3 m/sec) can be estimated to be about 7.1 hours [9]. Volatilization half-life from a model environmental pond can be estimated to be about 80 hours [14]. Based upon the vapor pressure, dibutyl ketone may be expected to evaporate from dry surfaces.

Dibutyl Ketone

Water Concentrations:

Effluent Concentrations: Dibutyl ketone has been qualitatively detected in trench leachates collected from commercially operated low-level radioactive waste disposal sites at Maxey Flats, KY and at West Valley, NY [6]. Dibutyl ketone has been identified in wastewater effluents from domestic sewage treatment [12]. Dibutyl ketone was detected in 2 of 63 industrial waters (concn less than 10 ppb) which were sampled from a wide range of chemical manufacturers in areas across the United States [11].

Sediment/Soil Concentrations:

Atmospheric Concentrations:

Food Survey Values:

Plant Concentrations:

Fish/Seafood Concentrations:

Animal Concentrations:

Milk Concentrations:

Other Environmental Concentrations:

Probable Routes of Human Exposure: Any exposure to dibutyl ketone in the workplace is expected to be by inhalation and by skin or eye contact.

Average Daily Intake:

Occupational Exposure:

Body Burdens:

Dibutyl Ketone

REFERENCES

1. Ambrose D et al; J Chem Therm 7: 272-90 (1975)
2. Atkinson R; Chem Rev 85: 69-201 (1985)
3. Atkinson R; Intern J Chem Kinet 19: 799-828 (1987)
4. CLOGP; GEMS-Graphic Exposure Modeling System CLOGP USEPA (1986)
5. Eisenreich SJ et al; Environ Sci Technol 15: 30-8 (1981)
6. Francis AJ et al; Nuclear Tech 50: 158-63 (1980)
7. Gerhold RM, Malaney GW; J Water Pollut Contr Fed 38: 562-79 (1966)
8. Hansch C et al; J Org Chem 33: 347-50 (1968)
9. Lyman WJ, Reehl WF, and Rosenblatt DH; Handbook of Chemical Property Estimation Methods. Environmental Behavior of Organic Compounds. Washington DC: American Chemical Society pp 4-9, 5-4, 5-10, 7-4, 7-5, 15-15 to 15-32 (1990)
10. Patty; Indus Hyg & Tox 3rd ed 2C: 4769 (1981)
11. Perry DL et al; Identification of Organic Compounds in Industrial Effluent Discharges. USEPA-600/4-79-016. p 44 (1979)
12. Shackelford WM, Keith LM; Frequency of Organic Compounds Identified in Water. USEPA-600/4-76-062. p 149 (1976)
13. Swann RL et al; Res Rev 85: 23 (1983)
14. US EPA; EXAMS II Computer Simulation (1987)
15. Weast RC et al. CRC Handbook of Chemistry and Physics. CRC Press, Inc: Boca Raton, FLA p.C-378 (1985)

1,2-Dichloroethylene

SUBSTANCE IDENTIFICATION

Synonyms:

Structure:

CAS Registry Number: 540-59-0

Molecular Formula: $C_2H_2Cl_2$

Wiswesser Line Notation: G1U1G

CHEMICAL AND PHYSICAL PROPERTIES

Boiling Point: 60.63 °C (cis-); 47.67 °C (trans-)

Melting Point: -80 °C (cis-); -49.8 °C (trans-)

Molecular Weight: 96.95

Dissociation Constants:

Log Octanol/Water Partition Coefficient: 1.86 (cis-1,2-dichloroethylene) and 2.09 (trans-1,2-dichloroethylene) [16]

Water Solubility: 3,500 mg/L (cis-) and 6,300 mg/L at 25 °C [29]

Vapor Pressure: 200 mm Hg (cis-) and 340 mm Hg (trans-) at 25 °C [29]

Henry's Law Constant: 0.00337 (cis-) and 0.00672 (trans-) atm-m³/mol [19]

167

1,2-Dichloroethylene

ENVIRONMENTAL FATE/EXPOSURE POTENTIAL

Summary: 1,2-Dichloroethylene may be released to the environment in emissions and wastewater during its production and use. Under anaerobic conditions existing in landfills, aquifers, or sediment, one is likely to find 1,2-dichloroethylenes that are formed as breakdown products from the reductive dehalogenation of common industrial solvents trichloroethylene, tetrachloroethylene, and 1,1,2,2-tetrachloroethane. The cis-1,2-dichloroethylene is apparently the more common isomer found, although it is often mistakenly reported as the trans isomer. The trans isomer, being a priority pollutant is more commonly analyzed for and the analytical procedures generally used do not distinguish between isomers. If 1,2-dichloroethylene is released on soil, it should evaporate and leach in soil very slowly. Very slow anaerobic biodegradation should occur in soil. If released into water, it will be lost mainly through volatilization (half-life 3 hr in a model river). Adsorption to sediment and suspended solids in water and bioconcentration in aquatic organisms should not be important. The results with respect to aerobic biodegradation in water are ambiguous, since some authors found 1,2-dichloroethylene to be recalcitrant while others found it to be biodegradable. Very slow anaerobic biodegradation in bottom sediments is likely. In the atmosphere, cis- and trans-1,2-dichloroethylene will be lost by reaction with photochemically produced hydroxyl radicals (half-lives 8 and 3.6 days, respectively) and scavenged by rain. Because it is relatively long-lived in the atmosphere, considerable dispersal from sources of emission should occur. The general population is exposed to 1,2-dichloroethylene by inhalation of air as well as ingestion of contaminated drinking water derived from ground water sources. Occupational exposure will be via dermal contact with the vapor and liquid or via inhalation.

Natural Sources:

Artificial Sources: 1,2-Dichloroethylene may be released to the environment in emissions and wastewater during its production and use as a solvent and extractant, in organic synthesis, and in the manufacture of perfumes, lacquers, and thermoplastics [18]. An assessment of the sources of the two isomers of 1,2-dichloroethylene is complicated by the fact that the trans isomer is a priority pollutant while the cis isomer is not and the standard USEPA methods of analysis do not allow the isomers to be differentiated [7]. This has resulted in monitoring reports erroneously listing the trans isomer when the cis isomer may be present [7]. The Michigan Department of Health occasionally find traces of the trans isomer but usually finds the cis isomer [7]. In an anaerobic, high-organic matrix, such as a landfill site, one is likely to find 1,2-dichloroethylenes as breakdown products due to reductive dehalogenation [7]. Degradation products are found in increasing proportions further from a source and where there are high concns of other degradable organic compounds [7]. Under simulated landfill conditions, it has been found that 1,2-dichloroethylene is formed from trichloroethene, tetrachloroethene, and 1,1,2,2-tetrachloroethane and therefore, common industrial solvents may be sources of dichloroethylenes in such environments [7,15]. Additionally, in muck and sediment microcosms tetrachloroethylene is converted into dichloroethylene although the relative amount of the trans isomer produced is much less than the cis [3,28]. Similarly, a source of 1,2-dichloroethylene in ground water is as a metabolite of the anaerobic degradation of more highly chlorinated ethenes such as trichloroethylene and tetrachloroethylene [7].

Terrestrial Fate: If 1,2-dichloroethylene is released to soil, it should evaporate readily and/or could leach into soil. 1,2-Dichloroethylene should undergo very slow anaerobic biodegradation in soil producing vinyl chloride.

Aquatic Fate: If released into water, 1,2-dichloroethylene will be lost mainly through volatilization (half-life 3 hr in a model river). At a site on a feeder canal downstream from a point where effluent was put into the canal, the concentration of 1,2-dichloroethylene ranged from <4 to 20.2 ppb [37]. However, at the point and beyond where the canal joined a river about 0.8 km away, the concn of 1,2-dichloroethylene was always below the detection limit of <4 ppb [37]. Since biodegradation and adsorption to sediment should not be significant, the loss is likely to be due to volatilization.

Atmospheric Fate: In the atmosphere 1,2-dichloroethylene will be lost by reaction with photochemically produced hydroxyl radicals with a half-life 8 and 3.6 days for the cis and trans isomers, respectively [12]. There is evidence that it will be scavenged by rain. The scavenging is expected since 1,2-dichloroethylene is significantly soluble in water.

1,2-Dichloroethylene

Biodegradation: Trans-1,2-dichloroethylene was recalcitrant in shake flask tests modified to accommodate volatile chemicals [23,24]. The concns examined in this studies ranged from 0.80 to 25 ppm. A 21-day acclimation period and the addition of a lactose cometabolite did not alter the biodegradability. Similarly, no biodegradation occurred in a river die-away test [24]. Contradictory results were obtained in a biodegradability screen test using a wastewater inoculum and 5 ppm of trans- 1,2-dichloroethylene [34]. A 67% of 1,2-dichloroethylene was lost in 7 days due to biodegradation, whereas 33% loss due to volatilization occurred in 10 days [34]. For the cis isomer, a 54% loss in 7 days was observed due to biodegradation, whereas a 34% loss due to volatilization occurred in 10 days [34]. Literature references to microbial degradation of low molecular weight chlorinated aliphatics generally find that they are not metabolized [6]. When trans-1,2-dichloroethylene was incubated with methanogenic aquifer material obtained adjacent to a landfill site in a serum bottle at 17 °C, at least 16 wk of incubation were required before disappearance began relative to autoclaved controls and the concn of the cis isomer was reduced to <2% of controls during the same period [38]. Vinyl chloride was identified as a degradation product after 1-2 weeks [38]. After 40 wk, the average concn of the trans isomer was reduced to 18% of controls, only traces of the cis isomer remained, and no vinyl chloride or other degradation product was found [38]. Another investigator found that when cis- or trans-1,2-dichloroethylene were incubated anaerobically using an inoculum from a municipal waste digester in order to simulate conditions in a landfill, vinyl chloride appeared within 6 weeks [15]. Biodegradation of 1,2-dichloroethylene was studied in microcosms prepared from uncontaminated organic sediment from the Everglades and allowed to sit to ensure oxygen depletion. Under these anoxic conditions, 50 and 73% of the cis and trans isomer were lost in 6 months with accompanying formation of vinyl chloride [2]. Ethyl chloride was formed in the case of the cis but not the trans isomer, indicating that there are different pathways other than the reductive dechlorination for the cis isomer [2].

Abiotic Degradation: In the atmosphere, cis- and trans-1,2-dichloroethylene react with photochemically produced hydroxyl radicals resulting in half-lives of 8 and 3.6 days, respectively [12]. The only product positively identified in this reaction was formyl chloride [12]. Chlorine substitution on alkenes markedly reduces their reactivity towards ozone; the half-lives resulting from ozone attack of the double bond are 129 and 44 days for the cis and trans isomer, respectively [35].

Bioconcentration: The recommended octanol/water partition coefficients for cis- and trans-1,2-dichloroethylene are 1.86 and 2.06, respectively [16], from which one can estimate a bioconcentration factor between 15 and 22 using a recommended regression equation [22]. Therefore 1,2-dichloroethylene should not bioconcentrate significantly in aquatic organisms.

Soil Adsorption/Mobility: The solubilities of cis- and trans-1,2-dichloroethylene in water are 3,500 and 6,300 mg/L, respectively [29], from which one can estimate a Koc between 36 and 49 using a recommended regression equation [22]. Therefore 1,2-dichloroethylene should have very low adsorption to soil or sediment and should show very high mobility in soil [33].

Volatilization from Water/Soil: The Henry's Law constants for cis- and trans-1,2-dichloroethylene are 0.00337 and 0.00672 atm-m^3/mol, respectively [19]. Based on these values and a recommended equation [22], the estimated half-lives for volatilization of cis and trans-1,2-dichloroethylene from a model river 1 m deep with a 1 m/sec current and a 3 m/sec wind are 3.1 and 3.0 hr, respectively. Diffusion through water controls volatilization [22]. The mean volatilization half-lives of cis- and trans-1,2-dichloroethylene from a slowly stirred beaker 6.5 cm deep were 19.4 and 24.0 minutes, respectively [9]. Based on a recommended equation [9], the experimental half-lives in the beaker correspond to half-lives of 5.0 and 6.2 hr in a body of water 1 m deep.

Water Concentrations: DRINKING WATER: In a 1982 USA ground water survey, 16 samples from a random of 466 contained 1,2-dichloroethylenes at concns above a quantitation limit of 0.2 ppb [8]. The median and maximum concn were 1.1 and 2 ppb, respectively [8]. A survey of 30 Canadian potable water treatment facilities showed that between 1 and 6 of the supplies contained 1,2-dichloroethylene in their finished drinking water [27]. The highest level measured was 32 ppb [27]. Of the 14 UK water supplies surveyed, two contained detectable quantities of dichloroethylene [10]. The source of both of these supplies was river water. Contaminated drinking water wells in Massachusetts and

171

New York contained 323, 294 and 91 ppb of 1,2-dichloroethylene whereas the highest reported concn in drinking water derived from surface sources was 9.8 ppb [5]. A Vero Beach, FL well water located 268 m from a leaking underground storage tank containing trichloroethylene contained 12 ppm of 1,2-dichloroethylene [37]. GROUND WATER: Ground water 1 mile from the site of an abandoned industrial landfill contained 138 ppb of 1,2-dichloroethylene [21]. At a Connecticut solvent recovery site, the concentration of 1,2-dichloroethylene in ground water was 30 and 2.7 ppm on-site at the water table and a deep well, respectively, while the corresponding levels decreased to not detected and 4.3 ppm 80 m downgradient [7]. Similar results were obtained at a solvent recovery site in Wisconsin. Here the analogous concn were 30 and 8.7 ppm on-site and not detected and 47 ppm 80 m downgradient [7]. An explanation offered for these results is that dichloroethylene is a degradation product and that degradation is faster closer to the surface where cometabolites are more abundant [7]. The Biscayne aquifer in the vicinity of the Miami Drum site contained 0-26 ppb of dichloroethylene [25]. 4.6% of the 315 wells waters sampled from the outcrop area of the Potomac-Raritan-Magothy aquifer system adjacent to the Delaware River contained trans-1,2-dichloroethylene [11]. The chemical was absent from wells downdip of the outcrop area [11]. A site study of a western Connecticut manufacturing plant that used large quantities of high quality trichloroethylene for degreasing found that 7 of 9 monitoring wells around the plant contained 1.2 - 320.9 ppb of trans-1,2-dichloroethylene [32]. SURFACE WATER: 6% of the 1144 stations in the USEPA STORET data base have detectable quantities of 1,2-dichloroethylene in the water [31]. The median concn was <5 ppb [31]. 1,2- Dichloroethylene was absent from samples taken from the Niagara River and open waters of Lake Ontario [14]. RAIN/SNOW: Rain water from a rain event in west Los Angeles contained 230 ng/L of 1,2-dichloroethylene [20].

Effluent Concentrations: 7.0% of the 1369 effluent samples in the USEPA STORET data base had detectable quantities of 1,2-dichloroethylene [31]. The median concn was <2.5 ppb [7]. Final effluent from the Los Angeles County wastewater treatment plant contained 5.2 ppb of 1,2- dichloroethylene [13]. 1,2-Dichloroethylene was detected emanating from a simulated landfill made from municipal waste and sewage sludge 8 months after landfilling was done [36]. 4 of 5 leachate samples from a landfill site that accepted both municipal and industrial wastes contained 1,2-dichloroethylene at concns ranging from 150 to 3200 ppb [7]. Based on site records of waste accepted, this is thought to be a breakdown product of more highly chlorinated compounds [7]. The average concn of 1,2-dichloroethylene in gas emerging from the surface of 9 landfills used for methane recovery was 200 ppb [39]. In a comprehensive survey of wastewater from 4000 industrial and publicly owned treatment works (POTWs) sponsored by the Effluent Guidelines Division of the USEPA, cis- or trans-1,2-dichloroethylene were identified in discharges of the following industrial category (frequency of occurrence; median concn in ppb; isomer): steam electric (1; 1.6; cis), leather tanning (1; 3.3; cis), iron and steel mfg (2; 1400.8; cis) (2; 2265.9; trans), nonferrous metals (1; 314.6; cis), organics and plastics (2; 121.5; cis) (3; 14.6; trans), inorganic chemicals (2; 3.9; trans), textile mills (1; 8.3; cis), plastics and synthetics (3; 20.1; cis), rubber processing (1; 712.0; cis) (2; 19.0; trans), auto and other laundries (1; 60.6; trans), explosives (1; 1.5; cis) (1; 3.9; trans), electronics (7; 140.7; trans), mechanical products (2; 13.7; trans), transportation equipment (1; 29.3; trans), publicly owned treatment works (63; 16.3; trans) [30].

Sediment/Soil Concentrations: 4% of the 361 samples in the USEPA STORET data base had detectable quantities of 1,2-dichloroethylene in sediment [31]. The median concn was <5 ppb [31]. Surficial sediment taken 6 km northwest of the discharge from the Los Angeles County wastewater treatment plant at a depth of 60 m contained <0.5 ppb (dry weight) of 1,2-dichloroethylene [13]. 1,2-Dichloroethylene has been detected, but not quantitated in sediment/soil/water samples at the Love Canal [17].

Atmospheric Concentrations: RURAL/REMOTE: Two samples from rural/remote areas in the USA contained no cis-1,2-dichloroethylene [4]. URBAN/SUBURBAN: US (669 site/samples) 68 ppt median, 3500 ppt maximum of the cis isomer [4]. SOURCE AREAS: USA (101 site/samples) 300 ppt median, 6700 ppt maximum of the cis isomer [4]. Traces of 1,2-dichloroethylene were detected in air samples outside of 3 of 9 houses tested at the Love Canal [1].

Food Survey Values:

1,2-Dichloroethylene

Plant Concentrations:

Fish/Seafood Concentrations: Samples of fish livers, shrimp muscle and invertebrates collected 6 km northwest of the discharge from the Los Angeles County wastewater treatment plant contained <0.3 ppb (wet weight) of 1,2-dichloroethylene [13]. None of the 95 stations in the USEPA STORET data base had detectable quantities (detection limit <5 ppb) of 1,2-dichloroethylene in fish [31].

Animal Concentrations:

Milk Concentrations:

Other Environmental Concentrations:

Probable Routes of Human Exposure: Occupational exposure to 1,2-dichloroethylene will be via inhalation and dermal contact with the vapor as well as by dermal contact with the liquid during its use as a solvent. The general population is exposed to 1,2-dichloroethylene in urban air as well as from contaminated drinking water sources.

Average Daily Intake: AIR INTAKE (assume air concn of 68 ppt and an inhalation rate of 20 m³/day): 5.4 ug/day. WATER INTAKE (assume water concn from contaminated sources of 1.1 ppb and a water intake of 20 L/day): 2.2 ug/day.

Occupational Exposure: Based on statistical estimates derived from the NIOSH survey conducted 1981-83 in the USA, 215 workers are potentially exposed to 1,2-dichloroethylene [26].

Body Burdens:

REFERENCES

1. Amoore JE, Hautala E; Appl Toxicol 3: 272-90 (1983)
2. Barrio-Lage G et al; Environ Sci Technol 20: 96-9 (1986)
3. Barrio-Lage GA et al; Environ Toxicol Chem 6: 571-8 (1987)
4. Brodzinsky R, Singh HB; Volatile Organic Chemicals In The Atmosphere Menlo Park,CA: SRI International Contract 68-02-3452 198 pp (1982)
5. Burmaster DE; Environ 24: 6-13, 33-36 (1982)

1,2-Dichloroethylene

6. Callahan MA et al; pp.51-1 to 51-10 in Water-Related Environmental Fate of 129 Priority Pollutants USEPA-440/4-79-029b Washington DC: Office of Water Planning and Standards (1979)
7. Cline PV, Viste DR; Waste Manage Res 3: 351-60 (1985)
8. Cotruvo JA; Sci Total Environ 47: 7-26 (1985)
9. Dilling WL; Environ Sci Technol 11: 405-9 (1977)
10. Fielding M et al; Organic Micropollutants In Drinking Water TR-159 Medmenham, Eng Water Res Cent 49 pp (1981)
11. Fusillo TV et al; Ground Water 23: 354-60 (1985)
12. Goodman MA et al; ACS Div Environ Chem 192nd Natl Mtg 26: 169-71 (1986)
13. Gossett RW et al; Mar Pollut Bull 14: 387-92 (1983)
14. Great Lakes Water Quality Board: p 195 in An Inventory Of Chemical Substances Identified In The Great Lakes Ecosystem Volume 1 - Summary Report To The Great Lakes Water Quality Board Windson Ontario, Canada (1983)
15. Hallen RT et al; ACS Div Environ Chem 192nd Natl Mtg 26: 344-6 (1986)
16. Hansch C, Leo AJ; Medchem Project Issue 26 Pomona, CA: Claremont College (1985)
17. Hauser TR et al; EPA's Monitoring Program at Love Canal 1980 Env Monit Assess 2: 249-72 (1982)
18. Hawley GG; p 335 in Condensed Chem Dictionary 10th ed Van Nostrand Reinhold NY (1981)
19. Hine J, Mookerjee PK; J Org Chem 40: 292-8 (1975)
20. Kawamura K, Kaplan IR; Environ Sci Tech 17: 497-501 (1983)
21. Lao RC et al; pp 107-18 in Analytical Techniques In Environmental Chemistry II Albaiges J ed New York NY: Pergamon Press Ltd (1982)
22. Lyman WJ et al; Handbook of Chem Property Estimation Methods. pp.4-1 to 4-33, 5-1 to 5-33, 15-1 to 15-34 McGraw-Hill NY (1982)
23. Mudder TI; Diss Abstr Int B 42: 1804 (1981)
24. Mudder TI, Musterman JL; Presented Before The Div Environ Chem Amer Chem Soc Kansas City, Mo Sept 1982 pp.52-3 (1982)
25. Myers VB; pp 354-7 in Florida Natl Conf Manage Uncontrolled Hazard Waste Sites (1983)
26. NIOSH: National Occupational Health Survey (1975)
27. Otson R et al; J Assoc Off Analyt Chem 65: 1370-4 (1982)
28. Parsons F et al; J Am Water Works Assoc 76: 56-9 (1984)
29. Riddick JA et al; Organic Solvents: Physical Properties and Methods of Purification Techniques of Chemistry 4th Ed New York, NY: Wiley-Interscience 2: 1325 (1986)
30. Shackelford WM et al; Analyt Chim Acta 146: 15-27 (1983)
31. Staples CA et al; Environ Toxicol Chem 4: 131-42 (1985)
32. Stuart JD; p 37 in Organics Transported Thru Selected Geological Media (NTIS PB83-224246 Comm Univ Storrs Inst of Water Resources (1983)
33. Swann RL et al; Res Rev 85: 17-28 (1983)
34. Tabak HH et al; J Water Pollut Contr Fed 53: 1503-18 (1981)
35. Tuazon EC et al; Arch Environ Contam Toxicol 13: 691-700 (1984)

1,2-Dichloroethylene

36. Vogt WG, Walsh JJ; Volatile Organic Compounds In Gases From Landfill Simulators Proc-APAC Annu Meet 78th (Vol 6) pp.17 (1985)
37. Wang T et al; Bull Environ Contam Toxicol 34: 578-86 (1985)
38. Wilson BH et al; Environ Sci Technol 20: 997-1002 (1986)
39. Zimmerman RE et al; Landfill Methane Trace Volatile Organic Constituent pp.230-9 in Proc Int Gas Res Conf (1983)

Dichlorofluoromethane

SUBSTANCE IDENTIFICATION

Synonyms: Freon 21

Structure:

$$H-\underset{\displaystyle Cl}{\overset{\displaystyle Cl}{C}}-F$$

CAS Registry Number: 75-43-4

Molecular Formula: CHCl$_2$F

Wiswesser Line Notation: GYGF

CHEMICAL AND PHYSICAL PROPERTIES

Boiling Point: 8.9 °C at 760 mm Hg

Melting Point: -135 °C

Molecular Weight: 102.92

Dissociation Constants:

Log Octanol/Water Partition Coefficient: 1.55 [12]

Water Solubility: 9,500 mg/L at 25 °C [20]

Vapor Pressure: 1216 mm Hg at 20 °C [22]

Henry's Law Constant: 0.025 atm-m^3/mol (estimated by the bond method) [14]

ENVIRONMENTAL FATE/EXPOSURE POTENTIAL

Summary: Dichlorofluoromethane may be released to the environment during its production, storage, transport, and use as a refrigerant, solvent, aerosol propellent, and fire extinguishing agent. When used for these purposes, dichlorofluoromethane will eventually be lost to the atmosphere unless it is captured and recycled. It is an extremely unreactive gas and losses due to photolysis, photooxidation, hydrolysis, or biodegradation in air, water, and soil will not be significant. If released on land, most of the dichlorofluoromethane will volatilize into the air. It is highly mobile in soil and therefore will have a potential for leaching into ground water. If released in water, it will be removed by volatilization. Its half-life in a model river is estimated to be 3.0 hr. In the atmosphere, dichlorofluoromethane is mainly removed by reaction with hydroxyl radicals. Its estimated half-life in the troposphere is about 2 yrs. As a result of long half-life, dichlorofluoromethane will accumulate and disperse all over the troposphere and diffuse into the stratosphere. The concentration of dichlorofluoromethane in the atmosphere will continue to increase as more is produced and released. Five percent of the dichlorofluoromethane that is released is expected to diffuse into the stratosphere where it will contribute in the destruction of the ozone layer. Dichlorofluoromethane will also be removed from the atmosphere by wet deposition. However, the dichlorofluoromethane removed in this manner will volatilize back into the atmosphere. The general population is exposed to very low levels of dichlorofluoromethane present in ambient air. Occupational exposure may occur via inhalation and dermal contact with the vapor or liquified gas.

Natural Sources: Emissions of dichlorofluoromethane have been reported from volcanos [11].

Artificial Sources: Dichlorofluoromethane may be released to the environment during its production, transport, storage, and use as a refrigerant, solvent, propellant, and fire extinguishing agent [6,13]. When used as a refrigerant, solvent, or aerosol propellant, dichlorofluoromethane will be eventually lost to the atmosphere unless it is captured and recycled.

Terrestrial Fate: Since dichlorofluoromethane is an inert gas with high vapor pressure at room temperature and has a low adsorption to soil (estimated Koc of 35), most of the chemical released on land will be lost by volatilization. Its low Koc also indicates that it is highly mobile in soil and therefore will have a high potential for leaching into ground water. Dichlorofluoromethane lacks suitable chromophoric group for absorption of light >290 nm. It also has a low rate of hydrolysis. Therefore, it is unlikely that photooxidation and hydrolysis will be significant in soil. Based on biodegradation of similar compounds, biodegradation of dichlorofluoromethane in soil would not be important.

Aquatic Fate: Dichlorofluoromethane has a very low rate of hydrolysis in water [10]. Its lack of suitable chromophoric group would preclude direct photolysis in water. Based on Henry's Law constant [14], and the recommended equation [18], its half-life in a model river 1 m deep flowing at a current speed of 1 m/sec and a wind speed of 3 m/sec is estimated to be 3.0 hr. Therefore, it will be removed from water predominantly by volatilization. Based on its estimated low Koc value [18], dichlorofluoromethane would not adsorb appreciably to sediment or suspended solids in water. The estimated low BCF value [18] suggests bioconcentration of the compound in aquatic organisms will not be important.

Atmospheric Fate: In the atmosphere, dichlorofluoromethane is mainly removed by reaction with hydroxyl radicals. Using a model that considers both vertical and latitudinal transport as well as reaction with atmospheric species, the half-life of dichlorofluoromethane in the troposphere was estimated to be 2.0 yr [9]. As a result of its long half-life, dichlorofluoromethane released to the atmosphere will accumulate and disperse all over the troposphere. Its background concentration in the northern hemisphere is predicted to be 2.4 times that in the southern hemisphere [9]. Five percent of dichlorofluoromethane released into the atmosphere is expected to diffuse slowly into the stratosphere where it will photolyze with low wavelength UV light producing chlorine atoms that would contribute to the destruction of the ozone layer. Due to its significant water solubility, dichlorofluoromethane will also be removed from the atmosphere by wet deposition. However, dichlorofluoromethane removed in this manner will volatilize back into the atmosphere.

Biodegradation: Since a structurally similar compound, dichlorodifluoromethane is not amenable to biodegradation [4], the biodegradation of dichlorofluoromethane is not likely to be important.

Dichlorofluoromethane

Abiotic Degradation: Dichlorofluoromethane is removed from the atmosphere by reaction with photochemically produced hydroxyl radicals by H-atom abstraction. The experimentally determined rate constant for this reaction is reported to vary between 2.6 to 3.54 x 10^{-14} cm^3/molecule-sec in the temperature range 20 to 25 °C [2]. Assuming this rate constant to be 3 x 10^{-14} cc/molecule-sec and the average tropospheric hydroxyl radical concentration of 5 x 10^{+5} [2], the half-life of dichlorofluoromethane in the atmosphere would be 1.46 yr. Utilizing a two-dimensional, tropospheric model that considers both vertical and latitudinal transport as well as reaction with atmospheric species, the half-life of dichlorofluoromethane in the troposphere is estimated to be 2.0 yr [9]. As a result of its long half-life, the dichlorofluoromethane released to the atmosphere will accumulate in the atmosphere and disperse all over the globe. About 5% of the dichlorofluoromethane is predicted to diffuse to the stratosphere [9] where it will be destroyed as a result of photodissociation and reaction with hydroxyl radicals and O(1D) atoms [1,8,9,15]. The photodissociation would produce free chlorine atoms which catalyze the destruction of stratospheric ozone layer. Due to its significant water solubility, dichlorofluoromethane will also be removed from the atmosphere by wet deposition. However, the dichlorofluoromethane removed in this manner will volatilize back into the atmosphere. Fluorocarbons are inert under environmental conditions [10]. The rate of hydrolysis of dichlorofluoromethane is very low, (<0.01 g/L of water-yr) at 30 °C [10]. This rate increases markedly, to 5.2 g/L of water-yr, in the presence of steel [10].

Bioconcentration: Based on its log octanol/water partition coefficient of 1.55 [12], a BCF of 8.9 for dichlorofluoromethane was estimated using a recommended regression equation [18]. This indicates that dichlorofluoromethane should not bioconcentrate in aquatic organisms.

Soil Adsorption/Mobility: Based on its water solubility, 9,500 mg/L at 25 °C [20], a Koc of 35 has been estimated for dichlorofluoromethane using a recommended regression equation [18]. Compounds with such low Koc value are expected to adsorb weakly to sediment and suspended solids in water and would be highly mobile in soil [21].

Volatilization from Water/Soil: Dichlorofluoromethane is a gas with a high vapor pressure, 1216 mm Hg at 20 °C [22], and would therefore be

Dichlorofluoromethane

expected to volatilize rapidly from water. The Henry's Law constant for dichlorofluoromethane estimated by the bond estimation method [14] is 0.025 atm-m^3/mole. Based on this Henry's Law constant and the recommended equation [18], the volatilization half-life of dichlorofluoromethane from a model river 1 m deep, flowing at 1 m/sec with a wind speed of 3 m/sec is estimated to be 3.0 hr. Hence, its volatilization from water will be very rapid; the volatilization rate will be controlled by diffusion of the compound through water [18].

Water Concentrations: DRINKING WATER: No dichlorofluoromethane was found in a survey of 1174 community wells and 617 private wells in Wisconsin in which dichlorofluoromethane was one of the target compounds [16].

Effluent Concentrations: Dichlorofluoromethane was found in gases emanating from a simulated landfill [23]. No quantitation was performed. The material in the simulated landfill consisted of municipal refuse and various loadings of municipal wastewater sludge. Dichlorofluoromethane was reported in landfill gas at a concn of 93 mg/m^3 [24]. The site was a clay pit that received both municipal, industrial, and liquid wastes that were at the sampling point >15 months.

Sediment/Soil Concentrations:

Atmospheric Concentrations: Based on measurements of concns in ambient air over the Red Sea and Indian Ocean, off the coast of Kenya, and at two locations in the Egyptian desert, the background concentration of dichlorofluoromethane ranges between 0 and 1 ppt with excursions up to 5-7 ppt in some instances [7]. In measurements of concns over deserts where samples were collected three times a day, no significant diurnal fluctuation was noted. The average concn. of dichlorofluoromethane measured in pristine atmospheres in Tasmania, the South Pole and the Pacific Northwest was 0.08 ppt [19]. Levels at a site in southern England averaged 1.6 ppt [19].

Food Survey Values:

Plant Concentrations:

Dichlorofluoromethane

Fish/Seafood Concentrations:

Animal Concentrations:

Milk Concentrations:

Other Environmental Concentrations:

Probable Routes of Human Exposure: The general population will be exposed to low levels of dichlorofluoromethane in ambient air. Occupation exposure may occur via inhalation and dermal contact with the liquified gas.

Average Daily Intake: AIR INTAKE (assume air concn of 1 ppt and an inhalation rate of 20 m^3/day): 0.086 ug/day; WATER INTAKE: insufficient data; FOOD INTAKE: insufficient data.

Occupational Exposure: If used as a refrigerant, people who service air conditioners (both automobile or building) and refrigerators are the most likely occupational group to be exposed to dichlorofluoromethane. If used as a fire extinguisher, firefighters may be another group who would also be susceptible to occasional exposure to this compound. Occupational exposure to dichlorofluoromethane among fire fighters has been studied [3].

Body Burdens: Dichlorofluoromethane was found in the expired air of 2 of 8 male volunteers; the amounts exhaled in these two subjects were 0.14 and 0.062 fg/hr [5]. Dichlorofluoromethane was again detected, but not quantified, in the expired air of a nonsmoking population of 62 males and female, suburban and urban residents [17]. The percentage of subjects whose breath contained dichlorofluoromethane was not reported.

REFERENCES

1. Altschuler AP; Adv Environ Sci Technol 10: 181-219 (1979)
2. Atkinson R; Chem Rev 85: 69-201 (1985)
3. Brandt-Rauf PW et al; Br J Ind Med 45: 606-12 (1988)

Dichlorofluoromethane

4. Callahan MA et al; Water-Related Environmental Fate of 129 Priority Pollutants Vol 2, USEPA Report No. 440/4-79-029b, Office of Water Planning and Standards, Washington, DC: U.S. Environmental Protection Agency p 62-1 to 62-6 (1979)
5. Conkle JP et al; Arch Environ Health 30: 290-5 (1975)
6. Council on Environmental Quality; Fluorocarbons and the environment. June 1975 (1975)
7. Crescentine G et al; Atmos Environ 20: 215-7 (1986)
8. Davidson JA et al; J Chem Phys 69: 4277-9 (1978)
9. Derwent RG, Eggleton AEJ; Atmos Environ 12: 1261-9 (1978)
10. Du Pont de Nemours Co; Freon Product Information B-2. Wilmington, DE: E.I. Du Pont de Nemours and Company (1980)
11. Graedel TE; Chemical Compounds in the Atmosphere NY: Academic Press (1978)
12. Hansch C, Leo AJ; MEDCHEM Project Claremont CA: Pomona College (1985)
13. Hawley GG; Condensed Chem Dictionary 10th ed NY: Van Nostrand Reinhold (1981)
14. Hine J, Mookerjee PK; J Org Chem 40: 292-98 (1975)
15. Hubrich C, Stahl F; J Photochem 12: 93-107 (1980)
16. Krill RM, Sonzogni WC; J Amer Water Works Assoc 78: 70-5 (1986)
17. Krotoszynski BK, O'Neill HJ; J Environ Sci Health A17: 855-83 (1982)
18. Lyman WJ et al; Handbook of Chem Property Estimation Methods NY: McGraw-Hill pp 4-1 to 4-33, 5-1 to 5-30, 15-1 to 15-34 (1982)
19. Penkett SA et al; Nature 286: 793-5 (1980)
20. Riddick JA et al; Organic Solvents New York: Wiley Interscience (1986)
21. Swann RL et al; Res Rev 85: 17-28 (1983)
22. Verschueren K; Handbook of Environmental Data on Organic Chemicals NY: Van Nostrand Reinhold Co. p 234 (1977)
23. Vogt WE, Walsh JJ; pp 1-17 in Proc. Air Pollut Control Assoc 78th Annual Meeting Vol 6 (1985)
24. Young P, Parker A; ASTM Spec Tech Publ 851 (Hazard Ind Waste Manage Test): 24-41 (1984)

1,1-Dichloropropane

SUBSTANCE IDENTIFICATION

Synonyms:

Structure:

$$CH_3 \overset{\displaystyle Cl}{\underset{\displaystyle Cl}{\diagup}}$$

CAS Registry Number: 78-99-9

Molecular Formula: $C_3H_6Cl_2$

Wiswesser Line Notation: GYG2

CHEMICAL AND PHYSICAL PROPERTIES

Boiling Point: 88.1 °C at 760 mm Hg

Melting Point:

Molecular Weight: 112.99

Dissociation Constants:

Log Octanol/Water Partition Coefficient: 2.307 (estimated) [10]

Water Solubility: 2156 mg/L (estimated from Kow) [6]. This value is compatible with the water solubility of 1,2-dichloropropane given as 2700 mg/L at 20 °C [5].

Vapor Pressure: 65.9 mm Hg (estimated from its boiling point) [6]. This estimated value is compatible with the vapor pressure of 49.67 mm Hg for 1,2-dichloropropane [8].

Henry's Law Constant: 0.0234 atm-m³/mole (estimated by the Bond method) [3].

1,1-Dichloropropane

ENVIRONMENTAL FATE/EXPOSURE POTENTIAL

Summary: There is no evidence of commercial production or sales of 1,1-dichloropropane in the U.S. either as the isolated chemical or in commercial mixtures. It is probably only used in small amounts possibly as a laboratory reagent. If released into soil or water during its production and use, it would be lost primarily by volatilization (half-life 3.1 hr from a model river). Bioconcentration in fish will not be significant. It is poorly adsorbed by soil and may leach into ground water. In the air, it will disperse and degrade primarily by reaction with photochemically produced hydroxyl radicals (estimated half-life 16 days). It will also be washed out by rain. Due to its limited production and use, exposure to the general public will be low.
Natural Sources:

Artificial Sources: No information could be found concerning the production and use of 1,1-dichloropropane or its presence as an impurity in a major product. It appears to be produced and used only in small quantities for research purposes and its input into the environment would be accordingly very low.

Terrestrial Fate: If released on soil, 1,1-dichloropropane will rapidly volatilize. It may also leach into the ground since it should not be strongly adsorbed by soil. There is no evidence as to its degradation in soil; however 1,2-dichloropropane shows little evidence of degradation in soil and the same may be true for 1,1-dichloropropane.

Aquatic Fate: If released into water, 1,1-dichloropropane will be lost rapidly primarily by volatilization with an estimated half-life 3.1 hr from a model river 1 m deep at a current speed of 1 m/sec and a wind speed of 3 m/sec. Adsorption to sediment and suspended solids in water should be low. Bioconcentration of 1,1-dichloropropane in aquatic organisms should not be important.

Atmospheric Fate: If released into the atmosphere 1,1-dichloropropane will primarily degrade by reaction with photochemically produced hydroxyl radicals with an estimated half-life 16 of days. As it is moderately long-lived, there is likely to be considerable dispersion.

1,1-Dichloropropane

1,1-Dichloropropane is moderately soluble water and therefore would be scavenged by rain.

Biodegradation: Based on the biodegradation potential of the structurally similar compound, namely 1,2-dichloropropane [2], the biodegradation rate of 1,1-dichloropropane in water and soil will be very slow compared to its rate of volatilization.

Abiotic Degradation: 1,1-Dichloropropane will react with photochemically produced hydroxyl radicals by H-atom abstraction. Based on an estimation method [1], the half-life for this reaction is about 16 days [1]. No data could be found concerning the hydrolysis of 1,1-dichloropropane. Alkyl halides are susceptible to neutral and basic hydrolysis, but additional halides on a carbon atom markedly decrease the hydrolysis rate [7]. The hydrolysis half-life of n-propyl chloride is 69 days at pH 7 and 25 °C [7]. Therefore, the hydrolysis half-life of 1,1-dichloropropane would be much longer and should not be very important in the environment.

Bioconcentration: Using the estimated value of the log octanol/water partition coefficient, 2.307 [10] and a recommended regression equation [6], one can estimate a bioconcentration factor of 33 for 1,1-dichloropropane. Therefore, bioconcentration of 1,1-dichloropropane in aquatic organisms should not be important.

Soil Adsorption/Mobility: Using the estimated value of the log octanol/water partition coefficient, 2.307 [10] and a recommended regression equation, one would estimate a K_{oc} of 125 for 1,1-dichloropropane. Adsorption to soil would therefore not be expected to be an important process and the compound should be highly mobile in soil [9].

Volatilization from Water/Soil: Based on an estimation method [6], the vapor pressure of 1,1-dichloropropane with a boiling point of 88.1 °C has been estimated to be 65.9 mm Hg. Since it also does not adsorb strongly and have moderately high vapor pressure, it would be expected to volatilize rapidly from soil. The estimated Henry's Law constant of 1,1 dichloropropane by the bond method [3] is 2.34×10^{-2} atm-m^3/mol. Based on this value for Henry's Law constant and a recommended equation [6], the estimated volatilization half-life of 1,1-dichloropropane from a river 1 m deep with a 1 m/sec current and 3 m/sec wind speed is 3.1 hr. The rate of volatilization should be controlled by diffusion through water [6].

Water Concentrations: SURFACE WATER: USGS survey of the Delaware River Estuary and selected New Jersey tributaries, May 1980 to Jan 1981: 1,1-dichloropropane was only detected in Sept 1980 samples in the vicinity of Philadelphia at 3.0-5.0 ppb [4].

Effluent Concentrations:

Sediment/Soil Concentrations:

Atmospheric Concentrations:

Food Survey Values:

Plant Concentrations:

Fish/Seafood Concentrations:

Animal Concentrations:

Milk Concentrations:

Other Environmental Concentrations:

Probable Routes of Human Exposure: Because of the significant vapor pressure, the most likely route of exposure to 1,1-dichloropropane should be inhalation of vapor.

Average Daily Intake:

Occupational Exposure:

Body Burdens:

1,1-Dichloropropane

REFERENCES

1. Atkinson R; Environ Toxicol Chem 7: 435-42 (1988)
2. Callahan MA et al; Water-Related Environmental Fate of 129 Priority Pollutants Vol 2, USEPA Report No. 440/4-79-029b, Office of Water Planning and Standards, Washington, DC: U.S. Environmental Protection Agency pp 54-1 to 54-5(1979)
3. Hine J, Mookerjee PK; J Org Chem 40: 292-8 (1975)
4. Hochreiter JJ Jr; Chem Quality Reconnaissance of the Water and Surficial Bed Material in the Delaware Estuary and Adjacent New Jersey Tributaries, 1980-81 USGS/WRI/NTIS 82-36 (1982)
5. Horvath AL; Halogenated Hydrocarbons - Solubility-Miscibility with Water NY: Mercel Dekker, Inc p 740 (1982)
6. Lyman WJ et al; Handbook of Chemical Property Estimation Methods. Environmental Behavior of Organic Compounds. McGraw-Hill NY pp 2-1 to 2-52, 4-1 to 4-33, 14-1 to 14-20, 15-1 to 15-34 (1982)
7. Mabey W, Mill T; J Phys Chem Ref Data 7: 383-415 (1978)
8. Riddick JA et al; Techniques of Chemistry Vol II, Organic Solvents - Physical Properties and Methods of Purification, 4th ed. NY: John Wiley & Sons p 499 (1986)
9. Swann RL et al; Res Rev 85: 17-28 (1983)
10. USEPA; Graphical Exposure Modeling System. CLOGP, Office of Toxic Substances, Washington, DC: USEPA (1985)

1,1-Dichloro-1,2,2,2-tetrafluoroethane

SUBSTANCE IDENTIFICATION

Synonyms: Frigen 114A

Structure:

$$\begin{array}{ccc} Cl & & F \\ | & & | \\ Cl - & C - C - & F \\ | & | \\ F & F \end{array}$$

CAS Registry Number: 374-07-2

Molecular Formula: $C_2Cl_2F_4$

Wiswesser Line Notation:

CHEMICAL AND PHYSICAL PROPERTIES

Boiling Point: 3.6 °C at 760 mm Hg

Melting Point: -94 °C

Molecular Weight: 170.92

Dissociation Constants:

Log Octanol/Water Partition Coefficient: 2.85 (estimated) [11]

Water Solubility: 60 mg/L at 21 °C [7]

Vapor Pressure: 760 mm Hg at 3.6 °C [12], 1455 mm Hg at 21 °C [8]

Henry's Law Constant: 1.69 atm cu-m/mol at 25 °C [6]

ENVIRONMENTAL FATE/EXPOSURE POTENTIAL

Summary: There is no evidence that 1,1-dichloro-1,2,2,2-tetrafluoroethane is manufactured or used commercially and therefore, little or none of the chemical may be released to the environment. If

189

released to soil, 1,1-dichloro-1,2,2,2- tetrafluoroethane would be expected to rapidly volatilize from the soil surface or leach into the soil. If released into water, 1,1-dichloro-1,2,2,2-tetrafluoroethane would rapidly volatilize (half-life 3.8 hr in a model river). Bioconcentration in aquatic organisms or adsorption to sediment would not be significant. 1,1-Dichloro-1,2,2,2-tetrafluoroethane is not expected to degrade in the troposphere and it would be slowly transported to the stratosphere by diffusion, a process that may take decades. In the stratosphere, photolysis of 1,1-dichloro-1,2,2,2-tetrafluoroethane may produce atomic chlorine which can then participate in a chain catalytic destruction of stratospheric ozone.

Natural Sources:

Artificial Sources: It is not clear that 1,1-dichloro-1,2,2,2-tetrafluoroethane (Fluorocarbon 114a) is used commercially or even manufactured in this country at the present time. It is not listed in Freon Products Information [4] or in the Aldrich catalog, suggesting that it is not used much commercially or for research purposes [1]. Therefore, significant releases to the environment would not be expected. Since its properties are similar to 1,2-dichloro-1,1,2,2-tetrafluoroethane (Fluorocarbon 114) [8], it might find similar uses as a solvent, aerosol propellent, refrigerant, blowing agent, dielectric fluid, and in fire extinguishers [5,8]. If used for such purposes, substantial emissions to the atmosphere will occur.

Terrestrial Fate: If released to soil, 1,1-dichloro-1,2,2,2-tetrafluoroethane would be expected to rapidly volatilize from the soil surface due to its high volatility. The estimated Koc suggests it would adsorb moderately to soil and possibly leach into ground water.

Aquatic Fate: If released into water, 1,1-dichloro-1,2,2,2-tetrafluoroethane would rapidly volatilize with an estimated half-life of 3.8 hr from a model river 1 m deep, flowing at a current speed of 1 m/sec and a wind speed of 3 m/sec. Adsorption to sediment would not be significant. The estimated bioconcentration factor suggest bioconcentration in aquatic organisms would not be important.

1,1-Dichloro-1,2,2,2-tetrafluoroethane

Atmospheric Fate: 1,1-Dichloro-1,2,2,2-tetrafluoroethane is not expected to degrade in the troposphere and it will disperse and slowly diffuse to the stratosphere, a process that may take decades [3]. In the stratosphere, it will slowly photolyze to produce chlorine atoms which can participate in chain catalytic destruction of ozone [10]. While some 1,1-dichloro-1,2,2,2-tetrafluoroethane may be lost from the atmosphere by wet deposition via rain and snow, any precipitated compound will be returned to the atmosphere by volatilization.

Biodegradation:

Abiotic Degradation: Hydroxyl radicals and ozone are the two most important atmospheric species affecting atmospheric persistence of most organic compounds. 1,1-Dichloro-1,2,2,2-tetrafluoroethane does not contain any groups that react with photochemically produced hydroxyl radicals. Similarly, it also does not contain any groups that would react with atmospheric ozone. In addition 1,1-dichloro-1,2,2,2-tetrafluoroethane does not contain any chromophores that absorb solar radiation >290 nm that would make it susceptible to direct photolysis in the troposphere. Therefore, 1,1-dichloro-1,2,2,2-tetrafluoroethane will be extremely persistent in the troposphere. In the stratosphere, 1,1-dichloro-1,2,2,2-tetrafluoroethane will slowly photolyze, releasing chlorine atoms which in turn are responsible for removing ozone [2]. Reaction with singlet oxygen should be an additional stratospheric sink [2]. Calculated stratospheric lifetimes for a structurally similar compound, that is, 1,2-dichloro-tetrafluoroethane is 50-65 yrs [2]. While no experimental data on hydrolysis of 1,1-dichloro-1,2,2,2-tetrafluoroethane could be found, the rate of hydrolysis of Freon compounds is extremely low [4]. The reaction is catalyzed by the presence of steel, but is still very low [4].

Bioconcentration: The calculated log octanol/water partition coefficient for 1,1-dichloro-1,2,2,2-tetrafluoroethane is 2.85 [11]. Using this log Kow value and a recommended regression equation [9] one estimates a BCF of 86. This indicates that bioconcentration in aquatic organism will not be important.

Soil Adsorption/Mobility: Using the water solubility data of 60 mg/L [7] and a recommended regression equation [9]one can estimate a Koc of 459 for 1,1-dichloro-1,2,2,2-tetrafluoroethane. Therefore, 1,1-dichloro-

1,1-Dichloro-1,2,2,2-tetrafluoroethane

1,2,2,2-tetrafluoroethane should adsorb moderately to soil or sediment.

Volatilization from Water/Soil: The Henry's Law constant for 1,1-dichloro-1,2,2,2- tetrafluoroethane is 1.69 atm-m^3/mol [6]. Using this value and a recommended equation [9], one estimates a half-life for volatilization from a model river 1 m deep with a 1 m/sec current and 3 m/sec wind to be 3.8 hr. Volatilization is controlled by diffusion in the liquid phase [9].

Water Concentrations:

Effluent Concentrations:

Sediment/Soil Concentrations:

Atmospheric Concentrations:

Food Survey Values:

Plant Concentrations:

Fish/Seafood Concentrations:

Animal Concentrations:

Milk Concentrations:

Other Environmental Concentrations:

Probable Routes of Human Exposure: Due to the high volatility, inhalation is the most likely route of exposure to 1,1-dichloro-1,2,2,2-tetrafluoroethane. If the compound is used as a refrigerant, dermal exposure is also likely during recharging of refrigerant in air conditioners or refrigerators.

Average Daily Intake:

Occupational Exposure: If used as a refrigerant, people who service air conditioners (both automobile or building) and refrigerators are the most

192

1,1-Dichloro-1,2,2,2-tetrafluoroethane

likely occupational group to be exposed to 1,1-dichloro-1,2,2,2-tetrafluoroethane. If used as a fire extinguisher, firefighters may be another group who would also be susceptible to occasional exposure to this compound. However, no evidence of exposure to 1,1-dichloro-1,2,2,2-tetrafluoroethane for either the general population or any occupational group in the United States is available.

Body Burdens:

REFERENCES

1. Aldrich; Catalog Handbook of Fine Chemicals Milwaukee, WI (1986)
2. Chou CC et al; J Phys Chem 82: 1-7 (1978)
3. Dilling WL; Atmospheric Environment pp 154-97 in Environmental Risk Analysis for Chemicals Conway RA ed New York: Van Nostrand Reinhold (1982)
4. Du Pont de Nemours Co; Freon Products Information B-2; A-98825 12/80 (1980)
5. Hawley GG; Condensed Chem Dictionary 10th ed Von Nostrand Reinhold NY p 339 (1981)
6. Hine J, Mookerjee PK; J Org Chem 40: 292-8 (1975)
7. Horvath AL; Halogenated Hydrocarbons. Solubility-Miscibility with Water. NY: Marcel Dekker p 699 (1982)
8. Kawan A, Flynn JB; Kirk-Othmer Encyclopedia of Chemical Technology 3rd ed, Grayson M ed, 1: 582-92 (1978)
9. Lyman WJ et al; Handbook of Chemical Property Estimation Methods New York: McGraw-Hill pp 4-1 to 4-30, 5-1 to 5-30, 15-1 to 15-34 (1982)
10. Makide Y, Rowland FS; Proc Natl Acad Sci 78: 5933-37 (1981)
11. USEPA; Graphical Exposure Modeling System. CLOGP, Office of Toxic Substances, Washington, DC: USEPA (1985)
12. Weast RC (ed); Handbook of Chemistry and Physics, 66th ed, Boca Raton, FL: CRC Press p C-264 (1985)

Diethylene Glycol Dimethyl Ether

SUBSTANCE IDENTIFICATION

Synonyms: Bis(2-methoxyethyl) ether

Structure:

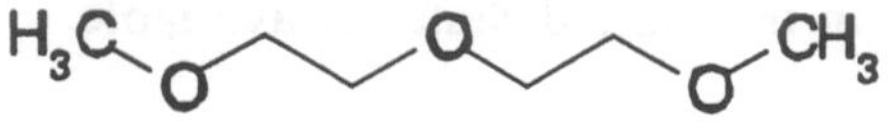

$$H_3C \diagdown O \diagup \diagdown O \diagup \diagdown O \diagdown CH_3$$

CAS Registry Number: 111-96-6

Molecular Formula: $C_6H_{14}O_3$

Wiswesser Line Notation:

CHEMICAL AND PHYSICAL PROPERTIES

Boiling Point: 162 °C

Melting Point: -68 °C

Molecular Weight: 134.18

Dissociation Constants:

Log Octanol/Water Partition Coefficient: -0.06 estimate [2]

Water Solubility: miscible [10]

Vapor Pressure: 2.96 mm Hg at 25 °C [3]

Henry's Law Constant: 2.28 x 10^{-9} atm-m^3/mole at 25 °C estimate [6]

ENVIRONMENTAL FATE/EXPOSURE POTENTIAL

Summary: Diethylene glycol dimethyl ether may be released to the environment via effluents at sites where it is produced or used as a solvent and an anhydrous reaction medium for organometallic synthesis. Diethylene glycol dimethyl ether is not expected to undergo hydrolysis

194

or direct photolysis in the environment. The complete miscibility of diethylene glycol dimethyl ether in water suggests that volatilization, adsorption and bioconcentration are not important fate processes. This is supported by the estimated Henry's Law constant, which indicates that volatilization of diethylene glycol dimethyl ether from natural waters and moist soil should be extremely slow. Yet, it may evaporate from dry surfaces, especially when present in high concentrations such as in spill situations. A low estimated log BCF suggests diethylene glycol dimethyl ether should not bioconcentrate among aquatic organisms. A low Koc indicates diethylene glycol dimethyl ether should not partition from the water column to organic matter contained in sediments and suspended solids, and it should be highly mobile in soil. Biodegradation may be an important removal mechanism of diethylene glycol dimethyl ether from aerobic soil and water; however, biodegradation data was not located in the available literature. In the atmosphere, diethylene glycol dimethyl ether is expected to exist almost entirely in the vapor phase and reactions with photochemically produced hydroxyl radicals should be important (estimated half-life of 14 hr). Physical removal of diethylene glycol dimethyl ether from air by precipitation and dissolution in clouds may occur; however, its short atmospheric residence time suggests that wet deposition is of limited importance. The most probable human exposure would be occupational exposure, which may occur through dermal contact or inhalation at workplaces where it is produced or used. Limited monitoring data indicate that nonoccupational exposures can occur from the ingestion of contaminated drinking water supplies.

Natural Sources:

Artificial Sources: Diethylene glycol dimethyl ether may be released to the environment via effluents at sites where it is produced or used as a solvent and an anhydrous reaction medium for organometallic synthesis [5].

Terrestrial Fate: Ethers are generally resistant to hydrolysis [8]. They do not absorb UV light in the environmentally significant range (>290 nm) [11]. Therefore, diethylene glycol dimethyl ether should not undergo hydrolysis in moist terrestrial environments, or direct photolysis on sunlit soil surfaces. The estimated Henry's Law constant indicates that volatilization of diethylene glycol dimethyl ether from moist soil should

195

not be an important fate process [8]. Yet, based upon the vapor pressure, diethylene glycol dimethyl ether should evaporate from dry surfaces, especially when present in high concentrations such as in spill situations. An estimated Koc of 20 [8] indicates diethylene glycol dimethyl ether should be highly mobile in soil [12]. Biodegradation may be an important removal mechanism of diethylene glycol dimethyl ether from aerobic soil; however, biodegradation data was not located in the available literature.

Aquatic Fate: Ethers are generally resistant to hydrolysis [8]. They do not absorb UV light in the environmentally significant range (>290 nm) [11]. Therefore, diethylene glycol dimethyl ether should not undergo hydrolysis or direct photolysis in aquatic environments. The complete miscibility of diethylene glycol dimethyl ether in water suggests that volatilization, adsorption and bioconcentration are not important fate processes. This is supported by the estimated Henry's Law constant, which indicates that volatilization of diethylene glycol dimethyl ether from natural waters should be extremely slow [8]. An estimated Koc of 20 [8] indicates diethylene glycol dimethyl ether should not partition from the water column to organic matter contained in sediments and suspended solids; an estimated bioconcentration factor (log BCF) of -0.28 [8] indicates diethylene glycol dimethyl ether should not bioconcentrate among aquatic organisms. Biodegradation may be an important removal mechanism of diethylene glycol dimethyl ether from aquatic systems; however, biodegradation data was not located in the available literature.

Atmospheric Fate: Ethers do not absorb UV light in the environmentally significant range (>290 nm) [11]. Therefore, diethylene glycol dimethyl ether should not undergo direct photolysis in the atmosphere. Based on the vapor pressure, diethylene glycol dimethyl ether is expected to exist almost entirely in the vapor phase in ambient air [4] where vapor phase reactions with photochemically produced hydroxyl radicals may be important. The rate constant for diethylene glycol dimethyl ether has been estimated to be 2.81×10^{-11} cm^3/molecule-sec at 25 °C, which corresponds to an atmospheric half-life of about 14 hr at an atmospheric concentration of $5 \times 10^{+5}$ hydroxyl radicals per cm^3 [1]. The complete miscibility of diethylene glycol dimethyl ether in water indicates that physical removal from air by precipitation and dissolution in clouds may occur; however, its short atmospheric residence time suggests that wet deposition is of limited importance.

Biodegradation:

Abiotic Degradation: Ethers are generally resistant to hydrolysis [8]. They do not absorb UV light in the environmentally significant range (>290 nm) [11]. Therefore, diethylene glycol dimethyl ether should not undergo hydrolysis or direct photolysis in the environment. The rate constant for the vapor-phase reaction of diethylene glycol dimethyl ether with photochemically produced hydroxyl radicals in air has been estimated to be 2.81×10^{-11} cm^3/molecule-sec at 25 °C, which corresponds to an atmospheric half-life of about 14 hr at an atmospheric concentration of $5 \times 10^{+5}$ hydroxyl radicals per cm^3 [1].

Bioconcentration: Because diethylene glycol dimethyl ether is miscible in water, bioconcentration in aquatic systems is not expected to be an important fate process. Based upon the estimated log Kow, a bioconcentration factor (log BCF) of -0.28 for diethylene glycol dimethyl ether has been calculated using a recommended regression-derived equation [8]. This BCF value also indicates diethylene glycol dimethyl ether should not bioconcentrate in aquatic organisms.

Soil Adsorption/Mobility: Because diethylene glycol dimethyl ether is miscible in water, soil adsorption is not expected to be an important fate process. Based on the estimated log Kow, a Koc of 20 for diethylene glycol dimethyl ether has been calculated using a recommended regression-derived equation [8]. This Koc value indicates diethylene glycol dimethyl ether will be highly mobile in soil [12], and it should not partition from the water column to organic matter contained in sediments and suspended solids.

Volatilization from Water/Soil: Because diethylene glycol dimethyl ether is miscible in water, and based upon the estimated Henry's Law constant, the volatilization of diethylene glycol dimethyl ether from natural bodies of water and moist soils is not expected to be an important fate process [8]. Yet, based upon the vapor pressure, diethylene glycol dimethyl ether should evaporate from dry surfaces, especially when present in high concentrations such as in spill situations.

Diethylene Glycol Dimethyl Ether

Water Concentrations: DRINKING WATER: Diethylene glycol dimethyl ether was listed as a contaminant found in drinking water [7].

Effluent Concentrations:

Sediment/Soil Concentrations:

Atmospheric Concentrations:

Food Survey Values:

Plant Concentrations:

Fish/Seafood Concentrations:

Animal Concentrations:

Milk Concentrations:

Other Environmental Concentrations:

Probable Routes of Human Exposure: The most probable route of human exposure to diethylene glycol dimethyl ether is by inhalation, dermal contact and ingestion. Drinking water supplies have been reported to contain diethylene glycol dimethyl ether [7].

Average Daily Intake:

Occupational Exposure: The most probable human exposure to diethylene glycol dimethyl ether would be occupational exposure, which may occur at places where it is produced or used. NIOSH (NOES Survey as of 3/28/89) has estimated that 207 workers are potentially exposed to diethylene glycol dimethyl ether in the USA [9]. Nonoccupational exposures may occur among populations with contaminated drinking water supplies [7].

Body Burdens:

Diethylene Glycol Dimethyl Ether

REFERENCES

1. Atkinson R; Intern J Chem Kin 19: 799-828 (1987)
2. CLOGP; PCGEMS Graphical Exposure Modeling System USEPA (1986)
3. Daubert TE, Danner RP; Data Compilation, Tables of Properties of Pure Cmpds, Design Inst for Phys Prop Data, Am Inst for Phys Prop Data, NY,NY (1989)
4. Eisenreich SJ et al; Environ Sci Technol 15: 30-8 (1981)
5. Hawley GG; Condensed Chemical Dictionary 10th ed Van Nostrand Reinhold NY p 375 (1981)
6. Hine J, Mookerjee PK; J Org Chem 40: 292-8 (1975)
7. Kool HJ et al; Crit Rev Env Control 12: 307-57 (1982)
8. Lyman WJ et al; Handbook of Chemical Property Estimation Methods NY: McGraw-Hill pp 4-9, 5-4, 6-3, 15-16 (1982)
9. NIOSH; National Occupational Exposure Survey (NOES) (1989)
10. Riddick JA et al; Organic Solvents NY: John Wiley & Sons Inc (1984)
11. Silverstein RM, Bassler GC; Spectrometric Id Org Cmpd NY: J Wiley & Sons Inc pp148-69 (1963)
12. Swann RL et al; Res Rev 85: 16-28 (1983)

Diethylene Glycol Monobutyl Ether

SUBSTANCE IDENTIFICATION

Synonyms: 2-(2-Butoxyethoxy)ethanol

Structure:

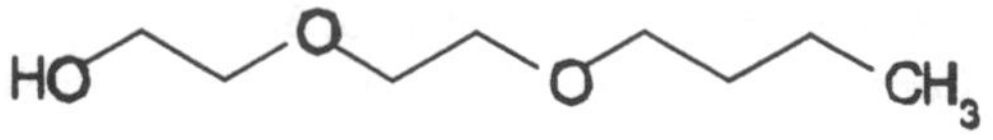

CAS Registry Number: 112-34-5

Molecular Formula: $C_8H_{18}O_3$

Wiswesser Line Notation: Q2O2O4

CHEMICAL AND PHYSICAL PROPERTIES

Boiling Point: 230.6 °C

Melting Point: -68.1 °C

Molecular Weight: 162.23

Dissociation Constants:

Log Octanol/Water Partition Coefficient: 0.91 estimate [5]

Water Solubility: miscible [7]

Vapor Pressure: 2.19 x 10^{-2} mm Hg at 25 °C [6]

Henry's Law Constant: 1.52 x 10^{-9} atm-m^3/mole at 25 °C estimate [10]

ENVIRONMENTAL FATE/EXPOSURE POTENTIAL

Summary: Diethylene glycol monobutyl ether may be released to the environment via a wide variety of effluents at sites where it is produced or used as a solvent or intermediate. Diethylene glycol monobutyl ether is not expected to undergo hydrolysis or direct photolysis in the environment. The complete miscibility of diethylene glycol monobutyl ether in water suggests that volatilization, adsorption and bioconcentration are not important fate processes. This is supported by the estimated Henry's Law constant, which indicates that volatilization of diethylene glycol monobutyl ether from natural waters and moist soil should be extremely slow. A low estimated log BCF suggests diethylene glycol monobutyl ether should not bioconcentrate among aquatic organisms. A low Koc indicates diethylene glycol monobutyl ether should not partition from the water column to organic matter contained in sediments and suspended solids, and it should be highly mobile in soil. Aqueous screening test data indicate that biodegradation may be an important removal mechanism of diethylene glycol monobutyl ether from aerobic soil and water. In the atmosphere, diethylene glycol monobutyl ether is expected to exist almost entirely in the vapor phase and reactions with photochemically produced hydroxyl radicals should be important (estimated half-life of 11 hr). Physical removal of diethylene glycol monobutyl ether from air by precipitation and dissolution is clouds may occur; however, its short atmospheric residence time suggests that wet deposition is of limited importance. The most probable human exposure would be occupational exposure, which may occur through dermal contact or inhalation at workplaces where it is produced or used. Workplace exposures have been documented. Limited monitoring data indicates that nonoccupational exposures can occur from the ingestion of contaminated drinking water supplies.

Natural Sources:

Artificial Sources: Diethylene glycol monobutyl ether may be released to the environment via effluents at sites where it is produced or used as a solvent for nitrocellulose, oils, dyes, gums, soaps and polymers, and as a plasticizer intermediate [9]. Diethylene glycol monobutyl ether may also be released to the environment via wastewater effluents from industries associated with paint and ink, printing and publishing, auto and other laundries, amusement and athletic goods, plastics and synthetics, timber products, pulp and paper, petroleum refining, coal mining, organic and inorganic chemicals, rubber processing, electronics, and from photographic and rum industries, foundries, textile mills, public-owned treatment works [4]. A tire fire also released diethylene glycol monobutyl ether to nearby surface waters [15].

Diethylene Glycol Monobutyl Ether

Terrestrial Fate: Alcohols and ethers are generally resistant to hydrolysis [12]. They do not absorb UV light in the environmentally significant range, >290 nm, and are commonly used as solvents for obtaining UV spectra [16]. Therefore, diethylene glycol monobutyl ether should not undergo hydrolysis in moist terrestrial environments, or direct photolysis on sunlit soil surfaces. The estimated Henry's Law constant indicates that volatilization of diethylene glycol monobutyl ether from moist soil should not be an important fate process [12]. An estimated Koc of 75 [12] indicates diethylene glycol monobutyl ether should be highly mobile in soil [17]. Aqueous screening test data [2,3,7] suggest that biodegradation may be an important removal mechanism of diethylene glycol monobutyl ether from aerobic soil.

Aquatic Fate: Alcohols and ethers are generally resistant to hydrolysis [12]. They do not absorb UV light in the environmentally significant range (>290 nm) and are commonly used as solvents for obtaining UV spectra [16]. Therefore, diethylene glycol monobutyl ether should not undergo hydrolysis or direct photolysis in aquatic environments. The complete miscibility of diethylene glycol monobutyl ether in water suggests that volatilization, adsorption and bioconcentration are not important fate processes. This is supported by the estimated Henry's Law constant, which indicates that volatilization of diethylene glycol monobutyl ether from natural waters should be extremely slow [12]. An estimated Koc of 75 [12] indicates diethylene glycol monobutyl ether should not partition from the water column to organic matter contained in sediments and suspended solids; and an estimated bioconcentration factor (log BCF) of 0.46 [12] indicates diethylene glycol monobutyl ether should not bioconcentrate among aquatic organisms. Aqueous screening test [2,3] data suggest that aerobic biodegradation may be an important removal mechanism of diethylene glycol monobutyl ether from aquatic systems.

Atmospheric Fate: Alcohols and ethers do not absorb UV light in the environmentally significant range (>290 nm) and are commonly used as solvents for obtaining UV spectra [16]. Therefore, diethylene glycol monobutyl ether should not undergo direct photolysis in the atmosphere. Based on the vapor pressure, diethylene glycol monobutyl ether is expected to exist almost entirely in the vapor phase in ambient air [8] where vapor phase reactions with photochemically produced hydroxyl

202

radicals may be important. The rate constant for diethylene glycol monobutyl ether has been estimated to be 3.62×10^{-11} cm^3/molecule-sec at 25 °C, which corresponds to an atmospheric half-life of about 11 hr at an atmospheric concentration of $5 \times 10^{+5}$ hydroxyl radicals per cm^3 [1]. The complete miscibility of diethylene glycol monobutyl ether in water indicates that physical removal from air by precipitation and dissolution in clouds may occur; however, its short atmospheric residence time suggests that wet deposition is of limited importance.

Biodegradation: Soil grab sample and river die-away test data pertaining to the biodegradation of diethylene glycol monobutyl ether in soil and natural waters were not located in the available literature. Although the rate cannot be determined, a few aerobic biological screening studies, which utilized settled waste water, sewage, or activated sludge for inocula, indicate that diethylene glycol monobutyl ether should biodegrade in the environment [2,3,7].

Abiotic Degradation: Alcohols and ethers are generally resistant to hydrolysis [16]. They do not absorb UV light in the environmentally significant range (>290 nm) and are commonly used as solvents for obtaining UV spectra [12]. Therefore, diethylene glycol monobutyl ether should not undergo hydrolysis or direct photolysis in the environment. The rate constant for the vapor-phase reaction of diethylene glycol monobutyl ether with photochemically produced hydroxyl radicals in air has been estimated to be 3.62×10^{-11} cm^3/molecule-sec at 25 °C, which corresponds to an atmospheric half-life of about 11 hr at an atmospheric concentration of $5 \times 10^{+5}$ hydroxyl radicals per cm^3 [1].

Bioconcentration: Because diethylene glycol monobutyl ether is miscible in water, bioconcentration in aquatic systems is not expected to be an important fate process. Based upon the estimated log Kow, a bioconcentration factor (log BCF) of 0.46 for diethylene glycol monobutyl ether has been calculated using a recommended regression-derived equation [12]. This BCF value also indicates diethylene glycol monobutyl ether should not bioconcentrate in aquatic organisms.

Soil Adsorption/Mobility: Because diethylene glycol monobutyl ether is miscible in water, soil adsorption is not expected to be an important fate process. Based on the estimated log Kow, a Koc of 75 for diethylene glycol monobutyl ether has been calculated using a recommended regression-derived equation [12]. This Koc value indicates diethylene glycol monobutyl ether will be highly mobile in soil [17], and it should not partition from the water column to organic matter contained in sediments and suspended solids.

Volatilization from Water/Soil: Because diethylene glycol monobutyl ether is miscible in water, and based on the estimated Henry's Law constant, the volatilization of diethylene glycol monobutyl ether from natural bodies of water and moist soils is not expected to be an important fate process [12].

Water Concentrations: DRINKING WATER: Diethylene glycol monobutyl ether was listed as a contaminant found in drinking water for a survey of US cities including Pomona, Escondido, Lake Tahoe and Orange Co, CA and Dallas, Washington, DC, Cincinnati, Philadelphia, Miami, New Orleans, Ottumwa, IA, and Seattle [11]. SURFACE WATER: In April 1980, diethylene glycol monobutyl ether was detected in Hayashida River water (the Matsubara area in Tatsuno City, Hyogo Prefecture) at concentrations of 35 and 240 ppb [19].

Effluent Concentrations: Diethylene glycol monobutyl ether was identified in 1 neutral fractions of 33 industrial wastewater effluents at concentrations of <10 ug/L [14]. For a national survey, diethylene glycol monobutyl ether was detected in 19 of 21 industrial categories of wastewater effluents [4]. Wastewater from the paint and ink industry contained diethylene glycol monobutyl ether at an average concentration of 35 mg/L; 47,933 for printing and publishing; 52 mg/L for textile mills; 244 mg/L for auto and other laundries; 5084 for amusement and athletic goods; 109 mg/L for plastics and synthetics; 7 mg/L for timber products; 34 mg/L for pulp and paper; 60 mg/L for petroleum refining; 7 mg/L for coal mining; 36 mg/L for inorganic chemicals; 84 mg/L for rubber processing; 3 mg/L for photographic industries; 3 mg/L for foundries; 214 mg/L for electronics; 147 mg/L for the rum industry; and 141 mg/L for unspecified industries; other categories included publicly owned treatment works and organics and plastics [4]. Diethylene glycol monobutyl ether was detected in surface waters downstream from a tire fire near Winchester, VA [15].

Diethylene Glycol Monobutyl Ether

Sediment/Soil Concentrations:

Atmospheric Concentrations:

Food Survey Values:

Plant Concentrations:

Fish/Seafood Concentrations:

Animal Concentrations:

Milk Concentrations:

Other Environmental Concentrations:

Probable Routes of Human Exposure: The most probable route of human exposure to diethylene glycol monobutyl ether is by inhalation, dermal contact and ingestion. Workplace exposures have been documented [18]. Drinking water supplies have been shown to contain diethylene glycol monobutyl ether [11].

Average Daily Intake:

Occupational Exposure: The most probable human exposure to diethylene glycol monobutyl ether would be occupational exposure, which may occur through dermal contact or inhalation at places where it is produced or used. NIOSH (NOES Survey as of 3/28/89) has estimated that 444,917 workers are potentially exposed to diethylene glycol monobutyl ether in the USA [13]. A study initiated in 1983, which surveyed the workplace atmospheres of 336 businesses in Belgium, showed that diethylene glycol monobutyl ether was present in 7 of 67 samples from sites where various materials such as varnishes, sterilization agents and cleaners are employed [18]. Nonoccupational exposures may occur among populations with contaminated drinking water supplies [11].

Body Burdens:

Diethylene Glycol Monobutyl Ether

REFERENCES

1. Atkinson R; Intern J Chem Kin 19: 799-828 (1987)
2. Babeu L, Vaishnav D; J Indust Microbiol 2: 107-15 (1987)
3. Bridie AL et al; Water Res 13: 627-30 (1979)
4. Bursey JT, Pellizzari ED; Analysis of Industrial Wastewater for Organic Pollutants in Consent Degree Survey. Contract No 68-03-2867. Athens, GA: USEPA Environ Res Lab (1982)
5. CLOGP; PCGEMS Graphical Exposure Modeling System USEPA (1986)
6. Daubert TE, Danner RP; Data Compilation, Tables of Properties of Pure Cmpds, Design Inst for Phys Prop Data, Am Inst for Phys Prop Data, NY, NY (1989)
7. Dow Chemical Co; The Glycol Ethers Handbook. Midland, MI (1981)
8. Eisenreich SJ et al; Environ Sci Technol 15: 30-8 (1981)
9. Hawley GG; Condensed Chemical Dictionary 10th ed Van Nostrand Reinhold NY p 375 (1981)
10. Hine J, Mookerjee PK; J Org Chem 40: 292-8 (1975)
11. Lucas SV; GC/MS Anal of Org in Drinking Water Concentrates and Advanced Treatment Concentrates Vol 1 USEPA-600/1-84-020A (NTIS PB85-128239) p 397 (1984)
12. Lyman WJ et al; Handbook of Chemical Property Estimation Methods NY: McGraw-Hill pp 15-15 to 15-29 (1982)
13. NIOSH; National Occupational Exposure Survey (NOES) (1989)
14. Perry DL et al; Ident of Org Compounds in Ind Effluent discharges USEPA-600/4-79-016 (NTIS PB-294794) p 230 (1979)
15. Peterson JC et al; Anal Chem 58: 70-4 (1986)
16. Silverstein RM, Bassler GC; Spectrometric Id Org Cmpd NY: J Wiley & Sons Inc pp 148-69 (1963)
17. Swann RL et al; Res Rev 85: 16-28 (1983)
18. Veulemans H et al; Am Indust Hyg Assoc J 48: 671-7 (1987)
19. Yasuhara, A et al; Environ Sci Technol 15: 570-3 (1981)

Diethylene Glycol Monoethyl Ether

SUBSTANCE IDENTIFICATION

Synonyms: Carbitol; 2-(2-ethoxyethoxy)ethanol

Structure:

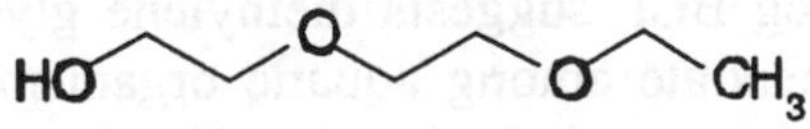

HO$\diagup\diagdown$O$\diagup\diagdown$O$\diagup$CH$_3$

CAS Registry Number: 111-90-0

Molecular Formula: C$_6$H$_{14}$O$_3$

Wiswesser Line Notation: Q2O2O2

CHEMICAL AND PHYSICAL PROPERTIES

Boiling Point: 195-202 °C

Melting Point: ND

Molecular Weight: 134.18

Dissociation Constants:

Log Octanol/Water Partition Coefficient: -0.15 estimate [5]

Water Solubility: miscible [7]

Vapor Pressure: 1.26 x 10^{-1} mm Hg at 25 °C [6]

Henry's Law Constant: 8.63 x 10^{-10} atm-m^3/mole at 25 °C estimate [11]

ENVIRONMENTAL FATE/EXPOSURE POTENTIAL

Summary: Diethylene glycol monoethyl ether may be released to the environment via a wide variety of effluents at sites where it is produced or used as a solvent. Diethylene glycol monoethyl ether is not expected to undergo hydrolysis or direct photolysis in the environment. The complete miscibility of diethylene glycol monoethyl ether in water suggests that volatilization, adsorption and bioconcentration are not important fate processes. This is supported by the estimated Henry's Law constant which indicates that volatilization of diethylene glycol monoethyl ether from natural waters and moist soil should be extremely slow. A low estimated log BCF suggests diethylene glycol monoethyl ether should not bioconcentrate among aquatic organisms. A low Koc indicates diethylene glycol monoethyl ether should not partition from the water column to organic matter contained in sediments and suspended solids, and it should be highly mobile in soil. Aqueous screening test data indicate that biodegradation is likely to be the most important removal mechanism of diethylene glycol monoethyl ether from aerobic soil and water. In the atmosphere, diethylene glycol monoethyl ether is expected to exist almost entirely in the vapor phase and reactions with photochemically produced hydroxyl radicals should be important (estimated half-life of 13 hr). Physical removal of diethylene glycol monoethyl ether from air by precipitation and dissolution in clouds may occur; however, its short atmospheric residence time suggests that wet deposition is of limited importance. The most probable human exposure would be occupational exposure, which may occur through dermal contact or inhalation at workplaces where it is produced or used. Atmospheric workplace exposures have been documented. Limited monitoring data indicates that nonoccupational exposures can occur from the ingestion of contaminated drinking water supplies.

Natural Sources:

Artificial Sources: Diethylene glycol monoethyl ether may be released to the environment via effluents at sites where it is produced or used as a solvent for nitrocellulose, resins, mineral oils, dyes, soaps, wood stains, textile printing, lacquers, and organic synthesis and as a diluent for brake fluid [10]. Diethylene glycol monoethyl ether may also be released to the environment via wastewater effluents from industries associated with iron and steel, printing and publishing, amusement and athletic goods, pulp and paper, and from public-owned treatment works [4].

Terrestrial Fate: Alcohols and ethers are generally resistant to hydrolysis [13]. They do not absorb UV light in the environmentally significant range (>290 nm) and are commonly used as solvents for obtaining UV spectra [16]. Therefore, diethylene glycol monoethyl ether should not undergo hydrolysis in moist terrestrial environments, or direct photolysis on sunlit soil surfaces. The estimated Henry's Law constant indicates that volatilization of diethylene glycol monoethyl ether from moist soil should not be an important fate process [13]. An estimated Koc of 20 [13] indicates diethylene glycol monoethyl ether should be highly mobile in soil [17]. Aqueous screening test data [2,3,7,15,18] indicate that biodegradation is likely to be the most important removal mechanism of diethylene glycol monoethyl ether from aerobic soil.

Aquatic Fate: Alcohols and ethers are generally resistant to hydrolysis [13]. They do not absorb UV light in the environmentally significant range (>290 nm) and are commonly used as solvents for obtaining UV spectra [16]. Therefore, diethylene glycol monoethyl ether should not undergo hydrolysis or direct photolysis in aquatic environments. The complete miscibility of diethylene glycol monoethyl ether in water suggests that volatilization, adsorption and bioconcentration are not important fate processes. This is supported by the estimated Henry's Law constant which indicates that volatilization of diethylene glycol monoethyl ether from natural waters should be extremely slow [13]. An estimated Koc of 20 [13] indicates diethylene glycol monoethyl ether should not partition from the water column to organic matter contained in sediments and suspended solids; and an estimated bioconcentration factor (log BCF) of -0.34 [13] indicates diethylene glycol monoethyl ether should not bioconcentrate among aquatic organisms. Aerobic screening test [2,3,7,15,18] data indicate that rapid aerobic biodegradation is likely to be the most important removal mechanism of diethylene glycol monoethyl ether from aquatic systems.

Atmospheric Fate: Alcohols and ethers do not absorb UV light in the environmentally significant range (>290 nm) and are commonly used as solvents for obtaining UV spectra [16]. Therefore, diethylene glycol monoethyl ether should not undergo direct photolysis in the atmosphere. Based on the vapor pressure, diethylene glycol monoethyl ether is expected to exist almost entirely in the vapor phase in ambient air [8] where vapor phase reactions with photochemically produced hydroxyl radicals should be important. The rate constant for diethylene glycol monoethyl ether has been estimated to be 2.93 x 10^{-11} cm^3/molecule-sec at 25 °C, which corresponds to an atmospheric half-life of about 13 hr at an atmospheric concentration of 5 x 10^{+5} hydroxyl radicals per cm^3 [1]. The complete miscibility of diethylene glycol monoethyl ether in water indicates that physical removal from air by precipitation and dissolution in clouds may occur; however, its short atmospheric residence time suggests that wet deposition is of limited importance.

Biodegradation: Soil grab sample and river die-away test data pertaining to the biodegradation of diethylene glycol monoethyl ether in soil and natural waters were not located in the available literature. Yet, aerobic biological screening studies, which utilized settled waste water, sewage, or activated sludge for inocula, indicate that diethylene glycol monoethyl ether should biodegrade rapidly in the environment [2,3,7,15,18]. After a 16-day acclimation period a 39.8% BODT was recorded using an 8-hr Warburg test [2]. A 5-day BOD test at 20 °C showed a 34.3% loss after 16 days of acclimation [2]. A second 5-day BOD test utilizing adapted seed had a 30% BODT [3]. Percent BODT at 5, 10 and 20 days incubation without acclimation were 5, 31 and 48%, respectively [7]. Percent BODT after 5, 10, 15, and 20 days incubation without acclimation were 17, 71, 75, and 87% respectively [15]. According to the Zahn-Wellens screening method, >90% loss of 400 ppm occurred in 28 days [15].

Abiotic Degradation: Alcohols and ethers are generally resistant to hydrolysis [16]. Likewise, they do not absorb UV light in the environmentally significant range (>290 nm) and are commonly used as solvents for obtaining UV spectra [13]. Therefore, diethylene glycol monoethyl ether should not undergo hydrolysis or direct photolysis in the environment. The rate constant for the vapor-phase reaction of diethylene glycol monoethyl ether with photochemically produced hydroxyl radicals in air has been estimated to be 2.93 x 10^{-11} cm^3/molecule-sec at 25 °C, which corresponds to an atmospheric half-life of about 13 hr at an atmospheric concentration of 5 x 10^{+5} hydroxyl radicals per cm^3 [1].

Bioconcentration: Because diethylene glycol monoethyl ether is miscible in water, bioconcentration in aquatic systems is not expected to be an important fate process. Based upon the estimated log Kow, a bioconcentration factor (log BCF) of -0.34 for diethylene glycol monoethyl ether has been calculated using a recommended regression-derived equation [13]. This BCF value also indicates diethylene glycol monoethyl ether should not bioconcentrate in aquatic organisms.

Soil Adsorption/Mobility: Because diethylene glycol monoethyl ether is miscible in water, soil adsorption is not expected to be an important fate process. Based on the estimated log Kow, a Koc of 20 for diethylene glycol monoethyl ether has been calculated using a recommended regression-derived equation [13]. This Koc value indicates diethylene glycol monoethyl ether will be highly mobile in soil [17], and it should not partition from the water column to organic matter contained in sediments and suspended solids.

Volatilization from Water/Soil: Because diethylene glycol monoethyl ether is miscible in water, and based upon the estimated Henry's Law constant, the volatilization of diethylene glycol monoethyl ether from natural bodies of water and moist soils is not expected to be an important fate process [13].

Water Concentrations: DRINKING WATER: Diethylene glycol monoethyl ether was listed as a contaminant found in drinking water for a survey of US cities including Pomona, Escondido, Lake Tahoe and Orange Co, CA and Dallas, Washington, DC, Cincinnati, Philadelphia, Miami, New Orleans, Ottumwa, IA, and Seattle [12].

Effluent Concentrations: For a national survey, diethylene glycol monoethyl ether was detected in 5 of 21 industrial categories of wastewater effluents [4]. Wastewater from the iron and steel industry contained diethylene glycol monoethyl ether at an average concentration of 497 mg/L; 52,189 mg/L for printing and publishing; 175 mg/L for amusement and athletic goods; and 40 mg/L for pulp and paper; other categories included public-owned treatment works [4].

Sediment/Soil Concentrations:

Atmospheric Concentrations:

Food Survey Values:

Plant Concentrations:

Diethylene Glycol Monoethyl Ether

Fish/Seafood Concentrations:

Animal Concentrations:

Milk Concentrations:

Other Environmental Concentrations:

Probable Routes of Human Exposure: The most probable route of human exposure to diethylene glycol monoethyl ether is by inhalation, dermal contact and ingestion. Atmospheric workplace exposures have been documented for French workers [9]. Drinking water supplies have been shown to contain diethylene glycol monoethyl ether [12].

Average Daily Intake:

Occupational Exposure: The most probable human exposure to diethylene glycol monoethyl ether would be occupational exposure, which may occur at places where it is produced or used. NIOSH (NOES Survey as of 3/28/89) has estimated that 311,819 workers are potentially exposed to diethylene glycol monoethyl ether in the USA [14]. Atmospheric workplace exposures have been documented for French workers [9]. Nonoccupational exposures may occur among populations with contaminated drinking water supplies [12].

Body Burdens:

REFERENCES

1. Atkinson R; Intern J Chem Kin 19: 799-828 (1987)
2. Bogan RH, Sawyer CN; Sew Ind Wastes 27: 917-28 (1955)
3. Bridie AL et al; Water Res 13: 627-30 (1979)
4. Bursey JT, Pellizzari ED; Analysis of Industrial Wastewater for Organic Pollutants in Consent Degree Survey. Contract No 68-03-2867. Athens, GA: USEPA Environ Res Lab (1982)
5. CLOGP; PCGEMS Graphical Exposure Modeling System USEPA (1986)
6. Daubert TE, Danner RP; Data Compilation, Tables of Properties of Pure Cmpds, Design Inst for Phys Prop Data, Am Inst for Phys Prop Data, NY,NY (1989)
7. Dow Chemical Co; The Glycol Ethers Handbook. Midland, MI (1981)
8. Eisenreich SJ et al; Environ Sci Technol 15: 30-8 (1981)
9. Ensminger A; CAH Notes DOC 131: 299-301 (1988)

10. Hawley GG; Condensed Chemical Dictionary 10th ed Van Nostrand Reinhold NY p 375 (1981)
11. Hine J, Mookerjee PK; J Org Chem 40: 292-8 (1975)
12. Lucas SV; GC/MS Anal of Org in Drinking Water Concentrates and Advanced Treatment Concentrates Vol 1 USEPA-600/1-84-020A (NTIS PB85-128239) p 397 (1984)
13. Lyman WJ et al; Handbook of Chemical Property Estimation Methods NY: McGraw-Hill p 7-4 (1982)
14. NIOSH; National Occupational Exposure Survey (NOES) (1989)
15. Price KS et al; J Water Pollut Contr Fed 46: 63-77 (1974)
16. Silverstein RM, Bassler GC; Spectrometric Org Cmpd NY: J Wiley & Sons Inc pp 148-69 (1963)
17. Swann RL et al; Res Rev 85: 16-28 (1983)
18. Zahn R, Wellens H; Z Wasser Abwasser Forsch 13: 1-7 (1980)

Diethylene Glycol Monomethyl Ether

SUBSTANCE IDENTIFICATION

Synonyms: Methyl Carbitol; 2-(2-Methoxyethoxy)ethanol

Structure:

CAS Registry Number: 111-77-3

Molecular Formula: $C_5H_{12}O_3$

Wiswesser Line Notation: Q2O2O1

CHEMICAL AND PHYSICAL PROPERTIES

Boiling Point: 193 °C

Melting Point: < -84 °C

Molecular Weight: 120.15

Dissociation Constants:

Log Octanol/Water Partition Coefficient: -0.68 estimate [4]

Water Solubility: miscible [14]

Vapor Pressure: 1.80×10^{-1} mm Hg at 25 °C [5]

Henry's Law Constant: 6.50×10^{-10} atm-m^3/mole at 25 °C estimate [9]

ENVIRONMENTAL FATE/EXPOSURE POTENTIAL

Summary: Diethylene glycol monomethyl ether may be released to the environment via effluents at sites where it is produced or used as a solvent, a component of brake fluid, and as a chemical intermediate. It may also be released to the environment via wastewater effluents from industries associated with paint and ink. Diethylene glycol monomethyl ether is not expected to undergo hydrolysis or direct photolysis in the environment. The complete miscibility of diethylene glycol monomethyl ether in water suggests that volatilization, adsorption and bioconcentration are not important fate processes. This is supported by the estimated Henry's Law constant, which indicates that volatilization of diethylene glycol monomethyl ether from natural waters and moist soil should be extremely slow. A low estimated log BCF suggests diethylene glycol monomethyl ether should not bioconcentrate among aquatic organisms. A low Koc indicates diethylene glycol monomethyl ether should not partition from the water column to organic matter contained in sediments and suspended solids, and it should be highly mobile in soil. Limited aqueous screening test data indicate that biodegradation may be an important removal mechanism of diethylene glycol monomethyl ether from aerobic soil and water. In the atmosphere, diethylene glycol monomethyl ether is expected to exist almost entirely in the vapor phase and reactions with photochemically produced hydroxyl radicals should be important (estimated half-life of 16 hr). Physical removal of diethylene glycol monomethyl ether from air by precipitation and dissolution in clouds may occur; however, its short atmospheric residence time suggests that wet deposition is of limited importance. The most probable human exposure would be occupational exposure, which may occur through dermal contact or inhalation at workplaces where it is produced or used. Limited monitoring data indicate that nonoccupational exposures can occur from the ingestion of contaminated drinking water supplies.

Natural Sources:

Artificial Sources: Diethylene glycol monomethyl ether may be released to the environment via effluents at sites where it is produced [10] or used as a solvent, a component of brake fluid and as a chemical intermediate [8]. Diethylene glycol monomethyl ether may also be released to the environment via wastewater effluents from industries associated with paint and ink [3].

Terrestrial Fate: Alcohols and ethers are generally resistant to hydrolysis [12]. They do not absorb UV light in the environmentally significant range (>290 nm) and are commonly used as solvents for obtaining UV spectra [15]. Therefore, diethylene glycol monomethyl ether should not undergo hydrolysis in moist terrestrial environments, or direct photolysis on sunlit soil surfaces. The estimated Henry's Law constant indicates that volatilization of diethylene glycol monomethyl ether from moist soil should not be an important fate process [12]. An estimated Koc of 10 [12] indicates diethylene glycol monomethyl ether should be highly mobile in soil [16]. Limited aqueous screening test data [4,5] suggests that biodegradation may be an important removal mechanism of diethylene glycol monomethyl ether from aerobic soil.

Aquatic Fate: Alcohols and ethers are generally resistant to hydrolysis [12]. They do not absorb UV light in the environmentally significant range (>290 nm) and are commonly used as solvents for obtaining UV spectra [15]. Therefore, diethylene glycol monomethyl ether should not undergo hydrolysis or direct photolysis in aquatic environments. The complete miscibility of diethylene glycol monomethyl ether in water suggests that volatilization, adsorption and bioconcentration are not important fate processes. This is supported by the estimated Henry's Law constant, which indicates that volatilization of diethylene glycol monomethyl ether from natural waters should be extremely slow [12]. An estimated Koc of 10 [12] indicates diethylene glycol monomethyl ether should not partition from the water column to organic matter contained in sediments and suspended solids and an estimated bioconcentration factor (log BCF) of -0.75 [12] indicates diethylene glycol monomethyl ether should not bioconcentrate among aquatic organisms. Limited aqueous screening test [2,6] data suggest that aerobic biodegradation may be an important removal mechanism of diethylene glycol monomethyl ether from aquatic systems.

Atmospheric Fate: Alcohols and ethers do not absorb UV light in the environmentally significant range (>290 nm) and are commonly used as solvents for obtaining UV spectra [15]. Therefore, diethylene glycol monomethyl ether should not undergo direct photolysis in the atmosphere. Based on the vapor pressure, diethylene glycol monomethyl ether is expected to exist almost entirely in the vapor phase in ambient air [7] where vapor-phase reactions with photochemically produced hydroxyl radicals may be important. The rate constant for diethylene glycol monomethyl ether has been estimated to be 2.44 x 10^{-11} cm^3/molecule-sec at 25 °C, which corresponds to an atmospheric half-life of about 16 hr at an atmospheric concentration of 5 x 10^{+5} hydroxyl radicals per cm^3 [1]. The complete miscibility of diethylene glycol monomethyl ether in water indicates that physical removal from air by precipitation and dissolution in clouds may occur; however, its short atmospheric residence time suggests that wet deposition is of limited importance.

Biodegradation: Soil grab sample and river die-away test data pertaining to the biodegradation of diethylene glycol monomethyl ether in soil and natural waters were not located in the available literature. Yet, aerobic biological screening studies, which utilized settled waste water, sewage or activated sludge for inocula, have demonstrated that diethylene glycol monomethyl ether should biodegrade in the environment [2,6]. Diethylene glycol monomethyl ether showed losses of 0, 21, and 66% when incubated at 20 °C for 5, 10, and 20 days, respectively [6]. Another 5-day BOD test at 20 °C lost 5% of the theoretic BOD [2].

Abiotic Degradation: Alcohols and ethers are generally resistant to hydrolysis [12]. They do not absorb UV light in the environmentally significant range (>290) nm and are commonly used as solvents for obtaining UV spectra [15]. Therefore, diethylene glycol monomethyl ether should not undergo hydrolysis or direct photolysis in the environment. The rate constant for the vapor-phase reaction of diethylene glycol monomethyl ether with photochemically produced hydroxyl radicals in air has been estimated to be 2.44 x 10^{-11} cm^3/molecule-sec at 25 °C, which corresponds to an atmospheric half-life of about 16 hr at an atmospheric concentration of 5 x 10^{+5} hydroxyl radicals per cm^3 [1].

Bioconcentration: Because diethylene glycol monomethyl ether is miscible in water, bioconcentration in aquatic systems is not expected to be an important fate process. Based upon the estimated log Kow, a bioconcentration factor (log BCF) of -0.75 for diethylene glycol monomethyl ether has been calculated using a recommended regression-derived equation [12]. This BCF value also indicates diethylene glycol monomethyl ether should not bioconcentrate in aquatic organisms.

Soil Adsorption/Mobility: Because diethylene glycol monomethyl ether is miscible in water, soil adsorption is not expected to be an important fate process. Based on the estimated log Kow, a Koc of 10 for diethylene glycol monomethyl ether has been calculated using a recommended regression-derived equation [12]. This Koc value indicates diethylene glycol monomethyl ether will be highly mobile in soil [16], and it should not partition from the water column to organic matter contained in sediments and suspended solids.

Volatilization from Water/Soil: Because diethylene glycol monomethyl ether is miscible in water, and based upon the estimated Henry's Law constant, the volatilization of diethylene glycol monomethyl ether from natural bodies of water and moist soils is not expected to be an important fate process [12].

Water Concentrations: DRINKING WATER: Diethylene glycol monomethyl ether was listed as a contaminant found in drinking water for a survey of US cities including Pomona, Escondido, Lake Tahoe and Orange Co, CA and Dallas, Washington, DC, Cincinnati, Philadelphia, Miami, New Orleans, Ottumwa, IA, and Seattle [11].

Effluent Concentrations: For a national survey, diethylene glycol monomethyl ether was detected in 1 of 21 industrial categories of wastewater effluents [3]. Wastewater from the paint and ink industries contained diethylene glycol monomethyl ether at an average concentration of 3571 mg/L [3]. Diethylene glycol monomethyl ether was listed as an volatile emission of glycol ether production [10].

Sediment/Soil Concentrations:

Atmospheric Concentrations:

Food Survey Values:

Plant Concentrations:

Fish/Seafood Concentrations:

Animal Concentrations:

Milk Concentrations:

Other Environmental Concentrations:

Diethylene Glycol Monomethyl Ether

Probable Routes of Human Exposure: The most probable route of human exposure to diethylene glycol monomethyl ether is by inhalation, dermal contact and ingestion. Drinking water supplies have been shown to contain diethylene glycol monomethyl ether [11].

Average Daily Intake:

Occupational Exposure: The most probable human exposure to diethylene glycol monomethyl ether would be occupational exposure, which may occur at places where it is produced or used. NIOSH (NOES Survey as of 3/28/89) has estimated that 104,248 workers are potentially exposed to diethylene glycol monomethyl ether in the USA [13]. Nonoccupational exposures may occur among populations with contaminated drinking water supplies [11].

Body Burdens:

REFERENCES

1. Atkinson R; Intern J Chem Kin 19: 799-828 (1987)
2. Bridie AL et al; Water Res 13: 627-30 (1979)
3. Bursey JT, Pellizzari ED; Analysis of Industrial Wastewater for Organic Pollutants in Consent Degree Survey. Contract No 68-03-2867. Athens, GA: USEPA Environ Res Lab (1982)
4. CLOGP; PCGEMS Graphical Exposure Modeling System USEPA (1986)
5. Daubert TE, Danner RP; Data Compilation, Tables of Properties of Pure Cmpds, Design Inst for Phys Prop Data, Am Inst for Phys Prop Data, NY, NY (1989)
6. Dow Chemical Co; The Glycol Ethers Handbook. Midland, MI (1981)
7. Eisenreich SJ et al; Environ Sci Technol 15: 30-8 (1981)
8. Hawley GG; Condensed Chemical Dictionary 10th ed Van Nostrand Reinhold NY p 375 (1981)
9. Hine J, Mookerjee PK; J Org Chem 40: 292-8 (1975)
10. Lovell RJ et al; Organic chemical manufacturing Vol 9 USEPA-450/3-80-028D p 545 (1980)
11. Lucas SV; GC/MS Anal of Org in Drinking Water Concentrates and Advanced Treatment Concentrates Vol 1 USEPA-600/1-84-020A (NTIS PB85-128239) p 397 (1984)
12. Lyman WJ et al; Handbook of Chemical Property Estimation Methods NY: McGraw-Hill pp 4-9, 5-4, 6-3, 15-16 (1982)
13. NIOSH; National Occupational Exposure Survey (NOES) (1989)
14. Riddick JA et al; Organic Solvents NY: John Wiley & Sons Inc (1984)

15. Silverstein RM, Bassler GC; Spectrometric Id Org Cmpd NY: J Wiley & Sons Inc pp 148-69 (1963)
16. Swann RL et al; Res Rev 85: 16-28 (1983)

Diethyl Ether

SUBSTANCE IDENTIFICATION

Synonyms:

Structure:

$$H_3C \diagdown O \diagdown CH_3$$

CAS Registry Number: 60-29-7

Molecular Formula: $C_4H_{10}O$

Wiswesser Line Notation: 2O2

CHEMICAL AND PHYSICAL PROPERTIES

Boiling Point: 34.6 °C

Melting Point: -116.3 °C (stable crystals); -123.3 °C (metastable crystals)

Molecular Weight: 74.12

Dissociation Constants:

Log Octanol/Water Partition Coefficient: 0.89 [13]

Water Solubility: 54950 mg/L at 25 °C [32]

Vapor Pressure: 537.1 mm Hg at 25 °C [4]

Henry's Law Constant: 0.00128 atm-m³/mol at 25 °C [17]

ENVIRONMENTAL FATE/EXPOSURE POTENTIAL

Summary: Diethyl ether may be released to the environment as a result of its industrial manufacture and use as a solvent and extractant. It has been detected in a variety of industrial effluents. Evaporation from solvent uses may be a major source of environmental release. If diethyl

221

ether is released to soil surfaces, it will volatilize rapidly. It is expected to be highly mobile in moist soil and subject to leaching to ground water. Its detection in various ground waters demonstrate that it will leach environmentally. If released to surface water, diethyl ether will be subject to rapid volatilization. Volatilization half-lives of 3.1 hr and 1.5 days have been estimated for a model river (one meter deep) and model pond (2 meters deep), respectively. Data from aqueous screening studies suggest that diethyl ether is resistent to biodegradation in environmental media. Aqueous hydrolysis, bioconcentration, photo-oxidation, direct photolysis and adsorption to sediment are not expected to be important. If released to the atmosphere, diethyl ether will exist almost entirely in the vapor phase where it will degrade via reaction with photochemically produced hydroxyl radicals (estimated half-life of 29 hr in average air). Direct photolysis will not be an important removal process since diethyl ether does not absorb light at wavelengths >290 nm. The most probable routes of exposure of the general population to diethyl ether are via inhalation of contaminated air and ingestion of contaminated drinking water. Exposure through dermal contact may occur in occupational settings. Inhalation and dermal exposure will be expected to be highest in workplaces where diethyl ether is made and used.

Natural Sources:

Artificial Sources: Diethyl ether may be released to the environment as a result of its use as an industrial solvent and extractant, in the production of smokeless gunpowder, and as a primer for gasoline [14,22]. Evaporation from solvent uses may be a major source of environmental release. Waste streams generated at manufacturing and use sites can release diethyl ether to the environment.

Terrestrial Fate: Due to its high vapor pressure, diethyl ether will volatilize rapidly from soil and terrestrial surfaces. Estimated Koc values of 11-73 [21] indicate that diethyl ether will be highly mobile in soil [31] and subject to significant leaching. Its detection in various ground waters demonstrate that it will leach environmentally [7]. Diethyl ether is not expected to chemically hydrolyze in soil. Aqueous screening test data from studies using activated sludge or sewage inocula suggest that diethyl ether is resistent to biodegradation in environmental media [16,33].

Diethyl Ether

Aquatic Fate: In surface water, diethyl ether will be subject to rapid volatilization. Using the measured Henry's Law constant, volatilization half-lives of 3.1 hr and 1.5 days have been estimated for a model river (one meter deep) and model pond (2 meters deep) respectively [21,36]. Screening test data from studies using activated sludge or sewage inocula suggest that diethyl ether is resistant to biodegradation in environmental media [16,33]. Based upon estimated Koc values of 11-73 [21] and estimated BCF values of 1.3-2.8 [21], adsorption to sediment or suspended particulate matter or bioconcentration in aquatic organisms should not be environmentally important. Analogy to similar aliphatic ethers indicates that aqueous hydrolysis should not be important [21]. Diethyl ether will not directly photolyze [5] nor significantly photooxidize via reaction with photochemically produced hydroxyl radicals in the water [2].

Atmospheric Fate: If diethyl ether is released to the atmosphere, it will exist almost entirely in the vapor phase based upon its relatively high vapor pressure [8]. It will degrade in the atmosphere by photooxidation via vapor-phase reaction with photochemically produced hydroxyl radicals. An atmospheric half-life of 29 hours at an atmospheric concentration of $5 \times 10^{+5}$ hydroxyl radicals per/cm^3 has been calculated for this process using the recommended measured rate constant at 25 °C [3]. Direct photolysis will not be an important removal process since diethyl ether does not absorb light at wavelengths >290 nm [5]. The relatively high water solubility of diethyl ether suggests that atmospheric washout is possible.

Biodegradation: Diethyl ether has been included in a list of compounds that were not biodegraded rapidly in screening tests utilizing sewage sludge inocula or soil inocula [1]. Many ethers are known to be resistant to biodegradation [1]. The 5-day biological oxygen demand measured for diethyl ether in screening tests ranged from 0% [16,33] to approximately 1.1% [16] of theoretical in studies using the standard dilution technique with sewage inocula and 0% of theoretical in a screening study using activated sludge inoculum [34]. A lag time of >10 days was observed in the latter study with activated sludge [34]. Zero percent theoretical BOD also was observed in a study which used the seawater dilution method with sewage inoculum over a 5-day period [33]. These screening test data suggest that diethyl ether is resistent to biodegradation in environmental

223

media; however, some of these screening test studies may not have been conducted for a long enough time period to allow determination of whether the biodegradation rate might increase after a longer lag period. No change was observed in the biological oxygen demand in tests of a semi-continuous activated sludge biological treatment simulator which tested for the removal of diethyl ether at a concentration of 200-800 ppm in a domestic sewage feed over a period of 24 hours [23].

Abiotic Degradation: The rate constant for the vapor-phase reaction of diethyl ether with photochemically produced hydroxyl radicals has been measured to be 13.4×10^{-12} cm^3/molecule-sec at 25 °C which corresponds to an atmospheric half-life of 29 hours at an atmospheric concentration of 5×10^5 hydroxyl radicals per cm^3 [3]. Reaction of diethyl ether with photochemically produced hydroxyl radicals in water should not be an important fate process based upon a half-life of 204 days which was calculated from a measured rate constant of 3.9×10^9 L/mol-sec [2]. Ethers are generally resistant the aqueous environmental hydrolysis [21]; therefore, hydrolysis is not expected to be significant under normal environmental conditions (pH 5-9) [21]. Direct photolysis will not be an important removal process since diethyl ether does not absorb light at wavelengths >290 nm [5].

Bioconcentration: Based upon the log octanol-water partition coefficient and the water solubility, the bioconcentration factor (BCF) for diethyl ether can be estimated to be 2.8 and 1.3, respectively, using recommended linear regression equations [21]. These estimated BCF values suggest that diethyl ether will not bioconcentrate significantly in aquatic organisms.

Soil Adsorption/Mobility: Based upon the log octanol-water partition coefficient and the water solubility, the Koc of diethyl ether can be estimated to be 73 and 11, respectively, using applicable linear regression equations [21]. These estimated Koc values suggest that diethyl ether will be high mobility in soil [31]. Therefore, diethyl ether may leach through soil to ground water.

Volatilization from Water/Soil: The value of the Henry's Law constant of diethyl ether indicates that volatilization is significant, and possibly rapid, from environmental waters [21]. Using the Henry's Law constant, the volatilization half-life from a river one meter deep flowing 1 m/sec with a wind velocity of 3 m/sec is estimated to be 3.1 hr at 25 °C [21]. The volatilization half-life from a model environmental pond (2 meters deep) has been estimated to be 1.5 days [36]. The relatively high vapor pressure of diethyl ether indicates that evaporation from soil surfaces will be rapid.

Water Concentrations: DRINKING WATER: Diethyl ether was detected, not quantified, in drinking water from the Torresdale Water Treatment Plant in Philadelphia, PA; it was found in samples taken one day out of seven days between Feb 1975 and January 1977 [30]. It was not detected in drinking water from 2 other treatment plants or in tap water from within the distribution system sampled between Aug 1975 and Sept 1976 [30]. It has been detected, not quantified, in the drinking water of 3 of 10 U.S. cities (Miami, FL, Philadelphia, PA, and Cincinnati, OH, were positive) in the National Organics Reconnaissance Survey that was initiated in 1974 [35]. It has been detected, but not quantified, in drinking water from unidentified sources [19]. GROUND WATER: Diethyl ether was detected at a concentration of 2.5 ug/L in 1 of 6 drinking water wells located downstream from a municipal and industrial solid waste landfill, Army Creek landfill - 60 miles southwest of Wilmington, DE [7]. It also was detected at 2.5 ug/L in the major recovery well that was removing the contaminated ground water [7]. Overall, diethyl ether was detected in 25% of the ground water samples analyzed in the study [7]. It was detected in 1982 at concentrations up to approx 1000 ug/L in an outwash aquifer contaminated by leachate from an organic chemical waste landfill near Ottawa, Canada, which accepted wastes between 1969 and 1980 [26]. Diethyl ether was detected, but not quantified, in ground water classified as uncontaminated (by inorganic indices) at 1 of 7 Minnesota municipal solid waste landfills sites tested [29]; it was not found in leachates from 6 landfill sites collected between 1958 and 1978 or in ground water samples classified as uncontaminated (by inorganic indices) at 1 of 13 landfills sites tested [29]. In the Netherlands, it was detected at concentrations ranging from 4 ppb to 15 ppm in ground water around a chemical water incineration site at which water was mass burned in massive piles of steel drums [15]. It was detected in the leachate plume at the Gloucester landfill, Ottawa, Canada at concentrations greater than 500 ppb [6]. SURFACE WATER: Diethyl ether has been detected, not quantified, in samples of water from the

Niagara River and the Cuyahoga River, a tributary of Lake Erie [11]. It has been found in the following Lake Michigan basin locations sampled in 1975 and 1976: Calumet - Sag Channel, 5 ug/L; Chicago Sanitary and Ship Channel, 3 ug/L [18]. Diethyl ether was found in samples of surface water from 9 of 204 sites near heavily industrialized areas across the USA sampled between Aug 1975 and Sept 1976 [10]; the concentrations ranged between 1 and 10 ppb in the samples where diethyl ether was found and the average concentration was 4.6 ppb [10].

Effluent Concentrations: Diethyl ether was detected in 9 of 63 samples of industrial effluents collected from a wide variety of industries across the U.S. (dates not reported) [28]. Six of the positive samples contained <10 ug/L diethyl ether, 2 samples contained 10-100 ug/L and 1 sample contained >100 ug/L [28]. Diethyl ether has been found at the following concentrations in the effluent of 3 out of 3 sewage treatment plants which discharge into Lake Michigan: West Side Sewage Treatment Plant, 1 ug/L; North Side Sewage Treatment Plant, 8 ug/L; Calumet Sewage Treatment Plant, 10 ug/L [11]. It also has been found in the effluent from the Chicago Central water works at 5 ug/L [11].

Sediment/Soil Concentrations:

Atmospheric Concentrations: URBAN: Diethyl ether was detected, but not quantified, in the ambient air from 1 of 5 sites in the Los Angeles Basin sampled in March and April 1975; Santa Monica was the site where 1 of 2 samples was positive [27]. SOURCE DOMINATED: Diethyl ether was found in the ambient air from 1 of 2 sites in the Kanawha Valley, WV sampled in Sept 1974 (S. Charleston, WV was the positive site), but was not found in any of the samples from 4 sites in the Houston, TX, area sampled in Nov 1974 [27]. It was not detected in any of the night ambient air samples analyzed during the study [27]. Diethyl ether was identified, but not quantified, at 4 of 7 sites in the Kanawha Valley, WV, sampled in Sept 1977; it was tentatively identified at another site [9]. Diethyl ether was identified, but not quantified, at 5 of 7 sites in the Shenandoah Valley, WA, sampled in Sept 1977; it was tentatively identified at another site [9].

Diethyl Ether

Food Survey Values: Diethyl ether has been tentatively identified, but not quantified, as a volatile component of raw chicken breast muscle and caecum [12].

Plant Concentrations:

Fish/Seafood Concentrations:

Animal Concentrations:

Milk Concentrations:

Other Environmental Concentrations:

Probable Routes of Human Exposure: Based upon ambient monitoring data, the most probable routes of general population exposure to diethyl ether are via inhalation of contaminated air [9,27] and ingestion of contaminated drinking water [19,30,35]. Inhalation and dermal exposure are expected to be highest in workplaces.

Average Daily Intake:

Occupational Exposure: NIOSH (NOES Survey 1981-1983) has statistically estimated that 175,489 workers are exposed to diethyl ether in the USA [25]. NIOSH (NOHS Survey 1972-1974) has statistically estimated that 1,672,606 workers are exposed to diethyl ether in the USA [24].

Body Burdens: Diethyl ether was detected in 37.5% of 387 samples of normal expired air from 54 carefully selected normal, healthy, nonsmoking, urban, adult human volunteers [20]. The geometric mean concentration of the positive samples was 0.331 ng/L [20].

REFERENCES

1. Alexander M; Biotechnol Bioeng 15: 611-47 (1973)
2. Anbar M, Neta P; Int J Appl Radiation Isotopes 18: 493-523 (1967)
3. Atkinson R; Chem Rev 85: 69-201 (1985)
4. Boublik T et al; The Vapour Pressures of Pure Substances. NY: Elsevier Sci Publ p 284 (1984)

Diethyl Ether

5. Calvert JG, Pitts JNJr; Photochemistry NY: John Wiley & Sons pp 441-2 (1966)
6. Devlin JF, Gorman WA; Water Pollut Res J Canada 22: 49-63 (1987)
7. DeWalle FB, Chian ESK; J Am Water Works Assoc 73: 206-11 (1981)
8. Eisenreich SJ et al; Environ Sci Technol 15: 30-8 (1981)
9. Erickson MD, Pellizzari ED; Analysis of Organic Air Pollutants in the Kanawha Valley, WV and the Shenandoah Valley, VA. USEPA-903/9-78-007 (NTIS PB286141) pp 46, 60 (1978)
10. Ewing BB et al; Monitoring to Detect Previously Unrecognized Pollutants in Surface Waters - Appendix: Organic Analysis Data. USEPA-560/6-77-015 p 299 (1977)
11. Great Lakes Water Quality Board; Inventory Chem Subst Id Great Lakes Ecos p 68 (1983)
12. Grey TC, Shrimpton DH; Brit Poultry Sci 8: 23-33 (1967)
13. Hansch C, Leo AJ; Medchem Project Issue No.26 Claremont, CA: Pomona College (1985)
14. Hawley GG; Condensed Chemical Dictionary 10th ed Van Nostrand Reinhold NY p 435 (1981)
15. Heida H; Contam Soil Int. TNO Conf 1st: 909-12 (1986)
16. Heukelekian H, Rand MC; J Water Pollut Control Assoc 29: 1040-53 (1955)
17. Hine J, Mookerjee PK; J Org Chem 40: 292-8 (1975)
18. Konasewich D et al; Status Report on Organic and Heavy Metal Contaminants in the Lakes Erie, Michigan, Huron and Superior basins, Great Lakes Qual Board p 291 (1978)
19. Kool HJ et al; Crit Rev Env Control 12: 307-57 (1982)
20. Krotoszynski BK et al; J Anal Toxicol 3: 225-34 (1979)
21. Lyman WJ et al; Handbook of Chem Property Estimation Methods Environ Behavior of Org Compounds Washington DC: Amer Chem Soc pp 4-9, 5-4, 5-10, 7-4, 15-16 to 15-29 (1990)
22. Merck; The Merck Index An Encyclopedia of Chemicals, Drugs, and Biologicals 10th ed Rahway, NJ: Merck & Co p 551 (1983)
23. Mills EJJr, Stack VTJr; Proc 8th Indust Waste Conf Eng Bull Purdue Univ, Eng Ext Ser pp 492-517 (1954)
24. NIOSH; The National Occupational Hazard Survey (NOHS) (1974)
25. NIOSH; The National Occupational Exposure Survey (NOES) (1983)
26. Patterson RJ et al; Wat Sci Technol 17: 57-69 (1985)
27. Pellizzari ED et al; Development of Analytical Techniques for Measuring Ambient Atmospheric Carcinogenic Vapors. USEPA-600/2-75-076. pp 75, 115 (1975)
28. Perry DL et al; Identification of Organic Compounds in Industrial Effluent Discharges. USEPA-600/4-79-016 (NTIS PB-294794) p 45 (1979)
29. Sabel GV, Clark TP; Waste Manag Res 2: 119-30 (1984)
30. Suffet IH et al; Water Res 14: 853-67 (1980)
31. Swann RL et al; Res Rev 85: 17-28 (1983)
32. Taft RW et al; Nature 313: 384-6 (1985)
33. Takemoto S et al; Suishitsu Odaku Kenkyu 4: 80-90 (1981)
34. Urano K, Kato Z; J Haz Materials 13: 147-59 (1986)

Diethyl Ether

35. USEPA; Preliminary Assessment of Suspected Carcinogens in Drinking Water. Interim Report to Congress, December 1975. Washington DC. p I-16 (1975)
36. USEPA; EXAMS II Computer Simulation (1987)

1,1-Difluoroethane

SUBSTANCE IDENTIFICATION

Synonyms: Freon 152A

Structure:

CAS Registry Number: 75-37-6

Molecular Formula: $C_2H_4F_2$

Wiswesser Line Notation: FYF

CHEMICAL AND PHYSICAL PROPERTIES

Boiling Point: -24.7 °C

Melting Point: -117 °C

Molecular Weight: 66.05

Dissociation Constants:

Log Octanol/Water Partition Coefficient: 0.75 [4]

Water Solubility: 2,800 mg/L at 25 °C and 1 atm partial pressure [6].

Vapor Pressure: 4,437.1 mm Hg at 25 °C [11]

Henry's Law Constant: 2.034×10^{-2} atm-m³/mole [5]

ENVIRONMENTAL FATE/EXPOSURE POTENTIAL

Summary: 1,1-Difluoroethane may be released to the environment as emissions during its use as a refrigerant and it may be released to the soil

1,1-Difluoroethane

from the disposal of refrigeration units containing this compound. If released to the soil, 1,1-difluoroethane may rapidly volatilize from soil surfaces to the atmosphere or leach through soil possibly into ground water. If released to water, volatilization would be the dominant fate process based on a half-life of 2.4 hours from a model river. If released to the atmosphere, 1,1-difluoroethane is expected to exist exclusively in the vapor phase. In the troposphere, 1,1-difluoroethane reacts slowly with photochemically generated hydroxyl radicals (half-life of 472 days). This relatively slow half-life in the lower atmosphere suggests that some 1,1-difluoroethane may gradually diffuse into the stratosphere. In the stratosphere, 1,1-difluoroethane is expected to slowly photolyze and contribute to the removal of stratospheric ozone. However, no data are available indicating 1,1-difluoroethane contributes to stratospheric ozone depletion. From its source of emissions, global atmospheric transport is expected to take place due to the stability of the chemical. The most probable route of human exposure to 1,1-difluoroethane by the general population is inhalation.

Natural Sources:

Artificial Sources: 1,1-Difluoroethane is used as a refrigerant [10]. This compound may be released to soil from the disposal of refrigeration units. Leaking and subsequent evaporation from refrigeration units can also release this compound in the atmosphere.

Terrestrial Fate: If released to soil, 1,1-difluoroethane may rapidly volatilize from soil surfaces or leach through soil, possibly into ground water.

Aquatic Fate: If released to surface water, essentially all 1,1-difluoroethane is expected to be lost by volatilization (half-life of approximately 2.4 hours from a model river) [7]. Bioaccumulation and adsorption to sediments are not significant fate processes in water.

Atmospheric Fate: Based on a vapor pressure of 4437.1 mm Hg at 25 °C [11], 1,1-difluoroethane is expected to exist exclusively in the vapor phase [3]. The relatively high water solubility of 1,1-difluoroethane [6] suggests that some loss by wet deposition occurs, but any loss by this mechanism is probably returned to the atmosphere by volatilization. In

231

1,1-Difluoroethane

the troposphere, 1,1-difluoroethane reacts slowly with photochemically generated hydroxyl radicals (half-life of 472 days) [1]. This relatively slow half-life in the lower atmosphere suggests that some 1,1-difluoroethane may gradually diffuse into the stratosphere. The half-life for tropospheric to stratospheric diffusion of compounds is on the order of 20 years [2]. In the stratosphere, 1,1-difluoroethane may slowly photolyze and contribute to the removal of stratospheric ozone. However, no data are available indicating 1,1-difluoroethane contributes to stratospheric ozone depletion. As a result of its persistence in the atmosphere, global atmospheric transport occurs.

Biodegradation:

Abiotic Degradation: The rate constant for the vapor-phase reaction of 1,1-difluoroethane with photochemically produced hydroxyl radicals has been experimentally measured to be 3.4×10^{-14} cm^3/molecule-sec at 22 °C [1] which corresponds to an atmospheric half-life of about 472 days at a daily average atmospheric hydroxyl radicals concn of $5 \times 10^{+5}$ per cm^3 [1]. 1,1-Difluoroethane is essentially inert to reaction with ozone molecules [1].

Bioconcentration: Based on a log Kow value of 0.75 [4], the bioconcentration factor (BCF) for 1,1-difluoroethane can be estimated to be 2.2 from a regression equation [7]. This BCF value suggests that the bioconcentration of 1,1-difluoroethane in aquatic organisms should not be important.

Soil Adsorption/Mobility: Based on a water solubility of 2,800 mg/L at 25 °C [6], a soil adsorption coefficient (Koc) for 1,1-difluoroethane of 50 was estimated using a linear regression equation [7]. This Koc value indicates that 1,1-difluoroethane would probably be moderately to highly mobile in soil and that it should be weakly adsorbed to suspended solids and sediments in water [12].

Volatilization from Water/Soil: The value of Henry's Law constant of 2.034×10^{-2} atm-m^3/mole [5] and the high volatility [11] of 1,1-difluoroethane suggests that the compound would volatilize rapidly from all bodies of water and soil surfaces [7]. Based on the value of Henry's Law constant, the volatilization half-life of 1,1-difluoroethane from a model river 1 m deep flowing at a current speed of 1 m/sec with a wind velocity of 3 m/sec has been estimated to be approximately 2.4 hours [7]. The volatilization half-life of 1,1-difluoroethane from a model pond has been estimated to be approximately 31 hours based on a water solubility of 2,800 mg/L, a Koc value of 50 and the Henry's Law constant of 2.034 x 10^{-2} atm-m^3/mole [13].

Water Concentrations:

Effluent Concentrations:

Sediment/Soil Concentrations:

Atmospheric Concentrations:

Food Survey Values:

Plant Concentrations:

Fish/Seafood Concentrations:

Animal Concentrations:

Milk Concentrations:

Other Environmental Concentrations:

Probable Routes of Human Exposure: The general population is exposed to 1,1-difluoroethane through inhalation of vapor containing the compound. In occupational settings, it is expected that exposure occurs by inhalation of contaminated air and dermal contact with this compound.

Average Daily Intake:

Occupational Exposure: Although a 1972-74 survey had statistically estimated that 17,880 workers were exposed to 1,1-difluoroethane [9], a 1981-83 National Exposure Survey recorded no worker exposures to 1,1-difluoroethane in the USA [8], indicating that the compound has not been produced or used since 1981.

1,1-Difluoroethane

Body Burdens:

REFERENCES

1. Atkinson R; Chem Rev 85: 69-201 (1985)
2. Dilling WL; In: Environmental Risk Analysis for Chemicals. Conway RA ed NY: Van Nostrand Reinhold p 154-97 (1982)
3. Eisenreich SJ et al; Environ Sci Technol 15: 30-38 (1981)
4. Hansch C, Leo AJ; MedChem Project, Issue No. 26, Claremont, CA: Pomona College (1985)
5. Hine J, Mookerjee PK; J Org Chem 40: 292-8 (1975)
6. Horvath AL; Halogenated Hydrocarbons - Solubility - Miscibility with Water, Marcel Dekker, Inc., NY p 726 (1982)
7. Lyman WJ et al; Handbook of Chemical Property Estimation Methods. NY: McGraw-Hill p 4-9, 5-5, 15-21 to 15-32 (1982)
8. NIOSH; National Occupational Exposure Survey (NOES), NIOSH, Cincinnati, OH (1989)
9. NIOSH; National Occupational Hazard Survey (NOHS), NIOSH, Cincinnati, OH (1974)
10. Nobis JF; Kirk-Othmer Encycl Chem Tech 3rd ed NY:Wiley 10: 866 (1980)
11. Riddick JA et al; Organic Solvents: Physical Properties and Methods of Purification 4th ed NY: John Wiley & Sons p 452 (1986)
12. Swann RL et al; Res Rev 85: 17-28 (1983)
13. USEPA; EXAMS II Computer Simulation, Environmental Research Lab, USEPA, Athens, GA (1987)

Di-iso-amyl Ether

SUBSTANCE IDENTIFICATION

Synonyms: 1,1'-Oxybis(3-methylbutane)

Structure:

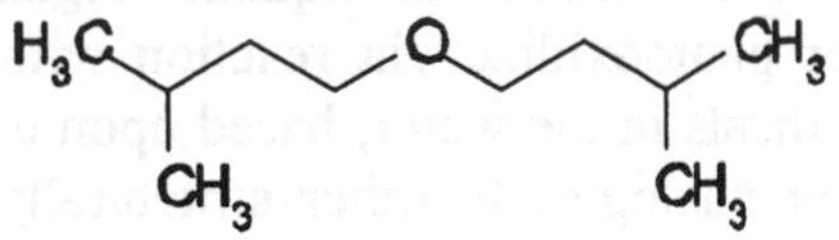

H3C
O
CH3
CH3
CH3

CAS Registry Number: 544-01-4

Molecular Formula: $C_{10}H_{22}O$

Wiswesser Line Notation:

CHEMICAL AND PHYSICAL PROPERTIES

Boiling Point: 173.4 °C

Melting Point:

Molecular Weight:: 158.28

Dissociation Constants:

Log Octanol/Water Partition Coefficient: 3.78 (estimated) [13]

Water Solubility: 200 mg/L at 25 °C [10]

Vapor Pressure: 1.4 mm Hg at 25 °C [10]

Henry's Law Constant: 1.46×10^{-3} atm-m^3/mole (calculated from vapor pressure and water solubility)

ENVIRONMENTAL FATE/EXPOSURE POTENTIAL

Summary: Di-iso-amyl ether may be released to the environment as a result of its use as a reaction solvent, lacquer solvent, and in the regeneration of rubber. If di-iso-amyl ether is released to soil, it will be subject to volatilization. It will be expected to exhibit moderate mobility in soil and, therefore, it may leach to ground water. It will not be expected to hydrolyze in soil. If di-iso-amyl ether is released to water, it will not be expected to significantly adsorb to sediment or suspended particulate matter, bioconcentrate in aquatic organisms, hydrolyze, directly photolyze, or photooxidize via reaction with photochemically produced hydroxyl radicals in the water, based upon estimated physical-chemical properties or analogies to other structurally related aliphatic ethers. Di-iso-amyl ether in surface water will be subject to rapid volatilization with estimated half-lives of 4.4 hr and 20 hr for volatilization from a river one meter deep flowing 1 m/sec with a wind velocity of 3 m/sec and a model pond, respectively. Di-iso-amyl ether may be resistent to biodegradation in environmental media based upon screening test data from studies using sewage inocula. Many ethers are known to be resistant to biodegradation. If di-iso-amyl ether is released to the atmosphere, it will be expected to exist almost entirely in the vapor phase based on its vapor pressure. It will be susceptible to photooxidation via vapor phase reaction with photochemically produced hydroxyl radicals with a half-life of 18 hours estimated for this process. Direct photolysis will not be an important removal process since aliphatic ethers do not absorb light at wavelengths >290 nm. The most probable route of general population exposure to di-iso-amyl ether is via inhalation of contaminated air, although exposure through dermal contact may also occur in occupational settings.

Natural Sources:

Artificial Sources: Di-iso-amyl ether may be released as a result of its use as a reaction solvent, lacquer solvent, and in the regeneration of rubber [6,9].

Terrestrial Fate: If di-iso-amyl ether is released to soil, it will be subject to volatilization based upon its estimated Henry's Law constant and vapor pressure. It will be expected to exhibit moderate mobility [11] in soil and, therefore, it may leach to ground water, based upon an estimated Koc of 237 [8]. It will not be expected to hydrolyze in soil [8]. Di-iso-amyl ether may be resistent to biodegradation in environmental media based upon aqueous screening test data from studies using sewage inocula [7]. Many ethers are known to be resistant to biodegradation [1].

Di-iso-amyl Ether

Aquatic Fate: If di-iso-amyl ether is released to water, it will not be expected to significantly adsorb to sediment or suspended particulate matter [8,10], bioconcentrate in aquatic organisms [8,10], hydrolyze [8], directly photolyze [4], or photooxidize via reaction with photochemically produced hydroxyl radicals in the water [2] based upon estimated physical-chemical properties or analogies to other structurally related aliphatic ethers [4,8,10]. Di-iso-amyl ether in surface water will be subject to rapid volatilization [8,10]. Using an estimated Henry's Law constant, a half-life for volatilization of di-iso-amyl ether from a river one meter deep flowing 1 m/sec with a wind velocity of 3 m/sec has been estimated to be 4.4 hr at 25 °C [8]. The volatilization half-life from a model pond, which considers the effect of adsorption has been estimated to be 20 hr [12]. Di-iso-amyl ether may be resistant to biodegradation in environmental media based upon screening test data from studies using sewage inocula [7]. Many ethers are known to be resistant to biodegradation [1].

Atmospheric Fate: If di-iso-amyl ether is released to the atmosphere, it will be expected to exist almost entirely in the vapor phase [5] based upon its vapor pressure. It will be susceptible to photooxidation via vapor phase reaction with photochemically produced hydroxyl radicals. An atmospheric half-life of 18 hours at an atmospheric concentration of 5 x 10^5 hydroxyl radicals per cm^3 has been calculated for this process based upon an estimated rate constant [3]. Direct photolysis will not be an important removal process since aliphatic ethers do not absorb light at wavelengths >290 nm [4].

Biodegradation: No data concerning the biodegradation of di-iso-amyl ether in environmental media were located. Many ethers are known to be resistant to biodegradation [1]. Screening test data from studies using activated sludge or sewage inocula suggest that di-iso-amyl ether is resistant to biodegradation in environmental media. Zero biological oxygen demand was measured for di-iso-amyl ether in aqueous screening tests which used sewage inocula for 5 days [7].

Abiotic Degradation: The rate constant for the vapor phase reaction of di-iso-amyl ether with photochemically produced hydroxyl radicals has been estimated to be 21 x 10^{-12} cm^3/molecule-sec at 25 °C [3] which corresponds to an atmospheric half-life of 18 hours at an atmospheric concentration of 5 x 10^5 hydroxyl radicals per cm^3. Hydrolysis is not expected to be significant under normal environmental conditions (pH 5-9) [8]. Direct photolysis will not be an important removal process since aliphatic ethers do not absorb light at wavelengths >290 nm [4].

Bioconcentration: Based upon its reported water solubility, a BCF of 31 has been estimated using a recommended regression equation [8]. Based upon this estimated BCF, di-iso-amyl ether will not be expected to bioconcentrate in aquatic organisms.

Soil Adsorption/Mobility: Based upon its reported water solubility, a Koc of 237 has been estimated using a recommended regression equation [8]. Based upon this estimated Koc, di-iso-amyl ether will be expected to exhibit moderate mobility in soil [11]. Di-iso-amyl ether, therefore, may slowly leach through soil to ground water.

Volatilization from Water/Soil: Based upon the calculated Henry's Law constant for di-iso-amyl ether, the volatilization half-life from a model river (1 meter deep flowing 1 m/sec with a wind speed of 3 m/sec) has been estimated to be 4.4 hr [8]. The volatilization half-life from a model pond, which considers the effect of adsorption, has been estimated to be 20 hr [4]. Based upon the Henry's Law constant and vapor pressure, di-iso-amyl ether will be subject to volatilization from surfaces and near-surface soil.

Water Concentrations:

Effluent Concentrations:

Sediment/Soil Concentrations:

Atmospheric Concentrations:

Food Survey Values:

Plant Concentrations:

Fish/Seafood Concentrations:

Di-iso-amyl Ether

Animal Concentrations:

Milk Concentrations:

Other Environmental Concentrations:

Probable Routes of Human Exposure: The most probable route of general population exposure to di-iso-amyl ether is via inhalation of contaminated air, although exposure through dermal contact may also occur in occupational settings.

Average Daily Intake:

Occupational Exposure:

Body Burdens:

REFERENCES

1. Alexander M; Biotechnol Bioeng 15: 611-47 (1973)
2. Anbar M, Neta P; Int J Appl Radiation Isotopes 18: 493-523 (1967)
3. Atkinson R; Intern J Chem Kinetics 19: 799-828 (1987)
4. Calvert JG, Pitts JNJr; Photochemistry John Wiley & Sons: New York pp 441-2 (1966)
5. Eisenreich SJ et al; Environ Sci Technol 15: 30-8 (1981)
6. Hawley GG; Condensed Chemical Dictionary 10th ed Van Nostrand Reinhold NY p 574 (1981)
7. Heukelekian H, Rand MC; J Water Pollut Control Assoc 29: 1040-53 (1955)
8. Lyman WJ et al; Handbook of Chem Property Estimation Methods NY: McGraw-Hill pp 4-9, 7-4 (1982)
9. Merck; The Merck Index An Encyclopedia of Chemicals, Drugs, and Biologicals 10th ed Rahway, NJ: Merck & Co p 738 (1983)
10. Riddick JA et al; Organic Solvents John Wiley & Sons Inc. NY (1984)
11. Swann RL et al; Res Rev 85: 17-28 (1983)
12. USEPA; EXAMS II Computer Simulation (1987)
13. USEPA; PC GEMS data base CLOGP3

Diisobutyl Ketone

SUBSTANCE IDENTIFICATION

Synonyms: 2,6-Dimethyl-4-heptanone

Structure:

CAS Registry Number: 108-83-8

Molecular Formula: $C_9H_{18}O$

Wiswesser Line Notation: 1Y1&1V1Y1&1

CHEMICAL AND PHYSICAL PROPERTIES

Boiling Point: 168.24 °C

Melting Point: -46.04 °C

Molecular Weight: 142.24

Dissociation Constants:

Log Octanol/Water Partition Coefficient: 2.65 (calculated) [6]

Water Solubility: 430 mg/L at 25 °C [13]

Vapor Pressure: 1.65 mm Hg at 25 °C [1]

Henry's Law Constant: 7.18×10^{-4} atm-m^3/mole (calculated from the vapor pressure and water solubility)

ENVIRONMENTAL FATE/EXPOSURE POTENTIAL

Summary: Diisobutyl ketone is released directly to the atmosphere by evaporation through its use as a solvent. If released to the atmosphere,

Diisobutyl Ketone

diisobutyl ketone is degraded relatively rapidly by reaction with photochemically produced hydroxyl radicals (estimated half-life of 14.2 hr). The results of several biodegradation screening studies suggest that biodegradation will be an important fate process in water and soil. If released to water, volatilization may be important. Volatilization half-lives of 4.9 and 57 hr have been estimated for a model river (one meter deep) and a model environmental pond, respectively. If released to soil, diisobutyl ketone may leach significantly based upon an estimated Koc value of 55. Diisobutyl can be expected to evaporate from dry surfaces. In occupational settings, exposure to diisobutyl ketone may occur through inhalation of vapors and through eye and skin contact.

Natural Sources:

Artificial Sources: Diisobutyl ketone is used as a solvent [11]; solvent evaporation will release diisobutyl ketone directly to the atmosphere. Diisobutyl ketone has been detected in wastewater effluents from waste treatment facilities [8].

Terrestrial Fate: Several biodegradation screening studies have observed significant biodegradation of diisobutyl ketone, particularly after periods of acclimation [3,12,16]; although these screening studies are not specific to the soil media, they suggest that biodegradation in soil is probably important. An estimated Koc value of 155 suggests medium mobility in soil and a potential for leaching [9,14]. Based upon the vapor pressure, diisobutyl ketone can be expected to evaporate from dry surfaces.

Aquatic Fate: Biodegradation and volatilization are important environmental fate processes for diisobutyl ketone in water. Several biodegradation screening studies have observed significant biodegradation of diisobutyl ketone, particularly after periods of acclimation [3,12,16]. Volatilization half-lives of 4.9 and 57 hr have been estimated for a model river (one meter deep) and a model environmental pond, respectively [9,15]. The chemical structure of diisobutyl ketone suggests that aquatic hydrolysis is not important [9]. An estimated Koc of 155 and BCF of 20 indicate that adsorption to sediment and bioconcentration in aquatic organisms are not important [9].

Diisobutyl Ketone

Atmospheric Fate: Based upon its relatively high vapor pressure, diisobutyl ketone can be expected to exist almost entirely in the vapor-phase in the ambient atmosphere [5]. Vapor-phase diisobutyl ketone is degraded relatively rapidly in the ambient atmosphere by reaction with photochemically produced hydroxyl radicals; the half-life for this reaction in air can be estimated to be about 14.2 hr [2].

Biodegradation: Using standard BOD techniques with acclimated microbial cultures, theoretical BODs of 37.4-46.7% were measured for diisobutyl ketone over 5-day incubation periods [3,16]. Theoretical BODs of 4, 39, 57, and 88% were measured over 5-, 10-, 15-, and 20-day inoculation periods, respectively, using standard BOD dilution water and a nonacclimated sewage inoculum [12]; theoretical BODs of 4, 9, and 18% were measured over 10-, 15-, and 20-day inoculation periods, respectively, using a synthetic salt BOD dilution water and a nonacclimated sewage inoculum [12].

Abiotic Degradation: The rate constant for the vapor-phase reaction of diisobutyl ketone with photochemically produced hydroxyl radicals has been experimentally measured to be 2.71×10^{-11} cm^3/molecule-sec at 25 °C [2], which corresponds to an atmospheric half-life of 14.2 hr at an atmospheric concentration of $5 \times 10^{+5}$ hydroxyl radicals/cm^3 [2]. Ketones are generally resistant to aqueous environmental hydrolysis [9]; therefore, diisobutyl ketone is not expected to hydrolyze in environmental waters.

Bioconcentration: Based upon the measured water solubility, the BCF for diisobutyl ketone can be estimated to be about 20 from a recommended linear regression-derived equation [9]. This BCF value suggests that bioconcentration in aquatic organisms is not important.

Soil Adsorption/Mobility: Based upon the measured water solubility, the Koc for diisobutyl ketone can be estimated to be 155 from a linear regression-derived equation [14]. This Koc value suggests that diisobutyl ketone has medium soil mobility [3].

Volatilization from Water/Soil: The value of the Henry's Law constant indicates that volatilization of diisobutyl ketone from environmental waters is probably significant, but may not be rapid [9]. Using the Henry's Law constant, the volatilization half-life from a model river (1 meter deep flowing 1 m/sec with a wind velocity of 3 m/sec) can be estimated to be about 4.9 hours [9]. The volatilization half-life from a model environmental pond (2 meters deep) can be estimated to be about 57 hours [15]. Based upon the vapor pressure, diisobutyl ketone can be expected to evaporate from dry surfaces.

Water Concentrations: DRINKING WATER: Diisobutyl ketone has been qualitatively detected in drinking water concentrates collected from Poplarville, MS in Mar 1979 and Philadelphia, PA in Feb 1976 [8]. GROUND WATER: Diisobutyl ketone has been detected at a maximum concentration of 0.3 ppb in ground water monitoring conducted in the Netherlands [17].

Effluent Concentrations: Diisobutyl ketone has been qualitatively detected in wastewater effluent concentrates collected from an advanced wastewater treatment facility in Pomona, CA in Sept 1974 [8].

Sediment/Soil Concentrations:

Atmospheric Concentrations:

Food Survey Values:

Plant Concentrations:

Fish/Seafood Concentrations:

Animal Concentrations:

Milk Concentrations:

Other Environmental Concentrations: Diisobutyl ketone has been detected at a concentration of 10 percent in a commercially available paint stripper solvent [7].

Probable Routes of Human Exposure: In occupational settings, exposure to diisobutyl ketone may occur through inhalation of vapors and through eye and skin contact [4].

Diisobutyl Ketone

Average Daily Intake:

Occupational Exposure: NIOSH (NOES Survey 1981-1983) has statistically estimated that 52,349 workers are potentially exposed to diisobutyl ketone in the USA [10]; 97% of these potential exposures result from products with trade names [10].

Body Burdens:

REFERENCES

1.   Ambrose D et al; J Chem Therm 7: 272-90 (1975)
2.   Atkinson R; Chem Rev 85: 69-201 (1985)
3.   Babeu L, Vaishnav DD; J Indust Microb 2: 107-15 (1987)
4.   Clayton GD, Clayton, FE; Patty's Industrial Hygiene and Toxicology, 3rd ed. NY: John Wiley & Sons, 2C: 4771 (1982)
5.   Eisenreich SJ et al; Environ Sci Technol 15: 30-8 (1981)
6.   GEMS; Graphic Exposure Modeling System. CLOGP. USEPA (1987)
7.   Hahn WJ, Werschulz PO; Evaluation of Alternatives to Toxic Organic Paint Strippers. USEPA-600/S2-86/063 (1986)
8.   Lucas, SV; GC/MS Analysis of Organics in Drinking Water Concentrates and Advanced Waste Treatment Concentrates: Vol 1. USEPA-600/1-84-020A. p 137 (1984)
9.   Lyman WJ et al; Handbook of Chemical Property Estimation Methods Washington DC: Amer Chem Soc pp 4-9, 5-10, 7-4, 15-15 to 15-29 (1990)
10. NIOSH; National Occupational Exposure Survey (NOES) (1983)
11. Papa AJ, Sherman PD Jr; Kirk-Othmer Encycl Chem Technol, 3rd ed, NY,NY: Wiley 13: 915-6 (1981)
12. Price KS et al; J Water Pollut Control Fed 46: 63-77 (1974)
13. Stross FH et al; J Amer Chem Soc 69: 1629-31 (1947)
14. Swann RL et al; Res Rev 85: 23 (1983)
15. US EPA; EXAMS II Computer Simulation (1987)
16. Vaishnav DD et al; Chemosphere 16: 695-703 (1987)
17. Zoeteman BCF et al; Sci Total Environ 21: 187-202 (1981)

Dimethyl Ether

SUBSTANCE IDENTIFICATION

Synonyms:

Structure:

CAS Registry Number: 115-10-6

Molecular Formula: C_2H_6O

Wiswesser Line Notation: 1O1

CHEMICAL AND PHYSICAL PROPERTIES

Boiling Point: -24 °C

Melting Point: -138.5 °C

Molecular Weight: 46.07

Dissociation Constants:

Log Octanol/Water Partition Coefficient:

Water Solubility: 35,300 mg/L at 25 °C [17]

Vapor Pressure: 4450 mm Hg at 25 °C [17]

Henry's Law Constant: 9.78×10^{-4} atm m^3/mole [10]

ENVIRONMENTAL FATE/EXPOSURE POTENTIAL

Summary: Dimethyl ether may be released to the environment as a result of its use as a refrigerant, solvent and extraction agent, propellant for sprays, and a catalyst and stabilizer for polymerization. If dimethyl ether is released to soil, it will volatilize when near the surface. It will, however, exhibit very high mobility in soil and may leach to ground water. Using estimated physical-chemical properties and/or analogies to other structurally related aliphatic ethers, if dimethyl ether is released to water, it will not significantly adsorb to sediment or suspended particulate matter, bioconcentrate in aquatic organisms, directly photolyze, or react with photochemically produced hydroxyl radicals in the water. It will not hydrolyze in water or soil. Dimethyl ether in surface water will be subject to rapid volatilization with estimated half-lives for volatilization of 2.6 hr from a river one meter deep flowing 1 m/sec with a wind velocity of 3 m/sec and 30 hr from a model pond. Dimethyl ether may be resistent to biodegradation in environmental media based on screening test data for diethyl ether using activated sludge or sewage inocula. Many ethers are known to be resistant to biodegradation. If dimethyl ether is released to the atmosphere, it will exist almost entirely in the vapor phase based on its vapor pressure. It will be susceptible to photooxidation by vapor-phase reaction with photochemically produced hydroxyl radicals with an estimated half-life of 5.4 days. It also will be susceptible to photooxidation by vapor-phase reaction with nitrate radicals in nighttime air with an estimated half-life of greater than or equal to 22 days. Direct photolysis will not be an important removal process since dimethyl ether does not absorb light at wavelengths >290 nm. The most probable routes of general population exposure to dimethyl ether are from inhalation of contaminated air and ingestion of contaminated drinking water. Exposure through dermal contact may occur in occupational settings. Inhalation and dermal exposure would be expected to be higher in workplaces where dimethyl ether is made and used. Exposure in infants also may occur through ingestion of contaminated mother's milk and some general population exposure may occur through the ingestion of contaminated food.

Natural Sources:

Artificial Sources: Dimethyl ether may be released as a result of its use as a refrigerant, solvent and extraction agent, propellant for sprays, and a catalyst and stabilizer for polymerization [8,12].

Terrestrial Fate: Using a reported Henry's Law constant of 9.78 x 10^{-4} atm m^3/mole [10], and a vapor pressure of 4450 mm Hg at 25 °C [17], if dimethyl ether is released to soil, it will volatilize rapidly. Since dimethyl ether has an estimated Koc of 14 [11,17], it will exhibit very high mobility [20] in soil and may leach to ground water. It is not be expected to hydrolyze in soil [11]. Aqueous screening test data from studies using activated sludge or sewage inocula for diethyl ether [9,21] suggest that dimethyl ether will be resistent to biodegradation in environmental media. Many ethers are known to be resistant to biodegradation [1].

Aquatic Fate: Using estimated physical-chemical properties or analogies to other structurally related aliphatic ethers [4,11,17], if dimethyl ether is released to water, it will not be expected to significantly adsorb to sediment or suspended particulate matter [11,17], bioconcentrate in aquatic organisms [11,17], hydrolyze [11], directly photolyze [4], or react with photochemically produced hydroxyl radicals in the water [2]. Dimethyl ether in surface water will rapidly volatilize [10,11]. Using a reported Henry's Law constant of 9.78×10^{-4} atm m^3/mole [10], a half-life for volatilization of dimethyl ether from a river one meter deep flowing 1 m/sec with a wind velocity of 3 m/sec has been estimated to be 2.6 hr at 25 °C [11]. The volatilization half-life from a model pond, which considers the effect of adsorption, has been estimated to be 30 hr [22]. Dimethyl ether may be resistent to biodegradation in environmental media based on screening test data for diethyl ether using activated sludge or sewage inocula [9,21]. Many ethers are known to be resistant to biodegradation [1].

Atmospheric Fate: If dimethyl ether is released to the atmosphere, given its vapor pressure of 4450 mm Hg at 25 °C [17], it will exist almost entirely in the vapor phase [6]. It will be susceptible to photooxidation by vapor-phase reaction with photochemically produced hydroxyl radicals and in nighttime air, with nitrate radicals. An atmospheric half-life of 5.4 days has been calculated for reaction with hydroxyl radicals during daytime at an atmospheric concentration of $5 \times 10^{+5}$ hydroxyl radicals/cm^{-3} using a measured rate constant of 2.98×10^{-12} cm^3/molecule-sec at 25 °C [3]. An atmospheric half-life of 22 days has been calculated for reaction with nitrate radicals at a nighttime atmospheric concentration of 2.4×10^{-8} nitrate radicals cm^{-3} using a measured rate constant of less than or equal to 3×10^{-15} cm^3/molecule-sec at 25 °C and 12 hr of nighttime air [19]. Direct photolysis will not be an important removal process since dimethyl ether does not absorb light at wavelengths >290 nm [4].

Biodegradation: No data concerning the biodegradation of dimethyl ether in environmental media were located. Many ethers are known to be resistant to biodegradation [1] and aqueous screening tests with the structurally similar diethyl ether suggest slow biodegradation [9].

Abiotic Degradation: The rate constant for the vapor-phase reaction of dimethyl ether with photochemically produced hydroxyl radicals has been measured to be 2.98×10^{-12} cm^3/molecule-sec at 25 °C [3], which corresponds to an atmospheric half-life of 5.4 days at an atmospheric concentration of $5 \times 10^{+5}$ hydroxyl radicals cm^{-3}. Hydrolysis is not expected to be significant under normal environmental conditions (pH 5-9) [11]. Direct photolysis will not be an important removal process since dimethyl ether does not absorb light at wavelengths >290 nm [4]. The rate constant for the vapor-phase reaction of dimethyl ether with nitrate radicals in nighttime air has been measured to be less than or equal to 3×10^{-15} cm^3/molecule-sec at 25 °C [19], which corresponds to an atmospheric half-life of greater than or equal to 22 days [each day was assumed to have 12 hr of nighttime air and a nighttime atmospheric concentration of 2.4×10^{-8} nitrate radicals cm^{-3}].

Bioconcentration: Using the reported water solubility of 35,300 mg/L at 25 °C [17] using a recommended regression equation [11], a BCF of 1.7 has been estimated. This estimated BCF suggests that dimethyl ether will not bioconcentrate in aquatic organisms.

Soil Adsorption/Mobility: Using a reported water solubility of 35,300 mg/L at 25 °C [17] and a recommended regression equation [11], a Koc of 14 has been estimated. This estimated Koc suggests that dimethyl ether will exhibit very high mobility in soil [20]. Dimethyl ether, therefore, may leach through soil to ground water if it does not volatilize or biodegrade first.

Volatilization from Water/Soil: The half-life for volatilization of dimethyl ether from a river one meter deep flowing 1 m/sec with a wind velocity of 3 m/sec has been estimated to be 2.6 hr at 25 °C [11] using a Henry's Law constant of 9.78×10^{-4} atm-m^3/mole [10]. The volatilization half-life from a model pond, which considers the effect of adsorption, has been estimated to be 30 hr [22]. The high Henry's Law constant and vapor pressure of 4450 Hg mm at 25 °C [17] suggest that dimethyl ether will volatilize from surfaces and near-surface soil.

Water Concentrations: DRINKING WATER: It has been detected, not quantified, in drinking water from 3 of 10 U.S. cities (Seattle, WA, Philadelphia, PA, and Cincinnati, OH, were positive) in the National Organics Reconnaissance Survey that was initiated in 1974 [23].

Effluent Concentrations: Dimethyl ether has been detected in 2 out of 63 samples of industrial effluents collected from a wide variety of industries across the U.S. at concentrations of >100 µg/L (dates not reported) [16]. It was qualitatively detected in primary treated municipal wastewater and sludge collected in 1983 from the Iona Island treatment plant in Vancouver, British Columbia, Canada [18].

Sediment/Soil Concentrations:

Atmospheric Concentrations: Dimethyl ether was identified, but not quantified, at 3 of 7 sites in the Kanawha Valley, WV, sampled in Sept. 1977; it was tentatively identified at another site [7].

Food Survey Values: Dimethyl ether was identified, but not quantified, as a volatile flavor component of baked Idaho Russet Burbank potatoes by one of two analytical techniques [5].

Plant Concentrations:

Fish/Seafood Concentrations:

Animal Concentrations:

Milk Concentrations: Dimethyl ether was detected in 1 of 12 samples of human breast milk from the cities of Bayonne, NJ, Jersey City, NJ, Bridgeville, PA, and Baton Rouge, LA [15].

Other Environmental Concentrations:

Dimethyl Ether

Probable Routes of Human Exposure: The most probable routes of general population exposure to dimethyl ether are by inhalation of contaminated air [7] and ingestion of contaminated drinking water [23]. Exposure through dermal contact may occur in occupational settings. Inhalation and dermal exposure are expected to be highest in workplaces where dimethyl ether is made and used. Exposure in infants also may occur through ingestion of contaminated mother's milk [15] and some general population exposure may occur through the ingestion of food [5].

Average Daily Intake:

Occupational Exposure: NIOSH (NOES Survey 1981-1983) has statistically estimated that 448 workers are exposed to dimethyl ether in the USA [13]. NIOSH (NOHS Survey 1972-1974) has statistically estimated that 50,009 workers are exposed to dimethyl ether in the USA [13].

Body Burdens: Dimethyl ether was detected in 1 of 12 samples of breast milk from the cities of Bayonne, NJ, Jersey City, NJ, Bridgeville, PA, and Baton Rouge, LA [15].

REFERENCES

1. Alexander M; Biotechnol Bioeng 15: 611-47 (1973)
2. Anbar M, Neta P; Int J Appl Radiation Isotopes 18: 493-523 (1967)
3. Atkinson R; Atmos Environ 24A: 1-41 (1990)
4. Calvert JG, Pitts JNJr; Photochemistry John Wiley & Sons NY pp 441-2 (1966)
5. Coleman EC et al; J Agric Food Chem 29: 42-8 (1981)
6. Eisenreich SJ et al; Environ Sci Technol 15: 30-8 (1981)
7. Erickson MD, Pellizzari ED; Analysis of Organic Air Pollutants in the Kanawha Valley, WV and the Shenandoah Valley, VA. USEPA-903/9-78-007 (NTIS PB286141) pp 46, 60 (1978)
8. Hawley GG; Condensed Chemical Dictionary 10th ed Van Nostrand Reinhold NY p 368 (1981)
9. Heukelekian H, Rand MC; J Water Pollut Control Assoc 29: 1040-53 (1955)
10. Hine J, Mookerjee PK; J Org Chem 40: 292-8 (1975)
11. Lyman WJ et al; Handbook of Chem Property Estimation Methods NY: McGraw-Hill p 4-9 (1982)
12. Merck; The Merck Index An Encyclopedia of Chemicals, Drugs, and Biologicals 10th ed Rahway, NJ: Merck & Co p 870 (1983)
13. NIOSH; The National Occupational Exposure Survey (NOES) (1983)
14. NIOSH; The National Occupational Hazard Survey (NOHS) (1974)

Dimethyl Ether

15. Pellizzari ED et al; Bull Environ Contam Toxicol 28: 322-8 (1982)
16. Perry DL et al; Identification of Organic Compounds in Industrial Effluent Discharges USEPA-600/4-79-016 (NTIS PB-294794) p 45 (1979)
17. Riddick JA et al; Organic Solvents 4th ed. Wiley-Interscience p 1325 (1986)
18. Rogers IH et al; Water Pollut Res J Canada 21: 187-204 (1986)
19. Sabljic A, Guestan H; Atmos Environ 24A: 73-8 (1990)
20. Swann RL et al; Res Rev 85: 17-28 (1983)
21. Takemoto S et al; Suishitsu Odaku Kenkyu 4: 80-90 (1981)
22. USEPA; EXAMS II Computer Simulation (1987)
23. USEPA; Preliminary Assessment of Suspected Carcinogens in Drinking Water. Interim Report to Congress, December 1975. Washington DC. p I-16 (1975)

N,N-Dimethylformamide

SUBSTANCE IDENTIFICATION

Synonyms: DMF

Structure:

CAS Registry Number: 68-12-2

Molecular Formula: C_3H_7NO

Wiswesser Line Notation: VHN1&1

CHEMICAL AND PHYSICAL PROPERTIES

Boiling Point: 153 °C

Melting Point: -61 °C

Molecular Weight: 73.09

Dissociation Constants:

Log Octanol/Water Partition Coefficient: -1.01 [11]

Water Solubility: miscible [14]

Vapor Pressure: 3.85 mm Hg at 25 °C [6]

Henry's Law Constant: 7.39×10^{-8} atm-m³/mole at 25 °C [18]

ENVIRONMENTAL FATE/EXPOSURE POTENTIAL

Summary: Dimethylformamide has been termed the universal organic solvent and is a widely used solvent for organic compounds where a low rate of evaporation is required. Consequently, dimethylformamide may be emitted to the environment by effluents from a variety of petrochemical industries. Dimethylformamide is expected to biodegrade rapidly in the environment. A calculated Koc of 7 indicates dimethylformamide should be highly mobile in soil. In aquatic systems, dimethylformamide is not expected to partition from the water column to organic matter contained in sediments and suspended solids or bioconcentrate in aquatic organisms. The Henry's Law constant suggests that volatilization of dimethylformamide from environmental waters will not be important. Dimethylformamide is expected to exist almost entirely in the gaseous phase in ambient air. The vapor-phase reaction with photochemically produced hydroxyl radicals (half-life of 2 hr) is likely to be an important fate process. The most probable human exposure would be occupational exposure, which may occur through dermal contact or inhalation at places where dimethylformamide is produced and used as a solvent of organic compounds.

Natural Sources:

Artificial Sources: Dimethylformamide has been termed the universal organic solvent [14] and is widely used as a solvent for organic compounds where a low rate of evaporation is required [14]. Dimethylformamide is used as a solvent for the formation of acrylic fibers, and sheets, films and coatings of other polymeric materials such as polyurethanes, specialty PVC, polyacrylonitrile, epoxy cellulose derivatives, and polyamides [15]. In the pharmaceutical industry, dimethylformamide is used as a crystallization medium for purification of vitamins, hormones and sulfonamides [15]. Dimethylformamide has commercial use as a solvent for electrolytes, particularly in high voltage capacitors, and in several electrolytic processes such as electroplating [15]. Consequently, dimethylformamide may be emitted to the environment by effluents from a variety of petrochemical industries.

Terrestrial Fate: Limited aqueous screening test data [4,19] and a river die-away test [7] suggest the biodegradation of dimethylformamide in soil will be rapid. A calculated Koc of 7 [13] indicates dimethylformamide will be highly mobile in soil [17]. The Henry's Law constant suggests that volatilization of dimethylformamide from moist soils will not be important [13].

N,N-Dimethylformamide

Aquatic Fate: River die-away test data suggests the biodegradation of dimethylformamide in aquatic systems should be rapid [7]. An estimated Koc ranging in the high mobility class for soil [17] indicates dimethylformamide will not partition from the water column to organic matter contained in sediments and suspended solids. Also, dimethylformamide is not expected to bioconcentrate in aquatic organisms. The Henry's Law constant suggests that volatilization of dimethylformamide from environmental waters will be not be important [13].

Atmospheric Fate: Based upon the vapor pressure, dimethylformamide is expected to exist almost entirely in the gaseous phase in ambient air [8]. The vapor-phase reaction of dimethylformamide with photochemically produced hydroxyl radicals is likely to be an important fate process. The rate constant for the vapor-phase reaction of dimethylformamide with photochemically produced hydroxyl radicals has been estimated to be 2.24×10^{-10} cm^3/molecule-sec at 25 °C, which corresponds to an atmospheric half-life of about 2 hours at an atmospheric concentration of $5 \times 10^{+5}$ hydroxyl radicals per cm^3 [2].

Biodegradation: Aqueous screening test data demonstrated that dimethylformamide was easily removed by sewage treatment facilities upon acclimation [19]. Wastewater from a polyamide synthesis operation at Kansas City, MO contained dimethylformamide at a concentration of 65,500 mg/L before entering a benchscale biological treatment system [4]. At feed rates of 90 lb/day/1000 cu ft, effluent from the biological reactor contained dimethylformamide at a concentration of less than 10 mg/L [4]. The concentration of dimethylformamide in the reactor sludge was not documented [4]. An aerobic unacclimated and acclimated river die-away test showed that dimethylformamide at an initial concentration of 30 mg/L completely disappeared within 6 and 3 days, respectively [7]. However, 24 to 48 hours was required before any degradation was observed among unacclimated samples [7]. Aerobic grab sample data for dimethylformamide in sea water showed a mineralization rate of less than 3% in 24 hours for initial concentrations of 10 ug/L and 100 ug/L [20]. However, 20% of dimethylformamide at a concentration of 0.1 ug/L was mineralized in 24 hr [20]. All samples were adjusted to sterilized controls [20].

N,N-Dimethylformamide

Abiotic Degradation: Dimethylformamide was classified in Class I: Noncategory with respect to ozone formation in smog chamber studies [5,10]. However, vapor phase reactions with photochemically produced hydroxyl radicals may be important in ambient air. The rate constant for the vapor-phase reaction of dimethylformamide with photochemically produced hydroxyl radicals has been estimated to be 2.24×10^{-10} cm^3/molecule-sec at 25 °C which corresponds to an atmospheric half-life of about 2 hours at an atmospheric concentration of $5 \times 10^{+5}$ hydroxyl radicals per cm^3 [2].

Bioconcentration: Based on the log Kow, a log BCF of -1.00 for dimethylformamide has been calculated from various regression-derived equations [13]. This log BCF value suggest dimethylformamide will not bioconcentrate in aquatic systems.

Soil Adsorption/Mobility: Based on the log Kow, a Koc of 7 for dimethylformamide has been calculated from various regression-derived equations [13]. This Koc value indicates dimethylformamide is highly mobile in soil [17].

Volatilization from Water/Soil: The Henry's Law constant suggests that volatilization of dimethylformamide from environmental waters will not be important [13].

Water Concentrations: DRINKING WATER: Dimethylformamide was listed as a contaminant found in drinking water for a survey of US cities including Pomona, Escondido, Lake Tahoe and Orange Co, CA and Dallas, TX, Washington, DC, Cincinnati, OH, Philadelphia, PA, Miami, FL, New Orleans, LA, Ottumwa, IA, and Seattle, WA [12]. SURFACE WATER: One of 204 samples contained dimethylformamide in a national survey of surface waters [9].

Effluent Concentrations: Dimethylformamide was detected in the air over a hazardous waste site in Lowell, MA and a neighboring industry at concentrations of 2.18 and greater than 50 ppb, respectively [1]. Dimethylformamide was detected in 1 of 63 industrial wastewater effluents at a concentration less than 10 ug/L [16]. Dimethylformamide was detected in the waste effluent of a plastics manufacturer at a concentration of 28,378 ng/uL of extract [3].

N,N-Dimethylformamide

Sediment/Soil Concentrations:

Atmospheric Concentrations: Dimethylformamide was detected in the air of Lowell, MA at a concentration of 8 ppb [1].

Food Survey Values:

Plant Concentrations:

Fish/Seafood Concentrations:

Animal Concentrations:

Milk Concentrations:

Other Environmental Concentrations:

Probable Routes of Human Exposure: Exposure of workers occurs mainly by inhalation of vapor and through skin contact.

Average Daily Intake:

Occupational Exposure: The most probable human exposure would be occupational exposure, which may occur through dermal contact or inhalation at places where dimethylformamide is produced and used as a solvent of organic compounds.

Body Burdens:

REFERENCES

1. Amster MB et al; Real Time Monitoring of Low Level Air Contaminants from Hazardous Waste sites. in Natl Conf Manage Uncontrolled Hazard Waste Sites, Silver Spring MD p 98-9 (1986)
2. Atkinson R; Intern J Chem Kin 19: 799-828 (1987)
3. Bursey JT, Pellizzari ED; Analysis of Industrial Wastewater for Organic Pollutants in Consent Degree Survey. Contract No 68-03-2867. Athens, GA: USEPA Environ Res Lab (1982)
4. Carter JL, Young DA; Proc Ind Waste Conf 38: 481-6 (1984)
5. Darnall KR et al; Environ Sci Technol 10: 692-6 (1976)

6. Daubert TE, Danner RP; Data Compilation, Tables of Properties of Pure Cmpds, Design Inst for Phys Prop Data, Am Inst for Phys Prop Data, NY, NY (1989)
7. Dojlido JR; Investigations of the Biodegradability and Toxicity of Organic Compounds USEPA 600/2-79-163 Cinn OH, Municipal Env Res Lab p 118 (1979)
8. Eisenreich SJ et al; Environ Sci Technol 15: 30-8 (1981)
9. Ewing BB et al; Monitoring to Detect Previously Unrecognized Pollutants in Surface Waters USEPA-560/6-77-015 p 75 (1977)
10. Farley FF; Photochemical Reactivity Classification of Hydrocarbons and other Organic Compounds USEPA-600/3-77-001B pp 713-27 (1977)
11. Hansch C, Leo AJ; Medchem Project Issue No 26. Claremont CA: Pomona College (1985)
12. Lucas SV; GC/MS Anal of Org in Drinking Water Concentrates and Advanced Treatment Concentrates Vol 1 USEPA-600/1-84-020A (NTIS PB85-128239) p 397 (1984)
13. Lyman WJ et al; Handbook of Chemical Property Estimation Methods NY: McGraw-Hill pp 4-9, 5-4, 5-10, 15-15 to 15-29 (1982)
14. Merck; The Merck Index 10th ed Rahway, NJ: Merck & Co p 391-2 (1983)
15. NIEHS; Dimethylformamide Draft Report Support for Chemical Nomination and Selection Process Natl Tox Prog NOI-ES-85218 p 6 (1988)
16. Perry DL et al; Ident of Org Compounds in Ind Effluent discharges USEPA-600/4-79-016 (NTIS PB-294794) p 230 (1979)
17. Swann RL et al; Res Rev 85: 16-28 (1983)
18. Taft RW et al; Nature 313: 384-6 (1985)
19. Thom NS, Agg AR; Proc R Soc Lond B 189: 347-57 (1975)
20. Ursin C; Chemosphere 14: 1539-50 (1985)

1,3-Dioxane

SUBSTANCE IDENTIFICATION

Synonyms:

Structure:

CAS Registry Number: 505-22-6

Molecular Formula: $C_4H_8O_2$

Wiswesser Line Notation:

CHEMICAL AND PHYSICAL PROPERTIES

Boiling Point: 105 °C at 755 mm Hg [9]

Melting Point: -42 °C [9]

Molecular Weight: 88.11

Dissociation Constants:

Log Octanol/Water Partition Coefficient: -0.419 (estimated) [6]

Water Solubility: Miscible [8]

Vapor Pressure: 39 mm Hg at 25 °C (estimated) [7]

Henry's Law Constant: approximately 4.88×10^{-6} atm-m^3/mole at 25 °C (experimental value for 1,4-dioxane) [4]

1,3-Dioxane

ENVIRONMENTAL FATE/EXPOSURE POTENTIAL

Summary: No data are available pertaining to either artificial or natural sources of environmental release of 1,3-dioxane. Limited prediction of environmental fate is based entirely on physical properties, chemical structure and analogy to similar compounds since experimental data are not available. If released to the atmosphere, 1,3-dioxane is expected to exist in the gas phase where it will be degraded relatively rapidly (estimated half-life of 2 days) by reaction with photochemically formed hydroxyl radicals. If released to soil, leaching may be possible since 1,3-dioxane is miscible in water. If released to water, volatilization is expected to be slow based on estimated half-lives of 7.2 days, 77.5 days, and 35.5 months from a shallow model river, a model environmental pond, and Lake Zurich, respectively. Aquatic hydrolysis, bioconcentration, and adsorption to sediment are not expected to be important. No data are available pertaining to the biodegradation of 1,3-dioxane in the environment. There are no data available to suggest that the general population of the USA is currently exposed to 1,3-dioxane.

Natural Sources:

Artificial Sources:

Terrestrial Fate: No experimental data are available pertaining to the chemical or biochemical degradation of 1,3-dioxane in soil. The miscibility of 1,3-dioxane in water [8] suggests that leaching through soil to associated ground waters may be possible.

Aquatic Fate: No experimental data are available pertaining to the chemical or biochemical degradation of 1,3-dioxane in natural water. The chemical structure of 1,3-dioxane suggests that aqueous hydrolysis will not occur [3]. The miscibility of 1,3-dioxane in water [8] suggests that bioconcentration in aquatic organisms and adsorption to sediments will not be important. Volatilization from environmental waters is expected to be slow based on estimated half-lives of 7.2 days, 77.5 days, and 35.5 months from a shallow model river, a model environmental pond, and Lake Zurich, respectively [3,5].

259

1,3-Dioxane

Atmospheric Fate: Based on the estimated vapor pressure, 1,3-dioxane can be expected to exist almost entirely in the vapor phase in the ambient atmosphere [2]. By analogy to 1,4-dioxane, vapor phase 1,3-dioxane will be degraded relatively rapidly in the atmosphere by reaction with photochemically formed hydroxyl radicals [1]; the half-life for this reaction in average air is approximately 2 days [1]. Due to its miscibility in water, physical removal of 1,3-dioxane from the atmosphere via wet deposition may be possible.

Biodegradation:

Abiotic Degradation: The rate constant for the vapor-phase reaction of 1,3,5-trioxane with photochemically produced hydroxyl radicals has been experimentally measured to be 7.9×10^{-12} cm^3/molecule-sec at 25 °C, which corresponds to an atmospheric half-life of about 2 days at an atmospheric concn of $5 \times 10^{+5}$ hydroxyl radicals per cm^3 [1]; 1,3-dioxane is expected to have a similar rate of reaction with atmospheric hydroxyl radicals because of the structural similarity. Ethers are generally resistant to aqueous environmental hydrolysis [3]; therefore, 1,3-dioxane is not expected to undergo aqueous hydrolysis in environmental waters.

Bioconcentration: Based upon the estimated log Kow, the BCF for 1,3-dioxane can be estimated to be 0.3 from a recommended regression-derived equation [3]. The estimated BCF and water solubility of 1,3-dioxane indicate that bioconcentration in aquatic organisms is not important.

Soil Adsorption/Mobility: 1,3-Dioxane has been reported to be miscible in water [8]; therefore, leaching through soil to associated ground waters may be possible.

Volatilization from Water/Soil: The Henry's Law constant for 1,4-dioxane has been experimentally measured to be 4.88×10^{-6} atm-m^3/mole at 25 °C [4]; the Henry's Law constant for 1,3-dioxane is expected to have a similar value. This value of Henry's Law constant indicates that volatilization from environmental waters is expected to be relatively slow [3]. Based on this Henry's Law constant, the volatilization half-life from a model river (1 m deep flowing 1 m/sec with a wind velocity of 3 m/sec) can be estimated to be about 7.2 days [3]; the volatilization half-lives from a model environmental pond and from Lake Zurich (Switzerland) can be estimated to be 77.5 days and 35.5 months, respectively [5].

Water Concentrations:

Effluent Concentrations:

Sediment/Soil Concentrations:

Atmospheric Concentrations:

Food Survey Values:

Plant Concentrations:

Fish/Seafood Concentrations:

Animal Concentrations:

Milk Concentrations:

Other Environmental Concentrations:

Probable Routes of Human Exposure: There are no data available to suggest that the general population of the USA is currently exposed to 1,3-dioxane.

Average Daily Intake:

Occupational Exposure:

Body Burdens:

REFERENCES

1.   Atkinson R; J Inter Chem Kinet 19: 799-828 (1987)
2.   Eisenreich SJ et al; Environ Sci Technol 15: 30-8 (1981)

1,3-Dioxane

3. Lyman WJ et al; Handbook of Chemical Property Estimation Methods. Environmental Behavior of Organic Compounds. Washington DC: American Chemical Society pp 4-9, 5-4, 5-10, 7-4, 7-5, 15-15 to 15-32. 1990.
4. Park JH et al; Anal Chem 59: 1970-6 (1987)
5. USEPA; EXAMS II Computer Simulation (1987)
6. USEPA; Graphical Exposure Modeling System (GEMS). CLOGP (1987)
7. USEPA; Graphical Exposure Modeling System (GEMS) (1988)
8. Weast RC; Handbook of Chemistry and Physics 60th Edition p C-277 (1979)
9. Weast RC et al. CRC Handbook of Chemistry and Physics. CRC Press, Inc: Boca Raton, FLA p.C-248 (1985)

Dipropylene Glycol

SUBSTANCE IDENTIFICATION

Synonyms:

Structure:

Mixture of isomers

CAS Registry Number: 25265-71-8

Molecular Formula: $C_6H_{14}O_3$

Wiswesser Line Notation:

CHEMICAL AND PHYSICAL PROPERTIES

Boiling Point: 233 °C

Melting Point:

Molecular Weight: 134.18

Dissociation Constants:

Log Octanol/Water Partition Coefficient: -1.07 estimate [6]

Water Solubility: miscible [4]

Vapor Pressure: 3.19 x 10^{-2} mm Hg at 25 °C [3]

Henry's Law Constant: 3.58 x 10^{-9} atm-m^3/mole at 25 °C estimate [8]

Dipropylene Glycol

ENVIRONMENTAL FATE/EXPOSURE POTENTIAL

Summary: Dipropylene glycol may be released to the environment via effluents at sites where it is produced or used as a solvent, and a plasticizer in polyester and alkyd resins, and in reinforced plastics. Dipropylene glycol is not expected to undergo hydrolysis or direct photolysis in the environment. The miscibility of dipropylene glycol in water suggests that volatilization, adsorption and bioconcentration are not important fate processes. This is supported by the estimated Henry's Law constant, which indicates that volatilization of dipropylene glycol from natural waters and moist soil should be extremely slow. A low estimated log BCF suggests dipropylene glycol should not bioconcentrate among aquatic organisms. A low Koc indicates dipropylene glycol should not partition from the water column to organic matter contained in sediments and suspended solids, and it should be highly mobile in soil. Although aerobic biodegradation screening test data suggests that the rate should be slow, biodegradation may still be an important removal mechanism of dipropylene glycol from aerobic soil and water. In the atmosphere, dipropylene glycol is expected to exist almost entirely in the vapor phase and reactions with photochemically produced hydroxyl radicals should be important (estimated half-life of 13 hr). Physical removal of dipropylene glycol from air by precipitation and dissolution in clouds may occur; however, its short atmospheric residence time suggests that wet deposition is of limited importance. The most probable human exposure would be occupational exposure, which may occur through dermal contact or inhalation at workplaces where it is produced or used. Limited monitoring data indicates that nonoccupational exposures can occur from the ingestion contaminated drinking water supplies.

Natural Sources:

Artificial Sources: Dipropylene glycol may be released to the environment via effluents at sites where it is produced or used as a solvent, and a plasticizer in polyester and alkyd resins, and in reinforced plastics [7].

Terrestrial Fate: Alcohols and ethers are generally resistant to hydrolysis [10]. They do not absorb UV light in the environmentally significant range (>290 nm) and are commonly used as solvents for

264

obtaining UV spectra [13]. Therefore, dipropylene glycol should not undergo hydrolysis in moist terrestrial environments, or direct photolysis on sunlit soil surfaces. The estimated Henry's Law constant indicates that volatilization of dipropylene glycol from moist soil should not be an important fate process [10]. An estimated Koc of 6 [10] indicates dipropylene glycol should be highly mobile in soil [14]. Although aerobic screening test data suggests that the rate should be slow [2,11], biodegradation may still be an important removal mechanism of dipropylene glycol from aerobic soil.

Aquatic Fate: Alcohols and ethers are generally resistant to hydrolysis [10]. They do not absorb UV light in the environmentally significant range (>290 nm) and are commonly used as solvents for obtaining UV spectra [13]. Therefore, dipropylene glycol should not undergo hydrolysis or direct photolysis in aquatic environments. The complete miscibility of dipropylene glycol in water suggests that volatilization, adsorption and bioconcentration are not important fate processes. This is supported by the estimated Henry's Law constant, which indicates that volatilization of dipropylene glycol from natural waters should be extremely slow [10]. An estimated Koc of 6 [10] indicates dipropylene glycol should not partition from the water column to organic matter contained in sediments and suspended solids; and an estimated bioconcentration factor (log BCF) of -1.04 [10] indicates dipropylene glycol should not bioconcentrate among aquatic organisms. Although aerobic screening test data suggests that the rate should be slow [2,11], biodegradation may still be an important removal mechanism of dipropylene glycol from aquatic systems.

Atmospheric Fate: Alcohols and ethers do not absorb UV light in the environmentally significant range (>290 nm) and are commonly used as solvents for obtaining UV spectra [13]. Therefore, dipropylene glycol should not undergo direct photolysis in the atmosphere. Based on the vapor pressure, dipropylene glycol is expected to exist almost entirely in the vapor phase in ambient air [5] where vapor phase reactions with photochemically produced hydroxyl radicals may be important. The rate constant for dipropylene glycol has been estimated to be 2.97×10^{-11} cm^3/molecule-sec at 25 °C, which corresponds to an atmospheric half-life of about 13 hr at an atmospheric concentration of $5 \times 10^{+5}$ hydroxyl radicals per cm^3 [1]. The miscibility of dipropylene glycol in water indicates that physical removal from air by precipitation and dissolution in clouds may occur; however, its short atmospheric residence time suggests that wet deposition is of limited importance.

Biodegradation: Soil grab sample and river die-away test data pertaining to the biodegradation of dipropylene glycol in soil and natural waters were not located in the available literature. Yet, a few aerobic biological screening studies, which utilized settled waste water, sewage, or activated sludge for inocula, indicate that dipropylene glycol should biodegrade slowly in the environment [2,11].

Abiotic Degradation: Alcohols and ethers are generally resistant to hydrolysis [10]. They do not absorb UV light in the environmentally significant range (>290 nm) and are commonly used as solvents for obtaining UV spectra [13]. Therefore, dipropylene glycol should not undergo hydrolysis or direct photolysis in the environment. The rate constant for the vapor-phase reaction of dipropylene glycol with photochemically produced hydroxyl radicals in air has been estimated to be 2.97×10^{-11} cm^3/molecule-sec at 25 °C, which corresponds to an atmospheric half-life of about 13 hr at an atmospheric concentration of $5 \times 10^{+5}$ hydroxyl radicals per cm^3 [1].

Bioconcentration: Because dipropylene glycol is miscible in water, bioconcentration in aquatic systems is nct expected to be an important fate process. Based upon the estimated log Kow, a bioconcentration factor (log BCF) of -1.04 for dipropylene glycol has been calculated using a recommended regression-derived equation [10]. This BCF value also indicates dipropylene glycol should not bioconcentrate in aquatic organisms.

Soil Adsorption/Mobility: Because dipropylene glycol is miscible in water, soil adsorption is not expected to be an important fate process. Based on an estimated log Kow, a Koc of 6 for dipropylene glycol has been calculated using a recommended regression-derived equation [10]. This Koc value indicates dipropylene glycol will be highly mobile in soil [14], and it should not partition from the water column to organic matter contained in sediments and suspended solids.

Dipropylene Glycol

Volatilization from Water/Soil: Because dipropylene glycol is miscible in water, and based upon the estimated Henry's Law constant, the volatilization of dipropylene glycol from natural bodies of water and moist soils is not expected to be an important fate process [10].

Water Concentrations: DRINKING WATER: Dipropylene glycol was listed as a contaminant found in drinking water for a survey of US cities including Pomona, Escondido, Lake Tahoe and Orange Co, CA and Dallas, Washington, DC, Cincinnati, Philadelphia, Miami, New Orleans, Ottumwa, IA, and Seattle [9].

Effluent Concentrations:

Sediment/Soil Concentrations:

Atmospheric Concentrations:

Food Survey Values:

Plant Concentrations:

Fish/Seafood Concentrations:

Animal Concentrations:

Milk Concentrations:

Other Environmental Concentrations:

Probable Routes of Human Exposure: The most probable route of human exposure to dipropylene glycol is by inhalation, dermal contact and ingestion. Drinking water supplies have been shown to contain dipropylene glycol [9].

Average Daily Intake:

Occupational Exposure: The most probable human exposure to dipropylene glycol would be occupational exposure, which may occur through dermal contact or inhalation at places where it is produced or

Dipropylene Glycol

used as a solvent, plasticizer and humectant. NIOSH (NOES Survey as of 3/28/89) has estimated that 218,354 workers are potentially exposed to dipropylene glycol in the USA [12]. Nonoccupational exposures may occur among populations with contaminated drinking water supplies [9].

Body Burdens:

REFERENCES

1. Atkinson R; Intern J Chem Kin 19: 799-828 (1987)
2. Bridie AL et al; Water Res 13: 627-30 (1979)
3. Daubert TE, Danner RP; Data Compilation, Tables of Properties of Pure Cmpds, Design Inst for Phys Prop Data, Am Inst for Phys Prop Data, NY, NY (1989)
4. Dow Chemical Co; The Glycol Ethers Handbook. Midland, MI (1981)
5. Eisenreich SJ et al; Environ Sci Technol 15: 30-8 (1981)
6. Hansch C, Leo AJ; Medchem Project Issue No 26. Claremont CA: Pomona College (1985)
7. Hawley GG; Condensed Chemical Dictionary 10th ed Van Nostrand Reinhold NY p 375 (1981)
8. Hine J, Mookerjee PK; J Org Chem 40: 292-8 (1975)
9. Lucas SV; GC/MS Anal of Org in Drinking Water Concentrates and Advanced Treatment Concentrates Vol 1 USEPA-600/1-84-020A (NTIS PB85-128239) p 397 (1984)
10. Lyman WJ et al; Handbook of Chemical Property Estimation Methods NY: McGraw-Hill pp 4-9, 5-4, 6-3, 15-16 (1982)
11. Niemi, GJ et al; Environ Toxicol Chem 6: 515-27 (1987)
12. NIOSH; National Occupational Exposure Survey (NOES) (1989)
13. Silverstein RM, Bassler GC; Spectrometric Id Org Cmpd NY: J Wiley & Sons Inc pp 148-69 (1963)
14. Swann RL et al; Res Rev 85: 16-28 (1983)

Ethyl Bromide

SUBSTANCE IDENTIFICATION

Synonyms:

Structure:

$$CH_3-CH_2-Br$$

CAS Registry Number: 74-96-4

Molecular Formula: C_2H_5Br

Wiswesser Line Notation: E2

CHEMICAL AND PHYSICAL PROPERTIES

Boiling Point: 38.4 °C [20]

Melting Point: -118.6 °C [20]

Molecular Weight: 108.98

Dissociation Constants:

Log Octanol/Water Partition Coefficient: 1.61 (recommended value) [9]

Water Solubility: 8939 at 25 °C [10]

Vapor Pressure: 467 mm Hg at 25 °C [3]

Henry's Law Constant: 7.49 x 10^{-3} atm-m^3/mole at 25 °C (estimated from vapor pressure and water solubility)

ENVIRONMENTAL FATE/EXPOSURE POTENTIAL

Summary: Ethyl bromide can be released to the environment in effluents from manufacturing and use facilities, in leachates from landfills, and through its use as a refrigerant. If released to the atmosphere, it will degrade relatively slowly by reaction with photochemically produced hydroxyl radicals (estimated half-life of 48 days). If released to water, ethyl bromide will be removed through hydrolysis and volatilization. The aqueous hydrolysis half-lives at 20 and 25 °C are 40 and 30 days, respectively. The volatilization half-lives from a model environmental river (1 meter deep) and model pond have been estimated to be 3.2 hr and 38.2 hr, respectively. If released to soil, ethyl bromide will be susceptible to hydrolysis under wet soil conditions. Its detection in landfill leachate demonstrates that environmental leaching can occur. Ethyl bromide's relatively high vapor pressure indicates that rapid evaporation from surfaces may occur. Insufficient data are available to predict the environmental importance of biodegradation in water or soil. Occupational exposure to ethyl bromide can occur through inhalation of vapor and dermal contact; it can enter the body through the skin or respiratory tract.

Natural Sources:

Artificial Sources: The detection of ethyl bromide in ambient air near bromine-chemical manufacturing facilities [4] suggests that the compound can be released to the environment from waste effluents at these facilities. Ethyl bromide has been detected in leachates from industrial and municipal landfills [2,7]. Ethyl bromide's use as a refrigerant [11] may eventually release the compound to the environment.

Terrestrial Fate: Based upon aqueous hydrolysis half-lives of 40 and 30 days at 20 and 25 °C [5,13], respectively, ethyl bromide should hydrolyze in moist soils. Estimated Koc values of 29 and 179 indicate that leaching in soil is possible [12]. The detection of ethyl bromide in landfill leachates has demonstrated that environmental leaching does occur [2]. Based upon the measured vapor pressure, ethyl bromide should evaporate rapidly from terrestrial surfaces. Insufficient data are available to predict the environmental importance of biodegradation in soil.

Aquatic Fate: Ethyl bromide hydrolyses in water; the aqueous hydrolysis half-lives at 20 and 25 °C are 40 and 30 days, respectively [5,13]. Insufficient data are available to predict the environmental importance of biodegradation or other degradation processes in water. Ethyl bromide can be expected to volatilize relatively rapidly from environmental waters. The volatilization half-lives from a model environmental river (1 meter deep) and model pond have been estimated to be 3.2 hr and 38.2 hr, respectively [12,19]. Aquatic bioconcentration and adsorption to sediment are not expected to be important.

Atmospheric Fate: Based upon the measured vapor pressure, ethyl bromide is expected to exist entirely in the vapor phase in the ambient atmosphere [6]. It is expected to degrade relatively slowly in an average ambient atmosphere (estimated half-life of about 48 days) by reaction with photochemically produced hydroxyl radicals [1]. Insufficient data are available to estimate other atmospheric degradation processes.

Biodegradation:

Abiotic Degradation: The rate constant for the vapor-phase reaction of ethyl bromide with photochemically produced hydroxyl radicals has been estimated to be 0.334×10^{-12} cm^3/molecule-sec at 25 °C which corresponds to an atmospheric half-life of about 48 days at an atmospheric concn of $5 \times 10^{+5}$ hydroxyl radicals per cm^3 [1]. The rate constant for the aqueous hydrolysis of ethyl bromide at 25 °C and pH 7 has been reported to be 2.64×10^{-7}/sec which corresponds to a half-life of 30 days [13]. At 20 °C, the aqueous hydrolysis rate constant has been reported to be 7.2×10^{-4}/hr, which corresponds to a half-life of about 40 days [5]. Ethyl bromide may react with naturally occurring nucleophiles in ground water (such as the HS ion found near sulfur and sulfide deposits) to form aliphatic sulfur-containing products [8,16].

Bioconcentration: Based upon the measured water solubility, the BCF for ethyl bromide can be estimated to be 3.7 from a regression-derived equation [12]. Based upon the measured log Kow, the Koc for ethyl bromide can be estimated to be 9.9 from a regression-derived equation [12]. These BCF values suggest that ethyl bromide will not bioconcentrate significantly in aquatic organisms.

Soil Adsorption/Mobility: Based upon the water solubility, the Koc for ethyl bromide can be estimated to be 29 from a regression-derived equation [12]. Based upon the measured log Kow, the Koc for ethyl bromide can be estimated to be 179 from a regression-derived equation

271

[12]. These BCF values suggest that ethyl bromide has a moderate to high soil mobility [18].

Volatilization from Water/Soil: Based upon the water solubility and vapor pressure values, the Henry's Law constant for ethyl bromide can be estimated to be 7.49×10^{-3} atm-m^3/mole. This value of Henry's Law constant indicates that volatilization from environmental waters is probably important [12]. Based on this Henry's Law constant, the volatilization half-life from a model river (1 m deep flowing 1 m/sec with a wind velocity of 3 m/sec) can be estimated to be about 3.2 hr [12]. Volatilization half-life from a model environmental pond can be estimated to be about 38.2 hr [19].

Water Concentrations: DRINKING WATER: Ethyl bromide was tentatively identified (but not quantified) in a drinking water sample collected from the Bellevue-Stratford Hotel in Philadelphia, PA in Aug 1976 [17].

Effluent Concentrations:

Sediment/Soil Concentrations:

Atmospheric Concentrations: Ethyl bromide was qualitatively detected in ambient air samples collected in the vicinity of the organic bromine chemical manufacturing areas in El Dorado and Magnolia, AR in 1977 [4].

Food Survey Values:

Plant Concentrations:

Fish/Seafood Concentrations:

Animal Concentrations:

Milk Concentrations:

Other Environmental Concentrations:

Ethyl Bromide

Probable Routes of Human Exposure: Occupational exposure to ethyl bromide can occur through inhalation of vapor and dermal contact [15]; it can enter the body through the skin or respiratory tract [15]. Since ethyl bromide has been detected in ambient air near bromine-chemical manufacturing facilities [4], populations living near manufacturing facilities may be exposed through inhalation.

Average Daily Intake:

Occupational Exposure: NIOSH (NOES Survey 1981-1983) has statistically estimated that 12,285 workers are potentially exposed to ethyl bromide in the USA [14].

Body Burdens:

REFERENCES

1. Atkinson R; J Inter Chem Kinet 19: 799-828 (1987)
2. Brown KW, Donnelly KC; Haz Waste Haz Mater 5: 1-30 (1988)
3. Daubert TE, Danner RP; Physical and Thermodynamic Properties of Pure Chemicals: Data Compilation, NY: Hemisphere Pub Corp (1989)
4. DeCarlo VJ; Ann NY Acad Sci 320: 678-81 (1979)
5. Ehrenberg L et al; Radiat Bot 15: 185-94 (1974)
6. Eisenreich SJ et al; Environ Sci Technol 15: 30-8 (1981)
7. Gould JP et al; In: Water Chlorination Environ Impact Health Eff 4: 525-39 (1983)
8. Haag WR; Mill T; Environ Toxicol Chem 7: 917-24 (1988)
9. Hansch C, Leo AJ; Medchem Project Issue No 26. Claremont CA: Pomona College (1985)
10. Horvath AL; Halogenated Hydrocarbons: Solubility-Miscibility With Water NY: Marcel Dekker p 490 (1982)
11. Kuney JH; Chemcyclopedia 1990, Washington DC: Amer Chem Soc 8: 81 (1989)
12. Lyman WJ et al; Handbook of Chemical Property Estimation Methods. Environmental Behavior of Organic Compounds. Washington DC: American Chemical Society pp 4-9, 5-4, 5-10, 7-4, 7-5, 15-15 to 15-32 (1990)
13. Mabey W, Mill T; J Phys Chem Ref Data 7: 383-415 (1978)
14. NIOSH; National Occupational Exposure Survey (NOES) (1983)
15. Parmeggiani L; Encyl Occup Health & Safety 3rd ed Geneva, Switzerland: International Labour Office pp 328-9 (1983)
16. Schwarzenbach RP et al; Environ Sci Technol 19: 322-7 (1985)
17. Suffet IH et al; Water Res 14: 853-67 (1980)
18. Swann RL et al; Res Rev 85: 23 (1983)

Ethyl Bromide

19. USEPA; EXAMS II Computer Simulation (1987)
20. Weast RC et al. CRC Handbook of Chemistry and Physics. CRC Press, Inc: Boca Raton, FLA p C-269 (1985)

Ethylene Glycol Diacetate

SUBSTANCE IDENTIFICATION

Synonyms:

Structure:

CAS Registry Number: 111-55-7

Molecular Formula: $C_6H_{10}O_4$

Wiswesser Line Notation: 1VO2OV1

CHEMICAL AND PHYSICAL PROPERTIES

Boiling Point: 190-191 °C

Melting Point: -31 °C

Molecular Weight: 146.14

Dissociation Constants:

Log Octanol/Water Partition Coefficient: 0.365 [11]

Water Solubility: 213,000 mg/L at 20 °C [9]

Vapor Pressure: 0.0439 mm Hg at 20 °C [4]

Henry's Law Constant: 3.98 x 10^{-8} atm-m^3/mole at 20 °C (calculated from vapor pressure and water solubility)

Ethylene Glycol Diacetate

ENVIRONMENTAL FATE/EXPOSURE POTENTIAL

Summary: Ethylene glycol diacetate is released directly to the atmosphere during the manufacture of resins, lacquers and printing inks. It is also released to the atmosphere through its use as a perfume fixative, as a solvent for oils, explosives and cellulose esters and ethers, and as a nondiscoloring plasticizer for ethyl and benzyl cellulose. If released to the atmosphere, ethylene glycol diacetate is degraded slowly by reaction with photochemically produced hydroxyl radicals (estimated half-life of 4.9 days in air). If released to soil, ethylene glycol diacetate can be expected to be very mobile based on an estimated Koc value of 37.63. A biodegradation study is available which suggests that ethylene glycol diacetate is readily biodegradable in sewage inocula which suggests that biodegradation of ethylene glycol diacetate in soil or water may be important. Based on an estimated vapor pressure of 4.39×10^{-2} mm Hg at 20 °C, ethylene glycol diacetate can be expected to slowly evaporate from dry soil surfaces. Based on a Henry's Law constant of 3.96×10^{-8} atm-m^3/mole at 20 °C, ethylene glycol diacetate is essentially nonvolatile in water. In occupational settings, exposure to ethylene glycol diacetate may occur through inhalation of vapors and through eye and skin contact.

Natural Sources:

Artificial Sources: Ethylene glycol diacetate is used in the manufacture of resins, lacquers and printing inks [5]. It is used as a perfume fixative and as a solvent for oils, explosives and cellulose esters and ethers [5,12]. It is also used as a nondiscoloring plasticizer for ethyl and benzyl cellulose [5]. During its use as an additive, fixative, solvent, or plasticizer, ethylene glycol diacetate may be released to the environment in waste streams.

Terrestrial Fate: An estimated Koc value of 37.63 suggests ethylene glycol diacetate has high mobility in soil and will leach significantly [6,10]. Based on an estimated vapor pressure of 4.39×10^{-2} mm Hg at 20 °C [4], ethylene glycol diacetate can be expected to evaporate from dry soil surfaces. Limited screening studies suggest that ethylene glycol diacetate may biodegrade in soil.

Ethylene Glycol Diacetate

Aquatic Fate: Based on a Henry's Law constant of 3.96×10^{-8} atm-m^3/mole, ethylene glycol diacetate is essentially nonvolatile in water [6]. An estimated Koc of 37.63 and BCF of 1.115 indicate that adsorption to sediment and bioconcentration in aquatic organisms are not significant [6]. Limited screening studies suggest that ethylene glycol diacetate may biodegrade in water.

Atmospheric Fate: Based on an estimated vapor pressure of 4.39×10^{-2} mm Hg at 20 °C [4], ethylene glycol diacetate can be expected to exist almost entirely in the vapor phase in the ambient atmosphere. Vapor-phase ethylene glycol diacetate is degraded slowly in the ambient atmosphere by reaction with photochemically formed hydroxyl radicals; the half-life for this reaction in typical air can be estimated to be about 4.9 days [1].

Biodegradation: Ethylene glycol diacetate was readily biodegradable as measured in the Hach respirometric and OECD Screening (die-away) tests using a sewage inocula [2].

Abiotic Degradation: The rate constant for the vapor-phase reaction of ethylene glycol diacetate with photochemically produced hydroxyl radicals has been estimated to be 3.24×10^{-12} cm^3/molecule-sec at 25 °C [1], which corresponds to an atmospheric half-life of about 4.9 days at an atmospheric concentration of $5 \times 10^{+5}$ hydroxyl radicals per cm^3 [1].

Bioconcentration: Based on a calculated log Kow of 0.365 [11], the BCF for ethylene glycol diacetate can be estimated to be 1.115 using a recommended regression-derived equation [6]. This BCF value suggests that ethylene glycol diacetate bioconcentration in aquatic organisms is not important.

Soil Adsorption/Mobility: Based on a calculated log Kow of 0.365(11), the Koc for ethylene glycol diacetate can be estimated to be approximately 37.6 [6], indicating that ethylene glycol diacetate has very high mobility in soil [10].

Volatilization from Water/Soil: Based on the Henry's Law constant, ethylene glycol diacetate is essentially nonvolatile in water [6]. Based on a vapor pressure of 4.39 x 10⁻² mm Hg at 20 °C [4], ethylene glycol diacetate can be expected to evaporate slowly from dry soil surfaces.

Water Concentrations:

Effluent Concentrations:

Sediment/Soil Concentrations:

Atmospheric Concentrations:

Food Survey Values:

Plant Concentrations:

Fish/Seafood Concentrations:

Animal Concentrations:

Milk Concentrations:

Other Environmental Concentrations:

Probable Routes of Human Exposure: In occupational settings, exposure to ethylene glycol diacetate may occur through inhalation of vapors and through eye and skin contact.

Average Daily Intake:

Occupational Exposures: NIOSH (NOHS Survey 1972-1974) has statistically estimated that 1,551 workers are potentially exposed to ethylene glycol diacetate in the USA [7]. NIOSH (NOES Survey 1981-1983) has statistically estimated that 1,417 workers are potentially exposed to ethylene glycol diacetate in the USA [8]. Ethylene glycol diacetate was quantitatively detected in an electrical cables insulation plant's extrusion area at a concentration range of 0.0-0.005 ppm [3].

Body Burdens:

Ethylene Glycol Diacetate

REFERENCES

1. Atkinson R; Environ Toxicol Chem 7: 435-42 (1988)
2. Cain RB; pp 325-70 in FEMS Symp 12, Microb. Degrad. Xenobiotics and "Builder" Recalcitrant Comps. (1981)
3. Cocheo V et al; Amer Ind Hyg Assoc J 44: 521-7 (1983)
4. Daubert TE, Danner DP; Physical & Thermodynamic Properties of Pure Chemicals Vol 3 NY, NY: Hemisphere Pub Corp (1989)
5. Hawley CG; The Condensed Chemical Dictionary 10th ed NY: Van Nostrand Reinhold Co (1981)
6. Lyman WJ et al; Handbook of Chemical Estimation Methods NY:Mcgraw-Hill pp 4-9 (1982)
7. NIOSH; National Occupational Hazard Survey (NOHS) (1974)
8. NIOSH; National Occupational Exposure Survey (NOES) (1989)
9. Riddick JA et al; Organic Solvents 4th ed; NY, NY: Wiley (1986)
10. Swann RL et al; Res Rev 85: 17-28 (1983)
11. USEPA; PCGEMS (1988)
12. Windholz M et al; The Merck Index 10th ed Rahway, NJ: Merck & Co Inc (1983)

Ethylene Glycol Monobutyl Ether

SUBSTANCE IDENTIFICATION

Synonyms: Butyl cellosolve; 2-Butoxyethanol

Structure:

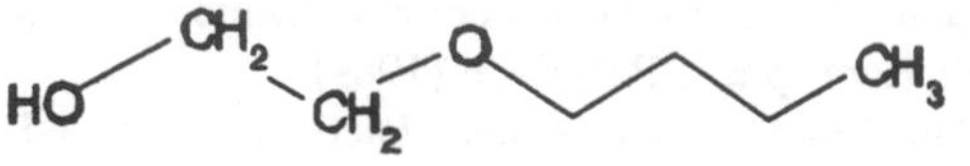

CAS Registry Number: 111-76-2

Molecular Formula: $C_6H_{14}O_2$

Wiswesser Line Notation: Q2O4

CHEMICAL AND PHYSICAL PROPERTIES

Boiling Point: 171-2 °C

Melting Point: -70 °C

Molecular Weight: 118.18

Dissociation Constants:

Log Octanol/Water Partition Coefficient: 0.83 [10]

Water Solubility: Miscible [19]

Vapor Pressure: 8.8 x 10^{-1} mm Hg at 25 °C [5]

Henry's Law Constant: 2.08 x 10^{-8} atm-m^3/mole at 25 °C estimate [12]

ENVIRONMENTAL FATE/EXPOSURE POTENTIAL

Summary: Ethylene glycol mono-n-butyl ether may be released to the environment via effluents at sites where it is produced or used as a solvent. Solvent-based building materials such as silicone caulk may release ethylene glycol mono-n-butyl ether to air as they dry. Leachate from municipal landfills and hazardous waste sites can also release ethylene glycol mono-n-butyl ether to ground waters. Ethylene glycol mono-n-butyl ether is not expected to undergo hydrolysis or direct photolysis in the environment. The complete miscibility of ethylene glycol mono-n-butyl ether in water suggests that volatilization, adsorption and bioconcentration are not important fate processes. This is supported by the estimated Henry's Law constant, which indicates that volatilization of ethylene glycol mono-n-butyl ether from environmental waters and moist soil should be extremely slow. A low estimated log BCF suggests ethylene glycol mono-n-butyl ether should not bioconcentrate among aquatic organisms. A low Koc indicates ethylene glycol mono-n-butyl ether should not partition from the water column to organic matter contained in sediments and suspended solids, and it should be highly mobile in soil. Limited monitoring data has shown it can leach to ground water. Aqueous screening test data indicate that biodegradation is likely to be the most important removal mechanism of ethylene glycol mono-n-butyl ether from aerobic soil and water. In the atmosphere, ethylene glycol mono-n-butyl ether is expected to exist almost entirely in the vapor phase and reactions with photochemically produced hydroxyl radicals should be important (estimated half-life of 17 hr). Physical removal of ethylene glycol mono-n-butyl ether from air by precipitation and dissolution in clouds may occur; however, its short atmospheric residence time suggests that wet deposition is of limited importance. The most probable human exposure would be occupational exposure, which may occur through dermal contact or inhalation at workplaces where it is produced or used. Workplace exposures have been documented. Limited monitoring data indicate that nonoccupational exposures can occur from the ingestion of contaminated drinking water supplies. Human exposures may also occur at construction sites and areas that have undergone remodelling.

Natural Sources:

Artificial Sources: Ethylene glycol mono-n-butyl ether may be released to the environment via effluents at sites where it is produced or used as a solvent for nitrocellulose resins, spray and quick-drying lacquers, varnishes, enamels, dry cleaning compounds, varnish and textile spot removers (in printing and dyeing), soaps solutions and emulsifying agents [11]. Ethylene glycol mono-n-butyl ether was listed as a volatile organic emission of silicone caulk [27]. Ethylene glycol mono-n-butyl ether is also released to the environment via leachate from municipal landfills and hazardous waste site [6,7,24].

Terrestrial Fate: Alcohols and ethers are generally resistant to hydrolysis [15]. They do not absorb UV light in the environmentally significant range, (>290 nm), and are commonly used as solvents for obtaining UV spectra [23]. Therefore, ethylene glycol mono-n-butyl ether should not undergo hydrolysis in moist terrestrial environments, or direct photolysis on sunlit soil surfaces. The estimated Henry's Law constant indicates that volatilization of ethylene glycol mono-n-butyl ether from moist soil should not be an important fate process [15]. An estimated Koc of 67 [15] indicates ethylene glycol mono-n-butyl ether should be highly mobile in soil [25]; limited monitoring data has shown it may leach to ground water [6,7,24]. Aqueous screening test data [2,5,18,20,26] suggest that biodegradation is likely to be the most important removal mechanism of ethylene glycol mono-n-butyl ether from aerobic soil.

Aquatic Fate: Alcohols and ethers are generally resistant to hydrolysis [15]. They do not absorb UV light in the environmentally significant range, >290 nm, and are commonly used as solvents for obtaining UV spectra [23]. Therefore, ethylene glycol mono-n-butyl ether should not undergo hydrolysis or direct photolysis in aquatic environments. The complete miscibility of ethylene glycol mono-n-butyl ether in water suggests that volatilization, adsorption and bioconcentration are not important fate processes. This is supported by the estimated Henry's Law constant, which indicates that volatilization of ethylene glycol mono-n-butyl ether from natural waters should be extremely slow [15]. An estimated Koc of 67 [1] indicates ethylene glycol mono-n-butyl ether should not partition from the water column to organic matter contained in sediments and suspended solids. An estimated bioconcentration factor (log BCF) of 0.40 [15] indicates ethylene glycol mono-n-butyl ether should not bioconcentrate among aquatic organisms. Aqueous screening test data [2,5,18,20,26] suggest that aerobic biodegradation is likely to be the most important removal mechanism of ethylene glycol mono-n-butyl ether from aquatic systems.

Atmospheric Fate: Alcohols and ethers do not absorb UV light in the environmentally significant range (>290 nm), and are commonly used as solvents for obtaining UV spectra [23]. Therefore, ethylene glycol mono-n-butyl ether should not undergo direct photolysis in the atmosphere. Based on the vapor pressure, ethylene glycol mono-n-butyl ether is expected to exist almost entirely in the vapor phase in ambient air [8] where vapor phase reactions with photochemically produced hydroxyl radicals may be important. The rate constant for ethylene glycol mono-n-butyl ether has been estimated to be 2.30×10^{-11} cm^3/molecule-sec at 25 °C, which corresponds to an atmospheric half-life of about 17 hr at an atmospheric concentration of $5 \times 10^{+5}$ hydroxyl radicals per cm^3 [1]. The complete miscibility of ethylene glycol mono-n-butyl ether in water indicates that physical removal from air by precipitation and dissolution in clouds may occur; however, its short atmospheric residence time suggests that wet deposition is of limited importance.

Biodegradation: Soil grab sample and river die-away test data pertaining to the biodegradation of ethylene glycol mono-n-butyl ether in soil and natural waters were not located in the available literature. Yet, a number of aerobic biological screening studies which utilized settled waste water, sewage, or activated sludge for inocula indicate that ethylene glycol mono-n-butyl ether should biodegrade rapidly in the environment [2,5,18,20,26]. Five-day BODT values ranged from 5% (without acclimation) [5] to 73% (with acclimation) [2]. Ten-day BODTs ranged from 57% [5] to 74% [18]. The maximum BODT reported was 88% for 20 days [18].

Abiotic Degradation: Alcohols and ethers are generally resistant to hydrolysis [15]. They do not absorb UV light in the environmentally significant range (>290 nm) and are commonly used as solvents for obtaining UV spectra [23]. Therefore, ethylene glycol mono-n-butyl ether should not undergo hydrolysis or direct photolysis in the environment. The rate constant for the vapor-phase reaction of ethylene glycol mono-n-butyl ether with photochemically produced hydroxyl radicals in air has been estimated to be 2.30×10^{-11} cm^3/molecule-sec at 25 °C, which corresponds to an atmospheric half-life of about 17 hr at an atmospheric concentration of $5 \times 10^{+5}$ hydroxyl radicals per cm^3 [1].

Bioconcentration: Because ethylene glycol mono-n-butyl ether is miscible in water, bioconcentration in aquatic systems is not expected to be an important fate process. Based upon the log Kow, a bioconcentration factor (log BCF) of 0.40 for ethylene glycol mono-n-butyl ether has been calculated using a recommended regression-derived equation [15]. This BCF value also indicates ethylene glycol mono-n-butyl ether should not bioconcentrate in aquatic organisms.

Soil Adsorption/Mobility: Because ethylene glycol mono-n-butyl ether is miscible in water, soil adsorption is not expected to be an important fate process. Based on the log Kow, a Koc of 67 for ethylene glycol mono-n-butyl ether has been calculated using a recommended regression-derived equation [15]. This Koc value indicates ethylene glycol mono-n-butyl ether will be highly mobile in soil [25], and it should not partition from the water column to organic matter contained in sediments and suspended solids.

Volatilization from Water/Soil: Because ethylene glycol mono-n-butyl ether is miscible in water, and based upon the estimated Henry's Law constant, the volatilization of ethylene glycol mono-n-butyl ether from natural bodies of water and moist soils is not expected to be an important fate process [15].

Water Concentrations: DRINKING WATER: Ethylene glycol mono-n-butyl ether was listed as a contaminant found in drinking water for a survey of US cities including Pomona, Escondido, Lake Tahoe and Orange Co, CA and Dallas, Washington, DC, Cincinnati, Philadelphia, Miami, New Orleans, Ottumwa, IA, and Seattle [14]. GROUND WATER: Ethylene glycol mono-n-butyl ether was detected at a concentration of 23 ug/L in 1 of 7 ground water samples collected near "The Valley of Drums", KY [24]. A ground water sample from an aquifer underlying a municipal landfill in Norman, OK contained ethylene glycol mono-n-butyl ether [6,7]. SURFACE WATER: In April 1980, ethylene glycol mono-n-butyl ether was detected in Hayashida River water (the Matsubara area in Tatsuno City, Hyogo Prefecture) at concentrations of 1310 and 5680 ppb [30].

Effluent Concentrations: Ethylene glycol mono-n-butyl ether was identified in 1 and 4 neutral fractions of 33 industrial wastewater effluents at concentrations of <10 and <100 ug/L, respectively [17]. Ethylene glycol mono-n-butyl ether was detected in ground water contaminated by leachate from a hazardous waste at "the Valley of

Drums", KY [23]. It was also detected, but not quantified, in the ground water leachate of a municipal landfill in Norman, OK [6,7]; it may be present in other landfill leachates [2,27]. Ethylene glycol mono-n-butyl ether was listed as a volatile organic emission of silicone caulk [17].

Sediment/Soil Concentrations:

Atmospheric Concentrations: INDOOR: According to the National Ambient Volatile Organic Compounds (VOCs) Database, the average daily indoor atmospheric concentration of ethylene glycol mono-n-butyl ether is 0.214 ppbV for 14 samples [21]. Ethylene glycol mono-n-butyl ether was detected at a concentration of 8 ug/m^3 in 1 of 6 samples of indoor air from 14 homes of northern Italy [4].

Food Survey Values:

Plant Concentrations:

Fish/Seafood Concentrations:

Animal Concentrations:

Milk Concentrations:

Other Environmental Concentrations: Ethylene glycol mono-n-butyl ether was contained in organic solvents with a frequency of occurrence of 0.4% [13]. A paint stripping formulation was comprised of 35% ethylene glycol mono-n-butyl ether [9]. Ethylene glycol mono-n-butyl ether was not detected in a machine cutting fluid prior to its use; however, the used fluid contained ethylene glycol mono-n-butyl ether at a concentration of 0.060 ug/g [29].

Probable Routes of Human Exposure: The most probable route of human exposure to ethylene glycol mono-n-butyl ether is by inhalation, dermal contact and ingestion. Workplace exposures have been documented [3,9,13,22,28]. Drinking water supplies have been shown to contain ethylene glycol mono-n-butyl ether [14].

Average Daily Intake:

Ethylene Glycol Monobutyl Ether

Occupational Exposure: NIOSH (NOES Survey as of 3/28/89) has estimated that 1,680,764 workers are potentially exposed to ethylene glycol mono-n-butyl ether in the USA [16]. According to the National Ambient Volatile Organic Compounds (VOCs) Database, the median workplace atmospheric concentration of ethylene glycol mono-n-butyl ether is 0.075 ppbV for 14 samples [22]. Workers at paint stripping operations that used stripping agents containing ethylene glycol mono-n-butyl ether were exposed to it [9]. Personal exposures to atmospheric ethylene glycol mono-n-butyl ether at a specialty chemical production facility in June of 1981 ranged from undetected levels to 0.1 ppm; indoor air concentrations within the facility were as high as 1.7 ppm [3]. A national survey of workplaces in Germany showed that workers were exposed to solvents containing ethylene glycol mono-n-butyl ether with a 0.4% frequency of occurrence [13]. A study initiated in 1983, which surveyed the workplace atmospheres of 336 businesses in Belgium, showed that ethylene glycol mono-n-butyl ether was present in 25 of 94 air samples taken from sites that utilize printing pastes; 10 of 81 samples from where painting took place; 1 of 20 samples from automobile repair shops; and 17 of 67 samples from sites where various materials such as varnishes, sterilization agents and cleaners are employed [28]. The geometric mean concentration of ethylene glycol mono-n-butyl ether in the air of printing shops was 4.1 mg/m^3 with a range of 1.5 to 17.7 mg/m^3; 18.8 mg/m^3 with a range of 3.4 to 93.6 mg/m^3 for painting areas; 5.9 mg/m^3 for car repair shops; and 8.5 mg/m^3 with a range of 0.2 to 1775 mg/m^3 for various industries [28]. Ethylene glycol mono-n-butyl ether was identified as a volatile emission from used machine cutting oils in a automobile manufacturing facility in Japan [29]. Nonoccupational exposures may occur among populations with contaminated drinking water supplies [14]. Because ethylene glycol mono-n-butyl ether is a component of solvent-based building materials such as silicone caulk [27], human exposures may occur at construction sites and areas that have undergone remodelling.

Body Burdens:

REFERENCES

1. Atkinson R; Intern J Chem Kin 19: 799-828 (1987)
2. Bridie AL et al; Water Res 13: 627-30 (1979)
3. Clapp DE et al; Environ Health Perspective 57: 91-5 (1984)

Ethylene Glycol Monobutyl Ether

4. DeBortoli M et al; Environ Int 12: 343-50 (1986)
5. Dow Chemical Co; The Glycol Ethers Handbook. Midland, MI (1981)
6. Dunlap WJ et al; Organic Pollutants contributed to groundwater by a Landfill USEPA-600/9-76-004 pp 96-110 (1976)
7. Dunlap WJ et al; Identif Anal Org Pollut 1: 453-77 (1976)
8. Eisenreich SJ et al; Environ Sci Technol 15: 30-8 (1981)
9. Hahn WJ, Werschulz PO; Evaluation of alternatives to toxic organic paint strippers. NTIS PB86 219-177/AS USEPA 600/S2-86/063 (1986)
10. Hansch C, Leo AJ; Medchem Project Issue No 26. Claremont CA: Pomona College (1985)
11. Hawley GG; Condensed Chemical Dictionary 10th ed Van Nostrand Reinhold NY p 375 (1981)
12. Hine J, Mookerjee PK; J Org Chem 40: 292-8 (1975)
13. Lehmann E et al; pp 31-41 in Safety and health aspects of organic solvents. Riihimaki V, Ulfvarson U eds Alan R Liss Inc. (1986)
14. Lucas SV; GC/MS Anal of Org in Drinking Water Concentrates and Advanced Treatment Concentrates Vol 1 USEPA-600/1-84-020A (NTIS PB85-128239) p 397 (1984)
15. Lyman WJ et al; Handbook of Chemical Property Estimation Methods NY: McGraw-Hill pp 4-9, 5-4, 6-3, 15-16 (1982)
16. NIOSH; National Occupational Exposure Survey (NOES) (1989)
17. Perry DL et al; Ident of Org Compounds in Ind Effluent discharges USEPA-600/4-79-016 (NTIS PB-294794) p 230 (1979)
18. Price KS et al; J Water Pollut Contr Fed 46: 63-77 (1974)
19. Riddick JA et al; Organic Solvents NY,NY: John Wiley & Sons Inc (1984)
20. Sasaki S; pp 283-98 in Aquat Pollutants: Transform & Biolog Effects. Hutzinger O et al ed Oxf: Pergamon Press (1978)
21. Shah JJ, Singh HB; Environ Sci Technol 22: 1381-8 (1988)
22. Shah JJ, Heyerdahl EK; National Ambient VOC Database Update USEPA-600/3-88/010 (1988)
23. Silverstein RM, Bassler GC; Spectrometric Id Org Cmpd J Wiley & Sons Inc pp 148-169 (1963)
24. Stonebreaker RD, Smith AJ; pp 1-10 in Contr Haz Mater Spills, Proc Natl Conf Louisville KY (1980)
25. Swann RL et al; Res Rev 85: 16-28 (1983)
26. Takemoto S et al; Suishitsu Odaku Kenkyu 4: 80-90 (1981)
27. Tichenor BA, Mason MA; JAPCA 38: 264-8 (1988)
28. Veulemans H et al; Am Indust Hyg Assoc J 48: 671-7 (1987)
29. Yasuhara, A et al; Agric Bio Chem 50: 1765-70 (1986)
30. Yasuhara, A et al; Environ Sci Technol 15: 570-3 (1981)

Ethylene Glycol Monoethyl Ether Acetate

SUBSTANCE IDENTIFICATION

Synonyms: 1-Acetoxy-2-ethoxyethane

Structure:

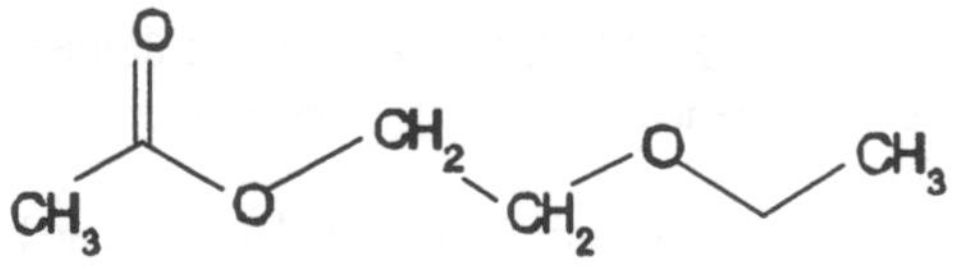

CAS Registry Number: 111-15-9

Molecular Formula: $C_6H_{12}O_3$

Wiswesser Line Notation: 2O2OV1

CHEMICAL AND PHYSICAL PROPERTIES

Boiling Point: 156 °C

Melting Point: -61.7 °C

Molecular Weight: 132.16

Dissociation Constants:

Log Octanol/Water Partition Coefficient: 0.65 estimate [5]

Water Solubility: 229,000 mg/L at 20 °C [18]

Vapor Pressure: 1.65 mm Hg at 20 °C [6]

Henry's Law Constant: 1.25×10^{-6} atm-m^3/mole at 20 °C (calculated from the vapor pressure and water solubility)

ENVIRONMENTAL FATE/EXPOSURE POTENTIAL

Summary: Ethylene glycol monoethyl ether acetate is directly released to the atmosphere through its use as a solvent in paints, lacquers, thinners, inks, stains, and varnishes. If released to the atmosphere, it will degrade primarily by reaction with photochemically produced hydroxyl radicals [estimated half-life of 1.2 days]. If released to soil or water, ethylene glycol monoethyl ether acetate is expected to degrade via biodegradation. Several biodegradation screening tests have demonstrated that ethylene glycol monoethyl ether acetate is readily biodegradable. It may leach readily in soils based upon its high water solubility. The major routes of occupational exposure to ethylene glycol monoethyl ether acetate are inhalation and skin absorption. Exposure to the general population can also occur through inhalation and dermal contact.

Natural Sources:

Artificial Sources: Ethylene glycol monoethyl ether acetate is directly released to the atmosphere through its use as a solvent in paints [21], lacquers, thinners, inks, stains, and varnishes.

Terrestrial Fate: The dominant degradation process for ethylene glycol monoethyl ether acetate in soil is expected to be biodegradation. Several screening biodegradation tests have demonstrated that ethylene glycol monoethyl ether acetate is readily biodegradable [3,16,20,24]. Based upon an estimated Koc of 5, ethylene glycol monoethyl ether acetate is expected to leach readily in soil [11]. The importance of leaching may be lessened by relatively rapid, concurrent biodegradation.

Aquatic Fate: The dominant removal process for ethylene glycol monoethyl ether acetate in water is expected to be biodegradation. Several biodegradation screening tests have demonstrated that ethylene glycol monoethyl ether acetate is readily biodegradable [3,16,20,24]. Volatilization from water is slow. The volatilization half-lives from a model environmental river (1 meter deep) and model pond have been estimated to be 34 days and 1 yr, respectively [11,25]. Aquatic hydrolysis, bioconcentration, and adsorption to sediment are not expected to be important.

Atmospheric Fate: Based upon the vapor pressure, ethylene glycol monoethyl ether acetate is expected to exist almost entirely in the vapor phase in the ambient atmosphere [9]. It is expected to degrade in an average ambient atmosphere by reaction with photochemically produced hydroxyl radicals with an estimated half-life of about 1.2 days [1]. Physical removal via wet deposition is likely since it is very soluble in water.

Biodegradation: A 5-day theoretical BOD of 41% was measured for ethylene glycol monoethyl ether acetate using a sewage inoculum and the standard dilution method [3]. Theoretical BODs of 36, 79, 82, and 80% were measured for 5, 10, 15, and 20 day inoculation periods, respectively, using a nonacclimated, settled wastewater seed [16]. Theoretical BODs of 10, 44, 59, and 69% were measured for 5, 10, 15, and 20 day inoculation periods, respectively, using a nonacclimated seawater seed with added raw wastewater [16]. Biodegradation of ethylene glycol monoethyl ether acetate was considered to be "well-biodegradable" (theoretical BOD > 30% after 14 days of inoculation) using the Japanese MITI protocol [20]. A 5-day theoretical BOD of 18.1% was measured using a sewage inoculum and the standard dilution method [24]; a 5-day theoretical BOD of 1.1% was measured using a seawater inoculum and a seawater dilution method [24].

Abiotic Degradation: The rate constant for the vapor-phase reaction of ethylene glycol monoethyl ether acetate with photochemically produced hydroxyl radicals has been experimentally determined to be 13.0×10^{-12} cm^3/molecule-sec at room temperature which corresponds to an atmospheric half-life of about 1.2 days at an atmospheric concentration of $5 \times 10^{+5}$ hydroxyl radicals per cm^3 [1]. Ethylene glycol monoethyl ether acetate is not reactive with water [26]; therefore, aqueous hydrolysis is not environmentally important.

Bioconcentration: Based upon the water solubility, the BCF for ethylene glycol monoethyl ether acetate can be estimated to be 0.6 from a regression-derived equation [11]. This BCF value suggests that ethylene glycol monoethyl ether acetate will not bioconcentrate significantly in aquatic organisms.

Soil Adsorption/Mobility: Based upon the water solubility, the Koc for ethylene glycol monoethyl ether acetate can be estimated to be 5 from a regression-derived equation [11]. This Koc value suggests that ethylene glycol monoethyl ether acetate has very high soil mobility [23].

Ethylene Glycol Monoethyl Ether Acetate

Volatilization from Water/Soil: The calculated Henry's Law constant for ethylene glycol monoethyl ether acetate suggests volatilization from environmental waters should be slow with the possible exception of very shallow rivers [11]. Based on this Henry's Law constant, the volatilization half-life from a model river (1 m deep flowing 1 m/sec with a wind velocity of 3 m/sec) can be estimated to be about 34 days [11]. Volatilization half-life from a model environmental pond can be estimated to be about 1 yr [25]. Ethylene glycol monoethyl ether acetate has a relative evaporation rate of 0.2 (relative to n-butyl acetate = 1) from nonadsorbing surfaces which classifies it as a slow-evaporating solvent [22]; the actual evaporation rate from a nonadsorbing surface at 25 °C (0-5% relative humidity, constant air-flow) is 50% evaporated in 25 min [7].

Water Concentrations: GROUND WATER: Ethylene glycol monoethyl ether acetate was qualitatively detected in ground water 30 meters beneath a paint factory in Milan, Italy [2].

Effluent Concentrations: The emission rate of ethylene glycol monoethyl ether acetate into the atmosphere from painting operations at an automotive assembly plant in Janesville, WI was estimated to be 37.9 gallons/hr [21].

Sediment/Soil Concentrations:

Atmospheric Concentrations: Ethylene glycol monoethyl ether acetate was qualitatively detected in ambient air samples collected near the Southern Black Forest in Germany during 1984-5 monitoring [10]. Ethylene glycol monoethyl ether acetate levels of 2-130 ug/m^3 were in indoor air of 5 residential apartments in Italy during 1983-4 monitoring [8].

Food Survey Values:

Plant Concentrations:

Fish/Seafood Concentrations:

Animal Concentrations:

Ethylene Glycol Monoethyl Ether Acetate

Milk Concentrations:

Other Environmental Concentrations: Inks, thinners and solvents commonly used in the screen printing industry may contain from 20-90% ethylene glycol monoethyl ether acetate [19].

Probable Routes of Human Exposure: The major routes of occupational exposure to ethylene glycol monoethyl ether acetate are inhalation and skin absorption [4]. Exposure to the general population can also occur through inhalation and dermal contact.

Average Daily Intake:

Occupational Exposure: NIOSH (NOES Survey 1981-1983) has statistically estimated that 221,841 workers are potentially exposed to ethylene glycol monoethyl ether acetate in the USA [14]. NIOSH (NOHS Survey 1972-1974) has statistically estimated that 321,125 workers are potentially exposed to ethylene glycol monoethyl ether acetate in the USA [13]. A total of 357 personal and workplace air samples were collected between 1982-1985 at 7 US semi-conductor manufacturing companies using ethylene glycol monoethyl ether acetate [15]; ethylene glycol monoethyl ether acetate concentrations ranged from 0.001-18.0 ppm with a mean concentration of 0.05 ppm for TWA exposure and a mean concentration of 1.56-2.82 ppm for short term exposure [15]. Levels of 27.0-166.5 mg/m^3 were detected in workplace air samples from two auto paint shops in Spain [12]. A survey of furniture factories in Finland detected ethylene glycol monoethyl ether acetate in 9% of all air samples with a mean concentration of 7 ppm [17]. The mean TWA concentration at various job sites in the screen print industry has been reported to be 5-18.5 ppm [19].

Body Burdens:

REFERENCES

1. Atkinson R; J Inter Chem Kinet 19: 799-828 (1987)
2. Botta D et al; Comm Eur Comm Eur 8518(Anal Org Micropollut Water): 261-75 (1984)
3. Bridie AL et al; Water Res 13: 627-30 (1979)
4. Clapp DE et al; Environ Health Perspec 57: 91-5 (1984)

Ethylene Glycol Monoethyl Ether Acetate

5. CLOGP; PCGEMS Graphical Exposure Modeling System USEPA (1986)
6. Daubert TE, Danner RP; Data Compilation, Tables of Properties of Pure Cmpds, Design Inst for Phys Prop Data, Am Inst for Phys Prop Data, NY,NY (1989)
7. Davis DS; Amer Perfumer Cosmet 81: 32-4 (1965)
8. DeBortoli M et al; Environ Intern 12: 343-50 (1986)
9. Eisenreich SJ et al; Environ Sci Technol 15: 30-8 (1981)
10. Juttner F; Chemosphere 15: 985-92 (1986)
11. Lyman WJ et al; Handbook of Chemical Property Estimation Methods NY: McGraw-Hill pp 4-9, 5-10, 15-15 to 15-29 (1982)
12. Medinilla J, Espigares M; Ann Occup Hyg 32: 509-13 (1988)
13. NIOSH; National Occupational Hazard Survey (NOHS) (1974)
14. NIOSH; National Occupational Exposure Survey (NOES) (1983)
15. Paustenbach DJ; J Toxicol Environ Health 23: 29-75 (1988)
16. Price KS et al; J Water Pollut Control Fed 46: 63-77 (1974)
17. Priha E et al; Ann Occup Hyg 30: 289-94 (1986)
18. Riddick JA et al; Organic Solvents NY: John Wiley & Sons Inc (1984)
19. Samini B; Amer Ind Hyg Assoc J 43: 858-62 (1982)
20. Sasaki S; pp 283-98 in Aquatic Pollutants Hutzinger O et al eds Oxford: Pergamon Press (1978)
21. Sexton K; Westberg H; Environ Sci Technol 14: 329-332 (1980)
22. Smith RL; Environ Health Perspect 57: 1-4 (1984)
23. Swann RL et al; Res Rev 85: 23 (1983)
24. Takemoto S et al; Suishitsu Odaku Kenkyu 4: 80-90 (1981)
25. USEPA; EXAMS II Computer Simulation (1987)
26. USEPA; Chemical Hazard Information Profile (CHIP): 2-Ethoxyethanol acetate. Washington, DC: USEPA (1982)

2-Ethylhexyl Acetate

SUBSTANCE IDENTIFICATION

Synonyms:

Structure:

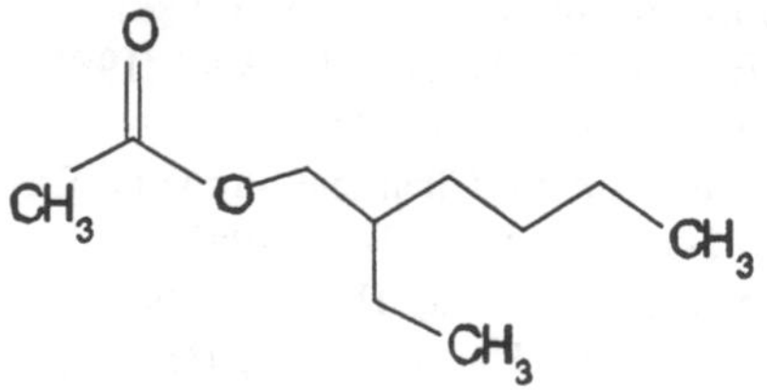

CAS Registry Number: 103-09-3

Molecular Formula: $C_{10}H_{20}O_2$

Wiswesser Line Notation: 4Y2&1OV1

CHEMICAL AND PHYSICAL PROPERTIES

Boiling Point: 199 °C

Melting Point: -93 °C

Molecular Weight: 172.27

Dissociation Constants:

Log Octanol/Water Partition Coefficient: 3.72, estimate [5]

Water Solubility: 98.4 mg/L at 25 °C, estimate [9]

Vapor Pressure: 0.23 mm Hg at 25 °C [2]

Henry's Law Constant: 2.81 x 10^{-3} atm-m^3/mole at 25 °C, estimate [6]

ENVIRONMENTAL FATE/EXPOSURE POTENTIAL

Summary: 2-Ethylhexyl acetate, which is used as a solvent, may be released in fugitive emissions during its manufacture, formulation, or use in commercial products. If released to soil, 2-ethylhexyl acetate is expected to display slight mobility. Volatilization is expected to occur from both moist and dry soils. Hydrolysis of 2-ethylhexyl acetate in soil is not expected to be a significant process except in highly basic soils with a pH >9. If released to water, 2-ethylhexyl acetate is expected to rapidly volatilize to the atmosphere. The half-life for volatilization from a model river is 4.2 h. 2-Ethylhexyl acetate may adsorb to sediment and suspended organic matter and it may bioconcentrate in fish and aquatic organisms. Hydrolysis of 2-ethylhexyl acetate in aquatic systems is not expected to be a significant process except under basic conditions of pH >9. In the atmosphere, 2-ethylhexyl acetate is expected to undergo a gas-phase reaction with photochemically produced hydroxyl radicals, with an estimated half-life on the order of 1.5 days. 2-Ethylhexyl acetate may undergo atmospheric removal by wet deposition processes. The probable routes of exposure to 2-ethylhexyl acetate are by inhalation and dermal contact during the production and use of this compound. The public may also be exposed to 2-ethylhexyl acetate by these routes and possibly contaminated drinking water.

Natural Sources: There are no data available to indicate that 2-ethylhexyl acetate is released to the environment from natural sources.

Artificial Sources: In 1989, one company was listed as a producer of ethylhexyl acetate [11]. Current production volumes are not available. In 1977, 1 million pounds of ethylhexyl acetate were produced in the United States [13]. Ethylhexyl acetate has been used as a high boiling solvent used to promote flow and retard blushing in lacquers, emulsions, and silk screen inks, and as a flow control agent in baking enamels [4]. It can also be used as a dispersant for vinyl organosols and as a coalescing agent for latex paints [4]. Ethylhexyl acetate may be released to the environment as a fugitive emission during its production, formulation and use.

Terrestrial Fate: If released to soil, a calculated soil adsorption coefficient of 2515 [9] obtained from the estimated octanol/water partition coefficient, suggests that 2-ethylhexyl acetate will display slight mobility in soil [12]. The estimated Henry's Law constant for 2-ethylhexyl acetate, and its vapor pressure, suggest that this compound will volatilize from both moist and dry soil. Hydrolysis of 2-ethylhexyl acetate in soil is not expected to be a significant process, except in highly basic soils with a pH >9, as in general, this process is too slow for alkyl esters to be environmentally significant under acidic, neutral, and slightly basic conditions [4].

Aquatic Fate: If released to water, 2-ethylhexyl acetate is expected to rapidly volatilize to the atmosphere. Based on the calculated Henry's Law constant, the half-life for volatilization from a model river is 4.2 hr [9]. From the estimated log octanol water partition coefficient, a calculated soil adsorption coefficient of 2515 [9] and a calculated bioconcentration factor of 395 [9] suggest that 2-ethylhexyl acetate may adsorb to sediment and suspended organic matter and is expected to bioconcentrate in fish and aquatic organisms. Hydrolysis of 2-ethylhexyl acetate in aquatic systems is not expected to be a significant process except under basic conditions of pH >9, as, in general, this process is too slow for alkyl esters to be environmentally significant under acidic, neutral, and slightly basic conditions [4].

Atmospheric Fate: In the atmosphere, a calculated rate constant for the gas phase reaction of 2-ethylhexyl acetate with photochemically produced hydroxyl radicals is 1.09×10^{-11} cm^3/mol-sec [1], which corresponds to an atmospheric half-life of 1.5 days. The water solubility, obtained from the estimated log octanol/water partition coefficient, suggests that this compound may be removed from the atmosphere by wet deposition.

Biodegradation:

Abiotic Degradation: In general, alkyl esters, especially those with branching on the carbon attached to the ester oxygen, are resistant to hydrolysis under acidic conditions except with highly acidic solutions or at elevated temperatures [3]. By analogy to propylacetate, hydrolysis of 2-ethylhexyl acetate at basic pHs <9 is expected to be too slow to be environmentally significant [4]. An estimated rate constant for the gas-phase reaction between 2-ethylhexyl acetate and photochemically produced hydroxyl radicals is 1.09×10^{-11} cu-cm/molecule-sec [1], which translates to a half-life of 1.5 days using an average atmospheric hydroxyl radical concentration of $5 \times 10^{+5}$ molecules/cu-cm [1].

Bioconcentration: From the estimated log octanol/water partition coefficient for 2-ethylhexyl acetate, a bioconcentration factor of 395 can be calculated using a regressional analysis [9]. This value suggests that 2-ethylhexyl acetate may bioconcentrate in fish and aquatic organisms.

Soil Adsorption/Mobility: From the estimated log octanol/water partition coefficient for 2-ethylhexyl acetate, a soil adsorption coefficient of 2515 can be calculated using a regressional analysis [9]. This value suggests that 2-ethylhexyl acetate will display slight mobility in soil [12].

Volatilization from Water/Soil: Based on the estimated Henry's Law constant at 25 °C for 2-ethylhexyl acetate, the half-life for volatilization from a model river 1 m deep, flowing at 1 m/sec and a wind speed of 3 m/sec is 4.2 hr [9]. The estimated Henry's Law constant of 2-ethylhexyl acetate and its vapor pressure at 25 °C, suggest that volatilization from both moist and dry soil to the atmosphere, respectively, will be significant processes.

Water Concentrations: DRINKING WATER: 2-Ethylhexyl acetate was listed as a compound detected in US drinking water supplies [7].

Effluent Concentrations:

Sediment/Soil Concentrations:

Atmospheric Concentrations:

Food Survey Values:

Plant Concentrations:

Fish/Seafood Concentrations:

Animal Concentrations:

Milk Concentrations:

Other Environmental Concentrations:

Probable Routes of Human Exposure: The probable routes of occupational exposure to 2-ethylhexyl acetate are by inhalation and dermal contact during the production and use of this compound. The general population may also be exposed to ethylhexyl acetate by inhalation or dermal contact.

Average Daily Intake:

Occupational Exposure: NIOSH (NOES Survey 1981-83) has statistically estimated that 2,499 workers are potentially exposed to 2-ethylhexyl acetate in the USA, 91% of which are exposed during the use of trade name compounds in which 2-ethylhexyl acetate is contained [10].

Body Burdens: 2-Ethylhexyl acetate was detected in 4.1% of 387 expired air samples obtained from 54 nonsmoking volunteers from Chicago, IL, at a mean concentration of 0.168 ng/L [8].

REFERENCES

1. Atkinson R; Chem Rev 85: 69-201 (1985)
2. Daubert TE, Danner RP; Physical and Thermodynamic Properties of Pure Chemicals: Data Compilation. Design Inst Phys Prop Data. Amer Inst Chem Eng Hemisphere Pub Corp, NY (1989)
3. Drossman H et al; Chemosphere 17: 1509-30 (1987)
4. Elam EU; Kirk-Othmer Encycl Chem Tech 3rd Ed. John-Wiley NY 9: 311-37 (1978)
5. GEMS; Graphic Exposure Modeling System CLOGP USEPA (1987)
6. Hine J, Mookerjee PK; J Org Chem 40: 292-8 (1975)
7. Kool HJ et al; CRC Crit Rev Env Control 12: 307-57 (1982)
8. Krotoszynski BK et al; J Anal Toxicol 3: 225-34 (1979)
9. Lyman WJ et al; Handbook of Chemical Property Estimation Methods NY: McGraw-Hill pp 4-1 to 4-33 and 5-1 to 5-30 and 15-1 to 15-34 (1982)
10. NIOSH; National Occupational Exposure Survey (NOES) (1989)
11. SRI International; Directory of Chemical Producers (1989)
12. Swann RL et al; Res Rev 85: 17-28 (1983)
13. USEPA; Nonconfidential Initial TSCA Inventory. Washington DC Off Toxic Subst (1977)

Ethyl Methyl Ether

SUBSTANCE IDENTIFICATION

Synonyms:

Structure:

$$CH_3\diagdown\diagup O\diagdown CH_3$$

CAS Registry Number: 540-67-0

Molecular Formula: C_3H_8O

Wiswesser Line Notation: 2O1

CHEMICAL AND PHYSICAL PROPERTIES

Boiling Point: 10.8 °C

Melting Point:

Molecular Weight: 60.10

Dissociation Constants:

Log Octanol/Water Partition Coefficient: 0.341 [7] (estimated)

Water Solubility: 205,000 mg/L [7,10] (estimated)

Vapor Pressure: 760 mm Hg at +7.5 °C [16]; 1506 mm Hg at 25 °C [5]

Henry's Law Constant: 1.14×10^{-3} atm-m^3/mole [11] (estimated from structure); 2.93×10^{-4} atm-m^3/mole (calculated from the water solubility and vapor pressure)

299

Ethyl Methyl Ether

ENVIRONMENTAL FATE/EXPOSURE POTENTIAL

Summary: Ethyl methyl ether may be released to the environment as a result of its manufacture and use. It has been used as an anesthetic. If ethyl methyl ether is released to soil, it will be subject to volatilization. It will be expected to exhibit very high mobility in soil and, therefore, it may leach to ground water. It will not be expected to hydrolyze in soil. If ethyl methyl ether is released to water, it will not be expected to significantly adsorb to sediment or suspended particulate matter, bioconcentrate in aquatic organisms, hydrolyze, directly photolyze, or photooxidize via reaction with photochemically produced hydroxyl radicals in the water, based upon estimated physical-chemical properties or analogies to other structurally related aliphatic ethers. In surface water, it will be subject to rapid volatilization with estimated half-lives of 4.6 and 2.2 days for volatilization from a river one meter deep flowing 1 m/sec with a wind velocity of 3 m/sec and a model pond, respectively. Ethyl methyl ether may be resistent to biodegradation in environmental media based upon screening test data for the structurally related diethyl ether from studies using activated sludge or sewage inocula. Many ethers are known to be resistant to biodegradation. If ethyl methyl ether is released to the atmosphere, it will be expected to exist almost entirely in the vapor phase based on its vapor pressure. It will be susceptible to photooxidation via vapor phase reaction with photochemically produced hydroxyl radicals with an estimated half-life of 2.6 days for this process. Direct photolysis will not be an important removal process since aliphatic ethers do not absorb light at wavelengths >290 nm. The most probable route of general population exposure to ethyl methyl ether is via inhalation of contaminated air. Exposure through dermal contact may occur in occupational settings.

Natural Sources:

Artificial Sources: Ethyl methyl ether may be released to the environment as a result of its manufacture and use. It has been used as an anesthetic [8].

Terrestrial Fate: If ethyl methyl ether is released to soil, it will be subject to volatilization based upon its calculated Henry's Law constant [10] and vapor pressure. It will be expected to exhibit very high mobility

300

[13] in soil and, therefore, it may leach to ground water, based upon an estimated Koc of 37 [10]. It will not be expected to hydrolyze in soil [10]. Aqueous screening test data from studies using activated sludge or sewage inocula for the structurally related diethyl ether [9,14] suggest that ethyl methyl ether may be resistent to biodegradation in environmental media. Many ethers are known to be resistant to biodegradation [1].

Aquatic Fate: If ethyl methyl ether is released to water, it will not be expected to significantly adsorb to sediment or suspended particulate matter [10], bioconcentrate in aquatic organisms [10], hydrolyze [10], directly photolyze [4], or photooxidize via reaction with photochemically produced hydroxyl radicals in the water [2], based upon estimated physical-chemical properties or analogies to other structurally related aliphatic ethers [4,10]. Ethyl methyl ether in surface water will be subject to rapid volatilization. Using a calculated Henry's Law constant [10], a half-life for volatilization of ethyl methyl ether from a river one meter deep flowing 1 m/sec with a wind velocity of 3 m/sec has been estimated to be 4.6 hr at 25 °C [10]. The volatilization half-life from a model pond, which considers the effect of adsorption, has been estimated to be 2.2 days [15]. Ethyl methyl ether may be resistent to biodegradation in environmental media based upon screening test data for the structurally related diethyl ether from studies using activated sludge or sewage inocula [9,14]. Many ethers are known to be resistant to biodegradation [1].

Atmospheric Fate: If ethyl methyl ether is released to the atmosphere, it will be expected to exist almost entirely in the vapor phase [6] since it is a gas at room temperature and pressure. It will be susceptible to photooxidation via vapor phase reaction with photochemically produced hydroxyl radicals. An atmospheric half-life of 2.6 days at an atmospheric concentration of 5×10^5 hydroxyl radicals per cm^3 has been calculated for this process based upon an estimated rate constant [3]. Direct photolysis will not be an important removal process since aliphatic ethers do not absorb light at wavelengths >290 nm [4].

Biodegradation: No data concerning the biodegradation of ethyl methyl ether in environmental media were located. Many ethers are known to be resistant to biodegradation [1] based mainly upon results from aqueous screening tests, including the structurally similar diethyl ether [9].

Abiotic Degradation: The rate constant for the vapor-phase reaction of ethyl methyl ether with photochemically produced hydroxyl radicals has been estimated to be 6.2×10^{-12} cm^3/molecule-sec at 25 °C [3] which corresponds to an atmospheric half-life of 2.6 days at an atmospheric concentration of 5×10^5 hydroxyl radicals per cm^3. Hydrolysis is not expected to be significant under normal environmental conditions (pH 5-9) [10]. Direct photolysis will not be an important removal process since aliphatic ethers do not absorb light at wavelengths >290 nm [4].

Bioconcentration: Based upon an estimated log Kow of 0.341 [7], a BCF of 1.1 has been estimated using a recommended regression equation [10]. Based upon this estimated BCF, ethyl methyl ether will not be expected to bioconcentrate in aquatic organisms.

Soil Adsorption/Mobility: Based upon an estimated octanol/water partition coefficient, a Koc of 37 has been estimated using a recommended regression equation [10]. Based upon this estimated Koc, ethyl methyl ether will be expected to exhibit very high mobility in soil [13]. Ethyl methyl ether, therefore, may leach through soil to ground water.

Volatilization from Water/Soil: Based upon an estimated Henry's Law constant for ethyl methyl ether, the volatilization half-life from a model river (1 meter deep flowing 1 m/sec with a wind speed of 3 m/sec) has been estimated to be 4.6 hr [10]. The volatilization half-life from a model pond, which considers the effect of adsorption, has been estimated to be 2.2 days [15]. Based upon the Henry's Law constant and vapor pressure, ethyl methyl ether will be subject to volatilization from surfaces and near-surface soil.

Water Concentrations:

Effluent Concentrations:

Sediment/Soil Concentrations:

Atmospheric Concentrations:

Ethyl Methyl Ether

Food Survey Values:

Plant Concentrations:

Fish/Seafood Concentrations:

Animal Concentrations:

Milk Concentrations:

Other Environmental Concentrations:

Probable Routes of Human Exposure: The most probable route of general population exposure to ethyl methyl ether is via inhalation of contaminated air. Exposure through dermal contact may occur in occupational settings.

Average Daily Intake:

Occupational Exposure:

Body Burdens:

REFERENCES

1. Alexander M; Biotechnol Bioeng 15: 611-47 (1973)
2. Anbar M, Neta P; Int J Appl Radiation Isotopes 18: 493-523 (1967)
3. Atkinson R; Intern J Chem Kinetics 19: 799-828 (1987)
4. Calvert JG, Pitts JNJr; Photochemistry John Wiley & Sons NY pp 441-2 (1966)
5. Daubert TE, Danner RP; Data Compilation Tables of Properties of Pure Compounds Am Inst Chem Eng (1989)
6. Eisenreich SJ et al; Environ Sci Technol 15: 30-8 (1981)
7. GEMS; Graphical Exposure Modeling System CLOGP3 (1986)
8. Hawley GG; Condensed Chemical Dictionary 10th ed Van Nostrand Reinhold NY p 440 (1981)
9. Heukelekian H, Rand MC; J Water Pollut Control Assoc 29: 1040-53 (1955)
10. Lyman WJ et al; Handbook of Chem Property Estimation Methods NY: McGraw-Hill pp 2-14, 4-9, 5-5, 7-4, 15-15 to 15-29 (1982)
11. Meylan WM, Howard PH; Environ Toxicol Chem 10:1283-93 (1991)
12. Riddick JA et al; Organic Solvents John Wiley & Sons Inc NY (1984)
13. Swann RL et al; Res Rev 85: 17-28 (1983)
14. Takemoto S et al; Suishitsu Odaku Kenkyu 4: 80-90 (1981)

Ethyl Methyl Ether

15. USEPA; EXAMS II Computer Simulation (1987)
16. Weast RC; Handbook of Chemistry and Physics 69th edition Boca Raton, FL: CRC Press (1988)

Formamide

Synonyms: Formic acid, amide

Structure:

$$H-\overset{\displaystyle O}{\underset{}{C}}-NH_2$$

CAS Registry Number: 75-12-7

Molecular Formula: CH_3NO

Wiswesser Line Notation: ZVH

CHEMICAL AND PHYSICAL PROPERTIES

Boiling Point: 210.5 °C at 760 mm Hg [17]

Melting Point: 2.55 °C [17]

Molecular Weight: 45.04

Dissociation Constants:

Log Octanol/Water Partition Coefficient: -1.51 [4]

Water Solubility: Miscible at 25 °C [2]

Vapor Pressure: 0.03 mm Hg at 25 °C (estimated using the Antoine equation and experimental data in T-range 70.5 to 210.5 °C) [17]

Henry's Law Constant: 1.531 x 10^{-8} atm-m^3/mole at 25 °C (estimated) [9]

ENVIRONMENTAL FATE/EXPOSURE POTENTIAL

Summary: Formamide may be released to the environment as a result of its manufacture and wide use as an intermediate and solvent. If released to the atmosphere, vapor-phase formamide is degraded relatively rapidly by reaction with photochemically produced hydroxyl radicals (estimated half-life of 2.1 hr in air). Based on the complete water solubility of formamide, removal by rainout may be important. Several biodegradation screening studies have observed biodegradation of formamide; therefore, if released to soil, formamide may be subject to biodegradation. Based on an estimated Koc value of 3.6, formamide is expected to leach. Based on a low vapor pressure, formamide may volatilize from dry surface soils; however, volatilization from moist soil surfaces is not an important fate process. If released to water, biodegradation is expected to be the predominant fate process. The chemical structure of formamide suggests that it may be susceptible to environmental hydrolysis although no rate data are available that indicate hydrolysis is an important aquatic fate process. Volatilization, adsorption to sediment and suspended materials, and bioconcentration (estimated BCF of 0.0417) are not expected to be important fate processes in aquatic environments. In occupational settings, exposure to formamide may occur through inhalation of vapors and through eye and skin contact. The most probable route of consumer exposure to formamide is probably as a result of its use in water-soluble ink formulations.

Natural Sources:

Artificial Sources: Formamide is used as a solvent in the manufacture processing of plastics, nonaqueous electrolysis, crystallization of pharmaceuticals such as penicillin and dihydrostreptomycin sulfate (82% of imported formamide is used by the pharmaceutical industry [16]), separation of chlorosilanes and purification of oils and fats [2]. It is also used in soil stabilization, as an ink solvent in fiber and felt-tip pens and markers, and in laboratory applications [16]. Formamide has been used or has the potential to be used as an additive to lube oil and hydraulic fluid, a component of deicing fluids for airport runways, a curing agent for epoxy resins, a plasticizer, an affinity enhancer for dyes, and a component of liquid fertilizers [2,16]. Formamide is utilized as an intermediate in the large-scale production of formic acid, as well as in

Formamide

the synthesis of hydrogen cyanide, imidazoles, pyrimidine, and 1,3,5-triazines [2]. Formamide is also used as a monomer in the production of polymers such as heat-resistant coatings [16]. Therefore, formamide is probably released to the environment as a result of its manufacture and wide use as an intermediate and solvent.

Terrestrial Fate: Several biodegradation screening studies have observed biodegradation of formamide [5,6,9,13]. Although these screening studies are not specific to soil media, they suggest that biodegradation in soil may be important. An estimated Koc value of 3.6 [8] indicates that formamide has very high mobility in soil [14] and leaching may occur. Based on the vapor pressure, formamide may evaporate from dry soil surfaces; however, volatilization from moist soils is not expected to be important.

Aquatic Fate: Based on the Henry's Law constant and the complete water solubility, formamide is not expected to volatilize from aquatic systems [8]. An estimated Koc of 3.6 and BCF of 0.0417 [8] indicate that adsorption to sediment and bioconcentration in aquatic organisms are not important removal processes in water. The chemical structure of formamide suggests that it may be susceptible to environmental hydrolysis [8] although no rate data are available that indicate hydrolysis is an important fate process in aquatic systems. Biodegradation is an important fate process in water based on its biodegradability in aqueous screening tests [5,6,9,13].

Atmospheric Fate: Based on the vapor pressure, formamide is expected to exist almost entirely in the vapor phase in the ambient atmosphere [3]. Vapor-phase formamide is degraded relatively rapidly in the ambient atmosphere by reaction with photochemically formed hydroxyl radicals; the half-life for this reaction in typical air can be estimated to be about 2.1 hr [1].

Biodegradation: Theoretical BODs were measured for formamide of 1.6, 4.7, and 11.8% over 6-, 12-, and 24-hr inoculation periods [9], respectively. Theoretical BODs greater than 30% over a 2-week incubation period [5,13], and 22.6 and 57.7% over a 2-week incubation period [6] were noted using the Japanese MITI standard BOD test.

Formamide

Abiotic Degradation: The rate constant for the vapor-phase reaction of formamide with photochemically produced hydroxyl radicals has been estimated to be 8.41×10^{-15} cm^3/molecule-sec at 25 °C which corresponds to an atmospheric half-life of about 2.1 hr at an atmospheric concentration of $5 \times 10^{+5}$ hydroxyl radicals per cm^3 [1].

Bioconcentration: Based on the measured log Kow, the bioconcentration factor (BCF) for formamide can be estimated to be 0.042 from a recommended regression-derived equation [8]. This BCF value is not indicative of bioconcentration in aquatic organisms.

Soil Adsorption/Mobility: Based on the measured log Kow, the Koc value for formamide can be estimated to be 3.6 from a recommended regression-derived equation [8]. This Koc value indicates that formamide has very high mobility in soil [14].

Volatilization from Water/Soil: The Henry's Law constant can be estimated to be 1.53×10^{-8} atm-m^3/mole at 25 °C based on a structure activity relationship [10]. This value of Henry's Law constant indicates that volatilization of formamide will not be a important fate process in water [8].

Water Concentrations:

Effluent Concentrations: Formamide has been detected in wastewater from a polyamide production plant [15]. Formamide was detected at 2.0 mg/L in gas condensate retort water in an oil-shale retort, but was not detected in the processed retort water [7].

Sediment/Soil Concentrations:

Atmospheric Concentrations:

Food Survey Values:

Plant Concentrations:

Fish/Seafood Concentrations:

Formamide

Animal Concentrations:

Milk Concentrations:

Other Environmental Concentrations:

Probable Routes of Human Exposure: Formamide is used as a solvent in the manufacture and processing of plastics, nonaqueous electrolysis, crystallization of pharmaceuticals such as penicillin and dihydrostreptomycin sulfate (82% of imported formamide is used by the pharmaceutical industry), separation of chlorosilanes and purification of oils and fats [2]. It is also used in soil stabilization, as an ink solvent in fiber and felt-tip pens and markers, and in laboratory applications [16]. Formamide has been used or has the potential to be used as an additive to lube oil and hydraulic fluid, a component of deicing fluids for airport runways, a curing agent for epoxy resins, a plasticizer, an affinity enhancer for dyes, and a component of liquid fertilizers [2,16]. Formamide is utilized as an intermediate in the large-scale production of formic acid, as well as in the synthesis of hydrogen cyanide, imidazoles, pyrimidine, and 1,3,5-triazines [2]. Formamide is also used as a monomer in the production of polymers such as heat-resistant coatings [16]. In occupational settings, exposure to formamide may occur through inhalation of vapors and through eye and skin contact. The most probable route of consumer exposure to formamide is probably a result of its use in water-soluble ink formulations [16].

Average Daily Intake:

Occupational Exposure: NIOSH (NOHS Survey 1972-1974) has statistically estimated that 6,512 workers are exposed to formamide in the USA [11]. NIOSH (NOES Survey 1981-1983) has statistically estimated that 2,724 workers are exposed to formamide in the USA [12].

Body Burdens:

REFERENCES

1. Atkinson R; Int J Chem Kinet 19: 799-828 (1987)
2. Eberling CL; Kirk-Othmer Encycl Chem Tech 3rd ed NY:Wiley Interscience 11: 258-63 (1980)

Formamide

3. Eisenreich SJ et al; Environ Sci Technol 15: 30-8 (1981)
4. Hansch C, Leo AJ; Medchem Project Issue No 26 Claremont, CA: Pomona College (1985)
5. Kawasaki M; Ecotox Environ Safety 4: 444-54 (1980)
6. Kitano M; Biodegradation and Bioaccumulation Test on Chemical Substances, OECD Tokyo Meeting, Reference Book TSU-No. 3 (1978)
7. Leenheer JA et al; Environ Sci Technol 16: 714-23 (1982)
8. Lyman WJ et al; Handbook of Chemical Property Estimation Methods. Environmental Behavior of Organic Compounds. Washington DC: American Chemical Society pp 4-9, 5-4, 5-10, 7-4, 7-5, 15-15 to 15-32 (1990)
9. Melaney GW, Gerhold RM; Ext Ser 112: 249-57 (1962)
10. Meylan WM, Howard PH; Environmental Toxicology and Chemistry 10:1283-1293 (1991)
11. NIOSH National Occupational Hazard Survey (NOHS) (1974)
12. NIOSH National Occupational Exposure Survey (NOES) (1983)
13. Sasaki S; pp 283-98 in Aquatic Pollutants Hutzinger O et al, eds Oxford: Pergamon Press (1978)
14. Swann RL et al; Res Rev 85: 23 (1983)
15. USEPA Chemical Hazard Information Profile; Draft Report: Formamide Sept. 18, Washington, DC: Office of Toxic Substances (1981)
16. USEPA; Fed Reg 51: 6929-33 (1986)
17. Windholz M et al; The Merck Index 10th ed Rahway, NJ: Merck & Co Inc p 605 (1983)

Furfural

Synonyms: 2-Furancarboxaldehyde

Structure:

CAS Registry Number: 98-01-1

Molecular Formula: $C_5H_4O_2$

Wiswesser Line Notation:

CHEMICAL AND PHYSICAL PROPERTIES

Boiling Point: 161.8 °C

Melting Point: -38.7 °C

Molecular Weight: 96.08

Dissociation Constants:

Log Octanol/Water Partition Coefficient: 0.41 [13]

Water Solubility: 86,000 mg/L at 25 °C [1]

Vapor Pressure: 2.5 mm Hg at 25 °C [34]

Henry's Law Constant: 3.7 x 10^{-6} atm m³/mol at 25 °C (calculated from the water solubility and vapor pressure)

ENVIRONMENTAL FATE/EXPOSURE POTENTIAL

Summary: Furfural is a naturally occurring compound which is also produced commercially. It may be released to the environment as a fugitive emission during its manufacture, formulation, or use in commercial products. Furfural may also be released to the environment in the smoke from burning wood. If released to soil, furfural is expected to display high mobility and it has the potential to leach into ground water. Limited data suggests that in may undergo biodegradation in soil. Volatilization from the soil surface to the atmosphere may occur; however it is not expected to be a rapid process. If released to water, furfural is expected to undergo microbial degradation, under both aerobic and anaerobic conditions. Acclimation has been found to increase the rate of biodegradation, and high concentrations of furfural inhibit biodegradation. Furfural is not expected to adsorb to sediment or suspended organic matter, nor is it expected to bioconcentrate in fish and aquatic organisms. Hydrolysis is not expected to be a significant fate process under environmental conditions. In the atmosphere, furfural is expected to exist predominately in the vapor phase. Destruction by the vapor-phase reaction with photochemically produced hydroxyl radicals is expected to be an important process with an estimated half-life of 0.44 days. Nighttime destruction by the vapor-phase reaction with nitrate radicals may be an important process in urban areas. Limited data suggests that direct photochemical degradation of furfural may occur in the atmosphere. Atmospheric removal by wet deposition may be a significant process. Occupational exposure to furfural may occur by inhalation or dermal contact during its production, formulation or use, Exposure to the general population may occur by the ingestion of contaminated drinking water or ingestion of food in which it is contained. The general population may also receive exposure to furfural by inhalation of smoke from wood fires, or by inhalation or dermal contact during the use of commercial products which contain this aldehyde.

Natural Sources: Found in several essential oils from plants of the pinaceae family, in the essential oil from cajenne linaloe, in the oil from leaves of trifolium pratense and trifolium incarnatum, in the distillation waters of several essential oils such as ambrette and angelica seeds, in ceylon cinnamon essential oil, in petitgrain oil, ylang-ylang, lavender, lemongrass, calamus, eucalyptus, neroli, sandalwood, tobacco leaves, and others [11].

Artificial Sources: Furfural has numerous commercial applications including the solvent refining of lubricating oils, butadiene, and other organic materials, as a solvent for nitrocellulose, cellulose and shoe dyes,

Furfural

as a synthetic intermediate, as an additive in phenolic and furan polymers, as a wetting agent for the manufacture of abrasives and brake linings, as a weed killer and fungicide, in road construction and metal refining, in flavorings, and as an analytical reagent [28,37,41]. It may be released to the environment as a fugitive emission during its production, formulation or use in commercial products. Furfural may also be released to the environment in the smoke of wood burning stoves or fireplaces.

Terrestrial Fate: If released to soil, furfural is expected to display high mobility and it has the potential to leach into ground water. Volatilization from the soil surface to the atmosphere is expected to occur; however, it is not expected to be a rapid process. Limited data suggests that furfural may undergo microbial degradation in soil.

Aquatic Fate: If released to water, furfural is expected to undergo microbial degradation under both aerobic and anaerobic conditions. The rate of microbial degradation has been found to increase as the microorganisms become acclimated, and high concentrations have been found to inhibit both the degradation of furfural and the degradation of other organic species present. Furfural is not expected to adsorb to sediment or suspended organic matter, nor is it expected to bioaccumulate in fish and aquatic organisms. Hydrolysis under environmental conditions is not expected to be a significant process. Volatilization from water to the atmosphere is expected to occur, although it is not expected to be a rapid process. Using the Henry's Law constant, the estimated half-life for the volatilization of furfural from a model river is 9.9 days [26].

Atmospheric Fate: In the atmosphere, furfural is expected to exist predominately in the vapor phase. Removal by wet deposition may be a significant process. Reaction of vapor-phase furfural with photochemically produced hydroxyl radicals is expected to be an important process. This reaction has an estimated half-life of 0.44 days [2]. Nighttime destruction by the vapor phase reaction with nitrate radicals may be an important process in urban areas. Limited data suggests that direct photochemical degradation of furfural may occur in the atmosphere.

Biodegradation: Furfural at an initial concn of 300 mg/L, in solution with phenol and N-methyl-2-pyrolidine, was found to undergo 98%

313

removal in a flow-through laboratory bioreactor using an activated sludge inoculum under aerobic conditions [35]. At an initial feed concentration of 1,000 mg/L, degradation was also found to occur. In a static reactor using unacclimated inocula, a lag period of 4 to 7 days was observed [35]. Furfural at an initial concn of 200 mg/L COD underwent 96.3% removal (time not given, but less than 120 hours) using a thickened adapted activated sludge under aerobic conditions [33]. In river die-away studies using water taken from the Great Miami, Little Miami, Ohio Rivers, 1.0 ppm of furfural was completely degraded within 3 days under aerobic conditions [9]. The rate of degradation at a higher initial furfural concn was found to be dependent upon the degree of acclimation, and fully acclimatized seeds were found to degrade 25 ppm within 5-12 days [9]. Furfural was listed as being confirmed to be well-biodegradable using the Japanese MITI test [17,36]. Furfural was shown to support the aerobic growth of organisms obtained from soil and grown on phenol [23], suggesting that this compound may biodegrade in soil. At an initial concn of 1.7 to 20, and 440 ppm, furfural underwent 46% and 17% theoretical biological oxygen demand, respectively, in 5 days using a sewage sludge seed under aerobic conditions [14]. Unactivated anaerobic organisms obtained from commercial wastewater reactors were found to completely utilize 580 mg/L of furfural in 30 days as measured by the production of methane and CO_2; however, it was found that at this furfural concentration the degradation rate of other carbon sources was inhibited [3]. At a furfural concn of 1160 g/L, no gas production was observed, although it was not clear if furfural was toxic to the microorganisms or merely inactivated them. Acclimated inocula from the same source were found to remove 99% of an initial 2320 mg/L of furfural in 32 days under anaerobic conditions [3].

Abiotic Degradation: An initial furfural concn of 1.5 ppm in distilled water was found to decrease by only 0.1 ppm after 30 days at 20 °C [9], suggesting that hydrolysis under neutral conditions is not an important process. High temperatures and long reaction times are necessary for the complete hydrolysis of furfural mineral acids [28]. The gas-phase concentration of furfural, obtained as a component of the smoke obtained from burning oak in a wood stove, was found to decrease when irradiated in an experimental chamber equipped both with sun lights and UV lights [20] suggesting that direct photooxidation of furfural may occur in the atmosphere. The rate of disappearance was greater when NOx was added to the system [20]. The half-life of the vapor-phase reaction of furfural with photochemically produced hydroxyl radicals is estimated at 0.44 days [2] which suggests that it will be a significant atmospheric fate process. The nighttime degradation of the aldehyde group of furfural by the vapor-phase reaction with nitrate radicals may be an important process in urban areas where higher concentrations of this oxidant ar expected [5].

Bioconcentration: Based on the water solubility and log octanol/water partition coefficient, respective bioconcentration factors in the range 0.008 - 1.2 can be calculated using an appropriate regression equation [26]. These values suggest that the bioconcentration of furfural in fish and aquatic organisms will be an insignificant process.

Soil Adsorption/Mobility: Based on the water solubility and log octanol/water partition coefficient, respective soil adsorption coefficients in the range of 1 to 40 can be calculated using appropriate regression equations [26]. These values suggest that furfural will display very high mobility in soil [39] and as such it has the potential to leach into ground water.

Volatilization from Water/Soil: Based on a calculated Henry's Law constant, the estimated half-life for the volatilization of furfural from a model river 1 m deep flowing at 1 m/sec and a wind speed of 3 m/sec is 9.9 days [26]. The vapor pressure and Henry's Law constant also suggests that volatilization from the soil surface to the atmosphere may occur, but it is not expected to be a rapid process.

Water Concentrations: SURFACE WATER: Furfural was detected in 1 out of 204 surface water samples taken near heavily industrialized sites across the USA (detection limit 1 ppb) at a concentration of 2 ppb [10]. Furfural was found in 1 of 13 samples taken in the Lake Michigan basin, 1977, at a concn of 2 ug/L [21]. DRINKING WATER: Furfural has been identified in the drinking water supplies of the United States and Europe [22]. It has been qualitatively detected in the drinking water of Ottumwa, IA [25].

Effluent Concentrations: Furfural was detected as an emission from the burning of jack pine, cedar, oak, and ash in a fireplace [24]. It was

315

identified in the smoke obtained from the burning of oak [20]. Furfural was detected in the wastewater of a synthetic rubber company discharging into the Calcasieu River (or its tributaries), LA, at an approximate concn of 1.7 ug/L [18]. Although furfural has been identified in vehicle exhaust emissions, it was not detected in samples taken from the Allegheny Mountain tunnel, PA, 1979 [12]. Furfural was identified in the spent chlorination liquor from the wood pulping industry which uses the sulfite process [4]. The mean concentration of five sulfite evaporator condensates (wastewater) from a paper plant in Bellingham, WA, ranged from 179-471 mg/L, mean 274 mg/L [3].

Sediment/Soil Concentrations:

Atmospheric Concentrations: Furfural was qualitatively detected in indoor air above a floor finished with a natural oil 4-5 months after its application [40]. Furfural was qualitatively detected in the air above the Southern Black Forest (Kalbelescheur), Germany, 1984-85 [16].

Food Survey Values: Furfural was identified as the dominant volatile flavor component of beef fried together with vegetables, soy sauce and sugar (Sukiyaki), and it was present in two of the ingredients, beef and soy sauce [38]. Furfural has been identified as a volatile flavor component of roasted filberts nuts [19], fried bacon [15], nectarines [8], baked potatoes [6], and the neutral fraction of clove essential oil [29]. Furfural, as a mixture with 3-hexanol, was identified as a volatile flavor component of mangos preserved for 14 months at -15 °C, but not in samples of the fresh fruit [27]. Identified in the aroma fraction of blue cheese fat [7].

Plant Concentrations:

Fish/Seafood Concentrations:

Animal Concentrations:

Milk Concentrations: Furfural was qualitatively detected in 2 of 8 samples of mothers milk (detection limits not given) from samples obtained from four urban sites in the USA [32].

Other Environmental Concentrations:

Probable Routes of Human Exposure: Occupational exposure to furfural may occur by inhalation or dermal contact during its production, formulation, and use in commercial products. Exposure to the general population may occur by inhalation or dermal contact if commercial products containing this compound are used in the home, or by the ingestion of contaminated water. Ingestion of food containing furfural is a probable route of exposure. Exposure to the general population may also occur by inhalation of the smoke from wood-burning fireplaces or stoves.

Average Daily Intake:

Occupational Exposure: NIOSH (NOES Survey 1981- 1983) has statistically estimated that 134,158 workers are exposed to furfural in the USA [31]. NIOSH (NOHS survey 1972-1974) has statistically estimated that 15,412 workers are exposed to furfural in the USA [30].

Body Burdens: Furfural was qualitatively detected in 2 of 8 samples of mothers milk (detection limits not given) from samples obtained from four urban sites in the USA [32].

REFERENCES

1. Amoore JE, Hautala E; J Appl Toxicol 3: 272-90 (1983)
2. Atkinson R; Chem Rev 85: 69-201 (1985)
3. Benjamin MM et al; Water Res 18: 601-7 (1984)
4. Carlberg GE et al; Sci Total Environ 48: 157-67 (1986)
5. Carter WPL et al; Environ Sci Tech 15: 829-31 (1981)
6. Coleman EC et al; J Agric Food Chem 29: 42-8 (1981)
7. Day EA, Anderson DF; J Agr Food Chem 13: 2-4 (1965)
8. Engel KH et al; J Agric Food Chem 36: 549-53 (1988)
9. Ettinger MB et al; Proc 8th Industrial Waste Conf: Purdue Univ Ext Ser (1954)
10. Ewing BB et al; Monitoring to Detect Previously Unrecognized Pollutants in Surface Waters, Appendix: USEPA-560/6-77-015A Washington DC USEPA (1977)
11. Fenaroli Handbook of Flavor Ingredients 2nd ed. Vol 2 (1975)
12. Hampton CV et al; Environ Sci Technol 16: 287-98 (1982)
13. Hansch C, Leo AJ; Medchem project Issue No. 26: Claremont CA Pomona College (1985)
14. Heukelekian H, Rand MC; J Water Pollut Contr Assoc 29: 1040-53 (1955)

Furfural

15. Ho CT et al; J Agric Food Chem 31: 336-42 (1983)
16. Juttner F; Chemosphere 15: 985-92 (1986)
17. Kawasaki M; Ecotox Environ Safety 4: 444-54 (1980)
18. Keith LH; Sci Total Environ 3: 87-102 (1974)
19. Kinlin TE et al; J Agr Food Chem 20: 1021-8 (1972)
20. Kleindienst TE et al; Environ Sci Tech 20: 493-501 (1986)
21. Konasewich D et al; Status Report on Organic and Heavy Metal Contaminants in the Lakes Erie, Michigan, Huron and Superior: Great Lakes Water Qual Board (1978)
22. Kool HJ et al; CRC Crit Rev Env Control 12: 307-57 (1982)
23. Kramer N, Doetsch RN; Arch Biochem Biophys 26: 401-5 (1950)
24. Lipart F et al; Environ Sci Technol 18: 326-30 (1984)
25. Lucas SV; GC/MS Analysis of Organics in Drinking Water Concentrates and Advanced Waste Treatment Concentrates: Vol 3 USEPA-600/1-84-020 Columbus, OH (1984)
26. Lyman WJ et al; Handbook of Chemical Property Estimation Methods NY: McGraw Hill pp 4-1 to 4-33, 5-1 to 5-30, 15-1 to 15-29 (1982)
27. MacLeod AJ, Snyder CH; J Agric Food Chem 36: 137-9 (1988)
28. Mckillip WP, Sherman E; Kirk-Othmer Encycl Chem Tech 3rd ed NY: John-Wiley 11: 499-527 (1980)
29. Muchalal M, Crouzet J; Agric Biol Chem 49: 1583-9 (1985)
30. NIOSH; National Occupational Hazard Survey (NOHS) (1983)
31. NIOSH; National Occupational Exposure Survey (NOES) (1974)
32. Pellizzari ED et al; Bull Environ Contam Toxicol 28: 322-8 (1982)
33. Pitter P; Water Res 10: 231-5 (1976)
34. Riddick JA et al; Organic Solvents Physical Properties and Methods of Purification 4th ed NY: Wiley Interscience (1986)
35. Rowe EH, Tullos LFJR; Hydrocarbon Process 59: 63-5 (1980)
36. Sasaki S; pp 283-98 in Aquatic Pollutants Hutzinger O et al Ed Pergamon Press Oxford (1978)
37. Sax NI, Lewis RJSR; Hawley's Condensed Chemical Dictionary: 11th ed NY: Van Nostrand Reinhold Co p 546 (1987)
38. Shibamato T et al; J Agric Food Chem 29: 57-63 (1981)
39. Swann RL et al; Res Rev 85: 17-28 (1983)
40. Van Nettin C et al; Bull Environ Contam Toxicol 40: 672-7 (1988)
41. Windholz M; The Merck Index: 10th ed Rahweh,NJ: Merck & Co Inc (1983)

Furfuryl Alcohol

SUBSTANCE IDENTIFICATION

Synonyms:

Structure:

CAS Registry Number: 98-00-0

Molecular Formula: $C_5H_6O_2$

Wiswesser Line Notation: T5OJ B1Q

CHEMICAL AND PHYSICAL PROPERTIES

Boiling Point: 170 °C

Melting Point: -14.6 °C

Molecular Weight: 98.10

Dissociation Constants:

Log Octanol/Water Partition Coefficient: 0.28 [8]

Water Solubility: miscible [13]

Vapor Pressure: 0.609 mm Hg at 25 °C [4]

Henry's Law Constant: 7.86×10^{-8} atm-m^3/mole at 25 °C, calculated from VP/Wsol

ENVIRONMENTAL FATE/EXPOSURE POTENTIAL

Summary: Furfuryl alcohol may be released to the environment as a result of its manufacture and use as a solvent for dyes and resins, as a flavoring agent and other uses. Furfuryl alcohol has been qualitatively

319

detected in mixtures of volatile compounds from numerous foods, which suggests that the compound may be formed naturally. If released to soil, it will be expected to exhibit very high mobility, based upon the reported infinite solubility of the compound in water and an estimated Koc of 34. It may, therefore, leach through soil to ground water if it does not biodegrade or otherwise decompose first. It may be subject to biodegradation in soil based upon results observed in a laboratory aqueous aerobic biodegradation screening test. It should not be subject to volatilization from moist near surface soil based upon an estimated Henry's Law constant, which has been calculated from the water solubility and vapor pressure. However, it may volatilize from dry near surface soil and other dry surfaces based upon its vapor pressure. If released to water, it will not be expected to adsorb to sediment or suspended particulate matter or to bioconcentrate in aquatic organisms based upon its estimated Koc and BCF, respectively, and its high solubility in water. It may directly photolyze in surface water based upon its absorption of light at wavelengths >290 nm. It may be subject to biodegradation in natural waters based upon results observed in a laboratory biodegradation aqueous aerobic screening test using a rigorous activated sludge inoculum. It should not be subject to volatilization from surface waters based upon the estimated Henry's Law constant. If furfuryl alcohol is released to the atmosphere, it can be expected to exist mainly in the vapor phase in the ambient atmosphere based upon its vapor pressure. The estimated atmospheric half-life for vapor-phase reaction with photochemically produced hydroxyl radicals half-life is 3.7 hr at an atmospheric concentration of $5 \times 10^{+5}$ hydroxyl radicals per cm^3. Furfuryl alcohol may be susceptible to direct photolysis in the atmosphere. Exposure to furfuryl alcohol will occur via ingestion of contaminated foods and human mother's milk. Minor exposure might occur via inhalation of air contaminated with the volatile compounds from certain foods which contain the compound, especially during cooking of the foods.

Natural Sources: Furfuryl alcohol was qualitatively detected in mixtures of volatile compounds from mountain Beaufort cheese [5], roasted filberts [11], nectarines [7], fried bacon, pork, and smoke flavors [10], heated sukiyaki mixture (beef, vegetables, and sugar and soy sauce seasonings) and individual sukiyaki ingredients (beef and soy sauce) [20], and fried

Furfuryl Alcohol

chicken [22], which suggests that the compound may be formed naturally.

Artificial Sources: Furfuryl alcohol may be released to the environment as a result of its manufacture and use as a solvent for dyes and resins, wetting agent, penetrant, and flavoring agent, and its use in the manufacture of resins, furan polymers, modified urea-formaldehyde polymers, corrosion-resistent sealants and cements, and foundry cores [9,23].

Terrestrial Fate: If furfuryl alcohol is released to soil, it will be expected to exhibit very high mobility [21], based upon the reported infinite solubility of the compound in water and an estimated Koc of 34 calculated [12] from the measured log Kow. The compound may, therefore, leach through soil to ground water if it does not biodegrade or otherwise decompose first. It may be subject to biodegradation in soil based upon results observed in a laboratory biodegradation aqueous aerobic screening test using a rigorous activated sludge inoculum [17]. No information was found regarding the rate of decomposition in water or the products thereof. It should not be subject to volatilization from moist near surface soil based upon an estimated Henry's Law constant, which has been calculated [12] from the measured water solubility and vapor pressure. However, it may volatilize from dry near surface soil and other dry surfaces based upon its vapor pressure.

Aquatic Fate: If furfuryl alcohol is released to water, it will not be expected to adsorb to sediment or suspended particulate matter or to bioconcentrate in aquatic organisms based upon its estimated Koc and BCF, respectively, which were calculated [12] from the measured log Kow, and its high solubility in water. It may directly photolyze based upon its absorption of light at wavelengths >290 nm [18]. It may be subject to biodegradation in natural waters based upon results observed in a laboratory aqueous aerobic biodegradation screening test using a rigorous activated sludge inoculum [17]. It should not be subject to volatilization from surface waters based upon an estimated Henry's Law constant, which has been calculated [12] from the measured water solubility and vapor pressure.

Atmospheric Fate: If furfuryl alcohol is released to the atmosphere, it can be expected to exist mainly in the vapor phase in the ambient atmosphere [6] based upon the measured vapor pressure. The estimated rate constant for vapor-phase reaction with photochemically produced hydroxyl radicals of 104.2 x 10^{-12} cm^3/molecule-sec at 25 °C [2] which corresponds to an atmospheric half-life of 3.7 hr at an atmospheric concentration of 5 x 10^{+5} hydroxyl radicals per cm^3. Furfuryl alcohol may be susceptible to direct photolysis in the atmosphere based upon its absorption of light at wavelengths >290 nm [4].

Biodegradation: A 97% removal of furfuryl alcohol was observed in 5 days in aerobic screening tests using a vigorous activated sludge system which was acclimated for 20 days prior to the experiments [17]. No information was found regarding biodegradation obtained under anaerobic conditions nor in natural water or soil.

Abiotic Degradation: Furfuryl alcohol may be susceptible to direct sunlight photolysis because it absorbs light at wavelengths >290 nm [18]. Reaction with photochemically generated singlet oxygen may be a significant removal process in surface water based upon a measured rate constant for this process of 1.2 x 10^{+8} L/mole-sec at 19 °C which corresponds to a half-life of 40 hr at a singlet oxygen concentration of 4 x 10^{-14} mole/L [19]. The rate constant for the vapor-phase reaction of furfuryl alcohol with photochemically produced hydroxyl radicals has been estimated to be 104.2 x 10^{-12} cm^3/molecule-sec at 25 °C which corresponds to an atmospheric half-life of 3.7 hr at an atmospheric concentration of 5 x 10^{+5} hydroxyl radicals per cm^3 [2].

Bioconcentration: An estimated BCF of 0.96 can be calculated from the measured log Kow using a recommended regression equation [12]. Based upon the estimated BCF and the reported infinite solubility of the compound in water, furfuryl alcohol will not be expected to significantly bioconcentrate in aquatic organisms.

Soil Adsorption/Mobility: An estimated Koc of 34 can be calculated from the measured log Kow using a recommended regression equation [12]. Based upon the estimated Koc and the reported infinite solubility of the compound in water, furfuryl alcohol will not be expected to strongly adsorb to sediment or suspended particulate matter [21], but it will be expected to exhibit very high mobility in soil [21].

Volatilization from Water/Soil: Based upon a water solubility of 1.0 x 10^{+6} mg/L [13] and the measured vapor pressure, the Henry's Law constant for furfuryl alcohol has been calculated to be 7.86 x 10^{-8} atm-m^3/mole [12]. Based upon this value for the Henry's Law constant, volatilization of furfuryl alcohol from surface water will not be expected to be an important transport process [12]. It may volatilize from dry near surface soil and other dry surfaces based upon its vapor pressure [2].

Water Concentrations:

Effluent Concentrations: Furfuryl alcohol was qualitatively detected in 4 out of >4000 samples of effluent from 4 of 46 industrial categories (pulp and paper, synfuels, publiclyowned treatment works, and rum industries) [3].

Sediment/Soil Concentrations:

Atmospheric Concentrations:

Food Survey Values: Furfuryl alcohol was detected at a concentration less than the quantification limit of 10 ppb in nectarine volatile components [7]. It was qualitatively detected (detection limit not specified) in mixtures of volatile components from roasted filberts [11], fried bacon, pork, and smoke flavors [10], heated sukiyaki mixture (beef, vegetables, and sugar and soy sauce seasonings) and individual sukiyaki ingredients (beef and soy sauce) [20], fried chicken [22] and mountain Beaufort cheese [5]. It has been listed as a constituent of the compound which contribute to the aroma of coffee [1].

Plant Concentrations:

Fish/Seafood Concentrations:

Animal Concentrations:

Furfuryl Alcohol

Milk Concentrations: Furfuryl alcohol was qualitatively detected (detection limit not specified) in 2 of 12 samples of human mother's milk from 4 U.S. cities [16].

Other Environmental Concentrations:

Probable Routes of Human Exposure: Exposure to furfuryl alcohol will occur via ingestion of contaminated foods [1,7,10,11,20,22] and human mother's milk [16]. Minor exposure might occur via inhalation of air contaminated with the volatile constituents from certain foods which contain the compound, especially during cooking of the foods.

Average Daily Intake:

Occupational Exposure: NIOSH (NOES Survey 1981-1983) has statistically estimated that 57,643 workers are potentially exposed to furfuryl alcohol in the USA [14]. NIOSH (NOHS Survey 1972-1974) has statistically estimated that 115,787 workers are potentially exposed to furfuryl alcohol in the USA [15].

Body Burdens: Furfuryl alcohol was qualitatively detected (detection limit not specified) in 2 of 12 samples of human mother's milk from 4 U.S. cities [16].

REFERENCES

1. Aeschbacher HU et al; Food Chem Toxicol 27: 227-32 (1989)
2. Atkinson R; Environ Toxicol Chem 7: 435-42 (1988)
3. Bursey JT, Pellizzari ED; Analysis of Industrial Wastewater for Organic Pollutants in Consent Degree Survey. Contract No. 68-03-2867. Athens,GA: USEPA Environmental Research Laboratories. p 88 (1982)
4. Daubert TE, Danner RP; Physical and Thermodynamic Properties of Pure Compounds. Data Compilation. Am Inst Chem Eng. (1989)
5. Dumont JP, Adda J; J Agric Food Chem 26: 364-7 (1978)
6. Eisenreich SJ et al; Environ Sci Technol 15: 30-8 (1981)
7. Engel KH et al; J Agric Food Chem 36: 549-53 (1988)
8. Hansch C, Leo AJ; Medchem Project Issue No 26. Claremont CA: Pomona College (1985)
9. Hawley GG; Condensed Chemical Dictionary 10th ed Van Nostrand Reinhold NY p 484 (1981)
10. Ho CT et al; J Agric Food Chem 31: 336-42 (1983)

Furfuryl Alcohol

11. Kinlin TE et al; J Agric Food Chem 20: 1021 (1972)
12. Lyman WJ et al; Handbook of Chem Property Estimation Methods NY: McGraw-Hill pp 4-9, 5-5, 7-4, 15-16 to 15-29 (1982)
13. McKillip WJ, Sherman E; Kirk-Othmer Encycl Chem Technol 3rd NY, NY: ed 11: 499-527 (1978)
14. NIOSH; The National Occupational Exposure Survey (NOES) (1983)
15. NIOSH; The National Occupational Hazard Survey (NOHS) (1974)
16. Pellizzari ED et al; Bull Environ Contam Toxicol 28: 322-8 (1982)
17. Pitter P; Water Res 10: 231-5 (1976)
18. Sadtler UV No. 5111. (19NA)
19. Scully FE Jr, Hoigne J; Chemosphere 16: 681-94 (1987)
20. Shibamoto T et al; J Agric Food Chem 29: 57-63 (1981)
21. Swann RL et al; Res Rev 85: 17-28 (1983)
22. Tang J et al; J Agric Food Chem 31: 1287-92 (1983)
23. Windholz M et al; Merck Index 10th ed. Rahway, NJ p 614 (1983)

Heptane

SUBSTANCE IDENTIFICATION

Synonyms:

Structure:

CAS Registry Number: 142-82-5

Molecular Formula: C_7H_{16}

Wiswesser Line Notation:

CHEMICAL AND PHYSICAL PROPERTIES

Boiling Point: 98.4 °C

Melting Point: -90.7 °C

Molecular Weight: 100.21

Dissociation Constants:

Log Octanol/Water Partition Coefficient: 4.66 [20]

Water Solubility: 2.93 mg/L at 25 °C [38]

Vapor Pressure: 45.8 mm Hg at 25 °C [38]

Henry's Law Constant: 2.06 atm-m³/mole at 25 °C, calculated from VP/Wsol

ENVIRONMENTAL FATE/EXPOSURE POTENTIAL

Summary: n-Heptane is a highly volatile constituent in the paraffin fraction of crude oil and natural gas. n-Heptane is released to the environment via the manufacture, use and disposal of many products

Heptane

associated with the petroleum and gasoline industries. Extensive data show release of n-heptane into the environment from printing pastes, paints, varnishes, adhesives and other coatings; landfills and waste incinerators; vulcanization and extrusion operations during rubber and synthetic production; and the combustion of gasoline fueled engines. Photolysis or hydrolysis of n-heptane is not expected to be important environmental fate processes. Biodegradation of n-heptane may occur in soil and water; however, volatilization and adsorption are expected to be far more important fate processes. A high Koc indicates n-heptane will be slightly mobile to immobile in soil. In aquatic systems n-heptane may partition from the water column to organic matter in sediments and suspended solids. The bioconcentration of n-heptane may be important in aquatic environments. The Henry's Law constant suggests rapid volatilization of n-heptane from environmental waters. The volatilization half-lives from a model river and a model pond, the latter considers the effect of adsorption, have been estimated to be 2.9 hr and 13 days, respectively. n-Heptane is expected to exist entirely in the vapor phase in ambient air. Reactions with photochemically produced hydroxyl radicals in the atmosphere have been shown to be important (estimated half-life of 2.4 days). Data also suggests that nighttime reactions with nitrate radicals may contribute to the atmospheric transformation of n-heptane, especially in urban environments. The most probable route of human exposure to n-heptane is by inhalation. Extensive monitoring data indicates n-heptane is a widely occurring atmospheric pollutant.

Natural Sources: n-Heptane is a constituent in the paraffin fraction of crude oil and natural gas [64].

Artificial Sources: n-Heptane is released to the environment via the manufacture, use and disposal of many products associated with the petroleum [2,21,49] and gasoline industries [30,49,57]. The combustion of gasoline fueled engines has been shown to release n-heptane into the atmosphere [19,43]. Vulcanization and extrusion operations during rubber and synthetic production as with shoes, tires and electrical insulation also emits n-heptane to the air [9]. Other well-documented materials responsible for the release of n-heptane to the environment include printing pastes, paints, varnishes, adhesives and other coatings [66,68,70]. Landfills [67] and waste incinerators [29] also release n-heptane into the environment.

327

Heptane

Terrestrial Fate: Photolysis or hydrolysis of n-heptane is not expected to be important in soils. The biodegradation of n-heptane may occur in soils; however, volatilization and adsorption are expected to be far more important fate processes. An estimated range for Koc from 2400 to 8200 [37] indicates n-heptane will be slightly mobile to immobile in most soils [61]. Based upon the estimated Henry's Law constant, n-heptane is also expected to rapidly volatilize from surface soils [2].

Aquatic Fate: Photolysis or hydrolysis [37] of n-heptane in aquatic systems is not expected to be important. The biodegradation of n-heptane may occur in aquatic environments; however, volatilization and adsorption are expected to be far more important fate processes. The log bioconcentration factor (log BCF) for n-heptane has been estimated to range from 2.53 to 3.31 [37], suggesting bioconcentration may be an important factor in aquatic systems. An estimated range for Koc from 2400 to 8200 [37] indicates n-heptane may strongly absorb to carbon [61] and may partition from the water column to organic matter contained in sediments and suspended solids. The estimated Henry's Law constant suggests rapid volatilization of n-heptane from environmental waters [37]. Based on this Henry's Law constant, the volatilization half-life from a model river has been estimated to be 2.9 hr [37]. The volatilization half-life from an model pond, which considers the effect of adsorption, can be estimated to be about 13 days [65].

Atmospheric Fate: Based on the vapor pressure, n-heptane is expected to exist entirely in the vapor phase in ambient air [14]. Direct photolysis of n-heptane in the atmosphere is not expected to be important [58]. However, vapor phase reactions with photochemically produced hydroxyl radicals in the atmosphere have been shown to be important. The rate constant for n-heptane was measured to be 7.18×10^{-12} cm^3/molecule-sec at 26 °C which corresponds to an atmospheric half-life of about 2.2 days at an atmospheric concentration of $5 \times 10^{+5}$ hydroxyl radicals per cm^3 [4,6]. Experimental data showed that 30.9% of the n-heptane fraction in a dark chamber reacted with NO_3 to form the corresponding alkyl nitrate [3,5], suggesting nighttime reactions with nitrate radicals may contribute to the atmospheric transformation of n-heptane, especially in urban environments.

Heptane

Biodegradation: The theoretical oxygen demand of benzene-acclimated activated sludge for n-heptane was 0.7, 4.3 and 23.4% after 6, 24 and 72 hr, respectively [39]. One mg of n-heptane and 1 mL of a 1:10 suspension of Hudson-Collamer silt-loam soil in mineral salt media and incubated in the dark at 25 °C [18]. Controls indicated no loss of oxygen whereas the degradability of n-heptane was assessed on the average oxygen consumption of 3.3, 7.4, 8.2 and 8.2 ug/ml for 2, 5, 10 and 20 days, respectively, for 2 trials [18]. At intervals of 6, 12 and 24 hr, endogenous respiration was greater than that of 3 preparations of n-heptane at a concentration of 500 mg/L in 20 mL volume solution of supernatant liquid with a suspended solids concentration adjusted to 2500 mg/L from differing aeration units of sewage treatment facilities [15]. n-Heptane incubated with sufficient oxygen in Florida ground water at 20 °C in the dark was reported to completely degrade within 7 days [11]. However, "experimental problems" were reported with the sterile controls [11]. In a similar study using a jet fuel mixture and freshwater from the Escambia River FL at 25 °C, a 99% loss of n-heptane in the controls was attributed to evaporation [60]. n-Heptane as a component of missile fuel was also lost to volatilization within 5 hr when incubated with water from the Range Point salt marsh FL [59].

Abiotic Degradation: Alkanes are generally resistant to hydrolysis [37]. Based on data for the structurally similar n-hexane and iso-octane, n-heptane is not expected to absorb UV light in the environmentally significant range, >290 nm [32]. Therefore, n-heptane probably will not undergo direct photolysis in the environment. The n-heptane concentration of 40 ppbC in an air sample was reduced by 43% within 6 hr of irradiation by natural sunlight in downtown Los Angeles, CA [34]. The rate constant for the vapor phase reaction of n-heptane with photochemically produced hydroxyl radicals was measured to be 7.18×10^{-12} cm^3/molecule-sec at 26 °C which corresponds to an atmospheric half-life of about 2.2 days at an atmospheric concentration of $5 \times 10^{+5}$ hydroxyl radicals per cm^3 [4,6]. At 27 °C the rate constant and half-life were reported to be 8.5×10^{-12} cm^3/sec and 1.9 days respectively [32]. At 39 °C the rate constant for the vapor phase reaction of n-heptane with photochemically produced hydroxyl radicals was determined to be 7.52×10^{-12} cm^3/molecule-sec, which corresponds to an atmospheric half-life of 2.1 days [46]. Experimental data showed that 30.9% of the n-heptane fraction in a dark chamber reacted with nitrate to form the corresponding alkyl nitrate [3,5], suggesting nighttime reactions with nitrate radicals may contribute to the atmospheric transformation of n-heptane, especially in urban environments.

Bioconcentration: Based upon the water solubility, and log Kow, the bioconcentration factor (log BCF) for n-heptane has been calculated to be 2.53 and 3.31, respectively, from recommended regression-derived equations [37]. These BCF values indicate bioconcentration may be important.

Soil Adsorption/Mobility: Based on the water solubility and a log Kow, the Koc of n-heptane has been calculated to range from 2400 to 8100 from various regression-derived equations [37]. These Koc values indicate n-heptane will be slightly mobile to immobile in soils [61].

Volatilization from Water/Soil: The Henry's Law constant indicates extremely rapid volatilization from environmental waters [37]. The volatilization half-life from a model river (1 meter deep flowing 1 m/sec with a wind speed of 3 m/sec) has been estimated to be 2.9 hr [37]. The volatilization half-life from a model pond, which considers the effect of adsorption, has been estimated to be 13 days [65].

Water Concentrations: DRINKING WATER: n-Heptane was reported as a compound identified in drinking water [33]. SEAWATER: Only trace quantities of n-heptane were detected in open surface waters and from an unpolluted coastal area of the north central Gulf of Mexico [52]. The n-heptane concentration ranged from 2 to 5 ng/L for 6 surface water samples collected at coastal areas of the north central Gulf of Mexico under anthropogenic influence [52]. n-Heptane was detected in 7 of 8 surface water samples in the Gulf of Mexico ranging in concentration from 0.1 to 4.1 ng/L with an average concentration of 1.9 ng/L [53]. GROUND WATER: n-Heptane was detected in the ground water downflow from an earthen disposal pit for produced water in the Duncan Oil Field, NM [13].

Effluent Concentrations: At a distance of 1 mile from its source, n-heptane was detected at a concentration of 6.0 ug/m^3 in the plume emitted from a General Motors plant located in Janesville, Wisconsin [56]. n-Heptane was identified as a stack emission from waste incinerators [29]. n-Heptane was also identified as a vapor emitted from landfills [67]. A refinery located in Tulsa OK was attributed with emissions to the surrounding atmosphere where the n-heptane concentration was measured to be 9.0 and 21.4 ppbC for two min before and after 1:33 PM [2]. The n-heptane content of the air downwind of a Mobil natural gas facility in Rio Blanco, CO was 50.7 ppbC [2]. The air outside of a shale oil wastewater treatment facility was found to contain n-heptane at a concentration of 10 ug/m^3 [21]. n-Heptane was detected in 1 of 63 industrial wastewater effluents at a concentration less than 10 ug/L [48]. Underwater hydrocarbon vents and formation water discharges from offshore oil production platforms were found to contain n-heptane concentrations in the vapor phase at 24 umol/L of gas and in the liquid state at 1330 ng/L, respectively [52]. A large septic tank serving 97 homes in Tacoma, WA received raw sewage influent containing n-heptane at a concentration of 6.2 ug/L and discharged n-heptane at a concentration less than 0.7 ug/L [12]. The average n-heptane concentration in the spent chlorination liquor of 3 differently treated sulfite pulps was 5 g/ton of pulp [8]. Data from Sept 2, 1979 identified n-heptane as a gaseous emission of the vehicle traffic through the Allegheny Mountain Tunnel of the Pennsylvania Turnpike [19]. The average exhaust from 67 gasoline fueled vehicles was found to contain n-heptane at a concentration 0.9% by weight in fuel [43].

Sediment/Soil Concentrations: n-Heptane was detected in the soil surrounding an earthen disposal pit for produced water in the Duncan Oil Field, NM [13]. n-Heptane was detected in 3 of 4 sediment samples from Walvis Bay of the Namibian shelf of SW Africa at concentrations of 0.12, 0.13 and 0.15 ng/g [2].

Atmospheric Concentrations: URBAN: The concentration of n-heptane in downtown Tuscaloosa, AL ranged from 16 to 60 ppb with an average of 11 ppb [24]. The n-heptane concentration of a rural site 35 miles from Tuscaloosa, the Talladega National Forest, ranged from trace amounts to 0.5 ppb with an average of 0.2 ppb [24]. In a 1979 study of 5 sites in NJ, n-heptane was identified in the air over the cities of Rutherford, Newark, South Amboy and Elizabeth [7]. The average n-heptane concentration in the air at the 6th floor of the Cooper Union Building in New York City, NY was 8 and 6 ppbC for 31 and 10 samples taken at 6:00-11:00 AM and 1:00-3:00 PM, respectively, in July 1978 [1]. The average n-heptane

Heptane

concentration in the air at the 82nd floor of the Empire State Building in New York City, NY was 3, 5 and 3 ppbC for 18, 21 and 17 samples taken at 6:00-9:00 AM 9:00-11:00 AM and 1:00-3:00 PM, respectively, in July 1978 [1]. At street level at the Empire State and World Trade Buildings in Manhattan, NY the average n-heptane concentration of 4 samples was 18 ppbC in July 1978 [1]. The ground level atmospheric concentration of n-heptane at 1:25 AM was 2.8 ppb and 15.0 ppb at 8:00 AM for Huntington Park, CA [54]. The n-heptane concentration ranged from 3 to 11 ppbV at a downtown Los Angeles location for the fall of 1981 [17]. The atmospheric n-heptane concentration ranged from 0.002 to 0.28 ppbV at Niwot Ridge Colorado [50]. The n-heptane concentration for 6 sites in Rio Blanco, CO ranged from 0.6 to 50.7 ppbC with an average of 9.8 ppbC [2]. n-Heptane was detected in the air of Boulder, CO at a concentration of 2 ug/m^3 and at a rural site outside of Boulder at a concentration of 1 ug/m^3 [21]. The average n-heptane concentration for 2 samples per 4 sites in Tulsa, OK was 8.6 ppbC with a range of 4.9 to 21.4 ppbC [2]. The 1977 maximum and average concentration of 690 points for n-heptane at a site in Houston, TX was 47 and 5 ppb of carbon, respectively [41]. The average n-heptane concentration from 6 to 9 AM in Houston, TX was 8 ppb of carbon [41]. n-Heptane was detected in 10 of 16 air samples from Houston, TX ranging in concentration from 16.2 to 138.9 ppm with an average of 59.1 ppm [36]. In 1983-4 the respective minimum, maximum and average outdoor air concentrations of n-heptane in northern Italy were 1, 17 and 5.1 ug/l [10]. The concentration of n-heptane in the downtown air of Zurich, Switzerland was 34.0 ppb [16]. At Deuselbach, Hunsruch in Germany the atmospheric n-heptane concentration was 0.044 ppb for October 23, 1983 [51]. n-Heptane was detected in the atmospheres of 6 industrialized cities of the USSR ranging in size of population from 0.4 to 4.5 million people [25,26,27]. The average n-heptane concentration (66 samples) in the air over Tokyo, Japan for 1980 was 0.5 ppb [63]. n-Heptane was identified in the ambient air of Sydney, Australia [42] with an average concentration of 0.7 ppbv [43]. RURAL: The respective median, minimum and maximum atmospheric concentrations of n-heptane for 5 rural locations in NC ranged from 0.1 to 3.5, 0.0 to 0.9 and 0.3 to 4.9 [55]. The n-heptane concentration of a rural site 35 miles from Tuscaloosa, AL ranged from 16 to 60 ppb with an average of 0.2 ppb [24]. At a rural site outside of Boulder, CO, n-heptane was detected at a concentration of 1 ug/cm [21]. REMOTE: The average n-heptane

Heptane

concentration for samples taken at altitudes of 2000, 2500 and 300 ft over Lake Michigan on Aug 27, 1976 was 0.3 ppb [40]. The average n-heptane concentration for samples taken at altitudes of 1000 and 1500 ft over Lake Michigan on Aug 28, 1976 was 0.2 ppbv [40]. SOURCE DOMINATED: At ground level the atmospheric concentration of n-heptane was 1.0, 1.5, 1.0 and 1.0 ug/m³ at 4, 7, 9 and 11 miles downwind of a General Motors plant in Janesville, Wisconsin [56]. INDOOR: n-Heptane occurred in 46% of the indoor air samples and 32% of the outdoor air samples taken in Chicago, Illinois [28]. The respective minimum, maximum and average indoor air concentrations of n-heptane for 14 homes and an office building were 1, 76 and 16 ug/L [10].

Food Survey Values: n-Heptane was identified as a volatile component of fried bacon [23], roasted filberts [31] and nectarines [62].

Plant Concentrations:

Fish/Seafood Concentrations: n-Heptane was detected in Walleye fish tissues from Lake St Clair, MI [22].

Animal Concentrations:

Milk Concentrations: n-Heptane was detected in 7 of 12 samples of mothers breast milk from the cities of Bayonne, NJ, Jersey City, NJ, Bridgeville, PA and Baton Rouge, LA [47].

Other Environmental Concentrations:

Probable Routes of Human Exposure: The most probable route of human exposure to n-heptane is by inhalation. Atmospheric workplace exposures have been documented [9,21,30,49,57,66,68,70]. n-Heptane is a highly volatile compound and monitoring data indicates that it is a widely occurring atmospheric pollutant.

Average Daily Intake:

Occupational Exposure: NIOSH (NOHS Survey 1972-1974) has statistically estimated that 1,192,459 workers are potentially exposed to n-heptane in the USA [44]. NIOSH (NOES Survey 1981-1983) has statistically estimated that 235,902 workers are potentially exposed to n-heptane in the USA [45]. A 1984 study showed n-heptane was emitted from gasoline exposing outside operators at the refineries to an average air concentration of 0.243 mg/m^3; n-heptane was detected in 39 of 56 samples [49]. Transport drivers were exposed to n-heptane at atmospheric concentration of 0.275 mg/m^3 and n-heptane was detected in 46 of 49 samples [49]. Gas station attendants were exposed to n-heptane at atmospheric concentration of 0.200 mg/m^3 and n-heptane was detected in 48 of 49 samples [49]. Tank contractors were exposed to vapors containing n-heptane at concentrations of 1.6 ppm at gasoline tank removal sites [57]. Attendants at a high volume service station in eastern PA were exposed to trace levels of n-heptane vapors in 15 of 18 air samples [30]. The indoor air of a shale oil wastewater treatment facility was found to contain n-heptane at a concentration of 24 ug/m^3 [21]. The atmospheric concentration of n-heptane ranged from 20 to 14000 ug/m^3 for the vulcanization area of a shoe sole manufacturing plant, and from 3 to 500 ug/m^3 for the vulcanization area and 0 to 70 ug/m^3 for the extrusion area of a tire retreading factory [9]. n-Heptane was detected in the workplace atmospheres of 336 businesses in Belgium with a frequency of occurrence of 11% for those that utilize printing pastes; 16% where painting took place; 65% of the automobile repair shops and 3% for sites where various materials such as varnishes are employed [68]. A total of 89 employees who spray paints or glues on their jobs for 3 companies were exposed to n-heptane at an average concentration of 0.1 ppm [70]. The workers of 7 out of 9 printing shops tested in Amsterdam were exposed to an average median concentration of n-heptane of 0.2 ppm with maximum concentrations ranging from 0.22 to 7.81 ppm [66].

Body Burdens: n-Heptane was detected in 7 of 12 samples of mothers breast milk from the cities of Bayonne, NJ, Jersey City, NJ, Bridgeville, PA and Baton Rouge, LA [47]. The air expired from humans contained n-heptane in 47.0% of the 387 samples collected from 54 subjects [35]. The average n-heptane concentration of 1.8 ng/L for all samples was expressed as the geometric mean with upper and lower limits of 4.7 and 0.69, respectively [35].

Heptane

REFERENCES

1. Altwicker ER et al; J Geophys Res 85: 7475-87 (1980)
2. Arnts RR, Meeks SA; Atmos Environ 15: 1643-51 (1981)
3. Atkinson R et al; J Phys Chem 86: 4563-9 (1982)
4. Atkinson R; Chem Rev 85: 69-201 (1985)
5. Atkinson R et al; Preprints Amer Chem Soc Div Environ Chem 23: 173-6 (1983)
6. Atkinson R et al; Internat J Chem Kin 14: 781-8 (1982)
7. Bozzelli JW et al; Analysis of Selected Toxic and Carcinogenic Substances in Ambient Air of NJ, State of NJ Dept Environ Protection (1980)
8. Carlberg GE et al; Sci Total Environ 48: 157-67 (1986)
9. Cocheo V et al; Am Ind Hyg Assoc J 44: 521-7 (1983)
10. DeBortoil M et al; Environ Int 12: 343-50 (1986)
11. Delfino JJ, Miles CJ; Fate and Transport of Agrochemicals in FL, Proc Soil Crop Soc Fl 44: 1-24 (1985)
12. Dewalle FB et al; Determination of Toxic Chemicals in Effluent from Household Septic Tanks. USEPA-600/S2-85-050 p 4 (1985)
13. Eiceman GA et al; Intern J Environ Anal Chem 24: 143-62 (1986)
14. Eisenreich SJ et al; Environ Sci Technol 15: 30-8 (1981)
15. Gerhold RM, Malaney GW; J Water Pollut Contr Fed 38: 562-79 (1966)
16. Grob K, Grob G; J Chromatogr 62: 1-13 (1971)
17. Grosjean D, Fung K; J Air Pollut Control Assoc 34: 537-43 (1984)
18. Haines JR, Alexander M; Appl Microbiol 28: 1084-5 (1974)
19. Hampton CV et al; Environ Sci Technol 16: 287-98 (1982)
20. Hansch C, Leo AJ; Medchem Project Issue No. 26 Claremont CA: Pomona College (1985)
21. Hawthorne SB, Sievers RE; Environ Sci Technol 18: 483-90 (1984)
22. Hesselberg RJ, Seelye JG; ADMEN Rep 82-1 US Fish Wildlife Soc Great L Fishery Lab p 49 (1982)
23. Ho CT et al; J Agric Food Chem 31: 336-42 (1983)
24. Holzer G et al; J Chromatog 142: 755-64 (1977)
25. Ioffe BV et al; Environ Sci Technol 13: 864-8 (1979)
26. Ioffe BV et al; Dokl Akad Nauk Sssr 243: 1186-9 (1978)
27. Ioffe BV et al; J Chromatogr 142: 787-95 (1977)
28. Jarke FM et al; Ashrae Trans 87: 153-66 (1981)
29. Junk GA, Ford CS; Chemosphere 9: 187-230 (1980)
30. Kearney CA, Dunham DB; Am Ind Hyg Assoc J 47: 535-9 (1986)
31. Kinlin TE et al; J Agric Food Chem 20: 1021 (1972)
32. Klopffer W et al; Ecotox Environ Safety 15: 298-319 (1988)
33. Kool HJ et al; Crit Rev Env Control 12: 307-57 (1982)
34. Kopczynski SL et al; Environ Sci Technol 6: 342-7 (1972)
35. Krotoszynski BK et al; J Anal Toxicol 3: 225-34 (1979)
36. Lonneman WA et al; Hydrocarbons in Houston Air. USEPA-600/3-79/018 p 44 (1979)

Heptane

37. Lyman WJ et al; Handbook of Chemical Property Estimation Methods NY:McGraw-Hill pp 4-9, 5-4, 10, 7-4 (1982)
38. Mackay D, Shiu WY; J Phys Chem Ref Data 19: 1175-99 (1981)
39. Malaney GW, McKinney RE; Water Sewage Works 113: 302-9 (1966)
40. Miller MM, Alkezweeny AJ; Ann NY Acad Sci 338: 219-32 (1980)
41. Monson PR et al; Pater 78-50.4 in 71st An Meeting Air Pollut Contr Assoc (1978)
42. Mulcahy MFR et al; Paper No. IV pp 17 in Occurrence Control Photochem Pollut. Proc Symp Workshop Sess (1976)
43. Nelson PF, Quigley SM; Atmos Environ 18: 79-87 (1984)
44. NIOSH; National Occupational Hazard Survey (NOHS) (1974)
45. NIOSH; National Occupational Exposure Survey (NOES) (1983)
46. Nolting F et al; J Atmos Chem 6: 47-59 (1988)
47. Pellizzari ED et al; Bull Environ Contam Toxicol 28: 322-8 (1982)
48. Perry DL et al; Ident of Org Compounds in Ind Effluent discharges. USEPA-600/4-79-016 (NTIS PB-294794) pp 230 (1979)
49. Rappaport SM et al; Appl Ind Hyg 2: 148-54 (1987)
50. Roberts JM et al; Atmos Environ 19: 1945-50 (1985)
51. Rudolph J, Khedim A; Int J Environ Anal Chem 290: 265-82 (1985)
52. Sauer TC Jr; Org Geochem 7: 1-16 (1981)
53. Sauer TC Jr et al; Mar Chem 7: 1-16 (1978)
54. Scott Research Labs Inc; Atmospheric Reactions Studies in the Los Angeles Basin. NTIS PB-194-058 p 86 (1969)
55. Seila RL et al; Atmospheric Volatile Hydrocarbon Composition at Five Remote Sites in NW NC. USEPA-600/D-84-092 (NTIS PB84-177930) (1984)
56. Sexton K, Westberg H; Environ Sci Technol 14: 329-32 (1980)
57. Shamsky S, Samimi B; App Ind Hyg 2: 242-5 (1987)
58. Silverstein RM, Bassler GC; Spectrometric Id of Org Cmpd, J Wiley and Sons Inc pp 148-169 (1963)
59. Spain JC, Somerville CC; Chemosphere 14: 239-48 (1985)
60. Spain JC et al; Degrad of Jet Fuel Hydrocarbons by Aquatic Microbial Communities NTIS AD-A139791/8 p 226 (1983)
61. Swann RL et al; Res Rev 85: 16-28 (1983)
62. Takeoka GR et al; J Agric Food Chem 36: 553-60 (1988)
63. Uno I et al; Atmos Environ 19: 1283-93 (1985)
64. USEPA; Drinking water Criteria Document for Gasoline. ECAO-CIN-D006, 8006-61-9 (1986)
65. USEPA; EXAMS II Computer Simulation (1987)
66. Verhoeff AP et al; Int Arch Occup Environ Health 60: 201-9 (1988)
67. Vogt WG, Walsh JJ; Volatile Organic Compounds in Gases from Landfill Simulators Proc APCA Annu Meet 78th 6: 2-17 (1985)
68. Vuelemans H et al; Am Indust Hyg Assoc J 48: 671-7 (1987)
69. Whelan JK et al; Geochim Cosmochim Acta 44: 1767-85 (1984)
70. Whitehead LW et al; Am Indust Hyg Assoc J 45: 767-72 (1984)

2-Hexanone

SUBSTANCE IDENTIFICATION

Synonyms: Methyl n-butyl ketone

Structure:

$$H_3C \overset{O}{\underset{}{\diagdown}} CH_3$$

CAS Registry Number: 591-78-6

Molecular Formula: $C_6H_{12}O$

Wiswesser Line Notation: 4V1

CHEMICAL AND PHYSICAL PROPERTIES

Boiling Point: 127.58 °C

Melting Point: -55.8 °C

Molecular Weight: 100.16

Dissociation Constants:

Log Octanol/Water Partition Coefficient: 1.38 [11]

Water Solubility: 16,000 mg/L at 25 °C [8]

Vapor Pressure: 11.62 mm Hg at 25 °C [27]

Henry's Law Constant: 9.57×10^{-5} atm-m^3/mole (calculated from the vapor pressure and water solubility)

ENVIRONMENTAL FATE/EXPOSURE POTENTIAL

Summary: The release of 2-hexanone to the environment is expected to occur through its manufacture, formulation, and use as a specialized organic solvent. Important among these is volatilization to the atmosphere

337

during the drying and curing of coatings in which it is used. If released to the atmosphere, reaction with photochemically produced hydroxyl radicals should be the most rapid method of degradation, with an estimated half-life on the order of 42 hours. The relatively high water solubility of 2-hexanone suggests that rain washout may be an important fate process. If released to soil, 2-hexanone is expected to display high mobility and is capable of undergoing rapid biodegradation. If released to water, 2-hexanone is not be expected to undergo aqueous hydrolysis, adsorption to sediment, or bioconcentration in aquatic organisms. Volatilization from both water and soil to the atmosphere is expected to be an important fate process. Human exposure to small quantities of 2-hexanone is expected to occur through the ingestion of natural and processed foods in which it has been detected. On a commercial basis, exposure is expected to occur through the inhalation of vapors during its manufacture, formulation, and use as a solvent. For the general population, inhalation of 2-hexanone vapors is expected to occur during the use, drying, and curing of coatings in which it is contained as a solvent.

Natural Sources: 2-Hexanone is known to occur in nature in only very low concentrations [16].

Artificial Sources: 2-Hexanone is used as a solvent for a wide variety of materials including lacquers, resins, oils, and the like. It is a medium evaporating solvent for nitrocellulose acrylates, vinyl, and alkyd coatings [23]. Release to the environment will most likely occur through volatilization during its manufacture and formulation, and through evaporation by the drying of commercial coatings in which it is contained.

Terrestrial Fate: 2-Hexanone compound is capable of rapid biodegradation in soil [18,25]. Several biodegradation screening studies have demonstrated that 2-hexanone biodegrades readily [2,28,35]. Estimated Koc values of 21 and 134 indicate that 2-hexanone will leach readily in soil [19]. The vapor pressure suggests that evaporation from soil surfaces to the atmosphere should be an important fate process.

Aquatic Fate: If released to water, 2-hexanone may undergo rapid biodegradation. Several biodegradation screening studies have

2-Hexanone

demonstrated that 2-hexanone biodegrades readily [2,28,35]. Volatilization from water may also be an important fate process. The estimated volatilization half-lives from a model river (1 meter deep) and a model environmental pond (2 meters deep) are 12.1 and 135 hours, respectively [19,34]. Aquatic hydrolysis, adsorption to sediment, and bioconcentration should not be an important fate processes.

Atmospheric Fate: The relatively high vapor pressure of 2-hexanone suggests that it will exist almost entirely in the vapor phase in the ambient atmosphere [7]. The estimated half-life for the vapor-phase reaction with photochemically produced hydroxyl radicals is 42 hours [1]. Although 2-hexanone is a candidate for direct photolysis, the reaction rate for this process should be much less than for other atmospheric chemical reactions (such as reaction with hydroxyl radicals), and thus should not be an important fate process [15]. The water solubility of 2-hexanone suggests that physical removal from air through wet deposition (dissolution into clouds, rainfall, etc.) may be an important fate process.

Biodegradation: Using an acclimated mixed microbial culture seed, a 59% theoretical biological oxygen demand for 2-hexanone was obtained after five days of inoculation [2]. 2-Hexanone was shown to undergo rapid (less than eight weeks) anaerobic biodegradation utilizing an anaerobic digester sludge inoculum [28]. In a five-day test using an activated microbial culture, 2-hexanone had a 61.4 % theoretical oxygen demand [35].

Abiotic Degradation: The rate constant for the vapor-phase reaction of 2-hexanone with photochemically produced hydroxyl radicals has been experimentally determined to be 9.1×10^{-12} cm^3/molecule-sec at 25 °C [1] which corresponds to an atmospheric half-life of 42 hr at an atmospheric concentration of $5 \times 10^{+5}$ hydroxyl radicals/cm^3 [1]. A review of literature photochemical degradation studies centering on the rate of ozone formation from the substrate showed that 2-hexanone can be considered to have low reactivity (2 on a 5 tier scale) [9]. Although 2-hexanone is capable of direct photochemical degradation, the rate of the reaction is not as rapid as decomposition by normal atmospheric chemical reactions (that is, reaction with hydroxyl radicals, nitrate radicals, and ozone) [15]. In a review of the photochemical smog reactivity of various organic

339

solvents, 2-hexanone was determined to be of low to moderate photochemical reactivity for all types of ensuing chemical reactions [17].

Bioconcentration: Based on the log octanol/water partition coefficient, a bioconcentration factor (BCF) for 2-hexanone can be estimated at 6 using a linear regression-derived equation [19]. This estimated BCF suggests that bioaccumulation in aquatic organisms should not occur.

Soil Adsorption/Mobility: Based on the log octanol/water partition coefficient and water solubility, Koc values of 21 and 134 can be estimated from linear regression-derived equations [19]. These Koc values suggest a high to very high soil mobility for 2-hexanone [29].

Volatilization from Water/Soil: The value of the Henry's Law constant indicates that volatilization of 2-hexanone from environmental waters is probably significant, but may not be rapid [19]. Using the Henry's Law constant, the volatilization half-life from a model river (1 meter deep flowing 1 m/sec with a wind velocity of 3 m/sec) can be estimated to be about 12.1 hours [19]. The volatilization half-life from a model environmental pond (2 meters deep) can be estimated to be about 135 hours [34]. Based upon the vapor pressure, 2-hexanone can be expected to evaporate from dry surfaces.

Water Concentrations: SURFACE WATER: 2-Hexanone was detected in Lake Erie at a concentration of 1 ug/L [14]. 2-Hexanone was found in an on-site lagoon at an unauthorized hazardous waste disposal site in Lang township, NJ, at a maximum concentration of 30 ug/L and average concentration of 20 ug/L [32]. GROUND WATER: 2-Hexanone was found in Biscayne Aquifer ground water in the vicinity of an inactive drum recycling facility in Dade County, FL in 1982 at a concentration of 7 ug/L [20]. It was found in two well water samples at an unauthorized hazardous waste disposal site in Lang township, NJ at a maximum concentration of 14000 ug/L and an average concentration of 7135 ug/L [33]. In 1984, 2-hexanone was detected in three of eleven on site wells at an abandoned landfill in Tybouts Corner, DE at a concentration range 6.1-91 ug/L [32].

Effluent Concentrations: 2-Hexanone was found in process water from in-situ coal gasification in Gillette, WY (7 ppm), in the aqueous

2-Hexanone

condensate from low-Btu gasification of Rosebud coal in Morgantown, WV (202 ppm), and in retort water from in-situ oil shale processing at Rock Springs, WY (53 ppm) [24]. It was found in the leachate of various municipal landfills at an average concentration of 0.148 mg/L [3]. It was identified in leachate discharge into a ditch near an abandoned landfill in Tybouts Corner, DE at concentrations up to 380 ug/L [32]. 2-Hexanone was detected in one out of sixty-three effluent and twenty-two intake waters from a wide range of chemical manufacturing in areas across the USA [26].

Sediment/Soil Concentrations: 2-Hexanone was found in soil samples at an unauthorized hazardous waste disposal site in Lang township, NJ, at a maximum concentration of 440 ug/kg (surficial soil) and 46 ug/kg (subsurface soil) [33]. It was not found in the sediment of an on-site lagoon that tested positive for the presence of 2-hexanone [33]. 2-Hexanone was detected in one out of three sediment samples in Red Lion creek near an abandoned landfill in Tybouts County, DE, in concentration less than 850 ug/L [32].

Atmospheric Concentrations: From Nov 1984-Jan 1985, 2-hexanone was qualitatively determined in air samples from the Southern Black Forest (Kalbelescheuer), and in suburban samples (Tubingen) West Germany [12].

Food Survey Values: 2-Hexanone has been identified as a component of the aromatic fraction of blue cheese [5]. It was identified as a component of Beaufort (Gruyere) cheese [6], as a component of raw chicken breast muscle [10], and as a volatile component of roasted filberts nuts [13]. 2-Hexanone was identified as a volatile flavor component of vacuum-distilled, blended nectarines, but not by direct headspace analysis of the fruit [30]. 2-Hexanone has been observed in white bread, toasted oats, milk (0.007-0.011 ppm) and cream (0.017-0.018 ppm) [16]. It was identified as a volatile component of cognac [31].

Plant Concentrations: 2-Hexanone has been identified as a volatile component of roasted filbert nuts [13] and as a volatile component of nectarines [30].

341

2-Hexanone

Fish/Seafood Concentrations:

Animal Concentrations: 2-Hexanone has been identified in raw chicken breast muscle [10].

Milk Concentrations: 2-Hexanone has been determined in milk at a concentration range of 0.007-0.011 ppm [16].

Other Environmental Concentrations:

Probable Routes of Human Exposure: For the general public, probable routes of exposure include the ingestion of natural and processed foods in which it has been identified, and through the inhalation of 2-hexanone vapors from commercial coatings which would release the solvent while drying and curing. The greatest hazards of 2-hexanone are by topical (eye and skin) and inhalation routes [4].

Average Daily Intake:

Occupational Exposure: NIOSH (NOHS Survey 1972-1974) has statistically estimated that 810 workers are exposed to 2-hexanone in the USA [21]. NIOSH (NOES Survey 1981-1983) has statistically estimated that 170,849 workers are exposed to 2-hexanone in the USA [22].

Body Burdens:

REFERENCES

1.   Atkinson R et al; Chem Rev 85: 69-201 (1985)
2.   Babeu L, Vaishnav DD; J Ind Microbiol 2: 107-15 (1987)
3.   Brown KW, Donnelly KC; Haz Wast Haz Water 5: 1-30 (1988)
4.   Clayton GD, Clayton, FE; Patty's Industrial Hygiene and Toxicology, 3rd ed. NY: John Wiley & Sons, 2C: 4747 (1982)
5.   Day EA, Anderson DF; J Agr Food Chem 13: 2-4 (1965)
6.   Dumont JP, Adda J; J Agr Food Chem 26: 364-7 (1978)
7.   Eisenreich SJ et al; Environ Sci Technol 15: 30-8 (1981)
8.   Erichsen R; Naturwissenschaften 39: 41-2 (1952)
9.   Farley FF; Int Conf Photochem Oxid Pollut Control Proc 2: 713-26 (1977)
10. Grey TC, Shrimpton DH; Brit Poultry Sci 8: 23-33 (1967)
11. Hansch C, Leo AJ; Medchem project Issue No. 26 Claremont, CA Pomona College (1985)

12. Juttner F; Chemosphere 15: 985-92 (1986)
13. Kinlin TE et al; J Agr Food Chem 20: 1021-8 (1972)
14. Konasewich D et al; Status Report on Organic Heavy Metal Contaminants in the Lakes Erie, Michigan, Huron and Superior Great Lakes Water Qual Board (1978)
15. Laity JL et al; Adv Chem Series 124: 95-112 (1973)
16. Lande SS et al; Investigation of Selected Environmental Contaminants: Ketonic Solvents USEPA-560/2-76-003 (1976)
17. Levy A; Adv Chem Ser 124: 70-94 (1973)
18. Lukins HB, Foster JW; J Bacteriol 85: 1074-87 (1963)
19. Lyman WJ et al; Handbook of Chemical Property Estimation Methods Washington DC: Amer Chem Soc pp 4-9, 5-4, 15-15 to 15-29 (1990)
20. Myers VB; Natl Conf Manage Uncontrolled Hazard Waste Site pp 354-7 (1983)
21. NIOSH; National Occupational Hazard Survey (NOHS)(1974)
22. NIOSH; National Occupational Exposure Survey (NOES)(1983)
23. Papa AJ, Sherman PD Jr; Kirk-Othmer Encycl Chem Tech 3rd Ed. John Wiley NY 13: 894-941 (1981)
24. Pellizzari ED et al; Amer Soc Test Mat. Spec Tech Publ STP 685: 256-74 (1979)
25. Perry JJ; Antonie van Leeuwenhoek 34: 27-36 (1968)
26. Perry DL et al; Identification of Organic Compounds in Industrial Efficient Discharges PB-291900 (1978)
27. Riddick JA et al; Organic Solvents: Physical Properties and Methods of Purification 4th ed Wiley Interscience NY (1986)
28. Shelton DR, Tiedje JM; App Env Microbiol 47: 85-7 (1984)
29. Swann RL et al; Res Rev 85: 17-28 (1983)
30. Takeoka GR et al; J Agric Food Chem 36: 553-60 (1988)
31. Terheide R et al; pp 249-81 in Anal Foods Beverages (Proc Symp), Chavalambous G Ed Academic Press New York (1978)
32. USEPA; Superfund Record of Decision USEPA/ROD/RO3-86/019 (1986)
33. USEPA; Superfund Record of Decision USEPA/ROD/RO3-86/024 (1987)
34. USEPA; Exams II Computer Simulation (1987)
35. Vaishnav DD et al; Chemosphere 16: 695-703 (1987)

2-Hydroxypropanenitrile

SUBSTANCE IDENTIFICATION

Synonyms:

Structure:

H₃C—CH(OH)—C≡N

CAS Registry Number: 78-97-7

Molecular Formula: C_3H_5NO

Wiswesser Line Notation: QY1&CN

CHEMICAL AND PHYSICAL PROPERTIES

Boiling Point: 182-4 °C (decomposes)

Melting Point: -40 °C

Molecular Weight: 71.08

Dissociation Constants:

Log Octanol/Water Partition Coefficient: -0.85 (estimated) [14]

Water Solubility: Miscible [4]

Vapor Pressure: 10 mm Hg at 74 °C [12]; 17 mm Hg at 90 °C [2]; 30 mm Hg at 102 °C [2]; 0.89 mm Hg at 25 °C (Antoine extrapolation)

Henry's Law Constant: 1.48×10^{-9} atm-m³/mol at 25 °C (estimated) [10]

ENVIRONMENTAL FATE/EXPOSURE POTENTIAL

Summary: 2-Hydroxypropanenitrile may be released to the environment in waste effluents generated at sites of its industrial manufacture or use. If released to the atmosphere, 2-hydroxypropanenitrile is expected to exist

344

almost entirely in the vapor phase where it will degrade relatively slowly (estimated half-life of 14.7 days) via reaction with photochemically produced hydroxyl radicals. Since 2-hydroxypropanenitrile is miscible in water, physical removal from air by wet deposition is likely to occur. If released to soil or water, microbial degradation may be an important removal process. One screening study has demonstrated that 2-hydroxypropanenitrile is readily biodegradable in Ohio River water. 2-Hydroxypropanenitrile is known to be unstable at high pHs; it reacts with alkali to form HCN. However, reaction rates in alkaline soil or natural water are not available. Leaching in soil is possible. Occupational exposure to 2-hydroxypropanenitrile can occur through inhalation of vapor and dermal contact. There are no data available to suggest that the general population is exposed to 2-hydroxypropanenitrile.

Natural Sources:

Artificial Sources: 2-Hydroxypropanenitrile is used as a solvent and chemical intermediate [6] and may be released to the environment in waste effluents generated at sites of its industrial manufacture or use.

Terrestrial Fate: One screening study has demonstrated that 2-hydroxypropanenitrile is readily biodegradable in Ohio River water [7]. Based on this study, microbial degradation in soil is likely to be an important removal mechanism. Insufficient data are available to predict the importance of other degradation processes in soil. 2-Hydroxypropanenitrile will degrade in alkaline media yielding HCN [3,11]; however, the rate of degradation at environmental pHs is not known. 2-Hydroxypropanenitrile will probably leach in soil based upon its miscibility in water and an estimated Koc value of 8.2 [9].

Aquatic Fate: One screening study has demonstrated that 2-hydroxypropanenitrile is readily biodegradable in Ohio River water [7]. Insufficient data are available to predict the importance of other degradation processes in water. 2-Hydroxypropanenitrile will degrade in alkaline media [3,11]; however, the rate of degradation at environmental pHs is not known. Aquatic bioconcentration, volatilization, and adsorption to sediment are not expected to be environmentally important.

Atmospheric Fate: Based upon the vapor pressure, 2-hydroxypropanenitrile is expected to exist almost entirely in the vapor phase in the ambient atmosphere [5]. It is expected to degrade in an average ambient atmosphere by reaction with photochemically produced hydroxyl radicals with an estimated half-life of about 14.7 days [1]. Physical removal via wet deposition (dissolution into clouds, rainfall, etc.) is likely to occur since it is miscible in water.

Biodegradation: Using Ohio River water as inoculum (with no special acclimation) and an aerobic test system, theoretical BODs of 50% and 70% were measured for 2-hydroxypropanenitrile after 2- and 5-day inoculation periods, respectively, at concentrations of 0.4-15 mg/L [7]; re-dosing the system resulted in a 60% theoretical BOD [7]. An activated sludge system that was acclimated to 2-hydroxypropanenitrile experienced an 87-98% BODT removal over a 4-week operation period while receiving influent 2-hydroxypropanenitrile levels that averaged 88 mg/L [8].

Abiotic Degradation: The rate constant for the vapor-phase reaction of 2-hydroxypropanenitrile with photochemically produced hydroxyl radicals can be estimated to be 1.09×10^{-12} cm^3/molecule-sec at 25 °C which corresponds to an atmospheric half-life of about 14.7 days at an atmospheric concentration of $5 \times 10^{+5}$ hydroxyl radicals/cm^3 [1]. Cyanohydrins, such as 2-hydroxypropanenitrile, are unstable at high pHs [3]; when 2-hydroxypropanenitrile is contacted with an alkali it evolves hydrocyanic acid [11]. Kinetic rate data pertaining to the hydrolysis of 2-hydroxypropanenitrile in water at environmental pHs were not located.

Bioconcentration: The miscibility of 2-hydroxypropanenitrile in water suggests that bioconcentration in aquatic organisms should not be environmentally important. Based upon the estimated log octanol-water partition coefficient, a BCF of 0.13 can be estimated from a recommended linear regression equation [9].

Soil Adsorption/Mobility: The miscibility of 2-hydroxypropanenitrile in water suggests that leaching in soil is likely to occur. Based upon the estimated log octanol-water partition coefficient, the Koc for 2-hydroxypropanenitrile can be estimated to be 8.2 from a linear regression equation [9]. This Koc value suggests that 2-hydroxypropanenitrile has very high soil mobility [13].

Volatilization from Water/Soil: The value of the Henry's Law constant indicates that 2-hydroxypropanenitrile is essentially nonvolatile from water [9]. Therefore, volatilization from environmental waters is not expected to be important.

Water Concentrations:

Effluent Concentrations:

Sediment/Soil Concentrations:

Atmospheric Concentrations:

Food Survey Values:

Plant Concentrations:

Fish/Seafood Concentrations:

Animal Concentrations:

Milk Concentrations:

Other Environmental Concentrations:

Probable Routes of Human Exposure: Occupational exposure to 2-hydroxypropanenitrile can occur through inhalation of vapor and dermal contact [11]. 2-Hydroxypropanenitrile is extremely toxic by skin absorption, eye contact, and ingestion; extreme care must be exercised during handling to ensure protection against personal contamination [11]. There are no data available to suggest that the general population is exposed to 2-hydroxypropanenitrile.

Average Daily Intake:

Occupational Exposure:

2-Hydroxypropanenitrile

Body Burdens:

REFERENCES

1. Atkinson R; J Inter Chem Kinet 19: 799-828 (1987)
2. Beilsteins Handbuch Der Organischen Chemie 3: 284 (1943)
3. Cholod MS; Kirk-Othmer Encycl Chem Technol 3rd ed NY: John Wiley & Sons 7: 385-96 (1979)
4. Dean JA; Lange's Handbook of Chem 13th ed NY: McGraw-Hill p 7-456 (1985)
5. Eisenreich SJ et al; Environ Sci Technol 15: 30-8 (1981)
6. Hawley GG; The Condensed Chemical Dictionary 10th ed NY: Van Nostrand Reinhold (1981)
7. Ludzack FJ et al; Sewage and Indust Wastes 31: 33-44 (1959)
8. Ludzack FJ et al; J Water Pollut Control Fed 33: 492-505 (1961)
9. Lyman WJ et al; Handbook of Chemical Property Estimation Methods Washington DC: Amer Chem Soc pp 4-9, 5-10, 15-15 to 15-29 (1990)
10. Meylan WM, Howard PH; Environ Toxicol Chem 10: 1283-93 (1991)
11. Parmeggiani L; Encycl Occup Health & Safety 3rd ed Geneva, Switzerland: International Labour Office pp 1445-6 (1983)
12. Sax NI; Dangerous Properties of Industrial Materials 6th ed NY: Van Nostrand Reinhold (1984)
13. Swann RL et al; Res Rev 85: 23 (1983)
14. USEPA; Graphical Exposure Modeling System. PC ver Apr 1987. PCLOGP v1.2 (1987)

Iodoform

SUBSTANCE IDENTIFICATION

Synonyms: Triiodomethane

Structure:

$$\begin{array}{c} \quad\; I \\ \quad\; | \\ H - C - I \\ \quad\; | \\ \quad\; I \end{array}$$

CAS Registry Number: 75-47-8

Molecular Formula: CHI_3

Wiswesser Line Notation: IYII

CHEMICAL AND PHYSICAL PROPERTIES

Boiling Point:

Melting Point: 120-123 °C

Molecular Weight: 393.78

Dissociation Constants:

Log Octanol/Water Partition Coefficient:

Water Solubility: 106 mg/L at 25 °C [6]

Vapor Pressure: 6.3 x 10^{-3} mm Hg at 25 °C (estimated from water solubility and Henry's Law constant)

Henry's Law Constant: 3.062 x 10^{-5} atm-m^3/mole at 25 °C (estimated) [5]

Iodoform

ENVIRONMENTAL FATE/EXPOSURE POTENTIAL

Summary: Iodoform can be released to the environment through its application as an antiseptic although present use is rather limited. If released to water, iodoform may evaporate into the atmosphere with a volatilization half-life of approximately 3.08 days from a model river. In water, some iodoform may also partition to sediment. In soil, its Koc of about 336 suggests moderate soil mobility. No data are available on other potential degradation processes in soil or water. If released to the atmosphere, vapor-phase iodoform is degraded slowly by reaction with photochemically produced hydroxyl radicals (estimated half-life of 325 days in air). There are no data which suggest that the general population is exposed to iodoform, but direct human exposure occurs through the application of the drug when applied to a wound. Workers involved in formulating and dispensing the drug may be exposed through dermal contact or inhalation of dust.

Natural Sources:

Artificial Sources: Iodoform is used as an antiseptic wound powder; however, today its use is rather limited [3,8].

Terrestrial Fate: The estimated Koc of approximately 336 [4] indicates that iodoform would probably be moderately mobile in soil [7]. No data are available regarding biodegradation or other chemical processes in soil.

Aquatic Fate: If released to water, iodoform may evaporate into the atmosphere with a volatilization half-life of approximately 3.08 days from a model river [4]. No data are available on other potential degradation processes in water. Based on the estimated Koc of about 336 [4], iodoform may partition from the water column to sediment. Based on an estimated BCF of 44.5 [4], iodoform is not expected to bioconcentrate significantly in aquatic organisms.

Atmospheric Fate: Vapor-phase iodoform is degraded slowly in the ambient atmosphere by reaction with photochemically formed hydroxyl radicals; the half-life for this reaction in typical air can be estimated to be about 325 days [1].

350

Iodoform

Biodegradation:

Abiotic Degradation: The rate constant for the vapor-phase reaction of iodoform with photochemically produced hydroxyl radicals has been estimated to be 4.94×10^{-14} cm^3/molecule-sec at 25 °C which corresponds to an atmospheric half-life of about 325 days at an atmospheric concn of $5 \times 10^{+5}$ hydroxyl radicals per cm^3 [1].

Bioconcentration: Based on the reported water solubility, the BCF for iodoform can be estimated to be 44.5 using a recommended regression-derived equation [4]. This BCF value suggests that iodoform would not bioconcentrate significantly in aquatic organisms.

Soil Adsorption/Mobility: Based on the reported water solubility, a soil adsorption coefficient (Koc) for iodoform of approximately 336 was estimated using a linear regression derived equation [4]. This Koc value indicates that iodoform would probably be moderately mobile in soil and that slight adsorption to suspended solids and sediments in water may take place [7].

Volatilization from Water/Soil: The estimated Henry's Law constant for iodoform suggests that volatilization from water may not be rapid but possibly significant [4]. Based on this value, the volatilization half-life of iodoform from a model river 1 m deep flowing 1 m/sec with a wind velocity of 3 m/sec has been estimated to be approximately 3.08 days [4].

Water Concentrations:

Effluent Concentrations:

Sediment/Soil Concentrations:

Atmospheric Concentrations:

Food Survey Values:

Plant Concentrations:

Fish/Seafood Concentrations:

351

Animal Concentrations:

Milk Concentrations:

Other Environmental Concentrations:

Probable Routes of Human Exposure: Iodoform is used as an antiseptic wound powder; however, today its use is rather limited [3,8]. Therefore, direct human exposure occurs through the application of the drug when applied to a wound. Workers involved in formulating and dispensing the drug may be exposed through dermal contact or inhalation of dust.

Average Daily Intake:

Occupational Exposure:

Body Burdens:

REFERENCES

1. Atkinson R; Int J Chem Kinet 19: 799-828 (1987)
2. Dean JA; Lange's Handbook of Chemistry 13th ed NY, NY: McGraw-Hill p 7-464 (1985)
3. Gump W; Kirk-Othmer Encycl Chem Tech 3rd ed. NY: John-Wiley 7: 803 (1979)
4. Lyman WJ et al; Handbook of Chemical Property Estimation Methods. Environmental Behavior of Organic Compounds. Washington DC: American Chemical Society pp 4-9, 5-4, 5-10, 7-4, 7-5, 15-15 to 15-32 (1990)
5. Meylan WM, Howard PH; Environmental Toxicology and Chemistry. 10: 1283-1293 (1991)
6. Seidell A; Solubilities of Organic Compounds; NY, NY: D Van Nostrand Co Inc (1941)
7. Swann RL et al; Res Rev 85: 17-28 (1983)
8. Windholz M; The Merck Index 10th ed; Merck & Co, Inc Rahway, NJ p 4906 (1983)

Iodomethane

Synonyms: Methyl iodide

Structure:

$$
\begin{array}{c}
\text{H} \\
\text{H}\!-\!\text{C}\!-\!\text{I} \\
\text{H}
\end{array}
$$

CAS Registry Number: 74-88-4

Molecular Formula: CH_3I

Wiswesser Line Notation: I1

CHEMICAL AND PHYSICAL PROPERTIES

Boiling Point: 42.5 °C

Melting Point: -66.5 °C

Molecular Weight: 141.94

Dissociation Constants:

Log Octanol/Water Partition Coefficient: 1.51 [4]

Water Solubility: 1,389 mg/L at 25 °C [7]

Vapor Pressure: 405.9 mm Hg at 25 °C [15]

Henry's Law Constant: 5.47×10^{-3} atm-m^3/mole at 25 °C [6]

ENVIRONMENTAL FATE/EXPOSURE POTENTIAL

Summary: Methyl iodide is produced by many marine photosynthetic organisms and therefore the ocean surface is thought to be a major natural source of methyl iodide. Some of the methyl iodide formed in ocean water will escape into the atmosphere by evaporation and some will react with chloride in seawater (half-life approximately 20 days at 19 °C) to form methyl chloride. In freshwaters, evaporation would be the most important process for the loss of methyl iodide. It will not be expected to adsorb to soil or sediment, or to bioconcentrate in aquatic organisms; however, there is some evidence of bioaccumulation in marine animals. The loss of methyl iodide by evaporation from both wet and dry soil surfaces may be the most important process, although the rate of photolysis on surficial soil remains unknown. In the atmosphere, methyl iodide shows reactivity, probably due to photoreaction. However, rate data are not available to estimate the tropospheric photolytic half-life of methyl iodide. Humans are exposed to methyl iodide from the ambient air and from ingesting seafood from the ocean.

Natural Sources: Formed in the sea as a natural product of marine algae with an estimated annual world production of $4 \times 10^{+10}$ kg [8].

Artificial Sources: Industrial emissions may result from its use as a methylating agent in organic synthesis [8]. Can be formed in the environment of nuclear reactors and vented in exhaust gases [8].

Terrestrial Fate: If spilled on land, methyl iodide would be expected to volatilize rapidly. Sorption to soils is not expected to be important. No studies could be found concerning its fate on land but it could photolyze or decompose by reacting with chlorides.

Aquatic Fate: Methyl iodide will be removed from the water by volatilization (half-life 3.7 hr in a typical river). In the ocean, methyl iodide will react with chloride to form methyl chloride (methyl iodide half-life 20 and 58 days at 19.2 and 10.8 °C, respectively). In placid bodies of fresh water, hydrolysis of methyl iodide may be significant (half-life 100-251 days). Photolysis may also be important but rate data necessary to assess the importance of the reaction remains unknown. It will not be expected to bind to sediments or to bioconcentrate in aquatic organisms; however, there is some evidence of bioaccumulation in marine animals.

Iodomethane

Atmospheric Fate: Methyl iodide has been found to be reactive in the troposphere. Photolysis in air may be an important process, but rate data necessary to evaluate the half-life is not known. The reaction of methyl iodide with photolytically produced hydroxyl radicals in the troposphere has a half-life of 371. Because of its significant water solubility, partial removal of atmospheric methyl iodide by wet deposition should occur.

Biodegradation:

Abiotic Degradation: Methyl iodide hydrolyzes slowly in water yielding methanol [13]. The half-life under neutral conditions is 110-251 days at 20-25 °C [13,20], increasing to 4 yr and 23 yr at 10 and 0 °C, respectively [20]. Acid does not catalyze the hydrolysis of alkyl halides and base catalysis is only important at higher pHs than are observed in the environment [13]. However, methyl iodide is unstable in seawater, reacting primarily with the chloride ion to form methyl chloride [20]. The half-life in seawater of 19.8 parts/thousand chlorine is 20 and 58 days at 19.2 and 10.8 °C, respectively [20]. Methyl iodide absorbs UV radiation up to approximately 340 nm and photolyses to form methyl radicals and iodine molecules [2]. Photolysis occurs in both the gas phase and in solution [9]. When vapor-phase methyl iodide at a concn 10 ppb in pure air at a relative humidity of 50% was irradiated with light of wavelength >290 nm, methyl iodide disappeared with a half-life of 7 hr [10]. Since no data on photolysis rate under natural sunlight irradiation conditions were found, the half-life for tropospheric photolysis of methyl iodide cannot be estimated. Based on an estimation method [1], the rate constant for the reaction of methyl iodide with photochemically produced hydroxyl radicals has been estimated to be 4.32×10^{-14} cm^3/molecule-sec. If the daily average concentration of hydroxyl radical in the atmosphere is assumed to be 5×10^5/cm^3 [1], the half-life of the reaction can be estimated to be 371 days.

Bioconcentration: Based on a log octanol/water partition coefficient of 1.51 [4] and a regression equation [12], the bioconcentration factor for methyl iodide in aquatic organisms has been estimated to be 8. Therefore, bioconcentration of the compound in aquatic organisms would not be important. However, a bioconcentration factor up to 240 was found for marine fish brain tissue and it is not clear whether the observed factors

355

were due to true bioconcentrations or due to synthesis of the compound by the fish [3].

Soil Adsorption/Mobility: Based on a water solubility of 1,389 mg/L [7] and a regression equation [12], log Koc for methyl iodide has been estimated to be 2.05. Therefore, it would be highly mobile in soil and would not adsorb strongly to suspended solids and sediments in water [19].

Volatilization from Water/Soil: Based on a Henry's Law constant of 5.47×10^{-3} atm-m^3/mole [6], the half-life of methyl iodide in a model river 1 m deep flowing at 1 m/sec with a wind velocity of 3 m/sec can be calculated to be 3.7 hr [12]. Due to its high vapor pressure of 405.9 torr [15] and its ability to evaporate rapidly from water, methyl iodide will evaporate rapidly from both dry and wet soil surfaces.

Water Concentrations: SEAWATER: Point Reyes, CA - nearshore 37 ppt [17]. Atlantic Ocean surface water 135 ppt [11].

Effluent Concentrations:

Sediment/Soil Concentrations:

Atmospheric Concentrations: REMOTE: US - 9 samples, 3 ppt (v/v) median [16]. RURAL: US - 3 samples, 7 ppt median [16]. SUBURBAN: US - 82 samples, 7 ppt median [16]. URBAN: US - 73 samples, 2 ppt median [16]. Global background 2 ppt [18]. Levels most frequently measurable in close proximity to the ocean are consistent with the suggestion that its primary origin is the ocean [10].

Food Survey Values:

Plant Concentrations:

Fish/Seafood Concentrations: Irish Sea - 3 species of mollusks 3-188 ppb, 10 ppb median [3]. Irish Sea - 5 species of fish 4-166 ppb, 17 ppb median [3].

Animal Concentrations:

Iodomethane

Milk Concentrations:

Other Environmental Concentrations:

Probable Routes of Human Exposure: The most probable route of general population exposure to methyl iodide is via ambient air since it is ubiquitous in the atmosphere. Exposure would also occur from ingesting seafood.

Average Daily Intake: Assuming an inhalation rate of 20 m^3 of air/day and the concn of methyl iodide in a typical ambient air as 3.5 ppt, the daily inhalation intake would be 0.41 ug.

Occupational Exposure: Since methyl iodide is used in organic synthesis and as a laboratory reagent [5], workers and laboratory personnel using this compound are occupationally exposed to this compound. In a 1981-83 National Occupational Exposure Survey (NOES), it has been statistically estimated that 2,182 workers are potentially exposed to methyl iodide [14].

Body Burdens:

REFERENCES

1. Atkinson R; Environ Toxicol Chem 7: 435-42 (1988)
2. Calvert JG, Pitts JM Jr; Photochemistry; John Wiley & Sons, New York NY pp 522-8 (1966)
3. Dickson AG, Riley JP; Ma Pollut Bull 7: 167-9 (1976)
4. Hansch C, Leo AJ; MedChem Project, Issue No. 26, Claremont, CA: Pomona College (1985)
5. Hawley GG (ed); The Condensed Chemical Dictionary, 10th ed, Van Nostrand Reinhold Co, NY p 681 (1981)
6. Hine J, Mookerjee PK; J Org Chem 40: 292-8 (1975)
7. Horvath, AL; Halogenated Hydrocarbons NY: Marcel Dekker Inc p 489 (1982)
8. IARC; Some Fumigants, the Herbicides, 2,4-D and 2,4,5-T, Chlorinated Dibenzodioxins and Miscellaneous Industrial Chemicals; 15: 245-54 (1977)
9. Lillian D et al; Environ Sci Technol 9: 1042-8 (1975)
10. Lillian D et al; Amer Chem Soc Symp Ser 17: 152-8 (1975)
11. Lovelock JE et al; Nature 241: 194-6 (1973)
12. Lyman WJ et al; Handbook of Chemical Property Estimation Methods Environmental Behavior of Organic Compounds; McGraw Hill Book Co, New York, NY pp 4-9, 5-5, 15-21 (1982)

357

Iodomethane

13. Mabey W, Mill T; J Chem Ref Data 7: 383-415 (1978)
14. NIOSH; National Occupational Exposure Survey (NOES), NIOSH, Cincinnati, OH, p 13 (1989)
15. Riddick JA et al; Techniques of Chemistry Vol II, Organic Solvents - Physical Properties and Methods of Purification 4th ed, NY: John Wiley and Sons p 546 (1986)
16. Shah JJ, Heyerdahl EK; National Ambient Volatile Organic (VOCs) Data Base Update, EPA Report No. 600/3-88/010, Atmospheric Sciences Res Lab, Research Triangle Park, NC (1988)
17. Singh HB et al; J Air Pollut Control Assoc 27: 332-6 (1977)
18. Singh HB et al; Atmospheric Distribution, Sources and Sink of Selected Hydrocarbons in the Atmosphere; p.134 USEPA-600/3-79-107 (1979)
19. Swann RL et al; Res Rev 85: 17-28 (1983)
20. Zafiriou OC; J Mar Res 33: 75-81 (1975)

Isopropyl Acetate

SUBSTANCE IDENTIFICATION

Synonyms:

Structure:

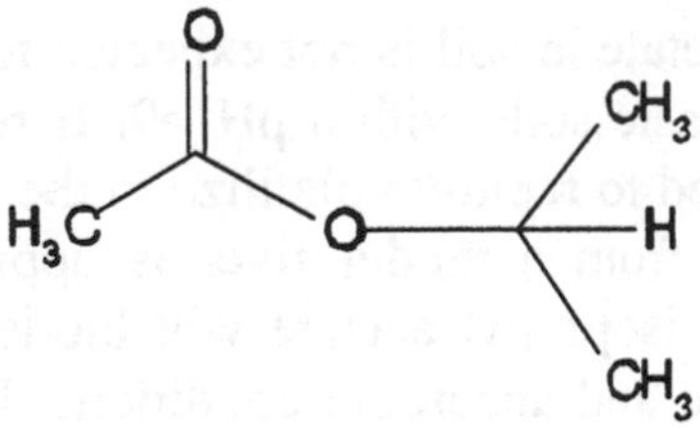

CAS Registry Number: 108-21-4

Molecular Formula: $C_5H_{10}O_2$

Wiswesser Line Notation: 1YOV1

CHEMICAL AND PHYSICAL PROPERTIES

Boiling Point: 90 °C

Melting Point: -73.4 °C

Molecular Weight: 102.13

Dissociation Constants:

Log Octanol/Water Partition Coefficient: 1.03, estimate [15]

Water Solubility: 3.09 x 10^4 mg/L at 20 °C [26],

Vapor Pressure: 59.2 mm Hg at 25 °C [1]

Henry's Law Constant: 2.81 x 10^{-4} atm cu-m/mole at 25 °C [20]

ENVIRONMENTAL FATE/EXPOSURE POTENTIAL

Summary: Isopropyl acetate, which is used as a solvent, may be released in fugitive emissions during its manufacture, formulation, or use in commercial products. Isopropyl acetate is also a naturally occurring compound. If released to soil, isopropyl acetate will display very high mobility and it has the potential to leach into ground water. Rapid volatilization is expected to occur from both moist and dry soils. Hydrolysis of isopropyl acetate in soil is not expected to be a significant process except in highly basic soils with a pH >9. If released to water, isopropyl acetate is expected to rapidly volatilize to the atmosphere. The half-life for volatilization from a model river is approximately 9 hr. Limited data suggests that isopropyl acetate will biodegrade in aquatic systems under both aerobic and anaerobic conditions. Isopropyl acetate will not significantly adsorb to sediment and suspended organic matter, nor will it bioconcentrate in fish and aquatic organisms. Hydrolysis of isopropyl acetate in aquatic systems is not expected to be a significant process except under basic conditions of pH >9. In the atmosphere, isopropyl acetate is expected to undergo a relatively slow gas-phase reaction with photochemically produced hydroxyl radicals with experimental half-lives on the order of 4-5 days. Isopropyl acetate may undergo atmospheric removal by wet deposition. The probable routes of exposure to isopropyl acetate are by inhalation and dermal contact during the production and use of this compound. The general public is likely to be exposed to isopropyl acetate by the ingestion of foods in which it is contained. Limited monitoring data indicate that the general population has the potential to be exposed to isopropyl acetate by the ingestion of contaminated drinking water.

Natural Sources: Isopropyl acetate may be released to the environment from natural sources. It has been detected as a volatile constituent of grape juice [27], nectarines [29] and apples [33].

Artificial Sources: In 1989 there were 3 companies listed as producers of isopropyl acetate [25]. Current production of isopropyl acetate in the United States amounted to 40.85 million pounds in 1988 [30]. Isopropyl acetate has been used as an active solvent for many synthetic resins such as ethyl cellusolve, cellulose acetate butyrate, cellulose nitrate, some vinyl copolymers, polystyrene methacrylate resins, and natural resins such as kauri, manila gums and pontianak [6]. It also finds use as a solvent for printing inks, and in organic synthesis and perfumery [23].

Isopropyl Acetate

Terrestrial Fate: If released to soil, a calculated soil adsorption coefficient of 14.8 [15] obtained from its water solubility suggests that isopropyl acetate will display very high mobility in soil [28] and has the potential to leach into ground water. The Henry's Law constant of isopropyl acetate, and its vapor pressure, suggest that this compound will rapidly volatilize from both moist and dry soil. Hydrolysis of isopropyl acetate in soil is not expected to be a significant process except in highly basic soils with a pH >9, as hydrolysis rate constants indicate that this process will be too slow to be environmentally significant under acidic, neutral, and slightly basic conditions [5,6].

Aquatic Fate: If released to water, isopropyl acetate is expected to rapidly volatilize to the atmosphere. Based on the measured Henry's Law constant, the half-life for volatilization from a model river is 6.1 hr [15]. The available data indicates that isopropyl acetate will biodegrade in aquatic systems under aerobic conditions [4,18]. From its water solubility, a calculated soil adsorption coefficient of 14.8 [15] and a calculated bioconcentration factor of 1.8 [15] indicate that isopropyl acetate will not significantly adsorb to sediment and suspended organic matter, nor will it bioconcentrate in fish and aquatic organisms. Hydrolysis of isopropyl acetate in aquatic systems is not expected to be a significant process except under basic conditions of pH >9, as hydrolysis rate constants indicate that this process will be too slow to be environmentally significant under acidic, neutral, and slightly basic conditions [5,6].

Atmospheric Fate: In the atmosphere, experimental rate constants for the gas phase reaction of isopropyl acetate with photochemically produced hydroxyl radicals in the range 3.0-3.72 x 10^{-12} cu-cm/mol-sec [2,3,10,32] correspond to an atmospheric half-life ranging from 4.2-5.2 days. In the presence of alkyl nitrates primary products from this reaction are acetone and methyl nitrate [10]. The relatively high water solubility of isopropyl acetate, suggests that this compound may be removed from the atmosphere by wet deposition.

Biodegradation: A screening test using an activated mixed microbial sewage inoculum indicated that isopropyl acetate had a 5-day BODT of 38% [18]. Isopropyl acetate, at an initial concentration of 0.4-3.2 ug/L, had a 52.3% 5-day BODT when incubated with an acclimated mixed microbial culture [4]. In a screening test using a settled sewage seed, 2.5

ppm isopropyl acetate was found to have a 5-day BODT of 12.7%, which increased to 40% after 10 days and 49.1% after 40 days [13]. Isopropyl acetate gave a 5 day BODT of 61% using a settled domestic wastewater seed, and BODTs of 72%, 74% and 76% after 10, 15 and 20 days, respectively [22]. When the same inoculum was added to synthetic sea water, the 5-, 10-, 15- and 20-day BODTs were 14%, 39%, 43%, 49%, respectively [22].

Abiotic Degradation: Experimental rate constants for the gas-phase reaction of isopropyl acetate with photochemically produced hydroxyl radicals range from 3.0-3.72 x 10^{-12} cu-cm/mol-sec at temperatures of 25-32 °C [2,3,10,32], which correspond to an atmospheric half-life ranging from 4.3-5.2 days using an average atmospheric hydroxyl radical concentration of 5 x 10^{+5} mol/cu-cm [2]. In the presence of alkyl nitrates, primary products from this reaction are acetone and, under atmospheric conditions, methyl nitrate [10]. The atmospheric photo-oxidation potential of isopropyl acetate, based on a 5-tiered rating system and determined in smog chamber studies, was assigned as Class II, low reactivity [8]. This ranking system assigns compound with negligible reactivity, such as methane, as class I and highly reactive dienes as class V [8]. The relative rate for the photo-oxidation of isopropyl acetate in a smog chamber was 0.4 times less reactive than toluene [12]. An experimental rate constant for the basic hydrolysis of isopropyl acetate in water at 25 °C, 0.030 L/mole-s [5], corresponds to half-lives of 267 days, 27 days, and 2.7 days at pH 8, 9 and 10, respectively. In general, alkyl esters, especially those with branching on the carbon attached to the ester oxygen, are resistant to hydrolysis under acidic conditions except with highly acidic solutions or at elevated temperatures [6]. An experimental rate constant for the aqueous reaction of isopropyl acetate with peroxy radicals at 30 °C is 2.4 x 10^{-3} L/mole-sec [9] which corresponds to a half-life of greater than 9000 years using an oxidant concentration of 1 x 10^{-9} mole/L [17].

Bioconcentration: From the water solubility of isopropyl acetate, a bioconcentration factor of 1.8 can be calculated by a recommended regression equation [15]. This value indicates that isopropyl acetate will not significantly bioconcentrate in fish and aquatic organisms.

Soil Adsorption/Mobility: From the water solubility of isopropyl acetate, a soil adsorption coefficient of 14.8 can be calculated by a regressional analysis [15]. This value indicates that isopropyl acetate will display very high mobility in soil [28].

Volatilization from Water/Soil: Based on the measured Henry's Law constant at 25 °C for isopropyl acetate, the half-life for volatilization from a model river 1 m deep, flowing at 1 m/sec and a wind speed of 3 m/sec is 6.1 hr [15]. The Henry's Law constant of isopropyl acetate and its vapor pressure at 25 °C, suggest that volatilization from both moist and dry soil to the atmosphere will be a significant fate process.

Water Concentrations: DRINKING WATER: isopropyl acetate was qualitatively detected in U.S. drinking water supplies [11]. Isopropyl acetate was detected, but not quantified, in drinking concentrates from Ottumwa, IA [14].

Effluent Concentrations:

Sediment/Soil Concentrations:

Atmospheric Concentrations: Isopropyl acetate was detected in 1 of 4 samples in the air surrounding Kin-Buc Waste Disposal site, NJ, 1976, at an estimated concentration of 6.5 ug/cu-m [21]. It was qualitatively detected in the air of the Netherlands and it was described as one of the principal compounds emitted to the air there [24]. Isopropyl acetate was found in air samples obtained in the industrialized Kanawha Valley, WV, 1977 [7].

Food Survey Values: Isopropyl acetate has been identified as a component of Concord grape juice essence [27]. It was detected in a headspace analysis of intact nectarines, but it was not found in samples of the blended fruit [29]. Isopropyl acetate was identified as a volatile flavor component of Kogyoke apples [33].

Plant Concentrations:

Fish/Seafood Concentrations:

Animal Concentrations:

Isopropyl Acetate

Milk Concentrations:

Other Environmental Concentrations:

Probable Routes of Human Exposure: The probable route of occupational exposure to isopropyl acetate are by inhalation and dermal contact during the production and use of this compound. The general public is likely to be exposed to isopropyl acetate by the ingestion of foods [27,29,33] in which it is contained. Limited monitoring data [11,14] indicate that the general population has the potential to be exposed to isopropyl acetate by the ingestion of contaminated drinking water.

Average Daily Intake:

Occupational Exposure: NIOSH (NOES Survey 1981-83) has statistically estimated that 125,376 workers are potentially exposed to isopropyl acetate in the USA, 94% of which are exposed during the use of trade name compounds in which isopropyl acetate is contained [19]. Isopropyl acetate was found in 4 air samples taken from 11 different auto paint shops in Spain at concentrations ranging 7.8-107 mg/m^3 [16]. It was qualitatively detected in 6% of air samples taken from printing industries in Belgium [31].

Body Burdens:

REFERENCES

1. Ambrose D et al; J Chem Therm 13: 795-802 (1981)
2. Atkinson R; Chem Rev 85: 69-201 (1985)
3. Atkinson R; Int J Chem Kinet 19: 799-828 (1987)
4. Babeu L, Vaishnav DD; J Ind Microbiol 2: 107-15 (1987)
5. Drossman H et al; Chemosphere 17: 1509-30 (1987)
6. Elam EU; Kirk-Othmer Encycl Chem Tech 3rd Ed. John-Wiley NY 9: 311-37 (1978)
7. Erickson MD, Pellizzari ED; Analysis of Organic Air Pollutants in the Kanawha Valley, WV and the Shenandoah Valley VA. USEPA-903/9-78-007 (1978)
8. Farley FF; Int Conf Photochem Oxid Pollut Control Proc 2: 713-26 (1977)
9. Hendry DG et al; J Phys Chem Ref Data 3: 937-78 (1974)
10. Kerr JA, Stocker DW; J Atmos Chem 4: 263-76 (1986)
11. Kool HJ et al; CRC Crit Rev Env Control 12: 307-57 (1982)

Isopropyl Acetate

12. Laity JL et al; Adv Chem Series 124: 95-112 (1973)
13. Lamb CB, Jenkins GF; Proc 8th Industrial Waste Conf, Purdue Univ: pp 326-9 (1952)
14. Lucas SV; GC/MS Analysis of Organics in Drinking Water Concentrates and Advanced Waste Treatment Concentrates: Vol. 3 USEPA-600/1-84-020 (NTIS PB85-128247), Columbus, OH (1984)
15. Lyman WJ et al; Handbook of Chemical Property Estimation Methods NY: McGraw-Hill pp 4-1 to 4-33, 5-1 to 5-30 and 15-1 to 15-34 (1982)
16. Medinilla J, Espigares M; Ann Occup Hyg 32: 509-13 (1988)
17. Mill T; Environ Toxicol Chem 1: 135-141 (1982)
18. Nieme GJ et al; Environ Toxicol Chem 6: 515-27 (1987)
19. NIOSH; National Occupational Exposure Survey (NOES) (1989)
20. Nirmalakhandan NN, Speece RE; Environ Sci Tech 22: 1349-57 (1988)
21. Pellizzari ED; Environ Sci Tech 16: 781-5 (1982)
22. Price KS et al; J Wat Pollut Contr Fed 46: 63-77 (1974)
23. Sax NI, Lewis RJSR; Hawley's Condensed Chemical Dictionary 11th ed Van Nostrand Reinhold Co. NY p 496 (1987)
24. Smeyers-Verbeke J et al; Atmosphere Environ 18: 2471-8 (1984)
25. SRI International; Directory of Chemical Producers (1989)
26. Stephan H, Stephan T; Solubilities of Inorganic and Organic Compounds in Binary Systems. Stephen H et al. eds, NY, NY 1: 1-79, 1604-43 (1963)
27. Stevens KL et al; J Food Sci 30: 1006-7 (1965)
28. Swann RL et al; Res Rev 85: 17-28 (1983)
29. Takeoka GR et al; J Agric Food Chem 36: 553-60 (1988)
30. USITC; Synthetic Organic Chemicals, United States Production and Sales, 1988: US Interl Trade Comm, Washington, DC. USITC publ #2219 (1989)
31. Veulemans H et al; Ind Hyg Assoc J 48: 671-6 (1987)
32. Wallington TJ et al.; J Phys Chem 92: 5024-8 (1988)
33. Yajima I et al; Agric Biol Chem 48: 849-5 (1984)

Isopropyl Ether

SUBSTANCE IDENTIFICATION

Synonyms:

Structure:

CAS Registry Number: 108-20-3

Molecular Formula: $C_6H_{14}O$

Wiswesser Line Notation: 1Y1&OY1&1

CHEMICAL AND PHYSICAL PROPERTIES

Boiling Point: 68-69 °C

Melting Point: -60 °C

Molecular Weight: 102.17

Dissociation Constants:

Log Octanol/Water Partition Coefficient: 1.56 [1]

Water Solubility: 2,000 mg/L at 20 °C [22]

Vapor Pressure: 149 mm Hg at 25 °C [22]

Henry's Law Constant: 9.78 x 10^{-3} atm-m³/mole [15]

ENVIRONMENTAL FATE/EXPOSURE POTENTIAL

Summary: Isopropyl ether may be released to the environment as a result of its use as an industrial solvent. If isopropyl ether is released to soil, it will be subject to volatilization. It will be expected to exhibit moderate mobility in soil and, therefore, it may leach to ground water.

Isopropyl Ether

It will not be expected to hydrolyze in soil. If isopropyl ether is released to water, it will not be expected to significantly adsorb to sediment or suspended particulate matter, bioconcentrate in aquatic organisms, hydrolyze, or directly photolyze, based upon estimated physical-chemical properties or analogies to other structurally related aliphatic ethers. It will not significantly photooxidize via reaction with photochemically produced hydroxyl radicals in the water. Isopropyl ether in surface water will be subject to rapid volatilization with estimated half-lives of 3.3 hr and 40 hr for volatilization from a river one meter deep flowing 1 m/sec with a wind velocity of 3 m/sec and a model pond, respectively. It may be resistent to biodegradation in environmental media based upon screening test data from studies using activated sludge or sewage inocula. Many ethers are known to be resistant to biodegradation. If isopropyl ether is released to the atmosphere, it will be expected to exist almost entirely in the vapor phase based on its vapor pressure. It will be susceptible to photooxidation via vapor-phase reaction with photochemically produced hydroxyl radicals with an estimated half-life of 17 hours for this process. Direct photolysis will not be an important removal process since aliphatic ethers do not absorb light at wavelengths >290 nm. The most probable routes of general population exposure to isopropyl ether are via inhalation of contaminated air and ingestion of contaminated drinking water. Exposure through dermal contact may occur in occupational settings. Inhalation and dermal exposure will be expected to be highest in workplaces where isopropyl ether is made and used.

Natural Sources:

Artificial Sources: Isopropyl ether may be released to the environment as a result of its use as an industrial solvent [16].

Terrestrial Fate: If isopropyl ether is released to soil, it will be subject to volatilization based upon the reported Henry's Law constant and the vapor pressure. It will be expected to exhibit moderate mobility [23] in soil and, therefore, it may leach to ground water, based upon an estimated Koc of 168 [18]. It will not be expected to hydrolyze in soil [18]. Aqueous screening test data from studies using activated sludge or sewage inocula [7,13] suggest that isopropyl ether may be resistent to biodegradation in environmental media. Many ethers are known to be resistant to biodegradation [2].

367

Isopropyl Ether

Aquatic Fate: If isopropyl ether is released to water, it will not be expected to significantly adsorb to sediment or suspended particulate matter, bioconcentrate in aquatic organisms, hydrolyze [18], or directly photolyze [8], based upon estimated physical-chemical properties or analogies to structurally related other aliphatic ethers [8,18]. It will not significantly photooxidize via reaction with photochemically produced hydroxyl radicals in the water [14]. Isopropyl ether in surface water will be subject to rapid volatilization [18]. Using the reported Henry's Law constant, a half-life for volatilization of isopropyl ether from a river one meter deep flowing 1 m/sec with a wind velocity of 3 m/sec has been estimated to be 3.3 hr at 25 °C [18]. The volatilization half-life from a model pond, which considers the effect of adsorption, has been estimated to be 40 hr [24]. Isopropyl ether may be resistent to biodegradation in environmental media based upon screening test data from studies using activated sludge or sewage inocula [7,13]. Many ethers are known to be resistant to biodegradation [2].

Atmospheric Fate: If isopropyl ether is released to the atmosphere, it will be expected to exist almost entirely in the vapor phase [9] based upon the reported vapor pressure. It will be susceptible to photooxidation via vapor-phase reaction with photochemically produced hydroxyl radicals. An atmospheric half-life of 17 hours at an atmospheric concentration of $5 \times 10^{+5}$ hydroxyl radicals per cm^3 has been calculated for this process based upon an estimated rate constant [6]. Direct photolysis will not be an important removal process since aliphatic ethers do not absorb light at wavelengths >290 nm [8].

Biodegradation: No data concerning the biodegradation of isopropyl ether in environmental media were located. An activated sludge aqueous screening study found that the compound was biodegraded slowly after a 15-day lag period with a 25% theoretical biological oxygen demand being measured after 25 days incubation [13]. A screening test study utilizing a sewage inoculum also indicated slow biodegradation had taken place based upon the 7% theoretical biological oxygen demand which was measured after 5 days [7]. These screening test results suggest that isopropyl ether may be resistent to biodegradation in the environment. Many ethers are known to be resistant to biodegradation [2].

Isopropyl Ether

Abiotic Degradation: The rate constant for the vapor-phase reaction of isopropyl ether with photochemically produced hydroxyl radicals has been estimated to be 23 x 10^{-12} cm^3/molecule-sec at 25 °C [6], which corresponds to an atmospheric half-life of 17 hours at an atmospheric concentration of 5 x 10^{+5} hydroxyl radicals per cm^3. Reaction of isopropyl ether with photochemically produced alkoxy radicals in water is not expected to be an important fate process [14]. Direct photolysis will not be an important removal process since aliphatic ethers do not absorb light at wavelengths >290 nm [8].

Bioconcentration: Based upon the reported log Kow, a BCF of 9.0 has been estimated using a recommended regression equation [18]. Based upon this estimated BCF, isopropyl ether will not be expected to bioconcentrate in aquatic organisms.

Soil Adsorption/Mobility: Based upon the reported log Kow, a Koc of 168 has been estimated using a recommended regression equation [18]. Based upon this estimated Koc, isopropyl ether will be expected to exhibit moderate mobility in soil [23]. Isopropyl ether, therefore, may slowly leach through some soils to ground water if it does not volatilize or biodegrade first.

Volatilization from Water/Soil: The half-life for volatilization of isopropyl ether from a river one meter deep flowing 1 m/sec with a wind velocity of 3 m/sec is estimated to be 3.3 hr at 20 °C [18] based on the reported Henry's Law constant. The volatilization half-life from a model pond, which considers the effect of adsorption, has been estimated to be 40 hr [24]. Based upon the Henry's Law constant and a reported vapor pressure, isopropyl ether will be subject to volatilization from surfaces and near-surface soil.

Water Concentrations: Isopropyl ether was detected in 1 of 14 samples of treated drinking water in England in 1978 and 1979; the drinking water with isopropyl ether was derived from river water [12]. GROUND WATER: Isopropyl ether has been found in ground water in the US from unspecified sites at a concentration range of 20-34 ppb [3]. Isopropyl ether has been detected at concentrations up to 160 ppb in the Old Bridge aquifer under an industrial plant in South Brunswick Township, NJ (no sampling dates specified) [4]. A contamination abatement system installed at this aquifer, including 7 extraction wells and a water treatment facility, reduced the isopropyl ether concentration by an estimated 92% [4]. It was detected at concentrations ranging from 13-128 ug/L in 12 of 12 samples of ground water taken from between March 1985 and Feb 1986 from an aquifer near Dijon, France which was contaminated by a nearby industrial area [5]. SURFACE WATER: Isopropyl ether has been found in the following Lake Michigan basin locations: Chicago Sanitary and Ship Channel, 1 ug/L [17]. Isopropyl ether was found at a concentration of 1 ppb in samples of surface water from 2 of 204 sites near heavily industrialized areas across the US sampled between Aug 1975 and Sept 1976 [11].

Effluent Concentrations: Isopropyl ether has been detected in 10 out of 63 samples of industrial effluents collected from a wide variety of industries across the USA (dates not reported) [21]. Six of the positive samples contained <10 ug/L isopropyl ether, 3 samples contained 10-100 ug/L and 1 sample contained >100 ug/L [21]. Samples of landfill gas from an unspecified landfill which had for more than 15 months accepted municipal and industrial solid wastes and unspecified liquid wastes contained 220 mg/m^3 isopropyl ether [25].

Sediment/Soil Concentrations:

Atmospheric Concentrations: Isopropyl ether was identified, but not quantified, at 4 of 7 sites in the Kanawha Valley, WV, sampled in Sept 1977 [10].

Food Survey Values:

Plant Concentrations:

Fish/Seafood Concentrations:

Animal Concentrations:

Milk Concentrations:

Other Environmental Concentrations:

Isopropyl Ether

Probable Routes of Human Exposure: The most probable routes of general population exposure to ether are via inhalation of contaminated air [10] and ingestion of contaminated drinking water [12]. Exposure through dermal contact may occur in occupational settings. Inhalation and dermal exposure will be expected to be highest in workplaces where isopropyl ether is made and used.

Average Daily Intake:

Occupational Exposure: NIOSH (NOES Survey 1981-1983) has statistically estimated that 3,253 workers are exposed to isopropyl ether in the USA [19]. NIOSH (NOHS Survey 1972-1974) has statistically estimated that 144,773 workers are exposed to isopropyl ether in the USA [20].

Body Burdens:

REFERENCES

1.   Abernethy SG et al; Environ Toxicol Chem 7: 469-81 (1988)
2.   Alexander M; Biotechnol Bioeng 15: 611-47 (1973)
3.   Aller L; Drastic: a standardized system for evaluating groundwater pollution potential using hydrologic settings. USEPA/600/2-85/018 (NTIS PB85-228146) p 147 (1985)
4.   Althoff WF et al; Groundwater 19: 495-504 (1981)
5.   Arnaud G; pp 465-74 Tech Sci Methods: Genie Urbain--Genie Rural (1986)
6.   Atkinson R; Intern J Chem Kinetics 19: 799-828 (1987)
7.   Bridie AL et al; Water Res 13: 627-30 (1979)
8.   Calvert JG, Pitts JNJr; Photochemistry John Wiley & Sons: NY pp 441-2 (1966)
9.   Eisenreich SJ et al; Environ Sci Technol 15: 30-8 (1981)
10.   Erickson MD, Pellizzari ED; Analysis of Organic Air Pollutants in the Kanawha Valley, WV and the Shenandoah Valley, VA USEPA- 903/9-78-007 (NTIS PB286141) p 46 (1978)
11.   Ewing BB et al; Monitoring to Detect Previously Unrecognized Pollutants in Surface Waters - Appendix: Organic Analysis Data. USEPA-560/6-77-015 p 299 (1977)
12.   Fielding M et al; Organic Pollutants in Drinking Water, TR-159 Water Res Cent p 49 (1981)
13.   Fujiwara Y et al; Yukagaku 33: 111-14 (1984)
14.   Hendry DG et al; J Phys Chem Ref Data 3: 944-78 (1974)
15.   Hine J, Mookerjee PK; J Org Chem 40: 292-8 (1975)
16.   Keeley DE; Kirk-Othmer Encycl Chem Tech 3rd ed Wiley Interscience NY 9: 391 (1980)

Isopropyl Ether

17. Konasewich D et al; Status Report on Organic and Heavy Metal Contaminants in the Lakes Erie, Michigan, Huron and Superior basins, Great Lakes Qual Board p 292 (1978)
18. Lyman WJ et al; Handbook of Chem Property Estimation Methods NY: McGraw-Hill pp 4-9, 5-5, 7-4, 15-16 to 15-29 (1982)
19. NIOSH; The National Occupational Exposure Survey (NOES) (1983)
20. NIOSH; The National Occupational Hazard Survey (NOHS) (1974)
21. Perry DL et al; Identification of Organic Compounds in Industrial Effluent Discharges. USEPA-600/4-79-016 (NTIS PB-294794) p 45 (1979)
22. Riddick JA et al; Organic Solvents. John Wiley & Sons Inc. NY (1984)
23. Swann RL et al; Res Rev 85: 17-28 (1983)
24. USEPA; EXAMS II Computer Simulation (1987)
25. Young P, Parker A; pp 24-41 in ASTM Special Technical Publication No. 851 (1984)

Isopropylamine

SUBSTANCE IDENTIFICATION

Synonyms: 2-Aminopropane

Structure:

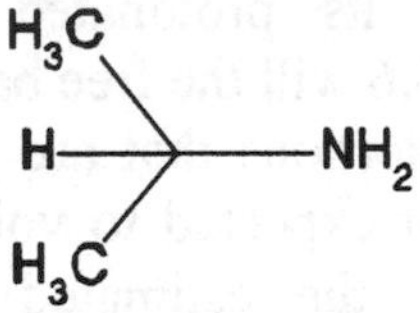

H_3C

H —— NH_2

H_3C

CAS Registry Number: 75-31-0

Molecular Formula: C_3H_9N

Wiswesser Line Notation: ZY1&1

CHEMICAL AND PHYSICAL PROPERTIES

Boiling Point: 33-34 °C

Melting Point: -101 °C

Molecular Weight: 59.11

Dissociation Constants: pKa = 10.6 [15]

Log Octanol/Water Partition Coefficient: 0.26 [6]

Water Solubility: miscible [18]

Vapor Pressure: 569.6 mm Hg at 25 °C [2]

Henry's Law Constant: 1.12×10^{-5} atm-m^3/mole, estimate [8]

ENVIRONMENTAL FATE/EXPOSURE POTENTIAL

Summary: Isopropylamine is a natural component of tobacco and maize, and it is probably released to the environment through their decomposition. Decomposing animal manure has been shown to release isopropylamine to the environment. Isopropylamine may also be released to the environment via effluents at sites where it is produced or used as a chemical intermediate or a solvent. In addition, isopropylamine is released to ambient air via cigarette smoke. With a pKa of 10.6, isopropylamine will mainly exist in its protonated form in the environment. Only in media with pHs >8.6 will the free base occur along with the conjugate acid in varying proportions that are pH dependent. The ionic form of isopropylamine is not expected to volatilize; for the unprotonated form of isopropylamine, the estimated Henry's Law constant indicates volatilization of isopropylamine from natural waters might be an important fate process under extreme alkaline conditions. Based on the vapor pressure, isopropylamine should rapidly evaporate from dry surfaces, especially when present in high concentrations such as in spill situations. The miscibility of isopropylamine in water suggests that adsorption and bioconcentration, in addition to volatilization, are not important fate processes. This is supported by low estimates for the bioconcentration factor (log BCF = 0.43) and soil adsorption coefficient (Koc = 33). However, cationic molecules generally adsorb to organic carbon and clays more strongly than their neutral counterparts. Hence, isopropylamine has the potential to bind to soil and partition from the water column to sediments and suspended solids. Limited data suggests isopropylamine should biodegrade rapidly in soil and water; however, it may be toxic to microorganisms at high concentrations. Isopropylamine is expected to exist almost entirely in the vapor phase in ambient air, where vapor-phase reactions with photochemically produced hydroxyl radicals may be important (estimated half-life of 10 hr). Physical removal from air by precipitation and dissolution in clouds may occur; however, the short atmospheric residence time of isopropylamine suggests that wet deposition is of limited importance. The most probable human exposure would be occupational exposure, which may occur through dermal contact or inhalation at workplaces where isopropylamine is produced or used. Atmospheric workplace exposures have been documented for farm and barnyard workers. The most common nonoccupational exposure is likely to result from either passive or active inhalation of cigarette smoke. Limited monitoring data indicates that other nonoccupational exposures can occur from the ingestion of certain foods.

Isopropylamine

Natural Sources: A study showed isopropylamine was emitted to the air from decomposing manure in animal feed lots [9]. Isopropylamine was identified as a natural component of tobacco leaves [9], maize and soybean residual [13].

Artificial Sources: Isopropylamine is used as a solvent, surface active agent, dehairing agent, solubilizer for 2,4-D acid, and as an intermediate in the synthesis of rubber accelerators, pharmaceuticals, dyes, insecticides, bactericides, textile specialties [7]. Isopropylamine may be released to the environment via effluents at sites where it is produced or used as a chemical intermediate or a solvent. In addition, isopropylamine is released to ambient air via cigarette smoke [9].

Terrestrial Fate: With a pKa of 10.6, isopropylamine will mainly exist in its protonated form in terrestrial environments [10]. Only in soils with pHs >8.6 will the free base occur along with the conjugate acid in varying proportions that are pH dependent [10]. The ionic form of isopropylamine is not expected to volatilize from most soils [10]. For the unprotonated form of isopropylamine, the estimated Henry's Law constant indicates volatilization of isopropylamine from moist soils might be an important fate process under extreme alkaline conditions [10]. Based on the vapor pressure, isopropylamine should rapidly evaporate from dry surfaces, especially when present in high concentrations such as in spill situations. The miscibility of isopropylamine in water suggests that adsorption is not an important fate process. This is supported by low estimates for the soil adsorption coefficient. A Koc of 33 [10] indicates isopropylamine should be highly mobile in soil [17]; however, cationic molecules generally adsorb to organic carbon and clays more strongly than their neutral counterparts [10]. Hence, isopropylamine has the potential to bind in soils. A single aerobic screening test, which utilized activated sludge for inocula, indicates isopropylamine should biodegrade rapidly in terrestrial environments [16]. A second screening test suggests isopropylamine may be toxic to microorganisms at high concentrations [11].

Aquatic Fate: With a pKa of 10.6, isopropylamine will mainly exist in its protonated form in aquatic systems [10]. Only in waters with pHs >8.6 will the free base occur along with the conjugate acid in varying proportions that are pH dependent [10]. The ionic form of isopropylamine is not expected to volatilize from water [10]. For the unprotonated form of isopropylamine, the estimated Henry's Law constant indicates volatilization of isopropylamine from natural waters might be an important fate process under extreme alkaline conditions [10]. The miscibility of isopropylamine in water suggests that adsorption and bioconcentration, in addition to volatilization, are not important fate processes. This is supported by low estimates for the bioconcentration factor (log BCF = 0.43) and soil adsorption coefficient (Koc = 33) [10]. However, cationic molecules generally adsorb to organic carbon and clays more strongly than their neutral counterparts [10]. Hence, isopropylamine has the potential to partition from the water column to sediments and suspended solids. A single aerobic screening test, which utilized activated sludge for inocula, indicates isopropylamine should biodegrade rapidly in the aquatic environment [16]. A second screening test suggests isopropylamine may be toxic to microorganisms at high concentrations [11].

Atmospheric Fate: Based on the vapor pressure, isopropylamine is expected to exist almost entirely in the vapor phase in ambient air [3]. In the atmosphere, vapor-phase reactions with photochemically produced hydroxyl radicals may be important. The rate constant for isopropylamine was estimated to be 3.87×10^{-11} cm^3/molecule-sec at 25 °C, which corresponds to an atmospheric half-life of about 10 hours at an atmospheric concentration of $5 \times 10^{+5}$ hydroxyl radicals per cm^3 [1]. The complete miscibility of isopropylamine in water suggests that physical removal from air by precipitation and dissolution in clouds may occur; however, the short atmospheric residence time of isopropylamine suggests that wet deposition is of limited importance.

Biodegradation: River die-away tests and grab sample data pertaining to the biodegradation of isopropylamine in natural waters and soil were not located in the available literature. A single aerobic screening test, which utilized activated sludge for inocula, suggests isopropylamine should biodegrade rapidly in the environment [16]. Isopropylamine was completely degraded within a 2 day period [16]. However, isopropylamine may be toxic to microorganisms at high concentrations [11]. After 8 days, an initial concentration of 500 ppm was reduced by

10% in a closed bottle inoculated with a microbial population of 5000 mg/L from activated sludge acclimated to aniline, and maintained at 20 °C [11].

Abiotic Degradation: The rate constant for the vapor-phase reaction of isopropylamine with photochemically produced hydroxyl radicals has been estimated to be 3.87×10^{-11} cm^3/molecule-sec at 25 °C, which corresponds to an atmospheric half-life of about 10 hours at an atmospheric concentration of $5 \times 10^{+5}$ hydroxyl radicals per cm^3 [1].

Bioconcentration: Because isopropylamine is miscible in water, bioconcentration in aquatic systems is not expected to be an important fate process. Based upon the log Kow, a bioconcentration factor (log BCF) of 0.43 for isopropylamine has been calculated using a recommended regression-derived equation [10]. This BCF value also indicates isopropylamine should not bioconcentrate among aquatic organisms.

Soil Adsorption/Mobility: Because isopropylamine is miscible in water, soil adsorption is not expected to be an important fate process. Based on the log Kow, a Koc of 33 for isopropylamine has been calculated using a recommended regression-derived equation [10]. This Koc value indicates isopropylamine will be very highly mobile in soil [17], and it should not partition from the water column to organic matter contained in sediments and suspended solids. However, with a pKa of 10.6, isopropylamine will mainly exist in its protonated form in the environment [10], and cationic molecules generally adsorb to organic carbon and clays more strongly than their neutral counterparts.

Volatilization from Water/Soil: With a pKa of 10.6, isopropylamine will mainly exist in its protonated form in the environment [10]. Only in environmental media with pHs >8.6 will the free base occur along with the conjugate acid in varying proportions that are pH-dependent [10]. The ionic form of isopropylamine is not expected to volatilize from water [10]. For the unprotonated form of isopropylamine, the Henry's Law constant suggests volatilization from natural waters might be an important fate process under extreme alkaline conditions [10]. Based upon the vapor pressure, isopropylamine should evaporate rapidly from dry

377

surfaces, especially when present in high concentrations such as in spill situations.

Water Concentrations:

Effluent Concentrations: Isopropylamine was detected in tobacco leaves and in cigarette smoke [9]. A study showed isopropylamine was emitted to the air from decomposing manure in animal feed lots [12].

Sediment/Soil Concentrations: Isopropylamine was identified as a constituent of soil from Moscow, USSR [5].

Atmospheric Concentrations:

Food Survey Values: Isopropylamine was identified as a volatile component of boiled beef [5]. Isopropylamine was detected in maize at a concentration of 2.3 mg/kg [13].

Plant Concentrations: Isopropylamine was identified as a component of tobacco leaves [9], maize at a concentration of 2.3 mg/kg, and soybean animal feedstuff in trace quantities [13].

Fish/Seafood Concentrations:

Animal Concentrations:

Milk Concentrations:

Other Environmental Concentrations:

Probable Routes of Human Exposure: The most probable route of human exposure to isopropylamine is by inhalation, dermal contact and ingestion. Cigarette smokers are likely to inhale isopropylamine [9], and atmospheric workplace exposures have been documented for farm and barnyard workers [12]. Certain foods have been shown to contain isopropylamine [5,13].

Average Daily Intake:

Isopropylamine

Occupational Exposure: The most probable human exposure to isopropylamine would be occupational exposure, which may occur through dermal contact or inhalation at places where it is produced or used. NIOSH (NOES Survey as of 3/28/89) has estimated that 7,527 workers are potentially exposed to isopropylamine in the USA [14]. A study showed isopropylamine was emitted to the air from decomposing manure in animal feed lots [12] exposing farm and barnyard workers. Nonoccupational exposures are likely to occur among cigarette smokers [9], or from the ingestion of certain foods [5,13].

Body Burdens:

REFERENCES

1. Atkinson R; Intern J Chem Kin 19: 799-828 (1987)
2. Daubert TE, Danner RP; Data Compilation, Tables of Properties of Pure Cmpds, Design Inst for Phys Prop Data, Am Inst for Prop Data, NY,NY (1989)
3. Eisenreich SJ et al; Environ Sci Technol 15: 30-8 (1981)
4. Golovnya RV, et al; Chem Senses Flavour 4: 97-105 (1979)
5. Golovnya RV et al; Amines in Soil as Possible Precursors of Nitroso Compounds. USSR Acad Med Sci pp 327-35 (1987)
6. Hansch C, Leo AJ; Medchem Project Issue No 26. Claremont CA: Pomona College (1985)
7. Hawley GG; Condensed Chemical Dictionary 10th ed Van Nostrand Reinhold NY p 375 (1981)
8. Hine J, Mookerjee PK; J Org Chem 40: 292-8 (1975)
9. Irvine WJ, Saxby MJ; Phytochemistry 8: 473-6 (1969)
10. Lyman WJ et al; Handbook of Chemical Property Estimation Methods NY: McGraw-Hill (1982)
11. Malaney GW; J Water Pollut Control Fed 32: 1300-11 (1960)
12. Mosier AR et al; Environ Sci Technol 7: 642-4 (1973)
13. Neurath GB et al; Food Cosmet Toxicol 15: 275-82 (1977)
14. NIOSH; National Occupational Exposure Survey (NOES) (1989)
15. Perrin DD; Dissociation Constants of Organic Bases in Aqueous Solution. IUPAC Chemical Data Series, Buttersworth London (1965)
16. Slave T et al; Rev Chim 25: 666-70 (1974)
17. Swann RL et al; Res Rev 85: 16-28 (1983)
18. Windholz M et al; Merck Index 10th ed. Rahway, NJ (1983)

Malononitrile

SUBSTANCE IDENTIFICATION

Synonyms: Propanedinitrile

Structure:

$$N \equiv C - CH_2 - C \equiv N$$

CAS Registry Number: 109-77-3

Molecular Formula: $C_3H_2N_2$

Wiswesser Line Notation: NC1CN

CHEMICAL AND PHYSICAL PROPERTIES

Boiling Point: 218-219 °C

Melting Point: 32 °C

Molecular Weight: 66.06

Dissociation Constants: pKa: 11.4 [6]

Log Octanol/Water Partition Coefficient: -1.20 (calculated) [5]

Water Solubility: 133,000 mg/L at room temperature [6]

Vapor Pressure: 11 mm Hg at 99 °C [2]; 20 mm Hg at 109 °C [2]; 0.015 mm Hg at 25 °C (Antoine extrapolation)

Henry's Law Constant: 1.27×10^{-8} atm-m^3/mol at 25 °C (estimated) [10]

ENVIRONMENTAL FATE/EXPOSURE POTENTIAL

Summary: Malononitrile may be released to the environment in wastewater effluents associated with its use in gold ore leaching and as a chemical intermediate. If released to the atmosphere, malononitrile will exist primarily in the vapor phase. Atmospheric degradation via photochemically produced hydroxyl radicals is slow (estimated half-life of 2.67 years). Removal from the atmosphere can occur through wet deposition and hydrolysis in cloud water. If released to water or moist soil, malononitrile will degrade through hydrolysis. At 25 °C, the aqueous hydrolysis half-life has been experimentally determined to vary from 21.4 days at pH 5 to 3.1 days at pH 9. Insufficient data are available to predict the importance of biodegradation in soil or water. Occupational exposure to malononitrile can occur through inhalation of vapor and dermal contact. There are no data available to suggest that the general population is exposed to malononitrile.

Natural Sources:

Artificial Sources: Malononitrile's uses in gold ore leaching and as a chemical intermediate [6] may result in releases to the environment via wastewater effluents.

Terrestrial Fate: The major degradation process for malononitrile in moist soil may be hydrolysis. At 25 °C, the aqueous hydrolysis half-life has been experimentally determined to vary from 21.4 days at pH 5 to 3.1 days at pH 9 [4]. Insufficient data are available to predict the importance of biodegradation. One microbial screening study has demonstrated that malononitrile can be toxic to activated sludge microorganisms at a relatively high concentration of 500 mg/L [7]. Based upon an estimated Koc value of 6.6 [8], malononitrile should leach readily in soil; the importance of leaching may be lessened, however, by concurrent hydrolysis.

Aquatic Fate: The major degradation process for malononitrile in water may be hydrolysis. At 25 °C, the hydrolysis half-life has been experimentally determined to vary from 21.4 days at pH 5 to 3.1 days at pH 9 [4]. Insufficient data are available to predict the importance of biodegradation. One microbial screening study has demonstrated that malononitrile can be toxic to activated sludge microorganisms at a relatively high concentration of 500 mg/L [7]. Aquatic volatilization, bioconcentration, direct photolysis, and adsorption to sediment are not expected to be environmentally important.

Malononitrile

Atmospheric Fate: Based upon the vapor pressure, malononitrile is expected to exist primarily in the vapor phase in the ambient atmosphere [3]. It is expected to degrade very slowly in an average ambient atmosphere by reaction with photochemically produced hydroxyl radicals with an estimated half-life of about 2.67 years [1]. Due to its high water solubility, malononitrile should be susceptible to wet deposition processes such as dissolution into clouds and physical removal via rainfall. Malononitrile that is dissolved into cloudwater may experience some hydrolysis. The half-life for the aqueous hydrolysis of malononitrile at 25 °C in acidic to neutral conditions is about 20-21 days [4].

Biodegradation: Using a Warburg respirometer, 72-hr incubation periods, and 3 different activated sludges, malononitrile was found to be toxic in all 3 sludges at a concentration of 500 mg/L [7].

Abiotic Degradation: Exposure of malononitrile to air and UV irradiation for a long time causes darkening and a release of hydrogen cyanide [6]. The neutral and base hydrolysis rate constants for malononitrile in water at 25 °C were experimentally determined to be 0.00135/hr and 806/M-hr, respectively, which corresponds to a half-life of 20.2 days at a pH of 7 [4]; at pHs of 5, 6, 8 and 9, the hydrolysis half-lives would be 21.4, 21.3. 13.4 and 3.1 days, respectively. The rate constant for the vapor-phase reaction of malononitrile with photochemically produced hydroxyl radicals can be estimated to be 1.64 x 10^{-14} cm^3/molecule-sec at room temperature, which corresponds to an atmospheric half-life of about 2.67 yrs at an atmospheric concentration of 5 x 10^{+5} hydroxyl radicals/cm^3 [1]. The UV absorption spectra of malononitrile in aqueous solution shows that malononitrile does not absorb UV light in the environmentally relevant spectra above 290 nm [9]; therefore, direct photolysis in the environment will not be important.

Bioconcentration: Based upon the water solubility, the BCF for malononitrile can be estimated to be 0.8 from a linear regression-derived equation [8]. This BCF value suggests that malononitrile will not bioconcentrate significantly in aquatic organisms.

Soil Adsorption/Mobility: Based upon the water solubility, the Koc for malononitrile can be estimated to be 6.6 from a linear regression-derived equation [8]. This Koc value suggests that malononitrile has very high soil mobility [13]; therefore, leaching to ground water is possible.

Volatilization from Water/Soil: The Henry's Law constant for malononitrile indicates that it is essentially nonvolatile from water [8]. Therefore, volatilization of malononitrile from environmental waters is not expected to be important.

Water Concentrations:

Effluent Concentrations:

Sediment/Soil Concentrations:

Atmospheric Concentrations:

Food Survey Values:

Plant Concentrations:

Fish/Seafood Concentrations:

Animal Concentrations:

Milk Concentrations:

Other Environmental Concentrations:

Probable Routes of Human Exposure: Occupational exposure to malononitrile can occur through inhalation of vapor and dermal contact [12]. Malononitrile is extremely toxic by skin absorption, eye contact, and ingestion; extreme care must be exercised during handling to ensure protection against personal contamination [12]. There are no data available to suggest that the general population is exposed to malononitrile.

Average Daily Intake:

Malononitrile

Occupational Exposure: NIOSH (NOES Survey 1981-1983) has statistically estimated that 1,235 workers are potentially exposed to malononitrile in the USA [11].

Body Burdens:

REFERENCES

1. Atkinson R; J Inter Chem Kinet 19: 799-828 (1987)
2. Beilsteins Handbuch Der Organischen Chemie 2: 589 (1943)
3. Eisenreich SJ et al; Environ Sci Technol 15: 30-8 (1981)
4. Ellington JJ et al; Measurement of Hydrolysis Rate Constants for Evaluation of Hazardous Waste Land Disposal: Vol 3 USEPA/600/3-88/028 p 18 (1988)
5. GEMS; Graphic Exposure Modeling System. CLOGP. USEPA (1987)
6. Hughes DW; Kirk-Othmer Encycl Chem Technol 3rd ed NY: John Wiley & Sons 14: 804-7 (1981)
7. Lutin PA; J Water Pollut Control Fed 42: 1632-42 (1970)
8. Lyman WJ et al; Handbook of Chemical Property Estimation Methods Washington DC: Amer Chem Soc pp 4-9, 5-10, 15-15 to 15-29 (1990)
9. Mendelson J et al; Science 120: 266-9 (1954)
10. Meylan WM, Howard PH; Environ Toxicol Chem 10: 1283-93 (1991)
11. NIOSH; National Occupational Exposure Survey (NOES) (1983)
12. Parmeggiani L; Encycl Occup Health & Safety 3rd ed Geneva, Switzerland: International Labour Office pp 1445-7 (1983)
13. Swann RL et al; Res Rev 85: 23 (1983)

Methyl Acetate

SUBSTANCE IDENTIFICATION

Synonyms: Acetic acid, methyl ester

Structure:

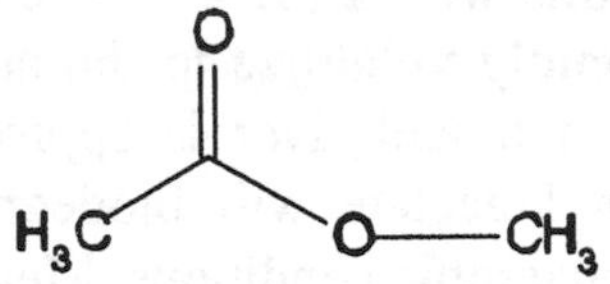

$$H_3C - C(=O) - O - CH_3$$

CAS Registry Number: 79-20-9

Molecular Formula: $C_3H_6O_2$

Wiswesser Line Notation: 1VO1

CHEMICAL AND PHYSICAL PROPERTIES

Boiling Point: 57 °C

Melting Point: -98.1 °C

Molecular Weight: 74.08

Dissociation Constants:

Log Octanol/Water Partition Coefficient: 0.18 [12]

Water Solubility: 2.435 x 10^5 mg/L at 20 °C [25]

Vapor Pressure: 216.2 mm Hg at 25 °C [1]

Henry's Law Constant: 1.15 x 10^{-4} atm-m^3/mole at 20 °C [5]

ENVIRONMENTAL FATE/EXPOSURE POTENTIAL

Summary: Methyl acetate, which is used as a solvent, may be released in fugitive emissions during its manufacture, formulation, or use in

385

commercial products. Methyl acetate is also a naturally occurring compound. If released to soil, methyl acetate will display very high mobility and it has the potential to leach into ground water. Rapid volatilization is expected to occur from both moist and dry soils. Hydrolysis of methyl acetate in soil is not expected to be a significant process except in highly basic soils with a pH >9. If released to water, methyl acetate is expected to rapidly volatilize to the atmosphere. The half-life for volatilization from a model river is approximately 9 hr. Limited data suggest that methyl acetate will biodegrade in aquatic systems under both aerobic and anaerobic conditions. Methyl acetate will not significantly adsorb to sediment and suspended organic matter, nor will it bioconcentrate in fish and aquatic organisms. Hydrolysis of methyl acetate in aquatic systems is not expected to be a significant process except under basic conditions of pH >9. In the atmosphere methyl acetate is expected to undergo a relatively slow gas-phase reaction with photochemically produced hydroxyl radicals, with experimental half-lives on the order of 50 to 100 days. However, laboratory experiments have indicated that the rate of sunlit photo-oxidations may be much faster. Methyl acetate may undergo atmospheric removal by wet deposition. The probable routes of exposure to methyl acetate are by inhalation and dermal contact during the production and use of this compound. The general public is likely to be exposed to methyl acetate by the ingestion of foods or distilled alcoholic beverages in which it is contained. Limited monitoring data indicate that the general population has the potential to be exposed to methyl acetate by the ingestion of contaminated drinking water.

Natural Sources: Methyl acetate may be released to the environment from natural sources. It has been detected as a volatile constituent of nectarines [27].

Artificial Sources: In 1989, three companies were listed as producers of methyl acetate [24]. Current production volumes are not available. In 1977, 110 million pounds of methyl acetate were produced in the United States [31]. Methyl acetate has been used as solvent for cellulose nitrate, cellulose acetates, resins and oils, in the manufacture of artificial leathers, and in organic synthesis [8,22]. Methyl acetate may be released to the environment as a fugitive emission during the production, formulation and use of these products.

Methyl Acetate

Terrestrial Fate: If released to soil, calculated soil adsorption coefficients for methyl acetate ranging from approximately 5 to 30 [17] obtained from its water solubility and log octanol/water partition coefficient, respectively, suggest that methyl acetate will display very high mobility in soil [26] and has the potential to leach into ground water. The Henry's Law constant and vapor pressure of methyl acetate, suggest that this compound will rapidly volatilize from both moist and dry soils to the atmosphere. Hydrolysis of methyl acetate in soil is not expected to be a significant process except in highly basic soils with a pH >9, as hydrolysis rate constants indicate that this process will be too slow to be environmentally significant under acidic, neutral, and slightly basic conditions [7,8].

Aquatic Fate: If released to water, methyl acetate is expected to rapidly volatilize to the atmosphere. Based on its measured Henry's Law constant, the half-life for volatilization from a model river is 9.1 hr [17]. Limited data suggest that methyl acetate will biodegrade in aquatic systems under both aerobic [13,18], and anaerobic [6,23] conditions. Calculated values for a soil adsorption coefficient ranging from approximately 5-30 [17] and calculated bioconcentration factors <1 [17] indicate that methyl acetate will not significantly adsorb to sediment and suspended organic matter, nor will it bioconcentrate in fish and aquatic organisms. Hydrolysis of methyl acetate in aquatic systems is not expected to be a significant process except under basic conditions of pH >9, as hydrolysis rate constants indicate that this process will be too slow to be environmentally significant under acidic, neutral, and slightly basic conditions [7,8].

Atmospheric Fate: In the atmosphere, experimental rate constants for the gas phase reaction of methyl acetate with photochemically produced hydroxyl radicals in the range 1.7-3.41 x 10^{-13} cm^3/mol-sec [3,4,11,32] correspond to an atmospheric half-life ranging between 47-94 days. Laboratory experiments have indicated that methyl acetate has the potential to undergo atmospheric oxidations at a rate equal to 55% removal in 24 hr when irradiated at 250 nm [10]. When irradiated at 360 nm, the rate increases to 90% removal after 6 hr [10]. The relatively high water solubility of methyl acetate suggests that this compound may be removed from the atmosphere by wet deposition.

Methyl Acetate

Biodegradation: Pure cultures of <u>Alcaligenes faecalis</u>, isolated from activated sludge, were found to oxidize methyl acetate after a short lag period [18]. A five-day BODT for methyl acetate was given as 26% using a sewage seed [13]. Methyl acetate was listed as a compound which should be amenable to biological degradation in wastewater treatment by anaerobic biotechnology [23]. Methyl acetate was found to undergo greater than 66% anaerobic biodegradation to methane in 90 days using an anaerobic digester seed acclimated to acetic acid [6].

Abiotic Degradation: Experimental rate constants for the gas-phase reaction of methyl acetate with photochemically produced hydroxyl radicals at 19-30 °C are 1.7-3.41×10^{-13} cm^3/mol-sec [3,4,11,32], which correspond to an atmospheric half-life ranging between 47-94 days using an average atmospheric hydroxyl radical concentration of 5×10^5 mole/cm^3 [2]. These values correspond to a removal rate of 0.06% per hr for this reaction [16]. Laboratory irradiation of 10 ppm methyl acetate and air in a quartz vessel at 250 nm resulted in 55% removal by photo-oxidation after 24 hr [10]. An identical experiment at 360 nm resulted in 90% removal after 6 hr [10]. A second order rate constant for the basic hydrolysis of methyl acetate in water at 25 °C, 0.182 L/mole-s [7] corresponds to half-lives of 44 days, 4.4 days, and 10.6 hr at pH 8, 9 and 10, respectively. In general, alkyl esters are resistant to hydrolysis under the acidic or neutral conditions which are typically found in the environment [8].

Bioconcentration: From the water solubility and log octanol/water partition coefficient of methyl acetate, bioconcentration factors of 0.57 and 0.81, respectively, can be calculated by a regressional analysis [17]. These values indicate that methyl acetate will not significantly bioconcentrate in fish and aquatic organisms.

Soil Adsorption/Mobility: From the water solubility and log octanol/water partition coefficient of methyl acetate, soil adsorption coefficients of 4.8 and 29.8, respectively, can be calculated by a recommended regression equation [17]. These values indicate that methyl acetate will display very high mobility in soil [26].

Volatilization from Water/Soil: Based on the measured Henry's Law constant for methyl acetate, the half-life for volatilization from a model river 1 m deep, flowing at 1 m/sec and a wind speed of 3 m/sec is 9.1 hr [17]. The Henry's Law constant of methyl acetate and its vapor pressure suggest that volatilization from both moist and dry soil to the atmosphere will be significant fate processes.

Water Concentrations: DRINKING WATER: Methyl acetate was detected, but not quantified, in the drinking water of Seattle, WA, in 1974 [30]. SURFACE WATER: Qualitatively detected in a waste pond from an abandoned pesticide plant located on the Rocky Mountain Arsenal, CO, 1981 [29].

Effluent Concentrations: Methyl acetate was detected as a trace organic constituent of the waste stream from industrial chemical processes, 1972-73 [15]. It was qualitatively identified in the effluent gas from refuse waste obtained from a food center in an experiment designed to determine the gases emitted from decaying waste matter at refuse sites, landfills, and trash transfer sites [14]. Methyl acetate was detected as an emission from the production of RDX at the Holston Army Ammunition Plant, TN, date not provided, at an emission rate of 733 lbs/day [21].

Sediment/Soil Concentrations:

Atmospheric Concentrations: Methyl acetate was qualitatively detected in the air of the industrialized Kanawha Valley, WV, 1977 [9].

Food Survey Values:

Plant Concentrations:

Fish/Seafood Concentrations:

Animal Concentrations:

Milk Concentrations:

Other Environmental Concentrations:

Probable Routes of Human Exposure: The probable routes of occupational exposure to methyl acetate are by inhalation and dermal

389

contact during the production and use of this compound. The general public is likely to be exposed to methyl acetate by the ingestion of foods [27] or distilled alcoholic beverages in which it is contained [28]. Limited monitoring data [30] indicates that the general population has the potential to be exposed to methyl acetate by the ingestion of contaminated drinking water.

Average Daily Intake:

Occupational Exposure: Of three US companies involved in spray painting and spray gluing, personal air samples from workers at only one company had methyl acetate residues at a mean concentration of 0.2 ppm [33]. Controls samples taken from workers who were not involved in these operations had no detectable amount of methyl acetate in their personal air samples [33]. Methyl acetate was found in 4 samples taken from 11 different auto paint shops in Spain at concentrations ranging 12.1-80.0 mg/cu-m [19]. NIOSH (NOES Survey 1981-83) has statistically estimated that 17,851 workers are exposed to methyl acetate in the USA, 92% of which are exposed during the use of trade name compounds in which methyl acetate is contained [20].

Body Burdens:

REFERENCES

1. Ambrose D et al; J Chem Therm 13: 795-802 (1981)
2. Atkinson R; Chem Rev 85: 69-201 (1985)
3. Atkinson R; Int J Chem Kinet 19: 799-828 (1987)
4. Atkinson R et al; Adv Photochem 11:375-488 (1979)
5. Buttery RG et al; J Agric Food Chem 17: 385-9 (1969)
6. Chou WL et al; Biotech Bioeng Symp 8: 391-414 (1979)
7. Drossman H et al; Chemosphere 17: 1509-30 (1987)
8. Elam EU; Kirk-Othmer Encycl Chem Tech 3rd Ed. John-Wiley NY 9: 311-37 (1978)
9. Erickson MD, Pellizzari ED; Analysis of Organic Air Pollutants in the Kanawha Valley, WV and the Shenandoah Valley, VA. USEPA-903/9-78-007 (1978)
10. Fujiki M et al; Simulation Studies of Degradation of Chemicals in the Environment. Chemical Res Report No. 1/1978. Environmental Agency of Japan, Office of Health Studies, Tokyo, Japan (1978)
11. Gusten H et al; J Atmos Chem 2: 83-96 (1984)
12. Hansch C, Leo AJ; Medchem project Issue No. 26 Claremont, CA Pomona College (1985)

Methyl Acetate

13. Heukelekian H, Rand MC; J Water Pollut Contr Assoc 29: 1040-53 (1955)
14. Koe LC, Ng WJ; Water, Air Soil Pollut 33: 199-204 (1987)
15. Leenheer JA et al; Environ Sci Tech 10: 445-51 (1976)
16. Lloyd AC; pp 27-48 in National Bureau of Standards, Wash DC: NBS-SP-577 (1978)
17. Lyman WJ et al; Handbook of Chemical Property Estimation Methods NY: McGraw-Hill pp 4-1 to 4-33, 5-1 to 5-30 and 15-1 to 15-34 (1982)
18. Marion CV, Malaney GW; J Water Pollut Control Fed 35: 1269-84 (1963)
19. Medinilla J, Espigares M; Ann Occup Hyg 32: 509-13 (1988)
20. NIOSH; National Occupational Exposure Survey (NOES) (1989)
21. Ryon MG et al; Database Assessment of the Health and Environmental Effects of Munitions Production Waste Products. Final Report. ORNL-6018 (NTIS DE84-016512) Oak Ridge Natl Labs, Oak Ridge, TN p 217 (1984)
22. Sax NI, Lewis RJSR; Hawley's Condensed Chemical Dictionary 11th ed Van Nostrand Reinhold Co. NY p 496 (1987)
23. Speece RE; Env Sci Tech 17: 416A-27A (1983)
24. SRI International; Directory of Chemical Producers (1989)
25. Stephan H, Stephan T; Solubilities of Inorganic and Organic Compounds in Binary Systems. Stephen H et al. eds, NY, NY 1: 1-79, 1604-43 (1963)
26. Swann RL et al; Res Rev 85: 17-28 (1983)
27. Takeoka GR et al; J Agric Food Chem 36: 553-60 (1988)
28. Terheide R et al; pp 249-81 in Anal Foods Beverages (Proc Symp), Chavalambous G Academic Press (1978)
29. Thoburn TW, Gunter BJ; Health Hazard Evaluation Report HETA 810176-968, Rocky Mt. Arsenal, Basin F, Commerce City, CO, NTIS PB-83-161257. Washington, DC (1981)
30. USEPA; Preliminary Assessment of Suspected Carcinogens in Drinking Water, Interim Report to Congress (1975)
31. USEPA; Nonconfidential Initial TSCA Inventory. Washington DC Off Tox Subst (1977)
32. Wallington TJ et al; J Phys Chem 92: 5024-8 (1988)
33. Whitehead LW et al; Am Ind Hyg Assoc 45: 767-72 (1984)

2-Methyl-1-butanol

SUBSTANCE IDENTIFICATION

Synonyms:

Structure:

CAS Registry Number: 137-32-6

Molecular Formula: $C_5H_{12}O$

Wiswesser Line Notation: Q1Y2&1

CHEMICAL AND PHYSICAL PROPERTIES

Boiling Point: 129 °C at 760 mm Hg [34]

Melting Point:

Molecular Weight: 88.15

Dissociation Constants:

Log Octanol/Water Partition Coefficient: 1.29 [31]

Water Solubility: 30,000 mg/L at 25 °C [3]

Vapor Pressure: 3.13 mm Hg at 25 °C [6]

Henry's Law Constant: 1.41×10^{-5} atm-m^3/mole at 25 °C [15]

ENVIRONMENTAL FATE/EXPOSURE POTENTIAL

Summary: 2-Methyl-1-butanol occurs naturally in the volatile components of many fruits. Volatile emissions from poultry manure contain 2-methyl-1-butanol. The use of 2-methyl-1-butanol as a solvent can release it to the atmosphere by evaporation. If released to air, it will degrade by reaction with photochemically produced hydroxyl radicals (estimated half-life of 47 hr). Physical removal from air via wet deposition is possible since it is relatively soluble in water. If released to soil or water, 2-methyl-1-butanol is expected to biodegrade. Leaching in soil is possible based upon its high water solubility; however, concurrent biodegradation should lessen the importance of leaching. Volatilization half-lives from a model environmental river (1 meter deep) and model pond have been estimated to be 61 hr and 28 days, respectively. The general population is exposed to 2-methyl-1-butanol through inhalation of aromas and ingestion of various foods, particularly fruits and fermented beverages.

Natural Sources: 2-Methyl-1-butanol has been identified as a volatile component of blue cheese aroma [7], Concord grape juice essence [24], nectarines [27], apples [5,35], papaya fruit [22], oranges [25], and tomatoes [23]. 2-Methyl-1-butanol is released in the volatile emissions from poultry manure [36].

Artificial Sources: 2-Methyl-1-butanol was identified as one of the odorous compounds emitted to the air from a sedimentation tank at a German water treatment facility [14]. Various solvent uses of 2-methyl-1-butanol will release 2-methyl-1-butanol to air through evaporation.

Terrestrial Fate: The only identifiable degradation process in soil is biodegradation. Although biodegradation data specific to 2-methyl-1-butanol are unavailable, analogy to similar aliphatic alcohols suggests that 2-methyl-1-butanol will readily biodegrade. Based upon an estimated Koc values of 15 and 120 [19], 2-methyl-1-butanol is expected to leach readily in soil [26]; however, the importance of leaching may be lessened by relatively rapid, concurrent biodegradation.

Aquatic Fate: The only important identifiable degradation process in water is biodegradation. Although biodegradation data specific to 2-methyl-1-butanol are unavailable, analogy to similar aliphatic alcohols suggests that 2-methyl-1-butanol will be readily biodegradable. The volatilization half-lives of 2-methyl-1-butanol from a model environmental river (1 meter deep) and model pond have been estimated to be 61 hr and 28 days, respectively [19,29]. Aquatic bioconcentration, hydrolysis, photodegradation and adsorption to sediment are not expected to be important.

Atmospheric Fate: Based the vapor pressure, 2-methyl-1-butanol is expected to exist almost entirely in the vapor phase in the ambient atmosphere [9]. It will degrade in an average ambient atmosphere by reaction with photochemically produced hydroxyl radicals (estimated half-life of 47 hr) [2]. Physical removal from air via wet deposition is possible since 2-methyl-1-butanol is relatively soluble in water.

Biodegradation: Data specific to the rate of environmental biodegradation of 2-methyl-1-butanol were not located. However, many biodegradation studies have demonstrated that the lower molecular weight aliphatic alcohols that are similar in structure to 2-methyl-1-butanol (such as 2-methyl-1-propanol) are readily biodegradable [4,8,11,13,16,20,21,28,30,33,37]. This analogy indicates that 2-methyl-1-butanol is also likely to be readily biodegradable in the environment.

Abiotic Degradation: The rate constant for the vapor-phase reaction of 2-methyl-1-butanol with photochemically produced hydroxyl radicals has been estimated to be 8.21×10^{-12} cm^3/molecule-sec at 25 °C which corresponds to an atmospheric half-life of about 47 hr at an atmospheric concn of $5 \times 10^{+5}$ hydroxyl radicals per cm^3 [1]. Alcohols are generally resistant to aqueous environmental hydrolysis [19]; therefore, 2-methyl-1-butanol is not expected to hydrolyze in the environment.

Bioconcentration: Based upon the water solubility, the BCF for 2-methyl-1-butanol can be estimated to be 1.8 from a regression-derived equation [19]. Based upon a measured log Kow, the BCF for 2-methyl-1-butanol can be estimated to be 5.6 from a regression-derived equation [19]. These BCF values suggest that 2-methyl-1-butanol will not bioconcentrate significantly in aquatic organisms.

Soil Adsorption/Mobility: Based upon the water solubility, the Koc for 2-methyl-1-butanol can be estimated to be 15 from a regression-derived equation [19]. Based upon the measured log Kow, the Koc for 2-methyl-1-butanol can be estimated to be 120 from a regression-derived equation [19]. These BCF values suggest that 2-methyl-1-butanol has high to very high soil mobility [26].

2-Methyl-1-butanol

Volatilization from Water/Soil: The value of Henry's Law constant indicates that volatilization from environmental waters is slow, but may be significant from shallow rivers [29]. Based on the Henry's Law constant, the volatilization half-life from a model river (1 m deep flowing 1 m/sec with a wind velocity of 3 m/sec) can be estimated to be about 61 hr [19]. Volatilization half-life from an model environmental pond can be estimated to be about 28 days [29].

Water Concentrations:

Effluent Concentrations: 2-Methyl-1-butanol was identified as one of the odorous compounds emitted to the air from a sedimentation tank at a German water treatment facility [14].

Sediment/Soil Concentrations:

Atmospheric Concentrations: 2-Methyl-1-butanol was qualitatively detected in indoor air samples from a residential dwelling that had received an oil floor finish known as "Glitsa" [32].

Food Survey Values: 2-Methyl-1-butanol has been identified as a volatile component of blue cheese aroma [7], Concord grape juice essence [24], nectarines [27], apples [5,35], roasted filbert nuts [17], papaya fruit [22], oranges [25], tomatoes [23], beer [10], and wine [12].

Plant Concentrations:

Fish/Seafood Concentrations:

Animal Concentrations:

Milk Concentrations:

Other Environmental Concentrations:

Probable Routes of Human Exposure: The general population is exposed to 2-methyl-1-butanol through inhalation of aromas and ingestion of various foods, particularly fruits and fermented beverages.

2-Methyl-1-butanol

Average Daily Intake:

Occupational Exposure:

Body Burdens: 2-Methyl-1-butanol was detected in 5.4% of 387 expired air samples collected from 54 human volunteers that were classified as healthy urban nonsmokers [18]; the 2-methyl-1-butanol concn ranged from 3.63 to 41.8 ng/L [18].

REFERENCES

1. Atkinson R; J Inter Chem Kinet 19: 799-828 (1987)
2. Atkinson R; J Phys Chem Ref Data, Monograph 1, p 145 (1989)
3. Barton AFM; Alcohols With Water. International Union of Pure And Applied Chemistry. Solubility Data Series. Vol 15 (1984)
4. Bridie AL et al; Water Res 13: 627-30 (1979)
5. Carelli A et al; J High Resol Chromatgr 12: 488-90 (1989)
6. Daubert TE; Danner RP; Physical and Thermodynamic Properties of Pure Chemicals: Data Compilation, NY: Hemisphere Pub Corp (1989)
7. Day EA, Anderson DF; J Agric Food Chem 13: 2-4 (1965)
8. Dias FF, Alexander M; Appl Microbial 22: 1114-8 (1971)
9. Eisenreich SJ et al; Environ Sci Technol 15: 30-8 (1981)
10. Feng X; Shipin Yu Fajiao Gongye 1986: 68-70 (1986)
11. Gerhold RM, Malaney GW; J Water Pollut Control Fed 38: 562-79 (1966)
12. Gonzalez CL, Bravo AF; Alimentaria (Madrid) 26: 55-60 (1989)
13. Hammerton C; J Appl Chem 5: 517-24 (1955)
14. Hangartner M; Intern J Environ Anal Chem 6: 161-9 (1979)
15. Hine J, Mookerjee PK; J Org Chem 40: 292-8 (1975)
16. Kawasaki M; Ecotoxic Environ Safety 4: 444-54 (1980)
17. Kinlin TE et al; J Agri Food Chem 20: 1021 (1972)
18. Krotoszynski BK et al; J Anal Toxicol 3: 225-34 (1979)
19. Lyman WJ et al; Handbook of Chemical Property Estimation Methods. Environmental Behavior of Organic Compounds. Washington DC: American Chemical Society pp 4-9, 5-4, 5-10, 7-4, 7-5, 15-15 to 15-32 (1990)
20. McKinney RE, Jeris JS; Sew Indust Wastes 27: 728-35 (1955)
21. Price KS et al; J Water Pollut Control Fed 46: 63-77 (1974)
22. Schwab W et al; J Agri Food Chem 37: 1009-12 (1989)
23. Sohn TW et al; Han'guk Nonghwa Hakhoechi 31: 292-7 (1988)
24. Stevens KL et al; J Food Sci 30: 1006-7 (1965)
25. Sugisawa H et al; Nippon Shokuhin Kogyo Gakkaishi 36: 455-62 1989)
26. Swann RL et al; Res Rev 85: 23 (1983)
27. Takeoka et al; J Agri Food Chem 36: 553-60 (1988)
28. Urano K, Kato Z; J Hazardous Materials 13: 147-59 (1986)
29. USEPA; EXAMS II Computer Simulation (1987)

2-Methyl-1-butanol

30. Vaishnav DD et al; Chemosphere 16: 695-703 (1987)
31. Valvani SC et al; J Pharm Sci 70: 502-7 (1981)
32. Van Netten C et al; Bull Environ Contam Toxicol 40: 672-7 (1988)
33. Wagner R; Vom Wasser 42: 271-305 (1974)
34. Weast RC et al. CRC Handbook of Chemistry and Physics. CRC Press, Inc: Boca Raton, FL p C-168 (1985)
35. Yajima I et al; Agric Biol Chem 48: 849-55 (1984)
36. Yasuhara A; J Chromatogr 387: 371-8 (1987)
37. Yonezawa Y, Urushigawa Y; Chemosphere 8: 139-42 (1979)

Methyl Formate

SUBSTANCE IDENTIFICATION

Synonyms: Formic acid, methyl ester

Structure:

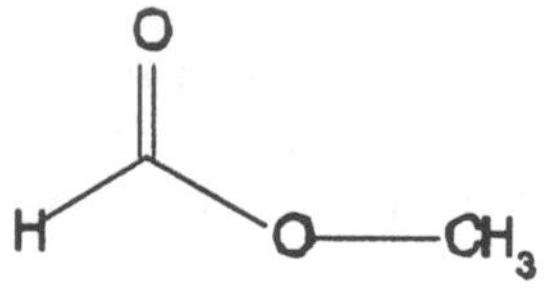

CAS Registry Number: 107-31-3

Molecular Formula: $C_2H_4O_2$

Wiswesser Line Notation: VHO1

CHEMICAL AND PHYSICAL PROPERTIES

Boiling Point: 31.5 °C [27]

Melting Point: -99 °C [27]

Molecular Weight: 60.05

Dissociation Constants:

Log Octanol/Water Partition Coefficient: -0.264 (estimated) [18]

Water Solubility: 230,000 mg/L at 25 °C [20]

Vapor Pressure: 585.7 mm Hg at 25 °C [2]

Henry's Law Constant: 2.23×10^{-4} atm-m^3/mole at 25 °C [6]

ENVIRONMENTAL FATE/EXPOSURE POTENTIAL

Summary: Methyl formate can be released to the environment in gases from the combustion of hydrocarbon fuels. Release can also occur at sites of production and use. Methyl formate may be formed in ambient air as

a by-product of the photochemical reaction of methane and hydroxyl radicals. It has been found to occur naturally in tomatoes and apples. If released to air, methyl formate will degrade relatively slowly by reaction with photochemically produced hydroxyl radicals (estimated half-life of 74 days). Physical removal from air via wet deposition is likely to occur since it is very soluble in water. If released to soil or water, hydrolysis and volatilization will be important removal mechanisms. Aqueous hydrolysis half-lives of 21.9 days, 2.19 days, 9.1 hr, and 0.91 hr can be predicted at respective pHs of 6, 7, 8, and 9 using an experimentally determined base-catalyzed rate constant at 25 °C. Volatilization half-lives from a model environmental river (1 meter deep) and model pond have been estimated to be 5.3 and 60 hr, respectively. Rapid evaporation from dry surfaces is likely to occur. Occupational exposure can occur through inhalation of vapor and dermal contact. Since methyl formate has been detected in tomatoes, apples, and coffee, exposure through ingestion will occur.

Natural Sources: Methyl formate was identified as a flavor component in tomato fruit [22] and as an aroma substance in apples [13].

Artificial Sources: Methyl formate is released in vent effluents during the commercial production of methanol [9]. Methyl formate can be released in waste streams from its commercial manufacture and use in the production of formamide [8]. Methyl formate's use in the commercial fumigant "Areginal" [12] will release the compound to the environment. Methyl formate can be formed as a minor by-product as a result of the photochemical reaction of methane and hydroxyl radicals [16]. Pyrolysis of biomass (such as wood) yielded methyl formate as a product [5]. Methyl formate has been detected in exhaust gases from the combustion of various hydrocarbon fuels [21].

Terrestrial Fate: Methyl formate can be expected to hydrolyze in moist soil; the rate of hydrolysis will increase with alkalinity. Aqueous hydrolysis half-lives of 21.9 days, 2.19 days, 9.1 hr, and 0.91 hr can be predicted at respective pHs of 6, 7, 8, and 9 using an experimentally determined base-catalyzed rate constant at 25 °C [11]. Based upon an estimated Koc of 5 [10], methyl formate is expected to leach readily in soil; however, the importance of leaching will be lessened by relatively rapid, concurrent hydrolysis. Methyl formate's relatively high vapor pressure indicates that rapid evaporation from dry surfaces is likely to occur. Insufficient biodegradation data specific to methyl formate are available to predict the importance of biodegradation in soil.

Aquatic Fate: Hydrolysis and volatilization can be important fate processes for methyl formate in water. Hydrolysis half-lives of 21.9 days, 2.19 days, 9.1 hr, and 0.91 hr can be predicted at respective pHs of 6, 7, 8, and 9 using an experimentally determined base-catalyzed rate constant at 25 °C [11]. The volatilization half-lives of methyl formate from a model environmental river (1 meter deep) and model pond have been estimated to be 5.3 and 60 hr, respectively [10,26]. Insufficient biodegradation data specific to methyl formate are available to predict the importance of biodegradation in water. Aquatic bioconcentration, direct photolysis and adsorption to sediment are not expected to be important.

Atmospheric Fate: Based upon the vapor pressure, methyl formate is expected to exist almost entirely in the vapor phase in the ambient atmosphere [3]. It will degrade relatively slowly (estimated half-life of 74 days) in an average ambient atmosphere by reaction with photochemically produced hydroxyl radicals [1]. Physical removal from air via wet deposition is likely to occur since methyl formate is very soluble in water.

Biodegradation: Methyl formate is listed as a compound that should be amenable to anaerobic biotechnology [23].

Abiotic Degradation: The rate constant for the vapor-phase reaction of methyl formate with photochemically produced hydroxyl radicals has been experimentally determined to be 2.27×10^{-13} cm^3/molecule-sec at 23 °C which corresponds to an atmospheric half-life of about 71 days at an atmospheric concn of $5 \times 10^{+5}$ hydroxyl radicals per cm^3 [1]. The base-catalyzed, aqueous hydrolysis rate constant for methyl formate at 25 °C has been reported to be 36.6/M-sec [11]; this rate constant corresponds to half-lives of 21.9 days, 2.19 days, 9.1 hr, and 0.91 hr at respective pHs of 6, 7, 8, and 9. Methyl formate vapor did not undergo direct photolysis when irradiated with UV light >300 nm [7].

Bioconcentration: Based upon the water solubility value, the BCF for methyl formate can be estimated to be 0.6 from a regression-derived equation [10]. This BCF value suggests that methyl formate will not bioconcentrate significantly in aquatic organisms.

Soil Adsorption/Mobility: Based upon the water solubility value, the Koc for methyl formate can be estimated to be 5 from a regression-derived equation [10]. This Koc value suggests that methyl formate has very high soil mobility [24].

Volatilization from Water/Soil: The experimental value of Henry's Law constant indicates that volatilization from environmental waters is probably significant, but may not be rapid [26]. Based on this Henry's Law constant, the volatilization half-life from a model river (1 m deep flowing 1 m/sec with a wind velocity of 3 m/sec) can be estimated to be about 5.3 hr [10]. Volatilization half-life from a model environmental pond can be estimated to be about 60 hr [26]. Methyl formate's relatively high vapor pressure indicates that rapid evaporation from dry surfaces is likely to occur.

Water Concentrations: DRINKING WATER: Methyl formate was qualitatively detected in drinking water samples collected in Seattle, WA [25].

Effluent Concentrations: The amount of methyl formate detected in exhaust gases from the combustion of various hydrocarbon fuels ranged from <0.1 to 0.7 ppm [21].

Sediment/Soil Concentrations:

Atmospheric Concentrations: Methyl formate was qualitatively detected in vapor-phase ambient air samples collected in the Kanawha Valley, WV in Sept 1977 [4].

Food Survey Values: Methyl formate was identified as a flavor component in tomato fruit [22]. Methyl formate was identified as a volatile constituent in brewed, roasted, and dried coffee [19]. It has also been detected as an aroma substance in apples [13].

Plant Concentrations: Methyl formate was identified as a flavor component in tomato fruit [22].

Fish/Seafood Concentrations:

Animal Concentrations:

Milk Concentrations:

Other Environmental Concentrations:

Probable Routes of Human Exposure: Occupational exposure to methyl formate can occur through inhalation of vapor and dermal contact [17]; it can be absorbed into the body through the respiratory and digestive systems and it may also penetrate the skin [17]. The general population can be exposed to methyl formate through inhalation of contaminated air. Since methyl formate has been detected in tomatoes, apples, and coffee [13,19,22], exposure through ingestion will occur.

Average Daily Intake:

Occupational Exposure: NIOSH (NOES Survey 1981-1983) has statistically estimated that 7,741 workers are potentially exposed to methyl formate in the USA [14]. NIOSH (NOHS Survey 1972-1974) has statistically estimated that 140 workers are potentially exposed to methyl formate in the USA [15].

Body Burdens:

REFERENCES

1. Atkinson R; J Phys Chem Ref Data, Monograph 1, p 145 (1989)
2. Daubert TE; Danner RP; Physical and Thermodynamic Properties of Pure Chemicals: Data Compilation, NY: Hemisphere Pub Corp (1989)
3. Eisenreich SJ et al; Environ Sci Technol 15: 30-8 (1981)
4. Erickson MD, Pellizzari ED; Analysis of Organic Air Pollutants in the Kanawha Valley, WV and the Shenandoah Valley, Va. USEPA-903/9-78-007 p 45 (1978)
5. Evans RJ, Milne TA; Energy Fuels 1: 123-37 (1987)
6. Hine J, Mookerjee PK; J Org Chem 40: 292-8 (1975)
7. Hustert K, Parlar H; Chemosphere 10: 1045-50 (1981)
8. Liepins R et al; Industrial Process Profiles for Environmental Use: Chpt 6. USEPA-600/2-77-023f, pp 6-494, 6-495 (1977)
9. Lovell RJ et al; Organic Chemical Manufacturing Volume 9: Selected Processes USEPA-450/3-80-028d p 143 (1980)

Methyl Formate

10. Lyman WJ et al; Handbook of Chemical Property Estimation Methods. Environmental Behavior of Organic Compounds. Washington DC: American Chemical Society pp 4-9, 5-4, 5-10, 7-4, 7-5, 15-15 to 15-32 (1990)
11. Mabey W, Mill T; J Phys Chem Ref Data 7: 383-414 (1978)
12. Meister RT et al; Farm Chemicals Handbook 1987, Willoughby, OH: Meister Publ Co p C-19 (1987)
13. Neubeller J, Buchloh G; Mitt Klosterneuburg 36: 34-46 (1986)
14. NIOSH; National Occupational Exposure Survey (NOES) (1983)
15. NIOSH; National Occupational Hazard Survey (NOHS) (1974)
16. Ogura K et al; J Mol Catal 43: 371-9 (1988)
17. Parmeggiani L; Encyl Occup Health & Safety 3rd ed Geneva, Switzerland: International Labour Office p 780-1 (1983)
18. PCGEMS; Personal Computer Graphical Exposure Modeling System. PCLOGP, Office of Toxic Substances, USEPA, Washington, DC (1988)
19. Rhoades JW; J Agri Food Chem 8: 136-41 (1960)
20. Riddick JA et al; Organic Solvents 4th ed; NY, NY: Wiley p 717 (1986)
21. Seizinger DE, Dimitriades B; J Air Pollut Control Assoc 22: 47-51 (1972)
22. Sohn TW et al; Han'guk Nonghwa Hakhoechi 31: 292-7 (1988)
23. Speece RE; Environ Sci Technol 17: 416A-27A (1983)
24. Swann RL et al; Res Rev 85: 23 (1985)
25. USEPA; Preliminary Assessment of Suspected Carcinogens in Drinking Water. Interim Report to Congress, June 1975 Washington, DC: USEPA p 10 (1975)
26. USEPA; EXAMS II Computer Simulation (1987)
27. Weast RC et al. CRC Handbook of Chemistry and Physics. CRC Press, Inc: Boca Raton, FLA p C-279 (1985)

N-Methylformamide

SUBSTANCE IDENTIFICATION

Synonyms:

Structure:

CAS Registry Number: 123-39-7

Molecular Formula: C_2H_5NO

Wiswesser Line Notation: VHM1

CHEMICAL AND PHYSICAL PROPERTIES

Boiling Point: 180-185 °C [14]

Melting Point:

Molecular Weight: 59.07

Dissociation Constants: pKa = -0.04 [10]

Log Octanol/Water Partition Coefficient: -0.624 [6] (estimated)

Water Solubility: Miscible [10]

Vapor Pressure: 3.08×10^{-2} mm Hg at 25 °C [10] (extrapolated using measured values)

Henry's Law Constant: 3.36×10^{-8} atm-m^3/mole at 25 °C [8] (estimated)

ENVIRONMENTAL FATE/EXPOSURE POTENTIAL

Summary: N-Methylformamide may be released to the environment from its manufacture and use, and from the use of dimethylformamide,

404

in which it is an impurity. N-Methylformamide may be released to the atmosphere as a result of the photolysis of dimethylamine or trimethylamine, and may occur in water as a result of photolysis of the aquatic herbicide fluridone. If released to the atmosphere, vapor-phase N-methylformamide is degraded relatively rapidly by reaction with photochemically produced hydroxyl radicals (estimated half-life of 1.7 hr in air). Based on the complete water solubility of N-methylformamide, removal by rainout may be important. If released to soil, N-methylformamide may be subject to biodegradation based on its biodegradation by microorganisms obtained through soil enrichment and its structural similarities to dimethylformamide. Based on an estimated Koc value of 10.9, N-methylformamide can be expected to leach significantly. Based on a low vapor pressure, N-methylformamide may volatilize from dry surface soils; however, volatilization from moist soil surfaces is not expected to be a significant fate process. If released to water, biodegradation is expected to be the most important fate process. Its hydrolysis is expected to be slight in aquatic systems based on hydrolysis data for dimethylformamide. Volatilization, adsorption to sediment and suspended materials, and bioconcentration (estimated BCF of 0.198) are not expected to be significant aquatic fate processes. In occupational settings, exposure to N-methylformamide may occur through inhalation of vapors and through eye and skin contact.

Natural Sources:

Artificial Sources: N-Methylformamide may be released to the environment from its manufacture and use, and from the use of dimethylformamide, in which it is an impurity [4]. N-Methylformamide has been detected in cigarette smoke [12] suggesting that it may be released to the environment as emissions from cigarette smoke. N-Methylformamide may be released to the atmosphere as a result of the photolysis of dimethylamine or trimethylamine [9], and may occur in water as a result of photolysis of the aquatic herbicide fluridone [11].

Terrestrial Fate: In soil, N-methylformamide may be subject to biodegradation based on its biodegradation by microorganisms obtained through soil enrichment [3] and its structural similarities to dimethylformamide which biodegrades in river water [2]. An estimated Koc value of 10.9 [7] indicates that N-methylformamide has very high mobility in soil and significant leaching may occur [13]. Based on the estimated vapor pressure, N-methylformamide may evaporate from dry soil surfaces; however, volatilization from moist soils is not expected to be important.

Aquatic Fate: If N-methylformamide is released to water, its hydrolysis is expected to be slight based on hydrolysis data for dimethylformamide [4]. Hydrolysis of N-methylformamide may be significant in highly acidic or alkaline waters based on the accelerated hydrolysis of dimethylformamide by acids and bases [4]. Based on the estimated Henry's Law constant and the complete water solubility, N-methylformamide is not expected to volatilize significantly in aquatic systems [7]. An estimated Koc of 3.6 [6,7] and BCF of 0.0417 [6,7] indicate that adsorption to sediment and bioconcentration are not significant fate processes in water. N-methylformamide may be subject to biodegradation based on its biodegradation by microorganisms obtained through soil enrichment [3] and its structural similarities to dimethylformamide which biodegrades in river water [2].

Atmospheric Fate: Based on the estimated vapor pressure, N-methylformamide is expected to exist almost entirely in the vapor phase in the ambient atmosphere [5]. Vapor-phase N-methylformamide is degraded relatively rapidly in the ambient atmosphere by reaction with photochemically formed hydroxyl radicals; the half-life for this reaction in typical air can be estimated to be about 1.7 hr [1].

Biodegradation: N-methylformamide has been shown to biodegrade by microorganisms obtained through soil enrichment [3].

Abiotic Degradation: The rate constant for the vapor-phase reaction of N-methylformamide with photochemically produced hydroxyl radicals has been estimated to be 2.22×10^{-10} cm^3/molecule-sec at 25 °C which corresponds to an atmospheric half-life of about 1.7 hr at an atmospheric concn of $5 \times 10^{+5}$ hydroxyl radicals per cm^3 [1].

Bioconcentration: Based on the estimated log Kow, the BCF for N-methylformamide can be estimated to be 0.20 using a recommended regression-derived equation [7]. This BCF value suggests that N-methylformamide would not bioconcentrate significantly in aquatic organisms.

Soil Adsorption/Mobility: Based on the estimated log Kow, the Koc for N-methylformamide can be estimated to be 10.9 using a recommended regression derived equation [7]. This Koc value indicates that N-methylformamide has very high mobility in soil and will leach [13].

Volatilization from Water/Soil: The estimated value of Henry's Law constant suggests that volatilization should not be a significant fate process in soil or water [7]. Based on this value, the volatilization half-life of N-methylformamide from a model river 1 m deep flowing 1 m/sec with a wind velocity of 3 m/sec has been estimated to be approximately 90.8 days [7]. Based on a measured vapor pressure of 0.4 mm Hg at 44 °C [10] and a measured vapor pressure of 1.5 mm Hg at 55 °C [10], a vapor pressure of 3.08×10^{-2} mm Hg at 25 °C can be extrapolated using the Antoine relationship. Based on this estimated vapor pressure, N-methylformamide may evaporate from dry soil surfaces; however, volatilization from moist soils is not expected to be significant.

Water Concentrations:

Effluent Concentrations:

Sediment/Soil Concentrations:

Atmospheric Concentrations:

Food Survey Values:

Plant Concentrations:

Fish/Seafood Concentrations: N-Methylformamide has been detected at concns of 0.09 and 4.06 ug/g in two samples of mussel collected at the Oarai Coast in Ibaraki, Japan on July 31, 1985 and July 31, 1986, respectively [15].

Animal Concentrations:

Milk Concentrations:

Other Environmental Concentrations: N-Methylformamide has been qualitatively detected in cigarette smoke [12].

Probable Routes of Human Exposure: N-Methylformamide has been listed as an impurity in dimethylformamide [4]; therefore, occupational exposure to N-methylformamide may occur during production and use of dimethylformamide. In occupational settings, exposure to formamide may occur through inhalation of vapors and through eye and skin contact.

Average Daily Intake:

Occupational Exposure:

Body Burdens:

REFERENCES

1. Atkinson R; Int J Chem Kinet 19: 799-828 (1987)
2. Dojlido JR; Investigations of Biodegradability and Toxicity of Organic Compounds USEPA-600/2-79-163 Cincinnati, OH: Municipal Environ Res Lab, Off Res Devel pp 2,4 (1979)
3. Doxtader KG, Alexander M; Soil Sci Soc Amer Proc 30: 351-56 (1966)
4. Eberling CL; Kirk-Othmer Encycl Chem Tech 3rd ed NY: Wiley Interscience 11: 263-8 (1980)
5. Eisenreich SJ et al; Environ Sci Technol 15: 30-8 (1981)
6. GEMS; Graphical Exposure Modeling System PCGEMS (1987)
7. Lyman WJ et al; Handbook of Chemical Property Estimation Methods. Environmental Behavior of Organic Compounds. Washington DC: American Chemical Society pp 4-9, 5-4, 5-10, 7-4, 7-5, 15-15 to 15-32 (1990)
8. Meylan WM and Howard PH; Environmental Toxicology and Chemistry 10:1283-1293 (1991)
9. Pitts JN et al; Environ Sci Technol 12: 946-53 (1978)
10. Riddick JA et al; Organic Solvents 4th ed NY: Wiley Interscience pp 655-656 (1986)
11. Saunders DG; J Agric Food Chem 31: 237-41 (1983)
12. Schumacher JN et al; J Agric Food Chem 25: 310-20 (1977)
13. Swann RL et al; Res Rev 85: 23 (1983)
14. Weast RC; CRC Handbook of Chemistry and Physics 66th ed Boca Raton, FL: CRC Press p C-279 (1985)
15. Yasuhara A, Morita M; Chemosphere 16: 2559-65 (1987)

5-Methyl-2-hexanone

SUBSTANCE IDENTIFICATION

Synonyms:

Structure:

CAS Registry Number: 110-12-3

Molecular Formula: $C_7H_{14}O$

Wiswesser Line Notation: 1Y1&2V1

CHEMICAL AND PHYSICAL PROPERTIES

Boiling Point: 144.9 °C

Melting Point: -73.9 °C

Molecular Weight: 114.19

Dissociation Constants:

Log Octanol/Water Partition Coefficient: 1.72 [26]

Water Solubility: 5400 mg/L at 20 °C [25]

Vapor Pressure: 4 mm Hg at 20 °C [25]

Henry's Law Constant: 1.11×10^{-4} atm-m^3/mol (calculated from the vapor pressure and water solubility)

ENVIRONMENTAL FATE/EXPOSURE POTENTIAL

Summary: 5-Methyl-2-hexanone could potentially be released to the environment in waste emissions generated at commercial manufacturing and use sites. Its use as a solvent can release it directly to the atmosphere

409

via evaporation. It could also be released in leachates and air emissions from landfills where wastes containing 5-methyl-2-hexanone have been disposed. If released to soil, 5-methyl-2-hexanone could potentially biodegrade, volatilize from surfaces or leach; concurrent biodegradation may minimize the importance of leaching. If released to water, 5-methyl-2-hexanone may be subject to biodegradation or volatilization (estimated volatilization half-life 11 hours from a model river). Bioconcentration in aquatic organisms, adsorption to suspended solids and sediments, aqueous hydrolysis, and reaction with naturally occurring oxidants are not expected to be significant fate processes. If released to the atmosphere, 5-methyl-2-hexanone is expected to exist almost entirely in the vapor phase. It will react in air with photochemically generated hydroxyl radicals (estimated half-life 1.9 days). The general population could potentially be exposed to 5-methyl-2-hexanone by inhalation of volatile flavor components or ingestion of certain foods, such as roasted filberts and fried bacon, which contain this compound.

Natural Sources:

Artificial Sources: 5-Methyl-2-hexanone could potentially be released to the environment in emissions from the one facility which manufactures this compound in the USA, from its use as a solvent for nitrocellulose, cellulose acetate butyrate, acrylics, and vinyl copolymers, during transport, during storage, and from sites where wastes containing this compound have been landfilled [8,14,23]. Potential also exists for release in wastewater from its manufacturing plant [14].

Terrestrial Fate: If released to soil, 5-methyl-2-hexanone could potentially biodegrade, leach or volatilize. Although data specific to 5-methyl-2-hexanone are not available, similar ketones are known to biodegrade readily [3,4,7,10,18,21,22,28]. Chemical hydrolysis and oxidation are not expected to be significant fate processes in soil; therefore, biodegradation is expected to be the only important degradation process in soil. An estimated Koc value of 39 indicates that 5-methyl-2-hexanone will leach readily [16]; however, concurrent biodegradation may minimize the importance of leaching. Surface evaporation and photolysis may also occur.

5-Methyl-2-hexanone

Aquatic Fate: If released to water, 5-methyl-2-hexanone may undergo volatilization and biodegradation. The estimated volatilization half-lives from a model river (1 meter deep) and model environmental pond (2 meters deep) are 11 hours and 5.5 days, respectively [16,27]. Although data specific to 5-methyl-2-hexanone are not available, similar ketones are known to biodegrade readily [3,4,7,10,18,21,22,28]. Bioconcentration in aquatic organisms, adsorption to suspended solids and sediments, aqueous hydrolysis, and reaction with commonly occurring oxidants found in natural waters are not expected to be significant fate processes.

Atmospheric Fate: Based upon its relatively high vapor pressure, 5-methyl-2-hexanone is expected to exist almost entirely in the vapor phase in the atmosphere [6]. If released to the atmosphere, 5-methyl-2-hexanone will degrade by reaction with photochemically generated hydroxyl radicals; the estimated half-life in typical air is 1.9 days [2]. Similar ketones are known to directly photolyze in sunlight [14]; therefore, direct photolysis may contribute to its atmospheric degradation. The water solubility of 5-methyl-2-hexanone suggests that physical removal from air via wet deposition (dissolution into clouds, rainfall, etc) is possible.

Biodegradation: Biodegradation data specific to the environmental biodegradation of 5-methyl-2-hexanone were not available. However, the results of many biodegradation screening studies performed on structurally similar ketones (such as 2-hexanone, diisobutyl ketone, 3-pentanone, methyl ethyl ketone) have demonstrated that these types of ketones biodegrade readily [3,4,7,10,18,21,22,28]. Therefore, biodegradation of 5-methyl-2-hexanone is probably an important (and perhaps dominant) removal process in soil and water.

Abiotic Degradation: The rate constant for the vapor-phase reaction of 5-methyl-2-hexanone with photochemically produced hydroxyl radicals can be estimated to be 8.2×10^{-12} cm^3/molecule-sec at 25 °C which corresponds to an atmospheric half-life of 1.9 days at an atmospheric concentration of $5 \times 10^{+5}$ hydroxyl radicals/cm^3 [2]. Ketones, particularly branched chain ketones, are susceptible to photochemical degradation in the environment [14]; although kinetic rates are not available, 5-methyl-2-hexanone may undergo some direct photolysis in sunlight. Ketones, in general, are resistant to aqueous chemical hydrolysis under environmental conditions [16]; therefore, chemical hydrolysis is not expected to be an important fate process for 5-methyl-2-hexanone. 5-Methyl-2-hexanone is stable to molecular oxygen, and as a result it is not expected to react with dissolved oxygen in water [14]. Based on the lack of reactivity of methyl ethyl ketone, a structurally similar compound, towards hydroxyl radicals (estimated half-life 2-4 years [1,5,17]) and alkyl peroxy radicals (estimated half-life approximately 400 years [9,17]) in natural sunlit water, chemical oxidation of 5-methyl-2-hexanone in the aquatic environment is not expected to be an important fate process.

Bioconcentration: Based upon the water solubility, the BCF for 5-methyl-2-hexanone can be estimated to be 5 from a recommended linear regression-derived equation [16]. This BCF value suggests that 5-methyl-2-hexanone will not bioconcentrate significantly in aquatic organisms.

Soil Adsorption/Mobility: Based upon the water solubility, the Koc for 5-methyl-2-hexanone can be estimated to be 39 from a linear regression-derived equation [16]. This Koc value suggests that 5-methyl-2-hexanone has high soil mobility [24]; therefore, leaching to ground water is possible.

Volatilization from Water/Soil: Based upon the Henry's Law constant, the volatilization half-life of 5-methyl-2-hexanone from a model river 1 m deep, flowing 1 m/sec with a wind speed of 3 m/sec can be estimated to be about 11 hours [16]. The volatilization half-life from a model environmental pond can be estimated to be 5.5 days [27]. The relatively high vapor pressure of 5-methyl-2-hexanone suggests that evaporation from dry surfaces will occur.

Water Concentrations:

Effluent Concentrations: 5-Methyl-2-hexanone was detected in concentrate obtained from water sample collected from an advanced waste treatment plant in Lake Tahoe, CA in Oct. 1974 [15].

Sediment/Soil Concentrations:

Atmospheric Concentrations:

5-Methyl-2-hexanone

Food Survey Values: 5-Methyl-2-hexanone was identified as a volatile component of roasted filberts [12]. 5-Methyl-2-hexanone was identified as a volatile flavor component of fried bacon [11].

Plant Concentrations:

Fish/Seafood Concentrations:

Animal Concentrations:

Milk Concentrations:

Other Environmental Concentrations:

Probable Routes of Human Exposure: The general population could potentially be exposed to 5-methyl-2-hexanone by inhalation of volatile flavor components and ingestion of certain foods, such as roasted filberts and fried bacon, which contain this compound [11,12]. Exposure to 5-methyl-2-hexanone in the workplace is most likely to occur by skin contact, eye contact, or inhalation, particularly at elevated temperatures [13]. Good warning properties of 5-methyl-2-hexanone should prevent overexposure [13].

Average Daily Intake:

Occupational Exposure: NIOSH (NOHS Survey 1972-1974) has statistically estimated that 25,326 workers are exposed to 5-methyl-2-hexanone in the USA [19]. NIOSH (NOES Survey 1981-1983) has statistically estimated that 6,837 workers are exposed to 5-methyl-2-hexanone in the USA [20].

Body Burdens:

REFERENCES

1. Anbar M, Neta P; J Appl Rad Iso 18: 493-523 (1967)
2. Atkinson R; Intern J Chem Kinet 19: 799-828 (1987)
3. Babeu L, Vaishnav DD; J Ind Microbiol 2: 107-15 (1987)
4. Bridie AL et al; Wat Res 13: 627-30 (1979)

5-Methyl-2-hexanone

5. Dorfman LM, Adams GE; Reactivity of the Hydroxyl Radical in Aqueous Solution; Washington, DC: Natl Bureau Stand NTIS COM-73-50623 (1973)
6. Eisenreich SJ et al; Environ Sci Tech 15: 30-8 (1981)
7. Ettinger MB; Ind Eng Chem 48: 256-9 (1956)
8. Hawley GG; The Condensed Chemical Dictionary 10th ed NY: Van Nostrand Reinhold p 682 (1981)
9. Hendry DG et al; J Phys Chem Ref Data 3: 944-7 (1974)
10. Heukelekian H, Rand MC; J Wat Pollut Control Assoc 29: 1040-53 (1955)
11. Ho CT et al; J Agric Food Chem 31: 336-42 (1983)
12. Kinlin TE et al; J Agric Food Chem 20: 1021-8 (1972)
13. Krasavage WJ et al; Patty's Industrial Hygiene and Toxicology 3rd ed Vol. IIC NY: Wiley (1982)
14. Lande SS et al; Investigation of Selected Potential Environmental Contaminants: Ketonic Solvent USEPA 560/2-76-003 pp 15, 25, 92-100, 106-7, 151 (1976)
15. Lucas SV; GC/MS Analysis of Organics in Drinking Water Concentrates and Advanced Waste Treatment Concentrates Vol. 1 USEPA-600/1-84-020A pp 321 (1984)
16. Lyman WJ et al; Handbook of Chemical Property Estimation Methods Washington DC: Amer Chem Soc pp 4-9, 5-10, 7-4, 15-1 to 15-32 (1990)
17. Mill T et al; Science 207: 886-7 (1980)
18. Mills EJ Jr, Stack VT Jr; Proc 8th Indust Waste Conf. Eng Bulletin Purdue Univ, Eng. Ext. Ser. pp 492-517 (1954)
19. NIOSH; National Occupational Hazard Survey (NOHS) (1974)
20. NIOSH; National Occupational Exposure Survey (NOES) (1983)
21. Price KS et al; J Water Pollut Control Fed 46: 63-77 (1974)
22. Shelton DR, Tiedje JM; App Env Microbiol 47: 85-7 (1984)
23. SRI; 1988 Directory of Chemical Producers. Menlo Park, CA: SRI International p 782 (1988)
24. Swann RL et al; Res Rev 85: 16-28 (1983)
25. Union Carbide; Ketones. Booklet No. F-41971 (1968)
26. USEPA; PCGEMS - Graphical Exposure Modeling System. CLOGP ver 3.32 (1986)
27. USEPA; EXAMS II computer simulation (1987)
28. Vaishnav DD et al; Chemosphere 16: 695-703 (1987)

3-Methylpyridine

SUBSTANCE IDENTIFICATION

Synonyms: 3-Picoline

Structure:

CAS Registry Number: 108-99-6

Molecular Formula: C_6H_7N

Wiswesser Line Notation:

CHEMICAL AND PHYSICAL PROPERTIES

Boiling Point: 143.9 °C

Melting Point: -18.3 °C

Molecular Weight: 93.12

Dissociation Constants: pKa = 5.63 at 25 °C [8]

Log Octanol/Water Partition Coefficient: 1.20 [11]

Water Solubility: miscible [8]

Vapor Pressure: 6.05 mm Hg at 25 °C [3]

Henry's Law Constant: 7.41 x 10^{-7} atm-m^3/mole at 25 °C [1]

ENVIRONMENTAL FATE/EXPOSURE POTENTIAL

Summary: 3-Methylpyridine may be released to the environment via effluents at sites where it is produced or used as a chemical intermediate in medicine, agriculture and other industries. It is also released to the environment with the manufacture and use of coal-derived liquid fuels and during the disposal of coal liquefication and gasification waste by-products. Due to dissociation, 3-methylpyridine and its conjugate acid will exist in environmental media in varying proportions that are pH dependent. Ions generally do not volatilize. The Henry's Law constant indicates that volatilization of 3-methylpyridine from environmental waters and moist soil should be extremely slow. Yet, 3-methylpyridine should evaporate from dry surfaces, especially when present in high concentrations such as in spill situations. In aquatic systems, the bioconcentration of 3-methylpyridine is not expected to be an important fate process. A low Koc indicates 3-methylpyridine should not partition from the water column to organic matter contained in sediments and suspended solids, and it should be highly mobile in soil. Limited monitoring data has shown it can leach to ground water. Biodegradation is likely to be the most important removal mechanism of 3-methylpyridine from aerobic soil and water. Both an aerobic river die-away test and aerobic soil grab sample study demonstrated rapid biodegradation of 3-methylpyridine after acclimation. Biodegradation is also expected to be fast in acclimated aerobic ground water; however, biodegradation may be quite slow under anaerobic conditions. In the atmosphere, 3-methylpyridine is expected to exist almost entirely in the vapor phase and reactions with photochemically produced hydroxyl radicals should be important (estimated half-life of 11 days). The complete miscibility of 3-methylpyridine in water suggests that physical removal from air by wet deposition (rainfall, dissolution in clouds, etc.) may occur. The most probable human exposure would be occupational exposure, which may occur through dermal contact or inhalation at workplaces where coal-derived fuels are produced or used. The most common nonoccupational exposure is likely to result from either passive or active inhalation of cigarette smoke. Limited monitoring data indicates that other nonoccupational exposures can occur from the ingestion of certain foods and contaminated drinking water supplies.

Natural Sources:

Artificial Sources: Methylpyridines are produced from the pyrolysis of coal or synthetically by reactions between aldehydes and ketones with ammonia [8]. Consequently, 3-methylpyridine may be released to the environment via effluents at sites where it is produced or used as an intermediate in medicine, agriculture and industry [8]. 3-Methylpyridine is also released to the environment via effluents from the manufacture and use of coal-derived liquid fuels and the disposal of coal liquefication and gasification waste by-products [4,7,15]. In addition, 3-methylpyridine is released to ambient air via cigarette smoke [10].

Terrestrial Fate: Due to dissociation, 3-methylpyridine and its conjugate acid will exist in soils in varying proportions that are pH dependent. Ions generally do not volatilize. The Henry's Law constant indicates that volatilization of 3-methylpyridine from moist soil should be extremely slow [14]. Yet, 3-methylpyridine should evaporate from dry surfaces, especially when present in high concentrations such as in spill situations. An estimated Koc of 107 [14] indicates 3-methylpyridine should be highly mobile in soil [20]; limited monitoring data has shown it may leach to ground water [19]. Biodegradation is likely to be the most important removal mechanism of 3-methylpyridine from aerobic soil. An aerobic soil grab sample study demonstrated rapid biodegradation of 3-methylpyridine [17]. However, biodegradation may be quite slow under anaerobic conditions [16].

Aquatic Fate: Due to dissociation, 3-methylpyridine and its conjugate acid will exist among environmental waters in varying proportions that are pH dependent. Under near-neutral and acidic conditions (pH <7.63), the ratio of 3-methylpyridine to its conjugate acid should increase with increasing pH [14]. Ions are not expected to volatilize from water. The miscibility of 3-methylpyridine in water suggests that volatilization, adsorption and bioconcentration are not important fate processes. This is supported by the Henry's Law constant which indicates that volatilization of 3-methylpyridine from environmental waters should be extremely slow [14]. An estimated Koc of 107 [14] indicates 3-methylpyridine should not partition from the water column to organic matter [20] contained in sediments and suspended solids; and an estimated bioconcentration factor (log BCF) of 5 indicates 3-methylpyridine should not bioconcentrate among aquatic organisms [14]. Aerobic biodegradation is likely to be the most important removal mechanism of 3-methylpyridine from aquatic systems. An aerobic river die-away test also showed that 3-methylpyridine biodegraded rapidly after acclimation in highly polluted natural waters [6]. Biodegradation is also expected to be fast in acclimated aerobic ground water; however, biodegradation may be quite slow under anaerobic conditions [16].

Atmospheric Fate: Based on the vapor pressure, 3-methylpyridine is expected to exist almost entirely in the vapor phase in ambient air [5]. In the atmosphere, vapor phase reactions with photochemically produced hydroxyl radicals may be important. The rate constant for 3-methylpyridine was estimated to be 1.43×10^{-12} cm^3/molecule-sec at 25 °C, which corresponds to an atmospheric half-life of about 11 days at an atmospheric concentration of $5 \times 10^{+5}$ hydroxyl radicals per cm^3 [2]. The complete miscibility of 3-methylpyridine in water suggests that physical removal from air by wet deposition (rainfall, dissolution in clouds, etc.) may occur.

Biodegradation: An aerobic biological screening study, which utilized a 10 mg/L yeast extract and an Aeric Ochraqualf soil for inocula, indicates that 3-methylpyridine is not readily biodegradable [18]. At 24 °C and a pH of 7, less than 1% of an initial 12.7 ppm of 3-methylpyridine was mineralized within 30 days as evidenced via the release of inorganic nitrogen [18]. However, an aerobic soil grab sample study demonstrated rapid biodegradation of 3-methylpyridine [17]. 3-Methylpyridine was added to Fincastle silt loam (Aeric Ochraqualf) with a pH of 6.7 and incubated at 25 °C [17]. Within 32 days, 69.3% of the available nitrogen was released to inorganic forms [17]. Sterilized controls lost 11.7% of the starting material to volatilization but did not release inorganic nitrogen [17]. An early aerobic river die-away test also showed that 3-methylpyridine biodegraded rapidly after acclimation in highly polluted natural waters maintained at 20 °C [6]. After 14- and 18-day acclimation periods, 100% of the original concentration of 1 ppm of 3-methylpyridine were removed within 2 and 4 days from the Ohio and Little Miami River waters, respectively [6]. For ground water that was taken from an aquifer contaminated by underground coal gasification, and to which soil was added as inocula, 3-methylpyridine degradation was also rapid [16]. Over 99% of 3-methylpyridine at an initial average concentration of 32.4 ppm was lost within 10 days when samples were incubated aerobically at 20 °C [16]. On the average, less than a 25% loss occurred within 31 days for sterilized, but otherwise identical controls [16]. For anoxic ground water samples with oxygen levels less than 0.1 to 0.4 ppm and without soil added, 3-methylpyridine degradation was shown to be much slower [16]. About 30% of the original average concentration of 24.8 ppm remained after 33 days [16].

Abiotic Degradation: The rate constant for the vapor-phase reaction of 3-methylpyridine with photochemically produced hydroxyl radicals has been estimated to be 1.43 x 10^{-12} cm^3/molecule-sec at 25 °C, which corresponds to an atmospheric half-life of about 11 days at an atmospheric concentration of 5 x 10^{+5} hydroxyl radicals per cm^3 [2].

Bioconcentration: Because 3-methylpyridine is miscible in water, bioconcentration in aquatic systems is not expected to be an important fate process. Based upon the log Kow, a bioconcentration factor (log BCF) of 5 for 3-methylpyridine has been calculated using a recommended regression-derived equation [14]. This BCF value also indicates 3-methylpyridine should not bioconcentrate among aquatic organisms.

Soil Adsorption/Mobility: Because 3-methylpyridine is miscible in water, soil adsorption is not expected to be an important fate process. Based on the log Kow, a Koc of 107 for 3-methylpyridine has been calculated using a recommended regression-derived equation [14]. This Koc value indicates 3-methylpyridine will be highly mobile in soil [20], and it should not partition from the water column to organic matter contained in sediments and suspended solids. However, due to dissociation, 3-methylpyridine and its conjugate acid will exist among environmental media in varying proportions that are pH dependent.

Volatilization from Water/Soil: Due to dissociation, 3-methylpyridine and its conjugate acid will exist among environmental media in varying proportions that are pH dependent. Ions are not expected to volatilize from water. Because 3-methylpyridine is miscible in water, and based upon the Henry's Law constant, volatilization of 3-methylpyridine from natural bodies of water and moist soils is also not expected to be an important fate process [14]. Yet, based upon the vapor pressure, 3-methylpyridine should evaporate from dry surfaces, especially when present in high concentrations such as in spill situations.

Water Concentrations: DRINKING WATER: 3-Methylpyridine was listed as a contaminant found in drinking water for a survey of US cities including Pomona, Escondido, Lake Tahoe and Orange Co, CA and Dallas, Washington, DC, Cincinnati, Philadelphia, Miami, New Orleans, Ottumwa, IA, and Seattle [13]. GROUND WATER: 3-Methylpyridine was detected in ground water samples near a coal gasification site near Hoe Creek in northeastern WY [19].

Effluent Concentrations: 3-Methylpyridine was detected in 2 of 6 wastewater effluents from energy related processes [15]. Retort water from a shale oil processing facility in DeBeque, CO contained 3-methylpyridine at an average concentration of 6 ppb; the process water at a coal gasification facility in Gillette, WY contained 3-methylpyridine at an average concentration of 16 ppb [15]. Wastewater from coal gasification at the Grand Fork's Energy Technology Center, ND was also reported to contain 3-methylpyridine [7]. In addition, wastewater effluent from a shale oil facility in Queensland, Australia was shown to contain 3-methylpyridine [4].

Sediment/Soil Concentrations:

Atmospheric Concentrations: URBAN: 3-Methylpyridine was not detected in the air of downtown Boulder, CO in Nov. 1982 [12]. REMOTE: 3-Methylpyridine was not detected in the air from a undeveloped location in CO in Nov. 1982 [12]. SOURCE DOMINATED: In Nov. 1982, 3-methylpyridine was detected in the air outside an oil shale wastewater facility of Occidental Oil Shale Inc. at Logan Wash, CO [12].

Food Survey Values: 3-Methylpyridine was identified as a volatile component of boiled beef [9].

Plant Concentrations:

Fish/Seafood Concentrations:

Animal Concentrations:

Milk Concentrations:

3-Methylpyridine

Other Environmental Concentrations: 3-Methylpyridine was detected in cigarette smoke at concentrations ranging from 12 to 36 ug/cigarette [10].

Probable Routes of Human Exposure: The most probable route of human exposure to 3-methylpyridine is by inhalation, dermal contact and ingestion. Cigarette smokers are likely to inhale 3-methylpyridine [10], and atmospheric workplace exposures have been documented [12]. Drinking water supplies have been shown to contain 3-methylpyridine [13].

Average Daily Intake:

Occupational Exposure: The most probable human exposure to 3-methylpyridine would be occupational exposure, which may occur through dermal contact or inhalation at places where it is produced or used. A 1982 study showed 3-methylpyridine was emitted to the air from wastewaters at a shale oil facility exposing inside workers [12]. Nonoccupational exposures are likely to occur among cigarette smokers [10] and populations with contaminated drinking water supplies [13]; or from the ingestion of certain foods [9].

Body Burdens:

REFERENCES

1.   Andon RJL et al; J Chem Soc, p 3188-96 (1954)
2.   Atkinson R; Intern J Chem Kin 19: 799-828 (1987)
3.   Chao J et al; J Phys Chem Ref Data 12: 1033-63 (1983)
4.   Dobson KR et al; Water Res 19: 849-56 (1985)
5.   Eisenreich SJ et al; Environ Sci Technol 15: 30-8 (1981)
6.   Ettinger LA et al; Indust Eng Chem 46: 791-3 (1954)
7.   Giabbai MF et al; Intern J Environ Anal Chem 20: 113-29 (1985)
8.   Goe GL; Kirk-Othmer Encycl Chem Tech 3rd NY: Wiley Interscience 19: 454-83 (1982)
9.   Golovnya RV et al; Chem Senses Flavour 4: 97-105 (1979)
10. Guerin MR, Buchanan MV; Environ Exposure to N-Aryl Compounds, Carcinogenic and mutagenic Responses to Aromatic Amines and Nitroarenes pp 37-45 (1988)
11. Hansch C, Leo AJ; Medchem Project Issue No 26. Claremont CA: Pomona College (1985)

3-Methylpyridine

12. Hawthorne SB, Sievers RE; Environ Sci Technol 18: 483-90 (1984)
13. Lucas SV; GC/MS Anal of Org in Drinking Water Concentrates and Advanced Treatment Concentrates Vol 1 USEPA-600/1-84-020A (NTIS PB85-128239) p 397 (1984)
14. Lyman WJ et al; Handbook of Chemical Property Estimation Methods NY: McGraw-Hill pp 4-9, 5-4, 6-3, 15-16 (1982)
15. Pelizzari E et al; ASTM Spec Tech Publ STP 686: 256-74 (1979)
16. Rodgers JE et al; Water Air Soil Pollut 24: 443-54 (1985)
17. Sims GK, Sommers LE; Environ Toxicol Chem 5: 503-9 (1985)
18. Sims GK, Sommers LE; Appl Environ Microbiol 51: 963-8 (1986)
19. Stuermer DH et al; Environ Toxicol Chem 16: 582-7 (1982)
20. Swann RL et al; Res Rev 85: 16-28 (1983)

4-Methylpyridine

SUBSTANCE IDENTIFICATION

Synonyms: 4-Picoline

Structure:

$$N$$

$$CH_3$$

CAS Registry Number: 108-89-4

Molecular Formula: C_6H_7N

Wiswesser Line Notation: T6NJ D1

CHEMICAL AND PHYSICAL PROPERTIES

Boiling Point: 144.9 °C

Melting Point: 3.7 °C

Molecular Weight: 93.12

Dissociation Constants: pKa = 5.98 at 25 °C [9]

Log Octanol/Water Partition Coefficient: 1.22 [10]

Water Solubility: Miscible [9]

Vapor Pressure: 5.77 mm Hg at 25 °C [4]

Henry's Law Constant: 7.07 x 10^{-7} atm-m^3/mole at 25 °C [1]

ENVIRONMENTAL FATE/EXPOSURE POTENTIAL

Summary: 4-Methylpyridine may be released to the environment via effluents at sites where it is produced or used as a chemical intermediate

423

in medicine, agriculture and other industries It is also released to the environment with the manufacture and use of coal-derived liquid fuels and during the disposal of coal liquefication and gasification waste by-products. Due to dissociation, 4-methylpyridine and its conjugate acid will exist in environmental media in varying proportions that are pH dependent. Ions generally do not volatilize. The Henry's Law constant indicates that volatilization of 4-methylpyridine from environmental waters and moist soil should be extremely slow. Yet, 4-methylpyridine should evaporate from dry surfaces, especially when present in high concentrations such as in spill situations. In aquatic systems, the bioconcentration of 4-methylpyridine is not expected to be an important fate process. A low Koc indicates 4-methylpyridine should not partition from the water column to organic matter contained in sediments and suspended solids, and it should be highly mobile in soil. Limited monitoring data has shown it can leach to ground water. Biodegradation is likely to be the most important removal mechanism of 4-methylpyridine from aerobic soil and water. Both an aerobic river die-away test and an aerobic soil grab sample study demonstrated rapid biodegradation to 4-methylpyridine after acclimation. In the atmosphere 4-methylpyridine is expected to exist almost entirely in the vapor phase and reactions with photochemically produced hydroxyl radicals should be important (estimated half-life of 11 days). The complete miscibility of 4-methylpyridine in water suggests that physical removal from air by wet deposition (rainfall, dissolution in clouds, etc.) may occur. The most probable human exposure would be occupational exposure, which may occur through dermal contact or inhalation at workplaces where coal-derived fuels are produced or used. The most common nonoccupational exposure is likely to result from either passive or active inhalation of cigarette smoke. Limited monitoring data indicates that other nonoccupational exposures can occur from the ingestion of certain foods and contaminated drinking water supplies.

Natural Sources:

Artificial Sources: Methylpyridines are produced from the pyrolysis of coal or synthetically by reactions between aldehydes and ketones with ammonia [9]. Consequently, 4-methylpyridine may be released to the environment via effluents at sites where it is produced or used as an intermediate in medicine, agriculture and industry [9]. 4-Methylpyridine is also released to the environment via effluents from the manufacture and use of coal-derived liquid fuels and the disposal of coal liquefication and gasification waste by-products [5,8,17]. In addition, 4-methylpyridine is released to ambient air via cigarette smoke [3].

Terrestrial Fate: Due to dissociation, 4-methylpyridine and its conjugate acid will exist in soils in varying proportions that are pH dependent. Ions generally do not volatilize. The Henry's Law constant indicates that volatilization of 4-methylpyridine from moist soil should be extremely slow [14]. Yet, 4-methylpyridine should evaporate from dry surfaces, especially when present in high concentrations such as in spill situations. An estimated Koc of 110 [14] indicates 4-methylpyridine should be highly mobile in soil [21]; and limited monitoring data has shown it may leach to ground water [20]. Biodegradation is likely to be the most important removal mechanism of 4-methylpyridine from aerobic soil. An aerobic soil grab sample study demonstrated rapid biodegradation of 4-methylpyridine [18].

Aquatic Fate: Due to dissociation, 4-methylpyridine and its conjugate acid will exist among environmental waters in varying proportions that are pH dependent. Under near-neutral and acidic conditions (pH < 7.98), the ratio of 4-methylpyridine to its conjugate acid should increase with increasing pH [14]. Ions are not expected to volatilize from water. The miscibility of 4-methylpyridine in water suggests that volatilization, adsorption and bioconcentration are not important fate processes. This is supported by the Henry's Law constant which indicates that volatilization of 4-methylpyridine from environmental waters should be extremely slow [14]. An estimated Koc of 110 [14] indicates 4-methylpyridine should not partition from the water column to organic matter [21] contained in sediments and suspended solids; and an estimated bioconcentration factor (log BCF) of 5 indicates 4-methylpyridine should not bioconcentrate among aquatic organisms [14]. Aerobic biodegradation is likely to be the most important removal mechanism of 4-methylpyridine from aquatic systems. An aerobic river die-away test also showed that 4-methylpyridine biodegraded rapidly after acclimation in highly polluted natural waters [7].

Atmospheric Fate: Based on the vapor pressure, 4-methylpyridine is expected to exist almost entirely in the vapor phase in ambient air [6]. In the atmosphere, vapor-phase reactions with photochemically produced hydroxyl radicals may be important. The rate constant for 4-methylpyridine was estimated to be 1.43 x 10^{-12} cm^3/molecule-sec at 25 °C, which corresponds to an atmospheric half-life of about 11 days at an atmospheric concentration of 5 x 10^{+5} hydroxyl radicals per cm^3 [2]. The complete miscibility of 4-methylpyridine in water suggests that physical removal from air by wet deposition (rainfall, dissolution in clouds, etc.) may occur.

Biodegradation: An aerobic biological screening study, which utilized a 10 mg/L yeast extract and an Aeric Ochraqualf soil for inocula, indicates that 4-methylpyridine is readily biodegradable [19]. At 24 °C and a pH of 7, 68% of an initial 15 ppm of 4-methylpyridine was mineralized within 24 days as evidenced via the release of inorganic nitrogen [19]. In addition, an aerobic soil grab sample study demonstrated rapid biodegradation of 4-methylpyridine [18]. 4-Methylpyridine was added to Fincastle silt loam (Aeric Ochraqualf) with a pH of 6.7 and incubated at 25 °C [18]. Within 16 and 32 days, 71.7% and 100%, respectively, of the available nitrogen was released to inorganic forms [18]. Sterilized controls lost 21.8% of the starting material to volatilization but did not release inorganic nitrogen [18]. An early aerobic river die-away test also showed that 4-methylpyridine biodegraded rapidly after acclimation in highly polluted natural waters maintained at 20 °C [7]. After 14- and 18-day acclimation periods, 100% of the original concentrations of 1 ppm of 4-methylpyridine were removed within 2 and 4 days from the Ohio and Little Miami River waters, respectively [7]. A screening test, which utilized 5 mL garden soil suspensions with glucose, yeast extract and mineral salts, compared aerobic and anaerobic biodegradation of 4-methylpyridine [15]. At concentrations of 93 and 94 ppm, 4-methylpyridine completely degraded within 66 to 170 days and 32 to 66 days under aerobic and anaerobic conditions, respectively [15].

Abiotic Degradation: The rate constant for the vapor-phase reaction of 4-methylpyridine with photochemically produced hydroxyl radicals has been estimated to be 1.43 x 10^{-12} cm^3/molecule-sec at 25 °C, which corresponds to an atmospheric half-life of about 11 days at an atmospheric concentration of 5 x 10^{+5} hydroxyl radicals per cm^3 [2].

4-Methylpyridine

Bioconcentration: Because 4-methylpyridine is miscible in water, bioconcentration in aquatic systems is not expected to be an important fate process. Based upon the log Kow, a bioconcentration factor (log BCF) of 5 for 4-methylpyridine has been calculated using a recommended regression-derived equation [14]. This BCF value also indicates 4-methylpyridine should not bioconcentrate among aquatic organisms.

Soil Adsorption/Mobility: Because 4-methylpyridine is miscible in water, soil adsorption is not expected to be an important fate process. Based on the log Kow, a Koc of 110 for 4-methylpyridine has been calculated using a recommended regression-derived equation [14]. This Koc value indicates the unionized 4-methylpyridine will be highly mobile in soil [21], and it should not partition from the water column to organic matter contained in sediments and suspended solids. However, due to dissociation, 4-methylpyridine and its conjugate acid will exist among environmental media in varying proportions that are pH dependent.

Volatilization from Water/Soil: Due to dissociation, 4-methylpyridine and its conjugate acid will exist among environmental media in varying proportions that are pH dependent. Ions are not expected to volatilize from water. Because 4-methylpyridine is miscible in water, and based upon the Henry's Law constant, volatilization of 4-methylpyridine from natural bodies of water and moist soils is also not expected to be an important fate process [14]. Yet, based upon the vapor pressure, 4-methylpyridine should evaporate from dry surfaces, especially when present in high concentrations such as in spill situations.

Water Concentrations: DRINKING WATER: 4-Methylpyridine was listed as a contaminant found in drinking water for a survey of US cities including Pomona, Escondido, Lake Tahoe and Orange Co, CA and Dallas, Washington, DC, Cincinnati, Philadelphia, Miami, New Orleans, Ottumwa, IA, and Seattle [13]. GROUND WATER: 4-Methylpyridine was detected in ground water samples near a coal gasification site near Hoe Creek in northeastern WY [20].

Effluent Concentrations: 4-Methylpyridine was detected in 1 of 6 wastewater effluents from energy-related processes [17]. The process water at a coal gasification facility in Gillette, WY contained 4-

427

methylpyridine in trace quantities [17]. Wastewater from coal gasification at the Grand Fork's Energy Technology Center, ND was also reported to contain 4-methylpyridine [8]. In addition, wastewater effluent from a shale oil facility in Queensland, Australia was shown to contain 4-methylpyridine [5].

Sediment/Soil Concentrations:

Atmospheric Concentrations: URBAN: 4-Methylpyridine was not detected in the air of downtown Boulder, CO in Nov 1982 [11]. RURAL: 4-Methylpyridine was not detected in the air from an undeveloped location in CO in Nov 1982 [11]. SOURCE DOMINATED: In Nov 1982, 4-methylpyridine was detected in the air outside an oil shale wastewater facility of Occidental Oil Shale Inc. at Logan Wash, CO [11].

Food Survey Values: 4-Methylpyridine was identified as a volatile component of fried bacon [12].

Plant Concentrations:

Fish/Seafood Concentrations:

Animal Concentrations:

Milk Concentrations:

Other Environmental Concentrations: The mainstream smoke of a popular 85-mm cigarette without filter tip contained 24.2 ug of 4-methylpyridine [3].

Probable Routes of Human Exposure: The most probable route of human exposure to 4-methylpyridine is by inhalation, dermal contact and ingestion. Cigarette smokers are likely to inhale 4-methylpyridine [3], and atmospheric workplace exposures have been documented [11]. Drinking water supplies have been shown to contain 4-methylpyridine [13].

Average Daily Intake:

4-Methylpyridine

Occupational Exposure: The most probable human exposure to 4-methylpyridine would be occupational exposure, which may occur through dermal contact or inhalation at places where it is produced or used. NIOSH (NOES Survey 1981-1983) has estimated that 9,577 workers are potentially exposed to 4-methylpyridine in the USA [16]. A 1982 study showed 4-methylpyridine was emitted to the air from wastewaters at a shale oil facility exposing inside workers [11]. Nonoccupational exposures are likely to occur among cigarette smokers [3] and populations with contaminated drinking water supplies [13]; or from the ingestion of certain foods [12].

Body Burdens:

REFERENCES

1. Andon RJL et al; J Chem Soc, p 3188-96 (1954)
2. Atkinson R; Intern J Chem Kin 19: 799-828 (1987)
3. Brunnemann KD et al; Anal Lett 11(7) 545 (1978)
4. Chao J et al; J Phys Chem Ref Data 12: 1033-63 (1983)
5. Dobson KR et al; Water Res 19: 849-56 (1985)
6. Eisenreich SJ et al; Environ Sci Technol 15: 30-8 (1981)
7. Ettinger LA et al; Indust Eng Chem 46: 791-3 (1954)
8. Giabbai, MF et al; Intern J Environ Anal Chem 20: 114-29 (1985)
9. Goe GL; Kirk-Othmer Encycl Chem Tech 3rd NY: Wiley Interscience 19: 454-83 (1982)
10. Hansch C, Leo AJ; Medchem Project Issue No 26. Claremont CA: Pomona College (1985)
11. Hawthorne SB, Sievers RE; Environ Sci Technol 18: 484-90 (1984)
12. Ho CT et al; J Agric Food Chem 31: 336-42 (1983)
13. Lucas SV; GC/MS Anal of Org in Drinking Water Concentrates and Advanced Treatment Concentrates Vol 1 USEPA-600/1-84-020A (NTIS PB85-128239) p 397 (1984)
14. Lyman WJ et al; Handbook of Chemical Property Estimation Methods NY: McGraw-Hill pp 4-9, 5-4, 6-3, 15-16 (1982)
15. Naik MN et al; Soil Biol Biochem 4: 313-23 (1972)
16. NIOSH; National Occupational Exposure Survey (NOES) (1989)
17. Pelizzari E et al; ASTM Spec Tech Publ STP 686: 256-74 (1979)
18. Sims GK, Sommers LE; Environ Toxicol Chem 5: 503-9 (1985)
19. Sims GK, Sommers LE; Appl Environ Microbiol 51: 963-8 (1986)
20. Stuermer DH et al; Environ Toxicol Chem 16: 582-7 (1982)
21. Swann RL et al; Res Rev 85: 16-28 (1983)

4-Methyl-2-pentanol

SUBSTANCE IDENTIFICATION

Synonyms:

Structure:

CAS Registry Number: 108-11-2

Molecular Formula: $C_6H_{14}O$

Wiswesser Line Notation: QY&1Y

CHEMICAL AND PHYSICAL PROPERTIES

Boiling Point: 131.7 °C at 760 mm Hg

Melting Point: -90 °C

Molecular Weight: 102.18

Dissociation Constants:

Log Octanol/Water Partition Coefficient: 1.43 [10]

Water Solubility: 1.64 x 10^4 mg/kg at 25 °C [22]

Vapor Pressure: 5.3 mm Hg at 25 °C [5]

Henry's Law Constant: 4.45 x 10^{-5} atm-m^3/mole [12]

ENVIRONMENTAL FATE/EXPOSURE POTENTIAL

Summary: 4-Methyl-2-pentanol may be released to the environment as a result of its manufacture and use as a solvent and its use in organic synthesis and brake fluids. It may be formed naturally as it has been found in the volatiles from mountain Beaufort cheese. If released to soil, it will be expected to exhibit high to very high mobility in soil based upon an estimated Koc; it may, therefore, leach through soil to ground water. It will not hydrolyze in moist soil, but it may be subject to volatilization from near-surface soil based upon estimated rates for its volatilization from water. It may be subject to biodegradation in soil based upon the rapid biodegradation observed in laboratory aqueous screening tests using sewage and activated sludge inoculum. If released to water, it will not be expected to adsorb to sediment and suspended particulate matter or to bioconcentrate in aquatic organisms based upon estimated Koc and BCF. It will not be expected to hydrolyze or directly photolyze. It may be subject to biodegradation in natural waters based upon the rapid biodegradation observed in laboratory aqueous screening tests using sewage and activated sludge inoculum. It will be subject to volatilization from surface waters based upon half-lives of 23 hr for volatilization from a model river one meter deep flowing 1 m/sec with a wind velocity of 3 m/sec and 10.6 days for volatilization from a model pond. If released to the atmosphere, it can be expected to exist mainly in the vapor phase in the ambient atmosphere based upon its vapor pressure. The estimated half-life for the reaction with photochemically produced hydroxyl radicals is 2.3 days at an atmospheric concentration of $5 \times 10^{+5}$ hydroxyl radicals per cm^3, based upon a measured rate constant. 4-Methyl-2-pentanol has been rated a moderately reactive compound with respect to ozone formations as observed in smog chamber tests. It will not be expected to directly photolyze in the atmosphere. Any general population exposure occurs mainly through the oral ingestion of contaminated drinking water and foods, such as certain cheeses, and through the inhalation of contaminated ambient air. Minor exposure may occur through dermal contact with contaminated water. Occupational exposure may occur through inhalation of contaminated air and dermal contact with solutions containing the compound.

Natural Sources: 4-Methyl-2-pentanol was qualitatively detected in the mixture of volatiles from mountain Beaufort cheese [6] which indicates that the compound may be formed naturally.

Artificial Sources: 4-Methyl-2-pentanol may be released to the environment as a result of its manufacture and use as a solvent for dyestuffs, oils, gums, resins, waxes, nitrocellulose and ethylcellulose; its use in organic synthesis, froth flotation; and its use in brake fluids.

4-Methyl-2-pentanol

Terrestrial Fate: If 4-methyl-2-pentanol is released to soil, it will be expected to exhibit high to very high mobility in soil [24] based upon estimated Koc of 143 and 21 estimated from log Kow and water solubility, respectively, using recommended regression equations [16]. It may, therefore, leach through soil. It will not hydrolyze in moist soil [16], but it may be subject to volatilization from near-surface soil based upon estimated rates for its volatilization from water [16,25]. It may be subject to biodegradation in soil based upon the rapid biodegradation observed in laboratory aqueous screening tests using sewage and activated sludge inoculum [21,26].

Aquatic Fate: If 4-methyl-2-pentanol is released to water, it will not be expected to adsorb to sediment and suspended particulate matter or to bioconcentrate in aquatic organisms based upon estimated Koc and BCF, respectively, estimated [16] from the log Kow and water solubility. It will not be expected to hydrolyze [16] or directly photolyze [23]. It may be subject to biodegradation in natural waters based upon the rapid biodegradation observed in laboratory aqueous screening tests using sewage and activated sludge inoculum [21,26]. It will be subject to volatilization from surface waters based upon half-lives of 23 hr for volatilization from a model river one meter deep flowing 1 m/sec with a wind velocity of 3 m/sec [16] and 10.6 days for volatilization from a model pond [25], respectively, estimated using the Henry's Law constant.

Atmospheric Fate: If 4-methyl-2-pentanol is released to the atmosphere, it can be expected to exist mainly in the vapor-phase in the ambient atmosphere [13] based upon its experimental vapor pressure. The measured rate constant for vapor-phase reaction with photochemically produced hydroxyl radicals of 7.1×10^{-12} cm^3/molecule-sec at 25 °C [20] corresponds to an atmospheric half-life of 2.3 days at an atmospheric concentration of $5 \times 10^{+5}$ hydroxyl radicals per cm^3. 4-Methyl-2-pentanol has been rated a moderately reactive (Class III) compound [8] with respect to ozone formation as observed in smog chamber tests [14]. 4-Methyl-2-pentanol will not be expected to directly photolyze in the atmosphere [23].

Biodegradation: A percent theoretical BOD of 84% was observed after 5 days in screening tests using the standard dilution technique and effluent from a biological sanitary waste treatment plant as inoculum [3].

4-Methyl-2-pentanol

A percent theoretical BOD of 43% was observed after 5 days in screening tests using the standard dilution technique and acclimated sewage as inoculum [17]. Tests using acclimated mixed microbial cultures as inoculum gave a percent theoretical BOD of 56% after 5 days [1]. In screening tests using filtered, settled domestic wastewater as inoculum, the observed percent theoretical BOD of 50%, 72%, 90% and 94% were observed after 5, 10, 15, 20 days, respectively [21]. In screening tests using activated sludge in a medium containing 100 ppm urea and approximately 16,000 ppm ethyl alcohol, the observed rate constant of disappearance of 4-methyl-2-pentanol was 0.432/hr which corresponds to a half-life of 17 hr [26]. The results of these laboratory screening tests indicate that 4-methyl-2-pentanol is readily biodegradable under the conditions used in the experiments. No information regarding biodegradation in natural media was found.

Abiotic Degradation: The rate constant for the vapor-phase reaction of 4-methyl-2-pentanol with photochemically produced hydroxyl radicals has been measured to be 7.1×10^{-12} cm^3/molecule-sec at 25 °C [20] which corresponds to an atmospheric half-life of 2.3 days at an atmospheric concentration of $5 \times 10^{+5}$ hydroxyl radicals per cm^3. 4-Methyl-2-pentanol was rated as a moderately reactive (Class III) compound [8] with respect to ozone formation as observed in smog chamber tests where the initial 4-methyl-2-pentanol concentration was 1.5 ppm (by volume) and the initial NOx concentration was 0.6 ppm [14]. Because alcohols do not absorb light at wavelengths above 290 nm [23], 4-methyl-2-pentanol will not be expected to directly photolyze in the environment. Hydrolysis is not expected to be a major pathway under normal environmental conditions because of the lack of hydrolysis functional groups (pH 5-9) [16].

Bioconcentration: Based upon the log Kow, a BCF of 7.2 has been estimated using a recommended regression equation [16]. Based upon the water solubility, a BCF of 2.6 has been estimated using a recommended regression equation [16]. Based upon these estimated BCF, 4-methyl-2-pentanol will not be expected to bioconcentrate in aquatic organisms.

Soil Adsorption/Mobility: Based upon the log Kow, a Koc of 143 has been estimated using a recommended regression equation [16]. Based upon the water solubility, a Koc of 21 has been estimated using a

4-Methyl-2-pentanol

recommended regression equation [16]. Based upon these estimated Koc, 4-methyl-2-pentanol will be expected to exhibit high to very high mobility in soil [24]. 4-Methyl-2-pentanol, therefore, may leach through soil to ground water if it does not volatilize or biodegrade first.

Volatilization from Water/Soil: The half-life for volatilization of 4-methyl-2-pentanol from a river one meter deep flowing 1 m/sec with a wind velocity of 3 m/sec is estimated to be 23 hr at 25 °C [16] based upon the Henry's Law constant. The volatilization half-life from a model pond, which considers the effect of adsorption, has been estimated to be 10.6 days [25].

Water Concentrations: DRINKING WATER: 4-Methyl-2-pentanol was qualitatively detected in 4 of 15 samples drinking water concentrate derived from large volume (>400 gallons) samples from 4 of 7 cities (Poplarville, MS, March 1979; Cincinnati, OH, Oct 1978; Philadelphia, PA, Feb 1976; Ottumwa, IA, Sept 1976) [15].

Effluent Concentrations: 4-Methyl-2-pentanol was detected at a concn of 136 ppb in a sample of influent water from an unspecified municipal wastewater treatment plant [19]. It was not indicated whether the compound was detected in final effluent from the plant [19]. It was detected at a concn of 45 ppb in effluent samples taken from the Los Angeles County wastewater treatment plant discharge, Palos Verdes, CA between Nov 1980 and Aug 1981 [9]. A concn of 140 ppm was found in samples of leachate or samples of ground water plume taken at an unspecified industrial landfill at an unspecified time [4]. 4-Methyl-2-pentanol was qualitatively detected in 4 of 16 samples of concentrate derived from large volume (>400 gallons) samples of advanced treatment concentrate from 4 of 6 cities (Lake Tahoe, CA, Oct 1974; Pomona, CA, 1975; Orange County, CA, Feb 1976; Blue Plains, Washington, DC, May 1975) [15].

Sediment/Soil Concentrations: 4-Methyl-2-pentanol was not detected (detection limit 0.5 ug/kg, dry weight basis) in two samples of marine sediment taken 6 km from the Los Angeles County wastewater treatment plant discharge, Palos Verdes, CA between Nov 1980 and Aug 1981 [9].

4-Methyl-2-pentanol

Atmospheric Concentrations: 4-Methyl-2-pentanol was qualitatively detected in ambient air samples taken Nov 20, 1979 at the BFI Landfill in New Jersey [2]. It was detected in ambient air during an air pollution peak at concn ranging from 0.9 to 7.0 ppb; date and location not specified [7].

Food Survey Values: 4-Methyl-2-pentanol was qualitatively detected in the mixture of volatiles from mountain Beaufort cheese [6].

Plant Concentrations:

Fish/Seafood Concentrations: 4-Methyl-2-pentanol was not detected (detection limit 0.3 ug/kg, dry weight basis) in tissue samples from aquatic organisms taken 6 km from the Los Angeles County wastewater treatment plant discharge, Palos Verdes, CA between Nov 1980 and Aug 1981 [9]. Samples analyzed included one sample each of liver tissue from Pacific Sanddab (Citharichthys xanthostigma), scorpion fish (Scrorpaena guttata), Dover sole (Microstomus pacificus), and white croaker (Genyonemus lineatus); muscle tissue from ridgeback prawn (Sicyonia ingentis); and a whole red pointer crab (Mursia guadichaudii) [9].

Animal Concentrations:

Milk Concentrations:

Other Environmental Concentrations:

Probable Routes of Human Exposure: General population exposure to 4-methyl-2-pentanol occurs mainly through the oral ingestion of contaminated drinking water [15] and foods, such as certain cheeses [6], and through the inhalation of contaminated ambient air [2]. Minor exposure may occur through dermal contact with contaminated water. Occupational exposure may occur through inhalation of contaminated air and dermal contact with solutions containing the compound.

Average Daily Intake:

4-Methyl-2-pentanol

Occupational Exposure: NIOSH (NOES Survey 1981-1983) has statistically estimated that 25,135 workers are potentially exposed to 4-methyl-2-pentanol in the USA [18].

Body Burdens:

REFERENCES

1. Babeu L, Vaishnav DD; J Indust Microbiol 2: 107-15 (1987)
2. Bozzelli JW et al; Analysis of Selected Toxic and Carcinogenic Substances in Ambient Air in NJ, State of NJ Dept Environ Protection (1980)
3. Bridie AL et al; Water Res 13: 627-30 (1979)
4. Brown KW, Donnelly KC; Haz Waste Haz Mater 5: 1-30 (1988)
5. Daubert TE, Danner RP; Physical and Thermodynamic Properties of Pure Compounds. Data Compilation. Am Inst Chem Eng. (1989)
6. Dumont JP, Adda J; J Agric Food Chem 26: 364-7 (1978)
7. Environ Dept Kanagawa Prefecture Government; Annual Report on Survey and Research of Air Pollution in Kanagawa Prefecture, Japan Vol 20. Kanagawa - Ken Saiki Osen Chose Kenkyu Hokoku 20: 86-90 (1978)
8. Farley FF; USEPA Intern Conf Photochemical Oxidant Poll Control Dimitriades B ed USEPA, Washington,DC USEPA-600/3-77-001b pp 713-27 (1977)
9. Gossett RW et al; Mar Pollut Bull 14: 387-92 (1983)
10. Hansch C, Leo AJ; Medchem Project Issue No. 26 Claremont, CA: Pomona College (1985)
11. Hawley GG; Condensed Chemical Dictionary 10th ed Van Nostrand Reinhold NY p 668 (1981)
12. Hine J, Mookerjee PK; J Org Chem 40: 292-8 (1975)
13. Eisenreich SJ et al; Environ Sci Technol 15: 30-8 (1981)
14. Laity JL et al; Advan Chem Series 124: 95-112 (1973)
15. Lucas SV; GC/MS Anal of Org in Drinking Water Concentrates and Advanced Treatment Concentrates Vol 1 USEPA-600/1-84-020a (NTIS-PB85-128221) pp 133, 166 (1984)
16. Lyman WJ et al; Handbook of Chem Property Estimation Methods NY: McGraw-Hill pp 4-9, 5-5, 7-4, 15-16 to 15-29 (1982)
17. Niemi GJ et al; Environ Toxicol Chem 6: 515-27 (1987)
18. NIOSH; The National Occupational Exposure Survey (NOES) (1983)
19. Norwood DL et al; Amer Water Works Assoc. Water Qual Technol Conf November 16-20, 1986. Technol Conf Proc pp 2-140 to 2-163 (1987)
20. Pitts JN Jr et al; USEPA Intern Conf Photochemical Oxidant Poll Control Dimitriades B ed. USEPA, Washington, DC USEPA-600/3-77-001b pp 687-704 (1977)
21. Price KS et al; J Water Pollut Control Fed 46: 63-77 (1974)
22. Riddick JA et al; Organic Solvents. John Wiley and Sons Inc. NY (1986)

4-Methyl-2-pentanol

23. Silverstein RM et al; Spectrometric Id of Org Cmpd, J Wiley and Sons Inc 3rd ed pp 238-55 (1974)
24. Swann RL et al; Res Rev 85: 17-28 (1983)
25. USEPA; EXAMS II Computer Simulation (1987)
26. Yonezawa Y, Urushigawa Y; Chemosphere 8: 139-42 (1979)

Nonane

Synonyms:

Structure:

$$H_3C \diagdown\diagup\diagdown\diagup\diagdown\diagup\diagdown CH_3$$

CAS Registry Number: 111-84-2

Molecular Formula: C_9H_{20}

Wiswesser Line Notation:

CHEMICAL AND PHYSICAL PROPERTIES

Boiling Point: 150.8 °C

Melting Point: -46.7 °C

Molecular Weight: 128.26

Dissociation Constants:

Log Octanol/Water Partition Coefficient: 5.46 estimate [4]

Water Solubility: 0.122 mg/L at 25 °C [69]

Vapor Pressure: 4.45 mm Hg at 25 °C [7]

Henry's Law Constant: 6.16 atm-m³/mole at 25 °C, calculated from VP/Wsol

ENVIRONMENTAL FATE/EXPOSURE POTENTIAL

Summary: n-Nonane is a constituent in the paraffin fraction of crude oil and natural gas. n-Nonane is released to the environment via the manufacture, use, and disposal of many products associated with the

438

petroleum and gasoline industries. Extensive data show release of n-nonane into the environment from solvent-based building materials, printing pastes, paints, varnishes, adhesives and other coatings; hazardous waste sites, landfills and waste incinerators; vulcanization and extrusion operations during rubber and synthetic production; and the combustion of gasoline, diesel fuels and plastics. Photolysis or hydrolysis of n-nonane is not expected to be environmentally important. Biodegradation of n-nonane may occur in soil and water; however, volatilization and adsorption are expected to be far more important environmental fate processes. A high Koc indicates n-nonane will be immobile in soil and may partition from the water column to organic matter contained in sediments and suspended solids. The bioconcentration of n-nonane may be important in aquatic systems. A Henry's Law constant of 6.16 atm-m^3/mole at 25 °C, suggests rapid volatilization of n-nonane from environmental waters. The volatilization half-lives from a model river and a model pond (the latter considers the effect of adsorption) have been estimated to be 3.3 hr and 78 days, respectively. n-Nonane is expected to exist almost entirely in the vapor phase in ambient air. Reactions with photochemically produced hydroxyl radicals in the atmosphere have been shown to be important (estimated half-life of 1.5 days). The most probable route of human exposure to n-nonane is by inhalation. Extensive monitoring data indicates n-nonane is a widely occurring atmospheric pollutant.

Natural Sources: n-Nonane is a constituent in the paraffin fraction of crude oil and natural gas [65].

Artificial Sources: n-Nonane is released to the environment via the manufacture, use, and disposal of many products associated with the petroleum [45,46,50] and gasoline industries [46,57]. The combustion of plastics [32], gasoline and diesel fuels have been shown to release n-nonane into the atmosphere [18,19,38]. Vulcanization and extrusion operations during rubber and synthetic production as with shoes, tires and electrical insulation also emit n-nonane to the air [5]. Other well-documented materials that are responsible for the release of n-nonane to the environment include solvent-based building materials, printing pastes, paints, varnishes, adhesives and other coatings [15,66,68]. Hazardous waste sites [30], landfills [67,70] and waste incinerators [24] also release n-nonane into the environment. Building materials such as petroleum-based solvents such as floor adhesives and waxes, wood stains, polyurethane finish and air fresheners emit n-nonane to indoor air [63].

Terrestrial Fate: Photolysis or hydrolysis [34] of n-nonane is not expected to be important in soils. The biodegradation of n-nonane may occur in soils; however, volatilization and adsorption are expected to be far more important fate processes. An estimated range for Koc from 13,900 to 22,250 [34] indicates n-nonane will be immobile in most soils [3]. Based upon the estimated Henry's Law constant, n-nonane is also expected to rapidly volatilize from moist surface soils [62].

Aquatic Fate: Photolysis and hydrolysis [34] of n-nonane in aquatic systems are not expected to be important. The log bioconcentration factor (log BCF) for n-nonane has been estimated to range from 3.31 to 3.92 [34], suggesting bioconcentration may be an important fate process in aquatic systems. Biodegradation of n-nonane may occur in aquatic environments; however, volatilization and adsorption are expected to be far more important fate processes. An estimated range for Koc from 13,900 to 22,250 [34] indicates n-nonane may partition from the water column to organic matter [62] contained in sediments and suspended solids. The estimated Henry's Law constant suggests rapid volatilization of n-nonane from environmental waters [34]. Based on the Henry's Law constant, the volatilization half-life from a model river has been estimated to be 3.3 hr [34]. The volatilization half-life from an model pond, which considers the effect of adsorption, has been estimated to be about 78 days [64].

Atmospheric Fate: Based on the vapor pressure, n-nonane is expected to exist almost entirely in the vapor phase in ambient air [11]. n-Nonane is not expected to undergo direct photolysis in ambient air. However, vapor-phase reactions with photochemically produced hydroxyl radicals in the atmosphere have been shown to be important. The rate constant for n-nonane was measured to be 10.7×10^{-12} cm^3/molecule-sec at 26 °C, which corresponds to an atmospheric half-life of about 1.5 days at an atmospheric concentration of $5 \times 10^{+5}$ hydroxyl radicals per cm^3 [3].

Biodegradation: At intervals of 6, 12, and 24 hr endogenous respiration was 0.2, 0.4, and 1.1 % of the TOD for n-nonane at a concentration of 500 mg/L in 20 mL volume soln of supernatant liquid with a suspended solids concentration adjusted to 2500 mg/L from differing aeration units of sewage treatment facilities [14]. A 27% loss of n-nonane loss occurred within 5 days and completely disappeared within 15 days from 1 mL of crude oil added to a 100 mL simulated seawater soln inoculated with sediment samples from Fukae of Kobe harbor, Japan and incubated at 20 °C [37]. Loss of 18 and 58% of n-nonane was observed within 5 and 15 days, respectively, from 1 mL of crude oil added to a 100 mL seawater soln collected at Fukae of Kobe harbor, Japan and incubated at 20 °C [37]. Complete recovery was reported for all the control samples [37].

Abiotic Degradation: Alkanes are generally resistant to hydrolysis [34]. Based on data for n-hexane and iso-octane, n-nonane is not expected to absorb UV light in the environmentally significant range, >290 nm [59]. Therefore, n-nonane probably will not undergo hydrolysis or direct photolysis in the environment. The rate constant for the vapor-phase reaction of n-nonane with photochemically produced hydroxyl radicals was measured to be 10.7×10^{-12} cm^3/molecule-sec at 26 °C which corresponds to an atmospheric half-life of about 1.5 days at an atmospheric concentration of $5 \times 10^{+5}$ hydroxyl radicals per cm^3 [3]. At 39 °C the rate constant for the vapor-phase reaction of n-nonane with photochemically produced hydroxyl radicals was determined to be 10.2×10^{-12} cm^3/molecule-sec, which corresponds to an atmospheric half-life of 1.3 days [42].

Bioconcentration: Based upon the water solubility and the estimated log Kow, the bioconcentration factor (log BCF) for n-nonane has been calculated, using recommended regression-derived equations, to be 2.31 and 3.92, respectively [34]. These bioconcentration factor values indicate bioconcentration may be an important fate process for n-nonane.

Soil Adsorption/Mobility: Based on the water solubility and the estimated log Kow, the Koc of n-nonane has been calculated, using various regression-derived equations, to range from 13,900 to 22,250 [34]. These Koc values indicate n-nonane will be immobile in soil [62].

Volatilization from Water/Soil: The Henry's Law constant for n-nonane indicates extremely rapid volatilization from natural waters [34]. The volatilization half-life from a model river (1 meter deep flowing 1 m/sec with a wind speed of 3 m/sec) has been estimated to be 3.3 hr [34]. The volatilization half-life from a model pond, which considers the effect of adsorption, has been estimated to be 78 days [64].

Water Concentrations: DRINKING WATER: n-Nonane was identified in 10 of 14 treated water supplies in England [13]. n-Nonane was listed as one of the many organic chemicals identified in drinking water in the USA as of 1974 [1] and as of 1982 [28]. The drinking water supply for the District of Columbia was found to contain n-nonane; the concentration was estimated to be 0.01 ppm [52]. n-Nonane was detected in New York City drinking water at a concentration of 0.02 ug/L [26]. SURFACE WATER: The average n-nonane concentration of 6 water samples from both Little Britain Lake and Welsh Harp Lake, England were 5.2 and 6.5 ppb, respectively [6]. The n-nonane average concentration of weekly samples taken over an approximate period of 1 year for Luton Brook, England was 7.2 ppb [6]. The average n-nonane concentration for water samples taken from the River Pinn at Brunel University, England was 3.0 ppb [6]. SEAWATER: Only trace quantities to 3 ng/L of n-nonane were detected in open surface waters of the north central Gulf of Mexico [51]. The n- nonane concentration of 3 surface water samples from an unpolluted coastal area of the north central Gulf of Mexico ranged from trace levels to 2 ng/L [51]. The n-nonane concentration also ranged from trace levels to 3 ng/L for 6 surface water samples collected at coastal areas of the north central Gulf of Mexico under anthropogenic influence [51]. n-Nonane was detected in 7 of 8 surface water samples in the Gulf of Mexico ranging in concentration from 0.2 to 2.3 ng/L with an average concentration of 1.0 ng/L [49]. RAIN WATER: The n-nonane concentration of rain water collected at Brunel University, England was 45.1 ppb [6].

Effluent Concentrations: n-Nonane was identified as a stack emission from waste incinerators [24]. One of five hazardous waste sites listed on the National Priorities List emitted gaseous n-nonane with a 25 to 50% frequency of occurrence [30]. n-Nonane was also identified as a vapor emitted from a simulated landfill [67] and an actual landfill at an average concentration of 0.9 ppmV [71]. A clay pit landfill in England that received municipal, industrial and liquid wastes emitted n-nonane gas at a concentration of 116 mg/L [70]. The combustion of plastics also emits n-nonane [32]. Building materials such as petroleum-based solvents such as floor adhesives and waxes, wood stains, polyurethane finish and air fresheners emit n-nonane to indoor air [63]. n-Nonane was detected in 4 of 63 industrial wastewater effluents at a concentration less than 10 ug/L [45]. Underwater hydrocarbon vents and formation water discharges from offshore oil production platforms were found to contain n-nonane concentrations in the vapor phase at 1 umol/L and in the liquid state at 50 ng/L, respectively [50]. Formation water contained n-nonane at a concentration of 520 ug/L [50]. Data from Sept 2, 1979 identified n-nonane as a gaseous emission of the vehicle traffic through the Allegheny Mountain Tunnel of the Pennsylvania Turnpike [18]. Data from Aug 25 to Sept 7, 1979 showed for a speed of 80 km/hr on straight and level highway, gasoline powered vehicles emitted n-nonane at an average rate of 3.6 mg/km and diesel trucks emitted n-nonane at an average of 5.1 mg/km [19]. The average exhaust from 67 gasoline fueled vehicles was found to contain n-nonane at a concentrations 0.2% by weight [38]. Motorboats emitted n-nonane to canal water with resultant concentrations ranging from 1 to 14 ng/L with an average of 6 ng/L for 8 samples [25].

Sediment/Soil Concentrations: The average n-nonane concentration of 5 samples for the sediments of Lake Pontchartrain, a shallow oligohaline estuary located in the deltaic plain of the Mississippi River near New Orleans, was 0.07 ppb [12]. n-Nonane was detected in the soil surrounding an earthen disposal pit for produced water in the Duncan Oil Field, NM [10].

Atmospheric Concentrations: URBAN: n-Nonane occurred in 46% of the indoor air samples and 38% of the outdoor air samples taken in Chicago, IL [23]. The average n-nonane concentration for 2 samples per 4 sites in Tulsa, OK was 2.2 ppbC with a range of 0.5 to 5.1 ppbC [2]. The n-nonane concentration for 6 sites in Rio Blanco, CO averaged 2.2 ppbC with a range from 0.2 to 8.8 [2]. The average concentration of n-nonane in 3 samples indoor and outdoor air from Neenah, WI were 0.21 and 0.07 ug/m^3, respectively [58]. The average concentration of n-nonane in 3 samples indoor and outdoor air from Newark, NJ were 1.38 and 0.73 ug/m^3, respectively [58]. The ground level atmospheric concentration of n-nonane at 1325 was 0.5 ppb and 6.7 ppb at 0800 for Huntington Park, CA [53]. At 1500 ft the n-nonane concentration was 0.2 ppb at 0743 and at 0807 at a height of 2,200 ft the n-nonane concentration was less than 0.1 ppb [53]. The n-nonane concentration ranged from 1 to 3 ppbV at a downtown Los Angeles location for the fall of 1981 [17]. The average atmospheric gas phase concentration of n-nonane was 489 ng/m^3 for 7 rain events in Portland, Oregon from Feb to Apr 1984 [31]. According to the National Ambient Volatile Organic Compounds (VOCs) Database, the median urban atmospheric concentration of n-nonane is 0.303 ppbV for 714 samples [56]. The average n-nonane concentration of air samples taken at Brunel University, England was 10.1 ppb [6]. n-Nonane was also identified in the ambient air of Paris, France at concentrations ranging from 2.2 to 6.5 ug/m^3 [47]. In 1983-4, the respective minimum, maximum and average outdoor air concentrations of n-nonane in northern Italy were less than 1, 9 and 2.1 ug/m^3 [8]. The same study determined the respective minimum, maximum and average indoor air concentration of n-nonane for 14 homes and an office building were less than 1, 165 and 27 ug/m^3 [8]. The concentration of n-nonane in the downtown air of Zurich, Switzerland was 1.7 ppb [16]. At Deuselbach, Hunsruck in Germany, the atmospheric n-nonane concentration was less than 0.01 ppb for October 23, 1983 [48]. n-Nonane was detected in the atmospheres of 6 industrialized cities of the USSR ranging in size of population from 0.4 to 4.5 million people [20,21,22]. The average n-nonane concentrations in the atmospheres of Pretoria, Johannesburg and Durban, South Africa were 0.7, 0.6, and 0.4 ppb, respectively [33]. n-Nonane was identified in the ambient air of Sydney, Australia [36] ranging in concentration from 0.1 to 1.6 ppbV with an average concentration of 0.4 ppbV [39]. n-Nonane was detected at an average concentration of 31.88 ppbC in the atmosphere over the British Columbia Research Council Laboratory at the University of British Columbia [61]. In the Netherlands, the average n-nonane concentrations in ambient air were 2.2 and 0.5 ppb for 20 and 24 samples, respectively [60]. SUBURBAN: According to the National Ambient Volatile Organic Compounds (VOCs) Database, the median suburban atmospheric concentration of n-nonane is 0.179 ppbV for 215 samples [56]. RURAL: At a rural site near Duren, Germany, the atmospheric n-nonane concentration was less than 0.15 ppb for March 1984 [48]. The respective median, minimum and maximum atmospheric concentrations of n-nonane for 5 rural locations in NC ranged from 0.1 to 1.3, 0.0 to 0.6, and 0.3 to 2.1 ppb [55]. The atmospheric concentration of n-nonane for Jones State Forest, TX ranged from 0.6 to 11.1 ppb with an average of 2.6 ppb for 10 samples [54]. According to the National Ambient Volatile Organic Compounds (VOCs) Database, the median rural atmospheric concentration of n-nonane is 0.032 ppbV for 12 samples [56]. REMOTE: For 9 samples collected over a 30-hour period, the average n-nonane concentration in the Smokey Mountains, NC was 0.7 ppbC with a range from 0.2 to 1.7 ppbC [2]. According to the National Ambient Volatile Organic Compounds (VOCs) Database, the median remote atmospheric concentration of n-nonane is 0.311 ppbV for 4 samples [56]. On Aug 27, 1976, the average n-nonane concentrations for air over Lake Michigan at altitudes of 2000, 2500 and 3000 ft was 0.7 ppbV [35]. On Aug 28, 1976, the average n-nonane concentrations for air over Lake Michigan at altitudes of 1000 and 1500 ft was 0.5 ppbV [35]. SOURCE DOMINATED: Construction activities accounted for an indoor air concentration of 85 ug/m^3 for n-nonane; whereas, 42 days after construction was completed, the indoor air concentration of n-nonane was 1.26 ug/m^3 [58]. According to the National Ambient Volatile Organic Compounds (VOCs) Database, the median source dominated and indoor atmospheric concentrations of n-nonane are 0.867 and 0.800 ppbV for 15 and 95 samples, respectively [56].

Food Survey Values: n-Nonane was identified as a volatile component of roasted filberts [27], and beaufort cheese [9].

Plant Concentrations:

Fish/Seafood Concentrations: The average n-nonane concentrations for the oyster and Rigolets clam populations of Lake Pontchartrain, a shallow oligohaline estuary located in the deltaic plain of the Mississippi River near New Orleans, were 25 and 24 ng/g of wet weight [12].

Animal Concentrations:

Milk Concentrations: n-Nonane was detected in 8 of 12 samples of mothers breast milk from the cities of Bayonne, NJ, Jersey City, NJ, Bridgeville, PA and Baton Rouge, LA [43].

Other Environmental Concentrations: n-Nonane was emitted from 8 adhesives used in building materials at an average rate of 740 ug/g/hr for a two-week drying period [15]. An air sample taken near an oil fire was found to contain n-nonane at a concentration of 0.16 mg/m^3 [44]. Two sets of 6 air samples both resulted in average n-nonane concentrations of 5.83 and 5.39 mg/m^3 for a room applied with alkyl resin paint diluted with white spirits [66]. The concentration of n-nonane vapors averaged 0.6 ppm at excavation sites for removing gasoline storage tanks [57].

Probable Routes of Human Exposure: The most probable route of human exposure to n-nonane is by inhalation. Atmospheric workplace exposures have been documented [5,15,46,57,66,68]. Extensive monitoring data indicates n-nonane is a widely occurring atmospheric pollutant and breath samples have demonstrated n-nonane exposure among urban residents [29].

Average Daily Intake:

Occupational Exposure: NIOSH (NOHS Survey 1972-1974) has estimated that 1,232 workers are exposed to n-nonane in the USA [41]. NIOSH (NOES Survey 1981-1983) has estimated that 427 workers are exposed to n-nonane in the USA [40]. The atmospheric concentration of n-nonane ranged from 0 to 60 ug/m^3 for the vulcanization area of a shoe sole manufacturing plant; from 0 to 10 ug/m^3 for the vulcanization area and 0 to 1 ug/m^3 for the extrusion area of a tire retreading factory; and from 0 to 5 ug/m^3 for the extrusion area of electrical insulation manufacturing plant [5]. A 1984 study showed n-nonane was emitted from gasoline exposing outside operators at the refineries to an average air concentration of 0.162 mg/m^3; n-nonane was detected in 25 of 56 samples [46]. Transport drivers were exposed to n-nonane at an average atmospheric concentration of 0.019 mg/m^3 and n-nonane was detected in 34 of 49 samples [46]. Gas station attendants were exposed to n-nonane at an average atmospheric concentration of 0.006 mg/m^3 and n-nonane was detected in 17 of 49 samples [46]. n-Nonane was detected in the workplace atmospheres of 336 businesses in Belgium with a frequency of occurrence of 6% for those that utilize printing pastes; 21% where painting took place; 35% of the automobile repair shops and 24% for sites where various materials such as varnishes are employed [68]. Painters were exposed to an average atmospheric concentration of n-nonane of 6.4 mg/m^3 when applying an alkyl resin paint diluted with white spirits [66]. Tank contractors were exposed to vapors containing n-nonane at concentrations up to 0.2 ppm with an average concentration of 0.1 ppm at gasoline tank removal sites [57]. According to the National Ambient Volatile Organic Compounds (VOCs) Database, the median

446

Nonane

workplace atmospheric concentration of n-nonane is 26 ppbV for 26 samples [3].

Body Burdens: n-Nonane was detected in 8 of 12 samples of mothers' breast milk from the cities of Bayonne, NJ, Jersey City, NJ, Bridgeville, PA and Baton Rouge, LA [43]. The air expired from humans contained n-nonane in 36.7% of the 387 samples collected from 54 subjects [29]. The average n-nonane concentration of 0.308 ng/L was expressed as the geometric mean with upper and lower limits of 1.37 and 0.069 ng/L, respectively [29].

REFERENCES

1. Abrams EF et al; Identification of Organic Compounds in Effluents from Industrial Sources EPA-560/3-75-002 (1975)
2. Arnts RR, Meeks SA; Atmos Environ 15: 1643-51 (1981)
3. Atkinson R et al; Inter J Chem Kin 14: 781-8 (1982)
4. CLOGP, PCGEMS Graphical Exposure Modeling System USEPA (1986)
5. Cocheo V et al; Am Ind Hyg Assoc J 44: 521-7 (1983)
6. Colenutt BA, Thornburn S; Int J Environ Stud 15: 25-32 (1980)
7. Daubert TE, Danner RP; Data Compilation, Tables of Properties of Pure Cmpds, Design Inst for Phys Prop Data, Am Inst for Phys Prop Data NY, NY (1985)
8. DeBortoli M et al; Environ Int 12: 343-50 (1986)
9. Dumont JP, Adda J; J Agric Food Chem 26: 364-7 (1978)
10. Eiceman GA et al: Intern J Environ Anal Chem 24: 143-62 (1986)
11. Eisenreich SJ et al; Environ Sci Technol 15: 30-8 (1981)
12. Ferrario JB et al; Bull Environ Contam Toxicol 34: 246-55 (1985)
13. Fielding M et al; Organic Pollutants in Drinking Water, TR-159 Water Res Cent p 49 (1981)
14. Gerhold RM, Malaney GW; J Water Pollut Contr Fed 38: 562-79 (1966)
15. Girman JR et al; Environ Int 12: 317-21 (1986)
16. Grob K, Grob G; J Chromatogr 62: 1-13 (1971)
17. Grosjean D, Fung K; J Air Pollut Control Assoc 34: 537-43 (1984)
18. Hampton CV et al; Environ Sci Technol 16: 287-98 (1982)
19. Hampton CV et al; Environ Sci Technol 17: 699-708 (1983)
20. Ioffe BV et al; Environ Sci Technol 13: 864-8 (1979)
21. Ioffe BV et al; Dokl Akad Nauk Sssr 243: 1186-9 (1978)
22. Ioffe BV et al; J Chromatogr 142: 787-95 (1977)
23. Jarke FH et al; Ashrae Trans 87: 153-66 (1981)
24. Junk GA, Ford CS; Chemosphere 9: 187-230 (1980)
25. Juttner F; Z Wasser-Abwasser-Forrsch 21: 36-9 (1988)
26. Keith LH et al; pp 329-73 in Ident Anal Org Pollut Water, Keith LH (ed) Ann Arbor MI (1976)
27. Kinlin TE et al; J Agric Food Chem 20: 1021 (1972)

Nonane

28. Kool HJ et al; Crit Rev Env Control 12: 307-57 (1982)
29. Krotoszynski BK et al; J Anal Toxicol 3: 225-34 (1979)
30. LaRegina J et al; Environ Prog 5: 18-27 (1986)
31. Ligocki MP et al; Atmos Environ 19: 1607-17 (1985)
32. Linak, WP et al; JAPCA 39: 836-46 (1989)
33. Louw CW et al; Atmos Environ 11: 703-17 (1977)
34. Lyman WJ et al; Handbook of Chemical Property Estimation Methods NY: McGraw-Hill pp 4-9, 5-4, 10, 7-4 (1982)
35. Miller MM, Alkezweeny AJ; Ann NY Acad Sci 338: 219-32 (1980)
36. Mulcahy MFR et al; Paper IV p 17 in Occurrence Contr Photochem Pollut, Proc Symp Workshop Sess (1976)
37. Nagata S, Kondo G; Photooxidation of Crude Oils, Proc 1977 Oil Spill Conf, Am Petrol Inst pp 617-20 (1977)
38. Nelson PF, Quigley SM; Atmos Environ 18: 79-87 (1984)
39. Nelson PF, Quigley SM; Environ Sci Technol 16: 650-5 (1982)
40. NIOSH; National Occupational Exposure Survey (NOES) (1989)
41. NIOSH; National Occupational Hazard Survey (NOHS) (1974)
42. Nolting F et al; J Atmos Chem 6: 47-59 (1988)
43. Pellizzari ED et al; Bull Environ Contam Toxicol 28: 322-8 (1982)
44. Perry R; Mass Spectroscopy in the Detection and Identification of Air Pollutants, Int Symp Ident Meas Environ Pollut pp 130-7 (1971)
45. Perry DL et al; Ident of Org Compounds in Ind Effluent discharges. USEPA-600/4-79-016 p 230 (1979)
46. Rappaport SM et al; Appl Ind Hyg 2: 148-54 (1987)
47. Raymond A, Guiochon G; Environ Sci Technol 8: 143-8 (1974)
48. Rudolph J, Khedim A; Int J Environ Anal Chem 290: 265-82 (1985)
49. Sauer TC Jr et al; Mar Chem 7: 1-16 (1978)
50. Sauer TC Jr; Org Geochem 7: 1-16 (1981)
51. Sauer TC Jr; Org Geochem 3: 91-101 (1981)
52. Scheiman MA et al; Biomed Mass Spectrom 4: 209-11 (1974)
53. Scott Research Labs Inc; Atmospheric Reaction Studies in the Los Angeles Basin, NTIS PB-194-058 p 86 (1969)
54. Seila RL; Non-urban Hydrocarbons Concentrations in Ambient Air No of Houston TX USEPA-500/3-79-010 p 38 (1979)
55. Seila RL et al; Atmospheric Volatile Hydrocarbon Composition at Five Remote Sites in NW NC, USEPA-600/D-84-092 (1984)
56. Shah JJ, Heyerdahl EK; National Ambient VOC Database Update USEPA-600/3-88/010 (1988)
57. Shamsky S, Samimi B; App Ind Hyg 2: 242-5 (1987)
58. Shields HC, Weschler CJ; J Air Pollut Control Fed 37: 1039-45 (1987)
59. Silverstein RM, Bassler GC; Spectrometric Id of Org Cmpd, J Wiley and Sons Inc pp 148-69 (1963)
60. Smeyers-Verbeke J et al; Atmos Environ 18: 2471-8 (1984)
61. Stump FD, Dropkin DL; Anal Chem 57: 2629-34 (1985)
62. Swann RL et al; Res Rev 85: 16-28 (1983)
63. Tichenor BA, Mason MA; JAPCA 38: 264-8 (1988)

Nonane

64. USEPA; EXAMS II Computer Simulation (1987)
65. USEPA; Drinking water Criteria Document for Gasoline ECAO-CIN-D006, 8006-61-9 (1986)
66. Van der Wal JF, Moerkerken A; Ann Occup Hyg 28: 39-47 (1984)
67. Vogt WG, Walsh JJ Volatile Organic Compounds in Gases from Landfill Simulators Proc APCA Annu Meet 78th 6: 2-17 (1985)
68. Vuelemans H et al; Am Indust Hyg Assoc J 48: 671-7 (1987)
69. Yalkowsky SH et al; Arizona Data Base of Water Solubility (1987)
70. Young P, Parker A; Vapors Odors and Toxic Gases from Landfills ASTM Spec Tech Publ 851: 24-41 (1984)
71. Zimmerman RE et al; Landfill Methane Trace Volatile Org Constituent, Proc Int Gas Res Conf pp 230-9 (1983)

n-Octane

SUBSTANCE IDENTIFICATION

Synonyms:

Structure:

$$H_3C\diagup\diagdown\diagup\diagdown\diagup\diagdown CH_3$$

CAS Registry Number: 111-65-9

Molecular Formula: C_8H_{18}

Wiswesser Line Notation: 8H

CHEMICAL AND PHYSICAL PROPERTIES

Boiling Point: 125.6 °C

Melting Point: -56.8 °C

Molecular Weight: 114.22

Dissociation Constants:

Log Octanol/Water Partition Coefficient: 5.18 at 22 °C [23]

Water Solubility: 0.66 mg/L at 25 °C [36]

Vapor Pressure: 14.1 mm Hg at 25 °C [36]

Henry's Law Constant: 3.21 atm-m³/mole at 25 °C, calculated from VP/Wsol

ENVIRONMENTAL FATE/EXPOSURE POTENTIAL

Summary: n-Octane is a highly volatile constituent in the paraffin fraction of crude oil and natural gas. n-Octane is released to the environment via the manufacture, use, and disposal of many products

450

associated with the petroleum and gasoline industries. Extensive data show release of n-octane into the environment from printing pastes, paints, varnishes, adhesives, and other coatings; hazardous waste sites, landfills, and waste incinerators; vulcanization and extrusion operations during rubber and synthetic production; and the combustion of gasoline fueled engines. Photolysis or hydrolysis of n-octane are not expected to be important environmental fate processes. Biodegradation of n-octane may occur in soil and water; however, volatilization and adsorption are expected to be far more important fate processes. A high Koc indicates n-octane will be immobile in soil and may partition from the water column to organic matter contained in sediments and suspended solids. The bioconcentration of n-octane may be important in aquatic systems. The Henry's Law constant suggests rapid volatilization of n-octane from environmental waters. The volatilization half-lives from a model river and a model pond (the latter considers the effect of adsorption) have been estimated to be 3.1 hr and 29.8 days, respectively. n-Octane is expected to exist entirely in the vapor phase in ambient air. Reactions with photochemically produced hydroxyl radicals in the atmosphere have been shown to be important (estimated half-life of 1.84 days). Data also suggests that nighttime reactions with nitrate radicals may contribute to the atmospheric transformation of n-octane, especially in urban environments. The most probable route of human exposure to n-octane is by inhalation. Extensive monitoring data indicates n-octane is a widely occurring atmospheric pollutant.

Natural Sources: n-Octane is a constituent in the paraffin fraction of crude oil and natural gas [66].

Artificial Sources: n-Octane is released to the environment via the manufacture, use, and disposal of many products associated with the petroleum [47,48,54] and gasoline industries [48,59]. The combustion of gasoline fueled engines has been shown to release n-octane into the atmosphere [21,22,41]. Vulcanization and extrusion operations during rubber and synthetic production as with shoes, tires and electrical insulation also emit n-octane to the air [8]. Other well-documented materials responsible for the release of n-octane to the environment include printing pastes, paints, varnishes, adhesives, and other coatings [17,68,69,71]. Hazardous waste sites [32], landfills [70,73] and waste incinerators [29] also release n-octane into the environment.

n-Octane

Terrestrial Fate: Photolysis or hydrolysis [35] are not expected to be important fate processes for n-octane in terrestrial environments. The biodegradation of n-octane may occur in soils; however, volatilization and adsorption are expected to be far more important fate processes. An estimated range for Koc from 5500 to 15,600 [35] indicates n-octane should be immobile in most soils [63]. Based upon the estimated Henry's Law constant, n-octane is also expected to rapidly volatilize from moist surface soils [35].

Aquatic Fate: Photolysis or hydrolysis [35] are not expected to be important fate processes for n-octane in aquatic systems. The biodegradation of n-octane may occur in natural bodies of water; however, volatilization and adsorption are expected to be far more important fate processes. The bioconcentration factor (log BCF) for n-octane has been estimated to range from 2.89 to 3.71 [35], suggesting bioconcentration may be an important factor in aquatic systems. An estimated range for Koc from 5500 to 15,600 [35] indicates n-octane will strongly absorb to carbon [63] and may partition from the water column to organic matter contained in sediments and suspended solids. The estimated Henry's Law constant suggests rapid volatilization of n-octane from environmental waters [35]. Based on the Henry's Law constant, the volatilization half-life from a model river has been estimated to be 3.1 hr [35]. The volatilization half-life from a model pond, which considers the effect of adsorption, can be estimated to be about 29.8 days [67].

Atmospheric Fate: Based on the measured vapor pressure, n-octane is expected to exist entirely in the vapor phase in ambient air [13]. Direct photolysis of n-octane in the atmosphere is not expected to be an important fate process. However, vapor-phase reactions with photochemically produced hydroxyl radicals in the atmosphere have been shown to be important. The rate constant for n-octane was measured to be 8.86×10^{-12} cm^3/molecule-sec at 26 °C, which corresponds to an atmospheric half-life of about 1.81 days at an atmospheric concentration of $5 \times 10^{+5}$ hydroxyl radicals per cm^3 [5]. At 25 °C the average rate constant for two well-correlated experimental determinations was 8.72×10^{-12} cm^3/molecule-sec (half-life of 1.84 days) [4]. Experimental data also showed that 33.2% of the n-octane fraction in a dark chamber reacted with nitrate radicals to form the corresponding alkyl nitrate [2,3], suggesting nighttime reactions with nitrate radicals may contribute to the atmospheric transformation of n-octane, especially in urban environments.

Biodegradation: The theoretical oxygen demand of benzene-acclimated activated sludge for n-octane was 1.0, 4.6 and 28.4% after 6, 24, and 72 hr, respectively [37]. One mg of n-octane and 1 mL of a 1:10 suspension of Hudson-Collamer silt loam soil in mineral salts media was incubated in the dark at 25 °C [20]. Controls without n-octane were used to determine the net oxygen consumed [20]. The average BODT of 2 trials for n-octane was 13, 58, 70 and 69% after 2, 5, 10 and 20 .days, respectively [20]. At intervals of 6, 12 and 24 hr, endogenous respiration was greater than that of 3 preparations of n-octane and activated sludge from differing aeration units of sewage treatment facilities [16]. A 49% loss of n-octane loss occurred within 5 days and completely disappeared within 15 days from 1 mL of crude oil added to a 100 mL simulated seawater solution inoculated with sediment samples from Fukae of Kobe Harbor, Japan and incubated at 20 °C [40]. Loss of 19 and 67% of n-octane was observed within 5 and 15 days, respectively, from 1 mL of crude oil added to a 100 mL seawater soln collected at Fukae of Kobe Harbor, Japan and incubated at 20 °C [40]. Although complete recovery was reported for the control samples, no account was made of volatilization losses. In a similar study using a jet fuel mixture and freshwater from the Escambia River FL at 25 °C, a 99% loss of n-octane in the controls was attributed to evaporation [61].

Abiotic Degradation: Alkanes are generally resistant to hydrolysis [35]. Based on data for n-hexane and iso-octane, n-octane is not expected to absorb UV light in the environmentally significant range, >290 nm [60]. Therefore, n-octane probably will not undergo hydrolysis or direct photolysis in the environment. The rate constant for the vapor-phase reaction of n-octane with photochemically produced hydroxyl radicals was measured to be 8.86 x 10^{-12} cm^3/molecule-sec at 26 °C which corresponds to an atmospheric half-life of about 1.81 days at an atmospheric concentration of 5 x 10^{+5} hydroxyl radicals per cm^3 [5]. At 25 °C the average rate constant for two well-correlated experimental determinations was 8.72 x 10^{-12} cm^3/molecule-sec, which corresponds to an atmospheric half-life of 1.84 days [4]. At 39 °C the rate constant for the vapor-phase reaction of n-octane with photochemically produced hydroxyl radicals was determined to be 8.8 x 10^{-12} cm^3/molecule-sec, which corresponds to an atmospheric half-life of 1.8 days [44]. Experimental data showed that 33.2% of the n-octane fraction in a dark chamber reacted with nitrate radical to form the corresponding alkyl nitrate [2,3], suggesting nighttime reactions with nitrate radicals may contribute to the atmospheric transformation of n-octane, especially in urban environments.

Bioconcentration: Based upon the water solubility and log Kow, the bioconcentration factor (log BCF) for n-octane has been calculated to be 2.89 and 3.71, respectively, from recommended regression-derived equations [35]. These bioconcentration factor values indicate bioconcentration may be important fate process for n-octane in the environment.

Soil Adsorption/Mobility: Based on the water solubility and the log Kow, the Koc of n-octane has been calculated to range from 5500 to 15,600 from various regression-derived equations [35]. These Koc values indicate n-octane will be immobile in soil [63].

Volatilization from Water/Soil: The Henry's Law constant for n-octane indicates extremely rapid volatilization from environmental waters [35]. The volatilization half-life from a model river (1 meter deep flowing 1 m/sec with a wind speed of 3 m/sec) has been estimated to be 3.1 hr [35]. The volatilization half-life from a model pond, which considers the effect of adsorption, has been estimated to be 29.8 days [67]. In a biodegradation study using a jet fuel mixture and freshwater from the Escambia River FL at 25 °C, a 99% loss of n-octane in the controls was attributed to evaporation [61].

Water Concentrations: DRINKING WATER: n-Octane was identified in 4 of 14 treated water supplies in England [15]. n-Octane was listed as one of the many organic chemicals identified in drinking water in the US as of 1974 [1]. SURFACE WATER: The average n-octane concentration of 6 water samples from Little Britain Lake, England was 3.7 ppb [9]. The average n-octane concentration of 6 water samples from Welsh Harp Lake, England was 10.8 ppb [9]. The n-octane average concentration of weekly samples taken over an approximate period of 1 year for Luton Brook, England was 10.7 ppb [9]. The average n-octane concentration for water samples taken from the River Pinn at Brunel University, England was 3.1 ppb [9]. n-Octane was identified in Delaware River water at Philadelphia, PA [7]. SEAWATER: Only trace quantities to 3 ng/L of n-octane were detected in open surface waters of the north central Gulf of Mexico [55]. The average n-octane concentration of 3 surface water samples from an unpolluted coastal area of the north central Gulf of Mexico was 9 ng/L [55]. The n-octane concentration ranged from 30 to 40 ng/L for 6 surface water samples collected at coastal areas of the north central Gulf of Mexico under anthropogenic influence [55]. n-Octane was detected in 6 of 8 surface water samples in the Gulf of Mexico ranging in concentration from 0.3 to 3.2 ng/L with an average concentration of 1.5 ng/L [53]. RAIN/SNOW: The n-octane concentration of rain water collected at Brunel University, England was 39.2 ppb [9].

Effluent Concentrations: At a distance of 1 mile from its source, n-octane was detected at a concentration of 46.0 ug/m^3 in the plume emitted from a General Motors plant located in Janesville, Wisconsin [58]. n-Octane was identified as a stack emission from waste incinerators [29]. One of five hazardous waste sites listed on the National Priorities List emitted gaseous n-octane with a 25 to 50% frequency of occurrence [32]. n-Octane was also identified as a vapor emitted from landfills [70]. A clay pit landfill in England that received municipal, industrial and liquid wastes emitted n-octane gas at a concentration of 95 mg/L [73]. n-Octane was detected in 1 of 63 industrial wastewater effluents at a concentration less than 10 ug/L [47]. Underwater hydrocarbon vents and formation water discharges from offshore oil production platforms were found to contain n-octane concentrations in the vapor phase at 6 umol/L and in the liquid state at 350 ng/L, respectively [54]. Data from Sept 2, 1979 identified n-octane as a gaseous emission of the vehicle traffic through the Allegheny Mountain Tunnel of the Pennsylvania Turnpike [21]. Data from Aug 25 to Sept 7, 1979 showed for a speed of 80 km/hr on straight and level highway, gasoline powered vehicles emitted n-octane at an average rate of 6.7 mg/km and diesel trucks emitted n-octane at an average of 4.4 mg/km [22]. The average exhaust from 67 gasoline fueled vehicles was found to contain n-octane at a concentration 0.4% by weight [41].

Sediment/Soil Concentrations: The average n-octane concentration of 5 samples for the sediments of Lake Pontchartrain, a shallow oligohaline estuary located in the deltaic plain of the Mississippi River near New

Orleans, was 0.07 ppb [14]. n-Octane was detected in the soil surrounding an earthen disposal pit for produced water in the Duncan Oil Field, NM [12].

Atmospheric Concentrations: URBAN: In a 1979 study of 5 sites in NJ, n-octane was identified in the air over the cities of Rutherford and Elizabeth [6]. At ground level the atmospheric concentration of n-octane was 5.0, 5.0, 4.5 and 3.5 ug/m³ at 4, 7, 9 and 11 miles downwind of a General Motors plant in Janesville, Wisconsin [58]. The 1977 maximum and average concentration of 690 points for n-octane at a site in Houston, Texas was 19 and 1 ppb of carbon, respectively [38]. The average n-octane concentration from 6 to 9 am in Houston, Texas was 2 ppb of Carbon [38]. The average outdoor air concentration (24 samples) for n-octane in Los Angeles, CA during Feb and May 1984 was 3.9 and 0.7 ug/m³, respectively [72]. The average outdoor air concentration (10 samples) for n-octane in Contra Costa, CA during June 1984 was 0.5 ug/m³ [72]. The ground level atmospheric concentration of n-octane at 1325 was 1.4 ppb and 10.4 ppb at 8:00 AM for Huntington Park, CA [56]. At 1500 ft the n-octane concentration was 3.3 ppb at 0743 PST and at 8:07 AM at a height of 2,200 ft the n-octane concentration was 0.4 ppb [56]. The n-octane concentration ranged from 1 to 5 ppbV at a downtown Los Angeles location for the fall of 1981 [19]. The average atmospheric gas phase concentration of n-octane was 774 ng/m³ for 7 rain events in Portland Oregon from Feb to Apr 1984 [33]. The atmospheric n-octane concentration ranged from 0.004 to 0.11 ppbV at Niwot Ridge, Colorado [51]. The average n-octane concentration of air samples taken at Brunel University, England was 13.5 ppb [9]. n-Octane was also identified in the ambient air of Paris, France [49]. In 1983-4 the respective minimum, maximum and average outdoor air concentrations of n-octane in northern Italy were 1, 15, and 2.4 ug/L [3]. The concentration of n-octane in the downtown air of Zurich, Switzerland was 3.4 ppb [18]. At Deuselbach, Hunsruck in Germany the atmospheric n-octane concentration was 0.014 ppb for October 23, 1983 [52]. n-Octane was detected in the atmospheres of 6 industrialized cities of the USSR ranging in size of population from 0.4 to 4.5 million people [25,26,27]. The average n-octane concentrations in the atmospheres of Pretoria, Johannesburg and Durban, South Africa were 1.4, 1.3, and 0.7 ppb, respectively [34]. The average n-octane concentration (66 samples) in the air over Tokyo Japan for 1980 was 0.3 ppb [65]. n-Octane was identified in the ambient air of Sydney, Australia [39] ranging in concentration from 0.1 to 2.1 ppbV with an average concentration of 0.5 ppbV [42]. n-Octane was detected in the atmosphere over the British Columbia Research Council Laboratory at the University of British Columbia [62]. RURAL: The respective median, minimum and maximum atmospheric concentrations of n-octane for 5 rural locations in NC ranged from 0.1 to 1.4, 0.0 to 0.7 and 0.2 to 2.3 ppb [57]. INDOOR: n-Octane occurred in 43% of the indoor air samples taken in Chicago Illinois [28]. The same study determined the respective minimum, maximum, and average indoor air concentrations of n-octane for 14 homes and an office building were 1, 65, and 14 ug/L [10].

Food Survey Values: n-Octane was identified as a volatile component of fried bacon [24], roasted filberts [30], nectarines [64], and beaufort cheese [11].

Plant Concentrations:

Fish/Seafood Concentrations: The average n-octane concentration for the oyster population of Lake Pontchartrain, a shallow oligohaline estuary located in the deltaic plain of the Mississippi River near New Orleans, was 22 ng/g of wet weight [14].

Animal Concentrations:

Milk Concentrations: Octane was detected in 7 of 12 samples of mothers' breast milk from the cities of Bayonne, NJ, Jersey City, NJ, Bridgeville, PA and Baton Rouge, LA [45].

Other Environmental Concentrations: n-Octane was emitted from 8 adhesives used in building materials at an average rate of 740 ug/g/hr for a two-week drying period [17]. An air sample taken near an oil fire was found to contain n-octane at a concentration of 0.21 mg/m^3 [46]. n-Octane was also emitted from a Swedish floor finish 49 mo after its application [69]. Two sets of 6 air samples both resulted in an average n-octane concentration of 12 mg/m^3 for a room applied with alkyl resin paint diluted with white spirits [68]. The concentration of n-octane vapors averaged 4.1 ppm at excavation sites for removing gasoline storage tanks [59].

Probable Routes of Human Exposure: The most probable route of human exposure to n-octane is by inhalation. Atmospheric workplace exposures have been documented [8,17,48,59,68,69,71]. Extensive monitoring data indicates n-octane is a widely occurring atmospheric pollutant and breath samples have demonstrated n-octane exposure among urban residents [72].

Average Daily Intake:

Occupational Exposure: NIOSH (NOHS Survey 1972-1974) has statistically estimated that 1,035,579 workers are exposed to n-octane in the USA [43]. The atmospheric concentration of n-octane ranged from 0 to 300 ug/m^3 for the vulcanization area of a shoe sole manufacturing plant; from 6 to 90 ug/m^3 for the vulcanization area and 0 to 10 ug/m^3 for the extrusion area of a tire retreading factory; and from 0 to 1 ug/m^3 for the extrusion area of electrical insulation manufacturing plant [8]. A 1984 study showed n-octane was emitted from gasoline exposing outside operators at the refineries to an average air concentration of 0.118 mg/m^3; n-octane was detected in 26 of 56 samples [48]. Transport drivers were exposed to n-octane at atmospheric concentration of 0.086 mg/m^3 and n-octane was detected in 39 of 49 samples [48]. Gas station attendants were exposed to n-octane at atmospheric concentrations of 0.023 mg/m^3 and n-octane was detected in 26 of 49 samples [48]. n-Octane was detected in the workplace atmospheres of 336 businesses in Belgium with a frequency of occurrence of 3% for those that utilize printing pastes; 11% where painting took place; 55% of the automobile repair shops and 2% for sites where various materials such as varnishes are employed [71]. Painters were exposed to atmospheric concentrations of n-octane of 12 mg/m^3 when applying an alkyl resin paint diluted with white spirits [68]. Tank contractors were exposed to vapors containing n-octane at concentrations of 0.7 ppm at gasoline tank removal sites [59]. The following list includes some common operations in which exposure to octane may occur: preparation of gasoline and rocket fuel; manufacture of polymers, benzene, toluene, printing inks, and xylene aromatics; laboratory procedures and studies; combustion studies; and preparation of liquid soaps and detergents. The health hazard from occupational exposure to air pollutants in the cabin of the Columbia spacecraft was evaluated. More than 100 contaminant gases, including n-octane were

n-Octane

detected in the Shuttle cabin. For the most part, the concentrations of these gases were maintained below a toxicity hazard level [50].

Body Burdens: n-Octane was detected in 7 of 12 samples of mothers' breast milk from the cities of Bayonne, NJ, Jersey City, NJ, Bridgeville, PA and Baton Rouge, LA [45]. The air expired from humans contained n-octane in 27.7% of the 387 samples collected from 54 subjects [31]. The average n-octane concentration of 0.296 ng/L was expressed as the geometric mean with upper and lower limits of 1.10 and 0.079, respectively [31]. The breath samples of 110 residents from Los Angeles, CA in Feb 1984, contained n-octane at an average concentration of 1.0 ug/m^3 [72]. The average octane concentration in the breath of 24 Los Angeles residents in May 1984 was 1.2 ug/m^3 [72]. For residents of Contra Costa, CA in June 1984 the average n-octane concentration of 67 breath samples was 0.6 ug/m^3 [72].

REFERENCES

1. Abrams EF et al; Identification of Organic Compounds in Effluents from Industrial Sources. USEPA-560/3-75-002 (1975)
2. Atkinson R et al; J Phys Chem 86: 4563-9 (1982)
3. Atkinson R et al; Preprints Amer Chem Soc Div Environ Chem 23: 173-6 (1983)
4. Atkinson R; Chem Rev 85: 69-201 (1985)
5. Atkinson R et al; Internat J Chem Kin 14: 781-8 (1982)
6. Bozzelli JW et al; Analysis of Selected Toxic and Carcinogenic Substances in Ambient Air of NJ, State of NJ Dept Environ Protection (1980)
7. Burnham AK et al; Amer Water Works Assoc J 65: 722-5 (1973)
8. Cocheo V et al; Am Ind Hyg Assoc J 44: 521-7 (1983)
9. Colenutt BA, Thornburn S; Int J Environ Stud 15: 25-32 (1980)
10. DeBortoil M et al; Environ Int 12: 343-50 (1986)
11. Dumont JP, Adda J; J Agric Food Chem 26: 364-7 (1978)
12. Eiceman GA et al: Intern J Environ Anal Chem 24: 143-62 (1986)
13. Eisenreich SJ et al; Environ Sci Technol 15: 30-8 (1981)
14. Ferrario JB et al; Bull Environ Contam Toxicol 34: 246-55 (1985)
15. Fielding M et al; Organic Pollutants in Drinking Water. Medmenham, UK: Water Res Cent TR-159 p 49 (1981)
16. Gerhold RM, Malaney GW; J Water Pollut Contr Fed 38: 562-79 (1966)
17. Girman JR et al; Environ Int 12: 317-21 (1986)
18. Grob K, Grob G; J Chromatogr 62: 1-13 (1971)
19. Grosjean D, Fung K; J Air Pollut Control Assoc 34: 537-43 (1984)
20. Haines JR, Alexander M; Appl Microbiol 28: 1084-5 (1974)
21. Hampton CV et al; Environ Sci Technol 16: 287-98 (1982)
22. Hampton CV et al; Environ Sci Technol 17: 699-708 (1983)

23. Hansch C, Leo AJ; Medchem Project Issue No 26. Claremont CA: Pomona College (1985)
24. Ho CT et al; J Agric Food Chem 31: 336-42 (1983)
25. Ioffe BV et al; Dokl Akad Nauk Sssr 243: 1186-9 (1978)
26. Ioffe BV et al; Environ Sci Technol 13: 864-8 (1979)
27. Ioffe BV et al; J Chromatogr 142: 787-95 (1977)
28. Jarke FH et al; Ashrae Trans 87: 153-66 (1981)
29. Junk GA, Ford CS; Chemosphere 9: 187-230 (1980)
30. Kinlin TE et al; J Agric Food Chem 20: 1021 (1972)
31. Krotoszynski BK et al; J Anal Toxicol 3: 225-34 (1979)
32. LaRegina J et al; Environ Prog 5: 18-27 (1986)
33. Ligocki MP et al; Atmos Environ 19: 1607-17 (1985)
34. Louw CW et al; Atmos Environ 11: 703-17 (1977)
35. Lyman WJ et al; Handbook of Chemical Property Estimation Methods NY: McGraw-Hill pp 4-9, 5-4, 10, 7-4 (1982)
36. Mackay D, Shiu WY; J Phys Chem Ref Data 19: 1175-99 (1981)
37. Malaney GW, McKinney RE; Water Sewage Works 113: 302-9 (1966)
38. Monson PR et al; Paper 78-50.4 in 71st Ann Meeting Air Pollut Contr Assoc (1978)
39. Mulcahy MFR et al; Paper IV Occurrence Contr Photochem Pollut, Proc Symp Workshop Sess p 17 in (1976)
40. Nagata S, Kondo G; pp 617-20 in Proc 1977 Oil Spill Conf, Am Petrol Inst (1977)
41. Nelson PF, Quigley SM; Atmos Environ 18: 79-87 (1984)
42. Nelson PF, Quigley SM; Environ Sci Technol 16: 650-5 (1982)
43. NIOSH; National Occupational Hazard Survey (NOHS) (1974)
44. Nolting F et al; J Atmos Chem 6: 47-59 (1988)
45. Pellizzari ED et al; Bull Environ Contam Toxicol 28: 322-8 (1982)
46. Perry R; pp 130-7 in Int Symp Ident Meas Environ Pollut (1971)
47. Perry DL et al; Ident of Org Compounds in Ind Effluent Discharges. USEPA-600/4-79-016 (NTIS PB-294794) p 230 (1979)
48. Rappaport SM et al; Appl Ind Hyg 2: 148-54 (1987)
49. Raymond A, Guiochon G; Environ Sci Technol 8: 143-8 (1974)
50. Rippstein WJ Jr, Coleman ME; Aviat Space Environ Med 54 (12): Sect 2, 60-7 (1983)
51. Roberts JM et al; Atmos Environ 19: 1945-50 (1985)
52. Rudolph J, Khedim A; Int J Environ Anal Chem 290: 265-82 (1985)
53. Sauer TC Jr et al; Mar Chem 7: 1-16 (1978)
54. Sauer TC Jr; Org Geochem 7: 1-16 (1981)
55. Sauer TC Jr; Org Geochem 3: 91-101 (1981)
56. Scott Research Labs Inc; Atmospheric Reaction Studies in the Los Angeles Basin. NTIS PB-194-058 p 86 (1969)
57. Seila RL et al; Atmospheric Volatile Hydrocarbon Composition at Five Remote Sites in NW NC. USEPA-600/D-84-092 (NTIS PB84-177930) (1984)
58. Sexton K, Westberg H; Environ Sci Technol 14: 329-32 (1980)
59. Shamsky S, Samimi B; App Ind Hyg 2: 242-5 (1987)

n-Octane

60. Silverstein RM, Bassler GC; Spectrometric Id of Org Cmpd, J Wiley and Sons Inc pp 148-169 (1963)
61. Spain JC et al; Degrad of Jet Fuel Hydrocarbons by Aquatic Microbial Communities. Tyndall AFB,FL: Air Force Eng Serv Ctr. AFESC/ESL-TR-83-26 (NTIS AD-A139791/8) p 226 (1983)
62. Stump FD, Dropkin DL; Anal Chem 57: 2629-34 (1985)
63. Swann RL et al; Res Rev 85: 16-28 (1983)
64. Takeoka GR et al; J Agric Food Chem 36: 553-60 (1988)
65. Uno I et al; Atmos Environ 19: 1283-93 (1985)
66. USEPA; Drinking water Criteria Document for Gasoline. ECAO-CIN-D006, 8006-61-9 (1986)
67. USEPA; EXAMS II Computer Simulation (1987)
68. Van der Wal JF, Moerkerken A; Ann Occup Hyg 28: 39-47 (1984)
69. Van Netten et al; Bull Environ Contam Toxicol 40: 672-7 (1988)
70. Vogt WG, Walsh JJ; pp 2-17 in Proc APCA Annu Meet 78th Vol 6 (1985)
71. Vuelemans H et al; Am Indust Hyg Assoc J 48: 671-7 (1987)
72. Wallace LA; Toxicol Environ Chem 12: 215-36 (1986)
73. Young P, Parker A; ASTM Spec Tech Publ 851: 24-41 (1984)

1-Octanol

SUBSTANCE IDENTIFICATION

Synonyms:

Structure:

$$H_3C\diagdown\diagup\diagdown\diagup\diagdown\diagup\diagdown CH_2-OH$$

CAS Registry Number: 111-87-5

Molecular Formula: $C_8H_{18}O$

Wiswesser Line Notation: Q8

CHEMICAL AND PHYSICAL PROPERTIES

Boiling Point: 194-195 °C

Melting Point: -16 TO -17 °C

Molecular Weight: 130.22

Dissociation Constants:

Log Octanol/Water Partition Coefficient: 2.97 [10]

Water Solubility: 540 mg/L at 25 °C [2]

Vapor Pressure: 7.94 x 10^{-2} mm Hg at 25 °C [5]

Henry's Law Constant:

ENVIRONMENTAL FATE/EXPOSURE POTENTIAL

Summary: 1-Octanol is released to the environment as a natural constituent of plants and microbes. It may also be released to the environment through effluents at sites where it is produced or used in perfumery, cosmetics, organic synthesis, solvent manufacture of high boiling esters, antifoaming agents and in food flavoring. Photolysis or hydrolysis of 1-octanol is not expected to be environmentally important. 1-Octanol should biodegrade rapidly in soil and water. Differing estimates of Koc indicate a wide range of adsorption characteristics for 1-octanol and the mobility class in soil may range from low to high; it may partition from the water column to organic matter in sediments and suspended solids. The potential for bioconcentration of 1-octanol in aquatic organisms is low. The Henry's Law constant indicates that volatilization of 1-octanol from natural waters may be important. The volatilization half-lives from a model river and a model pond (the latter considers the effect of adsorption) have been estimated to be about 1.8 and 82 days, respectively. 1-Octanol is expected to exist entirely in the vapor phase in ambient air. Vapor-phase reactions with photochemically produced hydroxyl radicals in the atmosphere may be important (estimated half-life of 1.3 days). Physical removal from air via precipitation has been shown to occur. The most probable human exposure to 1-octanol would be occupational exposure, which may occur through dermal contact or inhalation at places where it is produced or used. Common nonoccupational exposures would include the ingestion of foods containing it.

Natural Sources: 1-Octanol is released to the environment as a natural constituent of plants and microbes [9].

Artificial Sources: 1-Octanol may be released to the environment via effluents at sites where it is produced or used in perfumes, cosmetics, organic synthesis, solvent manufacture of high boiling esters, antifoaming agents and in food flavoring [12]. 1-Octanol is also released to the environment in landfill leachate [11].

Terrestrial Fate: Photolysis or hydrolysis [17] of 1-octanol is not expected to be important in soils. 1-Octanol should biodegrade rapidly in terrestrial environments based upon aqueous biodegradation screening tests. An estimated range for Koc from 137 to 983 [17] indicates a wide range of adsorption characteristics for 1-octanol and the mobility class in soil may range from low to high [23]. Based upon the estimated Henry's Law constant, the volatilization of 1-octanol from moist surface soils may be important [23].

1-Octanol

Aquatic Fate: Photolysis or hydrolysis [17] of 1-octanol in aquatic systems is not expected to be important. 1-Octanol should biodegrade rapidly in aquatic environments. The log bioconcentration factor (log BCF) has been estimated to range from 1.25 to 2.03 [17] indicating the potential for 1-octanol to bioconcentrate in aquatic organisms is low. An estimated range for Koc from 137 to 983 [17] indicates 1-octanol may partition from the water column to organic matter contained in sediments and suspended solids. The estimated Henry's Law constant suggests volatilization of 1-octanol from environmental waters may be important [17]. Based on the Henry's Law constant, the volatilization half-life from a model river has been estimated to be 1.8 days [17]. The volatilization half-life from an model pond, which considers the effect of adsorption, has been estimated to be about 82 days [25].

Atmospheric Fate: Based on the vapor pressure, 1-octanol is expected to exist entirely in the vapor phase in ambient air [8]. Vapor-phase reactions with photochemically produced hydroxyl radicals in the atmosphere may be important. The rate constant for 1-octanol was measured to be 1.195×10^{-11} cm^3/molecule-sec at 25 °C, which corresponds to an atmospheric half-life of about 1.3 days at an atmospheric concentration of $5 \times 10^{+5}$ hydroxyl radicals per cm^3 [1]. Physical removal from air via precipitation has been shown to occur [4].

Biodegradation: River die-away tests and grab sample data pertaining to the biodegradation of 1-octanol in natural waters and soil were not located in the available literature. Yet, a number of aerobic and anaerobic biological screening studies, which utilized settled waste water, sewage, or activated sludge for inocula, have demonstrated that 1-octanol is readily biodegradable [6,13,20,21,26,28,29]. Five-day BOD tests show BODTs of 33 [2], 37 [13] and 62.4% [26]. Other screening test data yielded half-lives of 0.07 [29] and 0.90 days for 1-octanol. 1-Octanol was completely removed and lost 75% of the starting material during 49- and 56-day anaerobic acclimated studies [20,21]. These studies indicate 1-octanol should degrade rapidly under most environmental conditions.

Abiotic Degradation: Alcohols are generally resistant to hydrolysis [17]. Alcohols absorb UV light at wavelengths <185 nm, which is not in the environmentally significant range, >290 nm [22]. Likewise, alcohols are commonly used as solvents for obtaining UV spectra [22]. Therefore 1-

octanol probably will not undergo hydrolysis or direct photolysis in the environment. The rate constant for the vapor-phase reaction of 1-octanol with photochemically produced hydroxyl radicals in air has been estimated to be 1.195×10^{-11} cm^3/molecule-sec at 25 °C, which corresponds to an atmospheric half-life of about 1.3 days at an atmospheric concentration of $5 \times 10^{+5}$ hydroxyl radicals per cm^3 [1].

Bioconcentration: Based upon the water solubility and log Kow, the bioconcentration factor (log BCF) for 1-octanol has been calculated to be 1.25 and 2.03, respectively, from recommended regression-derived equations [17]. These BCF values indicate the potential for 1-octanol to bioconcentrate in aquatic organisms is low.

Soil Adsorption/Mobility: Based upon the water solubility and log Kow, respective Koc values of 137 and 983 for 1-octanol have been calculated from various regression-derived equations [17]. These Koc values indicate a wide range of adsorption characteristics for 1-octanol and the mobility class in soil may range from low to high [23].

Volatilization from Water/Soil: The Henry's Law constant indicates that volatilization of 1-octanol from natural waters may be important [17]. The volatilization half-life from a model river (1 meter deep flowing 1 m/sec with a wind speed of 3 m/sec) has been estimated to be 1.8 days [17]. The volatilization half-life from a model pond, which considers the effect of adsorption, has been estimated to be about 82 days [25].

Water Concentrations: DRINKING WATER: 1-Octanol was listed as a contaminant found in drinking water for a survey of US cities including Pomona, Escondido, Lake Tahoe and Orange Co, CA and Dallas, Washington, DC, Cincinnati, Philadelphia, Miami, New Orleans, Ottumwa, IA, and Seattle [16]. RAIN/SNOW: The 1-octanol concentration of rain water collected at Brunel University, England was 39.2 ppb [4].

Effluent Concentrations: 1-Octanol was qualitatively identified as a component of leachate from a municipal waste landfill in the Netherlands [11].

Sediment/Soil Concentrations:

1-Octanol

Atmospheric Concentrations:

Food Survey Values: 1-Octanol was identified as a volatile component of fried bacon [14], roasted filberts [15], nectarines [24], apple juice [27], beaufort cheese [7] and cassava products [6].

Plant Concentrations:

Fish/Seafood Concentrations:

Animal Concentrations:

Milk Concentrations:

Other Environmental Concentrations: The spent chlorination liquor from bleaching of sulphite paper pulp contained 1-octanol at a concentration of 1 g/ton of pulp [3].

Probable Routes of Human Exposure: The most probable route of human exposure to 1-octanol is by inhalation, dermal contact or ingestion.

Average Daily Intake:

Occupational Exposure: NIOSH (NOES Survey as of 3/28/89) has estimated that 64,880 workers are potentially exposed to 1-octanol in the USA [18]. Nonoccupational exposures may include the ingestion of foods containing it [6,7,14,15,24,27].

Body Burdens: According the National Human Adipose Tissue Survey (NHATS) of 1982, 1-octanol was found in 12 of 46 composite samples; 4 samples each from subjects in the age groups of 0-14, 15-44 and 44+, respectively [19].

REFERENCES

1. Atkinson R; Intern J Chem Kin 19: 799-828 (1987)
2. Barton AFM; Alcohols with Water. International Union of Pure and Applied Chemistry. Solubility Data Series. Vol. 15 p 438 (1984)
3. Carlberg GE et al; Sci Total Environ 48: 157-67 (1986)

1-Octanol

4. Colenutt BA, Thornburn S; Int J Environ Stud 15: 25-32 (1980)
5. Daubert TE, Danner RP; Data Compilation, Tables of Properties of Pure Cmpds, Design Inst for Phys Prop Data, Am Inst for Phys Prop Data, NY, NY (1989)
6. Dougan J et al; J Sci Food Agric 34: 874-84 (1983)
7. Dumont JP, Adda J; J Agric Food Chem 26: 364-7 (1978)
8. Eisenreich SJ et al; Environ Sci Technol 15: 30-8 (1981)
9. Graedel TE; Chemical Compounds in the Atmosphere. Academic Press p 440 (1978)
10. Hansch C, Leo AJ; Medchem Project Issue No 26. Claremont CA: Pomona College (1985)
11. Harmsen J; Water Res 17: 699-705 (1983)
12. Hawley GG; Condensed Chemical Dictionary 10th ed Van Nostrand Reinhold NY p 375 (1981)
13. Heukelekian H, Rand MC; J Water Pollut Control Assoc 29: 1040-53 (1955)
14. Ho CT et al; J Agric Food Chem 31: 336-42 (1983)
15. Kinlin TE et al; J Agric Food Chem 20: 1021 (1972)
16. Lucas SV; GC/MS Anal of Org in Drinking Water Concentrates and Advanced Treatment Concentrates Vol 1 USEPA-600/1-84-020A (NTIS PB85-128239) p 397 (1984)
17. Lyman WJ et al; Handbook of Chemical Property Estimation Methods NY: McGraw-Hill pp 4-9, 5-4, 10, 7-4 (1982)
18. NIOSH; National Occupational Exposure Survey (NOES) (1989)
19. Onstot, JD et al; Characterization of HRGC/MS unidentified peaks from the board scan analysis of the FY82 NHATS Composites. Vol. I. MRI 8823-A01 USEPA p 56 (1987)
20. Shelton DR, Tiedje JM; Development of Tests for Determining Anaerobic Biodegradation Potential USEPA 560/5-81-013 (NTIS PB84-166495) p 92 (1981)
21. Shelton DR, Tiedje JM; Appl Environ Microbiol 47: 850-7 (1981)
22. Silverstein RM, Bassler GC; Spectrometric Id Org Cmpd, J Wiley & Sons Inc pp 148-69 (1963)
23. Swann RL et al; Res Rev 85: 16-28 (1983)
24. Takeoka GR et al; J Agric Food Chem 36: 553-60 (1988)
25. USEPA; EXAMS II Computer Simulation (1987)
26. Wagner R; Vom Wasser 42: 271-305 (1974)
27. Yajima I et al; Agric Biol Chem 48: 849-55 (1984)
28. Yonezawa Y et al; Kogai Shigen Kenkyusho Iho 11: 77-82 (1981)
29. Yonezawa Y, Urushigawa Y; Chemosphere 8: 139-42 (1979)

2-Octanol

SUBSTANCE IDENTIFICATION

Synonyms:

Structure:

CAS Registry Number: 123-96-6

Molecular Formula: $C_8H_{18}O$

Wiswesser Line Notation:

CHEMICAL AND PHYSICAL PROPERTIES

Boiling Point: 178.5 °C

Melting Point: 38.6 °C

Molecular Weight: 130.3

Dissociation Constants:

Log Octanol/Water Partition Coefficient: 2.72 [4]

Water Solubility: 1120 mg/L at 25 °C [1]

Vapor Pressure: 2.42 x 10^{-1} mm Hg at 25 °C [6]

Henry's Law Constant: 3.70 x 10^{-5} atm-m^3/mole at 25 °C, calculated from VP/Wsol

ENVIRONMENTAL FATE/EXPOSURE POTENTIAL

Summary: 2-Octanol may also be released to the environment through effluents at sites where it is produced or used in perfumery, organic synthesis, solvents, manufacture of plasticizers, antifoaming agents, petroleum additives, hydraulic oils and in masking industrial odors. Photolysis or hydrolysis of 2-octanol is not expected to be environmentally important. 2-Octanol should biodegrade rapidly in acclimated soil and water. Differing estimates of Koc indicate a wide range of adsorption characteristics for 2-octanol and the mobility class in soil may range from low to high; it may partition from the water column to organic matter in sediments and suspended solids. The potential for bioconcentration of 2-octanol in aquatic organisms is low. The estimated Henry's Law constant suggests volatilization of 2-octanol from environmental waters may be important. The volatilization half-lives from a model river and a model pond (the latter considers the effect of adsorption) have been estimated to be about 1.3 and 46 days, respectively. 2-Octanol is expected to exist entirely in the vapor phase in ambient air. Vapor-phase reactions with photochemically produced hydroxyl radicals in the atmosphere may be important (estimated half-life of 1.1 days). The most probable human exposure to 2-octanol would be occupational exposure, which may occur through dermal contact or inhalation at places were it is produced or used. Common nonoccupational exposures would include the ingestion of foods containing it.

Natural Sources: 2-Octanol has been identified as a constituent in the oil of reunion geranium; it also has been identified in a few species of mint and lavender.

Artificial Sources: 2-Octanol may be released to the environment via effluents at sites where it is produced or used in perfumery, organic synthesis, solvents, manufacture of plasticizers, antifoaming agents, petroleum additives, hydraulic oils and in masking industrial odors [13]. 2-Octanol is also released to the environment in landfill leachate [9] and wastewaters from sewage treatment facilities [12].

Terrestrial Fate: Photolysis or hydrolysis [16] of 2-octanol is not expected to be important in soils. 2-Octanol should biodegrade rapidly in acclimated terrestrial environments. An estimated range for Koc from 92 to 718 [16] indicates a wide range of adsorption characteristics for 2-octanol and the mobility class in soil may range from low to high [21]. Based upon the estimated Henry's Law constant, the volatilization of 2-octanol from moist surface soils may be important [21].

2-Octanol

Aquatic Fate: Photolysis or hydrolysis [16] of 2-octanol in aquatic systems is not expected to be important. 2-Octanol should biodegrade rapidly in acclimated aquatic environments. The log bioconcentration factor (log BCF) has been estimated to range from 1.07 to 1.84 [16] indicating the potential for 2-octanol to bioconcentrate in aquatic organisms is low. An estimated range for Koc from 92 to 718 [16] indicates 2-octanol may partition from the water column to organic matter contained in sediments and suspended solids. The estimated Henry's Law constant suggests volatilization of 2-octanol from environmental waters may be important [16]. Based on the Henry's Law constant, the volatilization half-life from a model river has been estimated to be 1.3 days [16]. The volatilization half-life from a model pond, which considers the effect of adsorption, has been estimated to be about 46 days [22].

Atmospheric Fate: Based on the vapor pressure, 2-octanol is expected to exist entirely in the vapor phase in ambient air [10]. Vapor-phase reactions with photochemically produced hydroxyl radicals in the atmosphere may be important. The rate constant for 2-octanol was measured to be 1.21×10^{-11} cm^3/molecule-sec at 25 °C, which corresponds to an atmospheric half-life of about 1.1 days at an atmospheric concentration of $5 \times 10^{+5}$ hydroxyl radicals per cm^3 [2]. 2-Octanol in air is not expected to undergo direct photolysis.

Biodegradation: River die-away tests and grab sample data pertaining to the biodegradation of 2-octanol in natural waters and soil were not located in the available literature. Yet, a number of aerobic and anaerobic biological screening studies, which utilized settled waste water, sewage, or activated sludge for inocula, have demonstrated that 2-octanol is biodegradable [3,11,18,19,23]. These studies indicate 2-octanol should degrade rapidly when acclimation has occurred.

Abiotic Degradation: Alcohols are generally resistant to hydrolysis [16]. Alcohols absorb UV light at wavelengths <185 nm, which is not in the environmentally significant range, >290 nm [20]. Likewise, alcohols are commonly used as solvents for obtaining UV spectra [20]. Therefore, 2-octanol probably will not undergo hydrolysis or direct photolysis in the environment. The rate constant for the vapor-phase reaction of 2-octanol with photochemically produced hydroxyl radicals in air has been estimated to be 1.21×10^{-11} cm^3/molecule-sec at 25 °C, which corresponds to an atmospheric half-life of about 1.1 days at an atmospheric concentration of $5 \times 10^{+5}$ hydroxyl radicals per cm^3 [2].

Bioconcentration: Based upon the water solubility and log Kow, the bioconcentration factor (log BCF) for 2-octanol has been calculated to be 1.07 and 1.84, respectively, from recommended regression-derived equations [16]. These BCF values indicate the potential for 2-octanol to bioconcentrate in aquatic organisms is low.

Soil Adsorption/Mobility: Based upon the water solubility and log Kow, respective Koc values of 92 and 718 for 2-octanol have been calculated from various regression-derived equations [16]. These Koc values indicate a wide range of adsorption characteristics for 2-octanol and the mobility class in soil may range from low to high [21].

Volatilization from Water/Soil: The Henry's Law constant indicates that volatilization of 2-octanol from natural waters may be important [16]. The volatilization half-life from a model river (1 meter deep flowing 1 m/sec with a wind speed of 3 m/sec) has been estimated to be 1.3 days [16]. The volatilization half-life from a model pond, which considers the effect of adsorption, has been estimated to be about 46 days [22].

Water Concentrations: GROUND WATER: 2-Octanol was qualitatively identified as a component of ground water below a municipal waste landfill in Norman, OK [9].

Effluent Concentrations: 2-Octanol was qualitatively identified as a component of leachate from a municipal waste landfill in Norman, OK [9]. The effluent of a Los Angeles County municipal sewage treatment facility contained 2-octanol at a concentration of 10 ug/L [12].

Sediment/Soil Concentrations:

Atmospheric Concentrations:

Food Survey Values: 2-Octanol was identified as a volatile component of mountain [8] and blue [7] cheeses.

2-Octanol

Plant Concentrations:

Fish/Seafood Concentrations:

Animal Concentrations:

Milk Concentrations:

Other Environmental Concentrations:

Probable Routes of Human Exposure: The most probable route of human exposure to 2-octanol is by inhalation, dermal contact or ingestion.

Average Daily Intake:

Occupational Exposure: NIOSH (NOES Survey as of 3/28/89) has estimated that 2,357 workers are potentially exposed to 2-octanol in the USA [17]. Nonoccupational exposures may include the ingestion of foods containing it [7,8].

Body Burdens: The air expired from humans contained n-octane [5,14] in 32.3% of the 387 samples collected from 54 subjects [15]. The average n-octanol concentration of 0.89 ng/L was expressed as the geometric mean with lower and upper limits of 0.26 and 3.1, respectively.

REFERENCES

1. Amidon GL et al; J Pharm Sci 63: 1858-66 (1974)
2. Atkinson R; Intern J Chem Kin 19: 799-828 (1987)
3. Babeu L, Vaishnav D; J Indust Microb 2: 107-15 (1987)
4. CLOGP; PCGEMS Graphical Exposure Modeling System USEPA (1986)
5. Conkle JP et al: Arch Environ Health 30: 290-5 (1975)
6. Daubert TE, Danner RP; Data Compilation, Tables of Properties of Pure Cmpds, Design Inst for Phys Prop Data, Am Inst for Phys Prop Data, NY, NY (1989)
7. Day EA, Anderson; J Agr Food Chem 13: 2-4 (1965)
8. Dumont JP, Adda J; J Agric Food Chem 26: 364-7 (1978)
9. Dunlap WJ et al; Identif Anal Org Pollut 1: 453-77 (1975)
10. Eisenreich SJ et al; Environ Sci Technol 15: 30-8 (1981)
11. Gerhold RM, Malaney GW; J Water Pollut Contr Fed 38: 562-79 (1966)

2-Octanol

12. Gossett RW et al; Mar Pollut Bull 14: 387-92 (1983)
13. Hawley GG; Condensed Chemical Dictionary 10th ed Van Nostrand Reinhold NY p 375 (1981)
14. Krotoszynski BK et al; J Chromat Sci 15: 239-44 (1977)
15. Krotoszynski BK et al; J Anal Toxicol 3: 225-34 (1979)
16. Lyman WJ et al; Handbook of Chemical Property Estimation Methods NY: McGraw-Hill pp 4-9, 5-4, 10, 7-4 (1982)
17. NIOSH; National Occupational Exposure Survey (NOES) (1989)
18. Shelton DR, Tiedje JM; Development of Tests for Determining Anaerobic Biodegradation Potential USEPA-560/5-81-013 (NTIS PB84-166495) p 92 (1981)
19. Shelton DR, Tiedje JM; Appl Environ Microbiol 47: 850-7 (1981)
20. Silverstein RM, Bassler GC; Spectrometric Id Org Cmpd, J Wiley & Sons Inc pp 148-69 (1963)
21. Swann RL et al; Res Rev 85: 16-28 (1983)
22. USEPA; EXAMS II Computer Simulation (1987)
23. Yonezawa Y, Urushigawa Y; Chemosphere 8: 139-42 (1979)

1,1'-Oxydi-2-propanol

SUBSTANCE IDENTIFICATION

Synonyms: 2,2'-Dihydroxydipropyl ether

Structure:

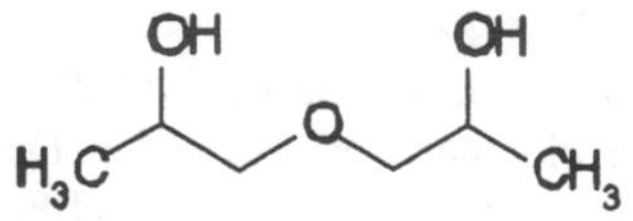

H₃C—CH(OH)—CH₂—O—CH₂—CH(OH)—CH₃

CAS Registry Number: 110-98-5

Molecular Formula: $C_6H_{14}O_3$

Wiswesser Line Notation:

CHEMICAL AND PHYSICAL PROPERTIES

Boiling Point: 233 °C

Melting Point:

Molecular Weight: 134.17

Dissociation Constants:

Log Octanol/Water Partition Coefficient: -0.687 [4] (estimated)

Water Solubility: Miscible [11]

Vapor Pressure: 0.01 mm Hg at 20 °C [10]

Henry's Law Constant: 3.58×10^{-9} atm-m³/mole at 25 °C [7] (estimated)

ENVIRONMENTAL FATE/EXPOSURE POTENTIAL

Summary: 1,1'-Oxydi-2-propanol is one of the isomers found in commercial dipropylene glycol (see p. 263), and therefore, may be released at sites where dipropylene glycol is produced or used as a solvent and plasticizer. 1,1'-Oxydi-2-propanol has been detected in groundwater near a landfill. If released to air, 1,1'-oxydi-2-propanol will degrade relatively rapidly by reaction with photochemically produced hydroxyl radicals (estimated half-life of 13 hr). Physical removal from air via wet deposition is possible since it is miscible in water. If released to soil, 1,1'oxydi-2-propanol is expected to leach readily. The rate or method of degradation in soil or water is currently uncertain due to lack of experimental data. There are no identifiable abiotic degradation processes. Insufficient data are available to predict the importance of biodegradation. Occupational exposure to 1,1'-oxydi-2-propanol can occur through dermal contact.

Natural Sources:

Artificial Sources:

Terrestrial Fate: The rate or method of 1,1'-oxydi-2-propanol degradation in soil is currently uncertain due to lack of experimental data. There are no identifiable abiotic degradation processes in soil. Insufficient data are available to predict the importance of biodegradation. 1,1'-Oxydi-2-propanol is miscible in water and is therefore expected to leach readily. It has been identified as a landfill leachate [2].

Aquatic Fate: The rate or method of 1,1'-oxydi-2-propanol degradation in water is currently uncertain due to lack of experimental data. There are no identifiable abiotic degradation processes in water. Insufficient data are available to predict the importance of biodegradation. Bioconcentration in aquatic organisms and adsorption to sediment are not expected to be important environmental fate processes because 1,1'-oxydi-2-propanol is miscible in water. Aquatic hydrolysis and volatilization are also not expected to be important.

Atmospheric Fate: Based upon the vapor pressure, 1,1'-oxydi-2-propanol is expected to exist primarily in the vapor phase in the ambient atmosphere [3]. Vapor-phase 1,1'-oxydi-2-propanol is expected to degrade readily in an average ambient atmosphere (estimated half-life of about 13 hr) by reaction with photochemically produced hydroxyl radicals [1]. 1,1'-Oxydi-2-propanol's miscibility in water suggests that physical removal from air via wet deposition is possible.

1,1'-Oxydi-2-propanol

Biodegradation:

Abiotic Degradation: The rate constant for the vapor-phase reaction of 1,1'-oxydi-2-propanol with photochemically produced hydroxyl radicals has been estimated to be 29.7 x 10^{-11} cm^3/molecule-sec at 25 °C which corresponds to an atmospheric half-life of about 13 hr at an atmospheric concn of 5 x 10^{+5} hydroxyl radicals per cm^3 [1]. Ethers and glycols are generally resistant to aqueous environmental hydrolysis [6]; therefore, 1,1'-oxydi-2-propanol is not expected to hydrolyze in environmental waters.

Bioconcentration: 1,1'-Oxydi-2-propanol is completely miscible in water which suggests that bioconcentration in aquatic organisms will not be important.

Soil Adsorption/Mobility: 1,1'-Oxydi-2-propanol is completely miscible in water which indicates that the compound will be readily mobile in soil.

Volatilization from Water/Soil: The Henry's Law constant for 1,1'-oxydi-2-propanol suggests that 1,1'-oxydi-2-propanol is essentially nonvolatile from water [6].

Water Concentrations: DRINKING WATER: 1,1'-Oxydi-2-propanol was qualitatively detected in drinking water concentrates collected in Cincinnati, OH on Oct 17, 1978 and Jan 14, 1980 [5]. GROUND WATER: 1,1'-Oxydi-2-propanol was qualitatively detected in ground water samples collected from a well adjacent to a landfill facility in Norman, OK [2]; the occurrence of 1,1'-oxydi-2-propanol in the well water is probably the result of leaching from the landfill [2].

Effluent Concentrations:

Sediment/Soil Concentrations:

Atmospheric Concentrations:

Food Survey Values:

Plant Concentrations:

476

1,1'-Oxydi-2-propanol

Fish/Seafood Concentrations:

Animal Concentrations:

Milk Concentrations:

Other Environmental Concentrations:

Probable Routes of Human Exposure: Occupational exposure to 1,1'-oxydi-2-propanol can occur through inhalation of vapor and dermal contact [9].

Average Daily Intake:

Occupational Exposure: NIOSH (NOES Survey 1981-1983) has statistically estimated that 2,961 workers are potentially exposed to 1,1'-oxydi-2-propanol in the USA [8].

Body Burdens:

REFERENCES

1. Atkinson R; Intern J Chem Kinet 19: 799-828 (1987)
2. Dunlap WJ et al; Identif Anal Org Pollut 1: 453-77 (1975)
3. Eisenreich SJ et al; Environ Sci Technol 15: 30-8 (1981)
4. GEMS; Graphical Exposure Modeling System PCGEMS (1987)
5. Lucas SV; GC/MS Analysis of Organics in Drinking Water Concentrates and Advanced Waste Treatment Concentrates Vol 1 USEPA-600/1-84-020A, NTIS PB85-128221 pp 45, 154 (1984)
6. Lyman WJ et al; Handbook of Chemical Property Estimation Methods. Environmental Behavior of Organic Compounds. Washington DC: American Chemical Society pp 4-9, 5-4, 5-10, 7-4, 7-5, 15-15 to 15-32 (1990)
7. Meylan WM and Howard PH; Environmental Toxicology and Chemistry 10:1283-1293 (1991)
8. NIOSH; National Occupational Exposure Survey (NOES) (1983)
9. Parmeggiani L; Encyl Occup Health & Safety 3rd ed Geneva, Switzerland: International Labour Office pp 973-4 (1983)
10. Sax NI, Lewis RJ Jr; Hawley's Condensed Chemical Dictionary 11th ed NY, NY: Van Nostrand Reinhold Co p.430 (1987)
11. Union Carbide; Chemicals and Plastics Physical Properties 1979-80, NY: Union Carbide Corp p 15 (1979)

n-Pentane

Synonyms:

Structure:

$$H_3C\diagup\diagdown\diagup CH_3$$

CAS Registry Number: 109-66-0

Molecular Formula: C_5H_{12}

Wiswesser Line Notation: 5H

CHEMICAL AND PHYSICAL PROPERTIES

Boiling Point: 36.1 °C

Melting Point: -129.7 °C

Molecular Weight: 72.15

Dissociation Constants:

Log Octanol/Water Partition Coefficient: 3.39 [20]

Water Solubility: 38.5 mg/L at 25 °C [34]

Vapor Pressure: 513 mm Hg at 25 °C [34]

Henry's Law Constant: 1.26 atm-m^3/mole at 25 °C, calculated from VP/Wsol

ENVIRONMENTAL FATE/EXPOSURE POTENTIAL

Summary: n-Pentane is a highly volatile constituent in the paraffin fraction of crude oil and natural gas. n-Pentane is released to the environment via the manufacture, use and disposal of many products

n-Pentane

associated with the petroleum and gasoline industries. Extensive data show release of n-pentane into the environment from hazardous waste disposal sites, landfills and waste incinerators and from the combustion of gasoline and diesel fueled engines. Photolysis, hydrolysis and bioconcentration of n-pentane are not expected to be important environmental fate processes. Biodegradation of n-pentane may occur in soil and water; however, volatilization and to some extent adsorption are expected to be far more an important environmental fate processes. A Koc range of 580 to 1600 indicates a low mobility class in soil for n-pentane. In aquatic systems, n-pentane may partition from the water column to organic matter contained in sediments and suspended materials. The Henry's Law constant suggests rapid volatilization of n-pentane from environmental waters. The volatilization half-lives from a model river and a model pond (the latter considers the effect of adsorption) have been estimated to be 2.5 hr and 3.5 days, respectively. n-Pentane is expected to exist entirely in the vapor phase in ambient air. Reactions with photochemically produced hydroxyl radicals in the atmosphere have been shown to be important (average half-life of 4.1 days). Data also suggests that nighttime reactions with nitrate radicals may contribute to the atmospheric transformation of n-pentane, especially in urban environments. The most probable route of human exposure to n-pentane is by inhalation. Extensive monitoring data indicates n-pentane is a widely occurring atmospheric pollutant.

Natural Sources: n-Pentane is a constituent in the paraffin fraction of crude oil and natural gas [69].

Artificial Sources: n-Pentane is released to the environment via the manufacture, use, and disposal of many products associated with the petroleum [4,48,50,52] and gasoline industries [4,6,50,58]. The combustion of gasoline and diesel fueled engines has been shown to release n-pentane into the atmosphere [10,19,40,60]. Hazardous waste disposal sites [31], landfills [14,70,72] and waste incinerators [11,26] also release n-pentane.

Terrestrial Fate: Photolysis [61] or hydrolysis [33] of n-pentane are not expected to be important in terrestrial environments. The biodegradation of n-pentane may occur in soils; however, primarily volatilization and to some extent adsorption are expected to be far more important fate processes. A calculated Koc range of 580 to 1600 [33] indicates a low mobility class for n-pentane in soils [65]. Based upon the estimated Henry's Law constant, n-pentane is expected to rapidly volatilize from surface soils [33].

Aquatic Fate: Photolysis or hydrolysis [33] of n-pentane are not expected to be important in aquatic systems. The biodegradation of n-pentane may occur in aquatic environments; however, primarily volatilization and to some extent adsorption are expected to be far more important fate processes. The log bioconcentration factor (log BCF) for n-pentane has been estimated to range from 1.90 to 2.35 [33] suggesting n-pentane will not bioconcentrate among aquatic organisms. An estimated range for Koc from 580 to 1600 [33] indicates n-pentane may absorb to carbon [65] and may partition from the water column to organic matter contained in sediments and suspended materials. The estimated Henry's Law constant suggests rapid volatilization of n-pentane from environmental waters [33]. Based on this Henry's Law constant, the volatilization half-life from a model river has been estimated to be 2.5 hr [33]. The volatilization half-life from a model pond, which considers the effect of adsorption, can be estimated to be about 3.5 days [68].

Atmospheric Fate: Based on the vapor pressure, n-pentane is expected to exist entirely in the vapor phase in ambient air [15]. n-Pentane does not absorb UV light in the environmentally significant range, >290 nm [61] and probably will not undergo direct photolysis in the atmosphere. However, vapor phase reactions with photochemically produced hydroxyl radicals in the atmosphere have been shown to be important. Rate constants for n-pentane were measured to be 4.06×10^{-12}, 5.30×10^{-12} and 3.51×10^{-12} cm^3/molecule-sec at 26 [5], 27 [14] and 27 [13] °C, respectively, which correspond to atmospheric half-lives of about 3.9 [5], 3.0 [14] and 4.5 [13] days at an atmospheric concentration of $5 \times 10^{+5}$ hydroxyl radicals per cm^3. Experimental data showed that 12.9% of the n-pentane fraction in a dark chamber reacted with NO_3 to form the corresponding alkyl nitrate [6,7], suggesting nighttime reactions with nitrate radicals may contribute to the atmospheric transformation of n-pentane, especially in urban environments.

Biodegradation: The theoretical oxygen demand of benzene-acclimated activated sludge for n-pentane was 0.3, 3.2, and 10.1% after 6, 24, and

72 hr, respectively [35]. The theoretical oxygen demand was 0.7, 0.5 and 0.7% at intervals of 6, 12, and 24 hr respectively, for 3 preparations of n-pentane and activated sludge from differing aeration units of sewage treatment facilities [16]. Within 24 hr n-pentane was oxidized to its corresponding methyl ketone, 2-pentanone [21,45], and the corresponding alcohol, 2-pentanol [21,46], by cell suspensions of over 20 methyltrophic organisms isolated from lake water and soil samples [21,45,46].

Abiotic Degradation: Alkanes are generally resistant to hydrolysis [33]. Based on data for iso-octane and n-hexane, n-pentane is not expected to absorb UV light in the environmentally significant range, >290 nm [61]. Therefore, n-pentane probably will not undergo hydrolysis or direct photolysis in the environment. An air sample's n-pentane concentration of 171 ppbC was reduced by 26% within 6 hr of irradiation by natural sunlight in downtown Los Angeles, CA [29]. The rate constants for the vapor-phase reaction of n-pentane with photochemically produced hydroxyl radicals was measured to be 4.06×10^{-12}, 5.30×10^{-12} and 3.51×10^{-12} cm^3/molecule-sec at 26 [5], 27 [14] and 27 [13] °C, respectively, which correspond to atmospheric half-lives of about 3.9 [5], 3.0 [14] and 4.5 [13] days at an atmospheric concentration of $5 \times 10^{+5}$ hydroxyl radicals per cm^3. At 39 °C the rate constant for the vapor-phase reaction of n-pentane with photochemically produced hydroxyl radicals was determined to be 4.29×10^{-12} cm^3/molecule-sec, which corresponds to an atmospheric half-life of 3.7 days [44]. Experimental data showed that 12.9% of the n-pentane fraction in a dark chamber reacted with nitrate radical to form the corresponding alkyl nitrate [6,7], suggesting nighttime reactions with nitrate radicals may contribute to the atmospheric transformation of n-pentane, especially in urban environments.

Bioconcentration: Based upon the water solubility and the log Kow, the bioconcentration factor (log BCF) for n-pentane has been calculated to be 1.90 and 2.35, respectively, from recommended regression-derived equations [33]. These BCF values are not indicative of important bioconcentration in aquatic organisms.

Soil Adsorption/Mobility: Based on the water solubility and the log Kow, the Koc of n-pentane has been calculated to range from 580 to 1600 from various regression-derived equations [33]. These Koc values indicate a slight soil mobility class for n-pentane in soils [65].

n-Pentane

Volatilization from Water/Soil: The Henry's Law constant indicates rapid volatilization from environmental waters [33]. The volatilization half-life from a model river (1 meter deep flowing 1 m/sec with a wind speed of 3 m/sec) has been estimated to be 2.5 hr [33]. The volatilization half-life from a model pond, which considers the effect of adsorption, has been estimated to be 3.5 days [68].

Water Concentrations: DRINKING WATER: n-Pentane was listed as one of the many organic chemicals identified in drinking water in the US as of 1974 [1,28,30]. SURFACE WATER: n-Pentane was identified in Delaware River water at Philadelphia, PA [17]. n-Pentane was listed as a contaminant present in the waters of Lake Ontario [17]. The Inner Harbor Navigation Canal of Lake Pontchartrain at New Orleans, LA was found to contain n-pentane at an average concentration for 8 samples of 2.4 ppb on May 6, 1980 [36]. SEAWATER: All 8 near-surface seawater samples from the intertropical Indian Ocean contained small amounts of n-pentane [8]. GROUND WATER: One of 11 ground water monitoring wells near the Granby Landfill,CT contained n-pentane at a concentration of 20 ppb [53].

Effluent Concentrations: n-Pentane was identified as a stack emission from waste incinerators [26]. Flue gases from a waste incinerator at Babylon, Long Island, NY was found to emit n-pentane at concentrations generally less than 0.7 ppm [11]. Three of five hazardous waste sites listed on the National Priorities List emitted gaseous n-pentane with a 25 to 75% frequency of occurrence [31]. n-Pentane was also identified as a vapor emitted from landfills with an average and maximum concentration of 0.4 and 5.0 ppm, respectively [70]. Land fill gas was found to contain n-pentane at a concentration of 0.4 ppmV [72]. One of 11 ground water monitoring wells near the Granby Landfill, CT contained n-pentane at a concentration of 20 ppb [53]. A refinery located in Tulsa, OK was attributed with emissions to the surrounding atmosphere where the n-pentane concentration was measured to be 46.4 and 81.7 ppbC for two min before and after 1:33 PM [4]. The n-pentane content of the air downwind of a Mobil natural gas facility in Rio Blanco, CO was 68.2 ppbC [4]. n-Pentane was detected in 1 of 63 industrial wastewater effluents at a concentration between 10 and 100 ug/L [48]. Underwater hydrocarbon vent discharges from offshore oil production platforms were found to contain n-pentane concentrations in the vapor phase at 260

n-Pentane

umol/L of gas [52]. Data from Sept 2, 1979 identified n-pentane as a gaseous emission of the vehicle traffic through the Allegheny Mountain Tunnel of the Pennsylvania Turnpike [19]. The average exhaust from 67 gasoline fueled vehicles was found to contain n-pentane at a concentration 3.3% by weight [41]. n-Pentane from car exhaust ranged in concentration from 0.03 to 0.07 ppmV with an average for 8 samples of 0.04 ppmV [40]. The average concentration of n-pentane for the exhaust of 46 automobiles was 1.7 weight % of total hydrocarbon according to both the federal test procedure and the New York City cycle [60].

Sediment/Soil Concentrations: n-Pentane was detected in 3 of 4 sediment samples from Walvis Bay of the Namibian shelf of SW Africa at concentrations of 0.12, 0.19 and 0.78 ng/g [71].

Atmospheric Concentrations: URBAN: The average n-pentane concentration in the air at the 6th floor of the Cooper Union Building in New York City NY was 31, 34 and 25 ppbC for 19, 12 and 10 samples taken at 6:00-9:00 AM, 9:00-11:00 AM and 1:00-3:00 PM, respectively, in July 1978 [9]. In a 1979 study of 5 sites in NJ, n-pentane was identified in the air over cities of Newark and Elizabeth [9]. The pentane concentration was 2.0, 1.0 and 1.0 ug/m^3 at 10, 15 and 40 mm, downwind of Janesville, WI on Aug 14, 1978 [2]. At Point Barrow, AK on Sept 2 and 3, 1967 n-pentane was detected in ambient air in 10 of 25 samples at concentrations ranging from 0.08 to 0.2 pp [12]. The average n-pentane concentration in the air at the 82 nd floor of the Empire State Building in New York City, NY was 12, 18 and 11 ppbC for 18, 21 and 17 samples taken at 6:00-9:00 AM, 9:00-11:00 AM and 1:00-3:00 PM, respectively, in July 1978 [2]. At street level at the Empire State and World Trade Buildings in Manhattan, NY the average n-pentane concentration of 4 samples was 63 ppbC in July 1978 [2]. In 1975 the average n-pentane concentration of 14 air samples taken between 5:30-8:30 AM and 12:30-3:30 PM at the World Trade Center in New York City, NY were 43 and 38 ppbC, respectively [2]. In 1975 the average n-pentane concentration of 11 and 8 air samples taken between 5:30-8:30 AM and 12:30-3:30 PM at the Interstate Sanitation Commission in New York City NY were 87 and 101 ppbC, respectively [2]. n-Pentane was detected at an average concentration of 27.7 ug/m^3 for 5 samples collected at the 82 nd floor of the World Trade Center in New York City between 5:00-5:30 PM, Aug 23, 1977 [3]. The ground level atmospheric concentration of n-pentane at 1:25 PM, was 9.1 ppb and 62.9 ppb at 0800 PST for Huntington Park, CA [54]. At 1500 ft the n-pentane concentration was 2.6 ppb at 7:43 AM, and at 8:07 AM at a height of 2,200 ft the n-pentane concentration was 1.1 ppb [54]. The n-pentane concentration ranged from 9 to 34 ppbV at a downtown Los Angeles location for the fall of 1981 [18]. From Aug 12, 1960 to Nov 18, 1960 the air of downtown Los Angeles ranged in n-pentane concentration from 0.012 to 0.064 ppm with an average for 16 samples of 0.034 ppm [40]. The ambient air of Riverside, CA was found to contain a n-pentane concentration of 22.4 ppb at 7:30 AM Sept 24, 1968 [63]. At 8:05-8:25 AM on Mar 3, 1966 the n-pentane concentration for a 2nd floor and roof top sample at the County Health and Finance Building in Riverside, CA were 18.6 and 17.0 ppb, respectively [62,63]. The n-pentane concentration for a 2nd floor sample at the County Health and Finance Building in Riverside, CA was 60.0 ppb at 7:40-8:00 AM December 22, 1965 [62]. The n-pentane concentration at 1100 ft just east of Antioch, CA was 3.5 ug/m^3, at 1000 ft near Pittsburgh, CA was 7.5 ug/m^3, at 1100 ft over Carquinez Strait, CA was 1.5 ug/m^3 and at 1000 ft over San Pablo Bay, CA was 0.5 ug/m^3 [57]. The n-pentane concentration for 6 sites in Rio Blanco, CO ranged from 1.3 to 68.2 ppbC with an average of 14.2 ppbC [4]. The average n-pentane concentration for 2 samples per 4 sites in Tulsa, OK was 44.6 ppbC with a range of 12.2 to 81.7 ppbC [4]. The 1977 maximum and average concentration of 676 points for n-pentane at a site in Houston, Texas was 190 and 27 ppb of carbon, respectively [55]. The atmospheric concentration of n-pentane at the Jones State Forest in rural TX was 11.1 ppbC [55]. The average n-pentane concentration from 6 to 9 AM in Houston, Texas was 42 ppb of Carbon [38]. n-Pentane was detected in all 16 air samples from Houston, TX ranging in concentration from 7.7 to 567.6 ppm with an average of 175.9 ppm [32]. n-Pentane was detected in trace amounts in 7 samples and ranged in concentration from 0.007 to 0.045 ppm for 7 samples out of a total of 15 samples taken in downtown Houston, TX 7-18-73 [59]. At Deuselbach, Hunsruck in Germany the atmospheric n-pentane concentration was 0.14 ppb for October 23, 1983 [51]. n-Pentane was detected in the atmospheres of 6 industrialized cities of the USSR ranging in size of population from 0.4 to 4.5 million people [23,24,25]. The average n-pentane concentration (66 samples) in the air over Tokyo, Japan for 1980 was 1.1 ppb [67]. n-Pentane was identified in the ambient air of Sydney, Australia [39] ranging in concentration from 0.5 to 23.5 ppbV with an average concentration of 6.6 ppbV [41]. n-Pentane was detected in the atmosphere over the British Columbia Research Council Laboratory at the University of British Columbia [64]. RURAL: The respective median, minimum and maximum atmospheric concentration of n-pentane for 5 rural locations in NC ranged from 0.7 to 20.5, 0.1 to 10.9, and 1.6 to 26.2 ppb [56]. The atmospheric concentration of n-pentane at the Jones State Forest in rural TX was 11.1 ppbC [55]. REMOTE: The average n-pentane concentration for samples taken at altitudes of 2000, 2500, and 300 ft over Lake Michigan on Aug 27, 1976 was 2.3 ppbV [37]. The average n-pentane concentration for samples taken at altitudes of 1000 and 1500 ft over Lake Michigan on August 28, 1976 was 0.5 ppbV [37]. The air over the Norwegian Arctic had an average n-pentane concentration for 5 samples from Bear Island, 2 from Hopen and 2 from Spitsbergen of less than 20 pptV on July 1982 and 339 ppt volume in the spring of 1983 [22].

Food Survey Values: n-Pentane was identified as a volatile constituent of freshly cooked fried chicken [66].

Plant Concentrations:

Fish/Seafood Concentrations:

Animal Concentrations:

Milk Concentrations: n-Pentane was detected in 8 of 12 samples of mothers breast milk from the cities of Bayonne, NJ, Jersey City, NJ, Bridgeville, PA and Baton Rouge, LA [47].

Other Environmental Concentrations: An air sample taken near an oil fire was found to contain n-pentane and n-pentene at a combined concentration of 0.51 mg/m^3 [49]. The concentration of n-pentane vapors averaged 88.6 ppm at excavation sites for removing gasoline storage tanks [58].

Probable Routes of Human Exposure: The most probable route of human exposure to n-pentane is by inhalation. Atmospheric workplace exposures have been documented [27,50,58]. n-Pentane is a highly volatile compound and monitoring data indicate that it is a widely occurring atmospheric pollutant.

Average Daily Intake:

Occupational Exposure: NIOSH (NOHS Survey 1972-1974) has statistically estimated that 44,379 workers are potentially exposed to n-pentane in the USA [43]. NIOSH (NOES Survey 1981-1983) has statistically estimated that 20,998 workers are potentially exposed to n-pentane in the USA [42]. A 1984 study showed n-pentane was emitted from gasoline exposing outside operators at the refineries to an average air concentration of 1.168 mg/m³; n-pentane was detected in 46 of 56 samples [50]. Transport drivers were exposed to n-pentane at atmospheric concentrations of 4.570 mg/m³ and n-pentane was detected in 48 of 49 samples [50]. Gas station attendants were exposed to n-pentane at atmospheric concentrations of 5.428 mg/m³ and n-pentane was detected in 48 of 49 samples [50]. Tank contractors were exposed to vapors containing n-pentane at concentrations of 37.4 ppm at gasoline tank removal sites [58]. Attendants at a high volume service station in eastern PA were exposed to levels of n-pentane ranging from 0.1 to 1.7 ppm for 18 of 18 air samples [27]. Common operations in which exposure to pentane may occur include: synthesis of amyl chlorides, polychlorocyclopentanes, or olefins; manufacture of low temperature thermometers, artificial ice, or polystyrene beads; and use as a blowing agent, solvent, or lighter and blowtorch fuel.

Body Burdens: n-Pentane was detected in 8 of 12 samples of mothers breast milk from the cities of Bayonne, NJ, Jersey City, NJ, Bridgeville, PA and Baton Rouge, LA [47].

REFERENCES

1. Abrams EF et al; Identification of Organic Compounds in Effluents from Industrial Sources USEPA-560/3-75-002 (1975)
2. Altwicker ER et al; J Geophys Res 85: 7475-87 (1980)
3. Altwicker ER, Whitby RA; in 72 Ann Meet Air Pollut Contr Assoc (1979)
4. Arnts RR, Meeks SA; Atmos Environ 15: 1643-51 (1981)
5. Atkinson R et al; Internat J Chem Kin 14: 781-8 (1982)
6. Atkinson R et al; Preprints Amer Chem Soc Div Environ Chem 23: 173-6 (1983)
7. Atkinson R et al; J Phys Chem 86: 4563-9 (1982)

n-Pentane

8. Bonsang B et al; J Atmos Chem 6: 3-20 (1988)
9. Bozzelli JW et al; Analysis of Selected Toxic and Carcinogenic Substances in Ambient Air of NJ. State of NJ Dept Environ Protection (2) Sexton K; Environ Sci Technol 17: 402-7 (1983)
10. Burnham AK et al; AWWA J 65: 722-5 (1973)
11. Carotti AA, Kaiser ER; J Air Pollut Contr Assoc 22: 224-53 (1972)
12. Cavanagh LA et al; Environ Sci Technol 3: 251-7 (1969)
13. Cox RA et al; Environ Sci Technol 14: 57-61 (1980)
14. Darnall KR et al; J Phys Chem 82: 1581-4 (1978)
15. Eisenreich SJ et al; Environ Sci Technol 15: 30-8 (1981)
16. Gerhold RM, Malaney GW; J Water Pollut Contr Fed 38: 562-79 (1966)
17. Great Lakes Water Quality Board; Inventory Chem Subst Id Great Lakes Ecos p 195 (1983)
18. Grosjean D, Fung K; J Air Pollut Control Assoc 34: 537-43 (1984)
19. Hampton CV et al; Environ Sci Technol 16: 287-98 (1982)
20. Hansch C, Leo AJ; Medchem Project Issue No 26. Claremont CA: Pomona College (1985)
21. Hou CT et al; Appl Environ Microbiol 46: 178-84 (1983)
22. Hov O et al; Geophys Res Lett 11: 425-8 (1984)
23. Ioffe BV et al; Environ Sci Technol 13: 864-8 (1979)
24. Ioffe BV et al; Dokl Akad Nauk Sssr 243: 1186-9 (1978)
25. Ioffe BV et al; J Chromatogr 142: 787-95 (1977)
26. Junk GA, Ford CS; Chemosphere 9: 187-230 (1980)
27. Kearney CA, Dunham DB; Am Ind Hyg Assoc J 47: 535-9 (1986)
28. Kool HJ et al; Crit Rev Env Control 12: 307-57 (1982)
29. Kopczynski SL et al; Environ Sci Technol 6: 342-7 (1972)
30. Kopfler FC et al; Adv Environ Sci Technol 8: 419-33 (1977)
31. LaRegina J et al; Environ Prog 5: 18-27 (1986)
32. Lonneman WA et al; Hydrocarbons in Houston Air USEPA-600/3-79/018 p 44 (1979)
33. Lyman WJ et al; Handbook of Chemical Property Estimation Methods NY: McGraw-Hill pp 4-9, 5-4, 10, 7-4 (1982)
34. Mackay D, Shiu WY; J Phys Chem Ref Data 19: 1175-99 (1981)
35. Malaney GW, McKinney RE; Water Sewage Works 113: 302-9 (1966)
36. McFall JA et al; Chemosphere 14: 1253-65 (1985)
37. Miller MM, Alkezweeny AJ; An NY Acad Sci 338: 219-32 (1980)
38. Monson PR et al; Paper 78-50.4 in 71st Ann Meeting Air Pollut Contr Assoc (1978)
39. Mulcahy MFR et al; Smog Forming Hydrocarbons in Urban Air, Occurrence Contr Photochem Pollut, Proc Symp Workshop Sess Paper IV p 17 (1976)
40. Neligan RE; Arch Environ Health 5: 581-91 (1962)
41. Nelson PF, Quigley SM; Atmos Environ 18: 79-87 (1984)
42. NIOSH; National Occupational Exposure Survey (NOES) (1983)
43. NIOSH; National Occupational Hazard Survey (NOHS) (1974)
44. Nolting F et al; J Atmos Chem 6: 47-59 (1988)
45. Patel RN et al; Appl Environ Microbiol 39: 727-33 (1980)

n-Pentane

46. Patel RN et al; Appl Environ Microbiol 39: 720-6 (1980)
47. Pellizzari ED et al; Bull Environ Contam Toxicol 28: 322-8 (1982)
48. Perry DL et al; Ident of Org Compounds in Ind Effluent discharges USEPA-600/4-79-016 (NTIS PB-294794) p 230 (1979)
49. Perry R; pp 130-7 in Int Symp Ident Meas Environ Pollut (1971)
50. Rappaport SM et al; Appl Ind Hyg 2: 148-54 (1987)
51. Rudolph J, Khedim A; Int J Environ Anal Chem 290: 265-82 (1985)
52. Sauer TC Jr; Org Geochem 7: 1-16 (1981)
53. Sawhney BL, Raabe JA; Ground Water Contamination Mvmt Org Pollut in Granby Landfill. New Haven, CT: CT Agric Exp Stat Bull 833 p 9 (1986)
54. Scott Research Labs Inc; Atmospheric Reaction Studies in the Los Angeles Basin. NTIS PB-194-058 p 86 (1969)
55. Seila RL; Non-urban Hydrocarbons Concentrations in Ambient Air No of Houston TX. USEPA-500/3-79-010 p 38 (1979)
56. Seila RL et al; Atmospheric Volatile Hydrocarbon Composition at Five Remote Sites in NW NC. USEPA-600/D-84-092 (NTIS PB84-177930) (1984)
57. Sexton K, Westberg H; Environ Sci Tech 14: 329-32 (1980)
58. Shamsky S, Samimi B; App Ind Hyg 2: 242-5 (1987)
59. Siddiqi AA, Worley FL; Atmos Environ 11: 131-43 (1977)
60. Sigsby JE et al; Environ Sci Technol 21: 466-75 (1987)
61. Silverstein RM, Bassler GC; Spectrometric Id of Org Cmpd, J Wiley and Sons Inc pp 148-169 (1963)
62. Stephens ER, Burleson FR; J Air Pollut Control Assoc 17: 147-53 (1967)
63. Stephens ER; Hydrocarbons in Polluted Air NTIS PB-230 993-/8 p 86 (1973)
64. Stump FD, Dropkin DL; Anal Chem 57: 2629-34 (1985)
65. Swann RL et al; Res Rev 85: 16-28 (1983)
66. Tang J et al; J Agric Food Chem 31: 1287-92 (1983)
67. Uno I et al; Atmos Environ 19: 1283-93 (1985)
68. USEPA; EXAMS II Computer Simulation (1987)
69. USEPA; Drinking water Criteria Document for Gasoline ECAO-CIN-D006, 8006-61-9 (1986)
70. Vogt WG, Walsh JJ; pp 2-17 in Proc APCA Annu Meet 78th Vol 6 (1985)
71. Whelan JK et al; Geochim Cosmochim Acta 44: 1767-85 (1984)
72. Zimmerman RE et al; pp 230-9 in Proc Int Gas Res Conf (1983)

3-Pentanone

SUBSTANCE IDENTIFICATION

Synonyms: Diethyl ketone

Structure:

$$H_3C \diagup \diagdown \overset{\displaystyle O}{\diagdown} \diagup \diagdown CH_3$$

CAS Registry Number: 96-22-0

Molecular Formula: $C_5H_{10}O$

Wiswesser Line Notation: 2V2

CHEMICAL AND PHYSICAL PROPERTIES

Boiling Point: 101.96 °C

Melting Point: -38.97 °C

Molecular Weight: 86.13

Dissociation Constants: pKa = 27.1 [24]

Log Octanol/Water Partition Coefficient: 0.99 [28]

Water Solubility: 48,100 mg/L at 25 °C [32]

Vapor Pressure: 37.0 mm Hg at 25 °C [5]

Henry's Law Constant: 8.72 x 10^{-5} atm-m^3/mol (calculated from the vapor pressure and water solubility)

ENVIRONMENTAL FATE/EXPOSURE POTENTIAL

Summary: 3-Pentanone is a naturally occurring chemical; it is found in plants as well as some foods such as chicken and fruits. It does not appear to have much industrial use. It has, however, been identified in

489

effluent from chemical manufacturing, mining, and the pulp and paper industry. It is also a combustion product, being found in auto exhaust and tobacco smoke and is produced during coal gasification. If released on soil, 3-pentanone will both evaporate into the atmosphere and leach into the ground. Leaching may not be important if concurrent biodegradation occurs rapidly. If released into surface water, 3-pentanone will be lost primarily by biodegradation and volatilization. The estimated volatilization half-life of 3-pentanone in a model river is 12 hr. It is readily biodegradable in screening tests, and it is probable that it will biodegrade in both water and soil. It may also photolyze, but no kinetic data on this degradation route are available. It would not be expected to bioconcentrate in fish and aquatic organisms. If released to the atmosphere, 3-pentanone will be degraded by reaction with photochemically produced hydroxyl radicals; the half-life of this process is 5.9 days. Direct photolysis and wet deposition may contribute to its atmospheric removal. The general population will be exposed to 3-pentanone in food, drinking water, and air. Monitoring data, however, are too limited to estimate the level of exposure. Occupational exposure to the chemical will most likely be by inhalation and dermal contact.

Natural Sources: 3-Pentanone is a plant volatile; it is emitted from European birch, European larch, European fir, Scots pine, Siberian pine, Evergreen cyprus, Northern white cedar, and ferns [16]. Sediment monitoring data has suggested that 3-pentanone is formed in various organic aquatic sediments via microbial activity [17,31].

Artificial Sources: 3-Pentanone may be released into the atmosphere from auto exhaust, wood pulping, and chemical manufacturing [11]. Wastewater from chemical manufacturing plants may also contain 3-pentanone. It is also found in tobacco smoke [11].

Terrestrial Fate: The primary degradation process for 3-pentanone in soil is expected to be biodegradation. Several biodegradation screening studies have demonstrated that 3-pentanone biodegrades readily [1,3,8,15]. It can also be expected to leach readily based upon estimated Koc values of 12 and 82 [20]; however, leaching may not be important if concurrent biodegradation occurs rapidly. 3-Pentanone released to surfaces will evaporate due to its high vapor pressure.

3-Pentanone

Aquatic Fate: When released into water, 3-pentanone will be primarily lost by biodegradation and volatilization. Several biodegradation screening studies have demonstrated that 3-pentanone biodegrades readily [1,3,8,15]. The estimated volatilization half-lives of 3-pentanone in a model river (1 m deep flowing at 1 m/sec with a 3 m/sec wind) and model environmental pond (2 m deep) are 12 hr [20] and 5.6 days [29], respectively. It may also photolyze, but no kinetic data on this degradation route are available. Adsorption to sediment and particulate matter in the water column should not be significant. Aquatic hydrolysis and bioconcentration are not expected to be important.

Atmospheric Fate: If released into the atmosphere, 3-pentanone can be expected to exist almost entirely in the vapor phase based upon its high vapor pressure [6]. It will degrade in the vapor phase by reaction with photochemically produced hydroxyl radicals; a half-life of 5.9 days can be estimated from an experimentally determined rate constant of 2.74×10^{-12} cm^3/molecule-sec [30]. No data concerning the direct photolysis rate of 3-pentanone are available; however, related alkylketones are expected to have photolysis half-lives of the order of days [4]; therefore, direct photolysis may contribute to its atmospheric degradation. The relatively high vapor pressure of 3-pentanone suggests that physical removal from air via wet deposition (dissolution into clouds, rainfall, etc.) is possible.

Biodegradation: Several investigators have shown that 3-pentanone readily biodegrades in screening tests using a sewage seed [1,3,8,15]. They report: 66.4% of theoretical BOD after 5 days incubation with acclimated cultures [1], 89% of theoretical BOD after 5 days [3] and 50.8% and 38% of theoretical BOD after 10 days [8,15]. In a semi-continuous activated sludge (SCAS) biological treatment simulation test, 38% of the theoretical BOD of 3-pentanone was lost during the 24 hr-treatment [21].

Abiotic Degradation: 3-Pentanone reacts with photochemically produced hydroxyl radicals by H-atom abstraction with an experimental rate constant of 2.74×10^{-12} cm^3/molecule-sec [30]; this corresponds to an atmospheric half-life of 5.9 days, assuming an average atmospheric concentration of $5 \times 10^{+5}$ hydroxyl radicals/cm^3. This photo-oxidation yields peroxyacetic acid as a stable end product [13]. 3-Pentanone has photochemical smog-forming potential as determined by its reactivity under simulated photochemical conditions [9,18]. All carbonyl compounds absorb UV radiation near the short-wavelength cutoff of the solar spectrum and therefore may potentially degrade as a result of direct photolysis [4]. While no photolysis rates are available for 3-pentanone, related alkylketones are estimated to photolyze at a rate of 12.5 x 10^{-6} sec^{-1} for a solar zenith angle of 30 °C [4]; this would result in a half-life of 1.3 days, assuming a quantum yield of 1 and a 12-hr sunlight day. The actual photolysis half-life should be much longer as quantum yields may be a fraction of unity and the assumed solar flux may not be realized. Therefore, both direct photolysis and photooxidation by hydroxyl radicals will contribute to 3-pentanone's photochemical degradation. Ketones are generally resistant to aqueous environmental hydrolysis [20]; therefore, 3-pentanone is not expected to hydrolyze in water.

Bioconcentration: Based upon the log octanol-water partition coefficient and the water solubility, the BCF (bioconcentration factors) for 3-pentanone can be estimated to be 3.3 and 1.4, respectively, from linear regression-derived equations [20]. This indicates that 3-pentanone will have a negligible tendency for bioconcentrating in fish.

Soil Adsorption/Mobility: Based upon the log octanol-water partition coefficient and the water solubility, the Koc for 3-pentanone can be estimated to be about 82 and 12, respectively, from linear regression-derived equations [20]. These Koc estimations indicate high to very high soil mobility [26]; therefore, a potential exists for leaching to ground water.

Volatilization from Water/Soil: The magnitude of the Henry's Law constant for 3-pentanone indicates that volatilization from environmental waters is possibly significant, but may not be rapid [20]. Based on the Henry's Law constant, the volatilization half-life from a model river (1 m deep flowing 1 m/sec with a wind velocity of 3 m/sec) can be estimated to be 12 hours [20]. Volatilization half-life from a model environmental pond (2 meters deep) can be estimated to be about 5.6 days [29]. The rate of volatilization will be sensitive to both liquid-phase mixing (current) and gas-phase mixing (wind) [23]. 3-Pentanone has a relatively high vapor pressure and a low soil adsorption, and therefore should readily volatilize from the soil surface.

3-Pentanone

Water Concentrations: DRINKING WATER: In a survey of 14 treated drinking water supplies of varied sources in England, 3-pentanone was detected in 6 supplies that were derived from both ground and surface sources [10]. GROUND WATER: 3-pentanone was not found in ground water at a coal gasification site in Hanna, WY although it was found in process water [22].

Effluent Concentrations: In a comprehensive survey of wastewater from 4000 industrial and publicly owned treatment works (POTWs) sponsored by the Effluent Guidelines Division of the USA EPA, 3-pentanone was identified in discharges of the following industrial category (positive occurrences; median concentration in ppb): nonferrous metals (1;26.7), ore mining (3;2.1), organics and plastics (1;47.3), plastics and synthetics (1; 2491.1), pulp and paper (1; 27.5) [25]. The highest effluent concentration was 2491.1 ppb in the plastics and synthetics industry [25]. 3-Pentanone was detected in effluent from an advanced wastewater treatment plant in Lake Tahoe, CA [19].

Sediment/Soil Concentrations: 3-Pentanone has been detected, but not quantified, in recently deposited sediment beneath a reed bed in Lake Constance, Switzerland [17]. The maximum concentrations were found adjacent to zones of former reed growth, suggesting that 3-pentanone was formed by microbial processes. Sediment cores at 3 shallow sites from Walvis Bay off the arid coast of Southwest Africa were analyzed for 3-pentanone and other chemicals [31]. This is an area where organic matter in the sediment is considered to be of marine origin due to a lack of on-shore vegetation and the fact that no major rivers discharge into the bay. At two sites the concentration of 3-pentanone ranged from not detected to 0.46 ppb and not detected to 0.75 ppb (dry weight); it was absent from the third site. Abrupt changes in sediment concentration with depth suggest that the chemical stayed close to its generation sites [31].

Atmospheric Concentrations: 3-Pentanone was detected in 1 of 7 air samples from the Kanawha Valley in WV (St Albans) but not in a similar study in the Shenandoah Valley, VA [7]. Levels of 0.57 ppb 3-pentanone were found at an oil shale facility; however, it was not detected in nearby rural and urban areas [14]. At a chemical manufacturing site, the concentration of 3-pentanone was 2 ppb [11].

3-Pentanone

Food Survey Values: 3-Pentanone has been identified as a volatile component of raw chicken breast [12], nectarines [27] and kiwi fruit [2].

Plant Concentrations: 3-Pentanone has been identified as a volatile component of nectarines [27] and kiwi fruit [2].

Fish/Seafood Concentrations:

Animal Concentrations:

Milk Concentrations:

Other Environmental Concentrations: 3-Pentanone has been detected in tobacco smoke [11].

Probable Routes of Human Exposure: The general population will be exposed to 3-pentanone as it occurs naturally in some foods [12,27] as well as cigarette smoke [11]. It is also found in drinking water [10]. Occupational exposure to the chemical will most likely be by inhalation and dermal contact.

Average Daily Intake:

Occupational Exposure:

Body Burdens:

REFERENCES

1. Babeu L, Vaishnav DD; J Indust Microbiol 2: 107-15 (1987)
2. Bartley JP, Schwede AM; J Agric Food Chem 37: 1023-5 (1989)
3. Bridie AL et al; Wat Res 13: 627-30 (1979)
4. Cox RA et al; Environ Sci Technol 15: 587-92 (1981)
5. Daubert TE, Danner RP; Physical and Thermodynamic Properties of Pure Chemicals: Data Compilation, NY: Hemisphere Pub Corp (1989)
6. Eisenreich SJ et al; Environ Sci Technol 15: 30-8 (1981)
7. Erickson MD, Pellizzari ED; Analysis of Organic Air Pollutants in the Kanawha Valley, WV and the Shenendoah Valley, VA. USEPA-903/9-78-007 (1978)
8. Ettinger MB; Ind Eng Chem 48: 256-9 (1956)
9. Farley FF; pp 713-26 in Inter Conf Photochem Oxidant Pollut Control Proc USEPA-600/377-001b (1977)

3-Pentanone

10. Fielding M et al; Organic micropollutants in drinking water TR-159. Medmenham, England: Water Resources Center (1981)
11. Graedel TE; Chemical Compounds in the Atmosphere. NY, NY: Academic Press p 184 (1978)
12. Grey TC, Shrimpton DH; Brit Poultry Sci 8: 23-33 (1967)
13. Hanst PL, Gay BW Jr; Atmos Environ 17: 2259-65 (1983)
14. Hawthorne SB, Sievers RE; Environ Sci Technol 18: 483-90 (1984)
15. Heukelekian H, Rand MC; J Wat Pollut Control Assoc 29: 1040-53 (1955)
16. Isadorov VA et al; Atmos Environ 19: 1-8 (1985)
17. Juettner F, Schroeder R; Arch Hydrobiol 94: 1269-75 (1982)
18. Laity JL et al; pp 95-112 in Adv Chem Series 124. Amer Chem Soc Washington, DC (1973)
19. Lucas SV; GC/MS analysis of organics in drinking water concentrates and advanced waste treatment concentrates. Vol 1. USEPA 600/1-84-020A (1984)
20. Lyman WJ et al; Handbook of Chemical Property Estimation Methods Washington, DC: Amer Chem Soc pp 5-4, 5-10 (1990)
21. Mills EJ Jr, Stack VT Jr; Proc 8th Indust Waste Conf. Eng Bulletin Purdue Univ, Eng. Ext. Ser. pp 492-517 (1954)
22. Pellizzari ED et al; ASTM Spec Tech Publ STP 686: 256-74 (1979)
23. Rathbun RE, Tai DY; Water, Air, Soil Pollut 17: 281-93 (1982)
24. Riddick JA et al; Organic Solvents NY: Wiley Interscience (1986)
25. Shackelford WM et al; Analyt Chim Acta 146: 15-27 (1983)
26. Swann RL et al; Res Rev 85: 23 (1983)
27. Takeoka GR et al; J Agric Food Chem 36: 553-60 (1988)
28. Tewari YB et al; J Chem Eng Data 27: 451-4 (1982)
29. US EPA; EXAMS II Computer Simulation (1987)
30. Wallington TJ, Kurylo MJ; J Phys Chem 91: 5050-4 (1987)
31. Whelen JK et al; Geochim Cosmochim Acta 44: 1767-85 (1980)
32. Yalkowsky SH; Arizona Database of Aqueous Solubilities. College of Pharmacy, Univ of Arizona (1989)

2-Phenoxyethanol

SUBSTANCE IDENTIFICATION

Synonyms: Ethylene glycol monophenyl ether

Structure:

CAS Registry Number: 122-99-6

Molecular Formula: $C_8H_{10}O_2$

Wiswesser Line Notation: Q2OR

CHEMICAL AND PHYSICAL PROPERTIES

Boiling Point: 245.2 °C at 760 mm Hg [17]

Melting Point: 14 °C [17]

Molecular Weight: 138.16

Dissociation Constants:

Log Octanol/Water Partition Coefficient: 1.16 [5]

Water Solubility: 26,940 mg/L at 25 °C [16]

Vapor Pressure: 0.03 mm Hg at 25 °C [3]

Henry's Law Constant: 2.0×10^{-7} atm-m^3/mole at 25 °C (estimated from water solubility and vapor pressure)

ENVIRONMENTAL FATE/EXPOSURE POTENTIAL

Summary: 2-Phenoxyethanol is released to the environment in wastewater effluents from various chemical, plastics, photographic, and mechanical product industries. Its use as a solvent will release it directly to air by evaporation. If released to air, vapor-phase 2-phenoxyethanol will degrade relatively rapidly by reaction with photochemically produced hydroxyl radicals (estimated half-life of 11.6 hr). Physical removal from air via wet deposition is possible since it is relatively soluble in water. If released to soil or water, 2-phenoxyethanol is expected to degrade through biodegradation. Leaching in soil is possible. 2-Phenoxyethanol will evaporate slowly from terrestrial surfaces. Occupational exposure to 2-phenoxyethanol occurs through inhalation of vapor and dermal contact. Its use as solvent for inks, resins and cellulose acetate and its use as a perfume fixative can expose the general population through dermal contact and inhalation of vapor.

Natural Sources:

Artificial Sources: 2-Phenoxyethanol has been detected in effluents from sewage treatment plants and in effluents from chemical manufacturing facilities [14]. 2-Phenoxyethanol's use as solvents for inks, resins and cellulose acetate [13] will release the compound to air through evaporation.

Terrestrial Fate: The only identifiable degradation process for 2-phenoxyethanol in soil is biodegradation. The results of one biological screening study have indicated that 2-phenoxyethanol is readily biodegradable [3]. 2-Phenoxyethanol may leach readily in soil based upon estimated Koc values of 16 and 102 [8]. It will evaporate slowly from surfaces; the evaporation rate from a solid surface to air at 77 °F (15% relative humidity) is about 0.4 wt % in 60 minutes [3].

Aquatic Fate: The only major identifiable degradation process for 2-phenoxyethanol in water is biodegradation. The results of one biological screening study have indicated that 2-phenoxyethanol is readily biodegradable [3]. Aquatic hydrolysis, volatilization, direct photolysis, bioconcentration, and adsorption to sediment are not expected to be important.

Atmospheric Fate: Based upon the vapor pressure, 2-phenoxyethanol is expected to exist almost entirely in the vapor phase in the ambient atmosphere [4]. It will degrade relatively rapidly in an average ambient atmosphere by reaction with photochemically produced hydroxyl radicals (estimated half-life of 11.6 hr) [1]. Physical removal from air via wet deposition is possible since 2-phenoxyethanol is relatively soluble in water.

Biodegradation: Theoretical BODs of 21% (5-day), 66% (10-day), and 75% (20-day) have been measured [3]; these BODTs indicate that 2-phenoxyethanol will largely be removed during biological waste treatment [3].

Abiotic Degradation: The rate constant for the vapor-phase reaction of 2-phenoxyethanol with photochemically produced hydroxyl radicals has been estimated to be 33.2 x 10^{-12} cm^3/molecule-sec at 25 °C which corresponds to an atmospheric half-life of about 11.6 hr at an atmospheric concn of 5 x 10^{+5} hydroxyl radicals per cm^3 [1]. The UV spectrum of an aqueous solution of 2-phenoxyethanol does not show any absorbance above 290 nm [6] which indicates that 2-phenoxyethanol will not directly photolyze in the environment; the addition of titanium dioxide to the aqueous solution as a photosensitizer allows absorbance above 290 nm [6]. Ether, alcohol and glycol functional groups are generally resistant to aqueous environmental hydrolysis [8]; therefore, 2-phenoxyethanol is not expected to hydrolyze in water.

Bioconcentration: Based upon the water solubility value, the BCF for 2-phenoxyethanol can be estimated to be 2 from a regression-derived equation [8]. Based upon the measured log Kow, the BCF for 2-phenoxyethanol can be estimated to be 4.5 from a regression-derived equation [8]. These BCF values suggest that 2-phenoxyethanol will not bioconcentrate in aquatic organisms.

Soil Adsorption/Mobility: Based upon the water solubility value, the Koc for 2-phenoxyethanol can be estimated to be 16 from a regression-derived equation [8]. Based upon the measured log Kow, the Koc for 2-phenoxyethanol can be estimated to be 102 from a regression-derived equation [8]. These Koc values suggest that 2-phenoxyethanol has a high to very high soil mobility [15].

Volatilization from Water/Soil: The value of Henry's Law constant indicates that 2-phenoxyethanol is essentially nonvolatile from environmental waters [8]. The evaporation rate of 2-phenoxyethanol from a solid surface is classified as <0.01 relative to butyl acetate=1 [3]; the actual evaporation loss of 2-phenoxyethanol from a surface at an air temperature of 77 °F and 15% relative humidity is 0.4 wt % in 60 min [3].

Water Concentrations: DRINKING WATER: 2-Phenoxyethanol was qualitatively detected in drinking water concentrates collected in Cincinnati, OH on Oct 17, 1978 [7]. GROUND WATER: 2-Phenoxyethanol concentrations of less than 5 ppm were detected in well waters collected in the vicinity of two industrial factories in Spain in 1984 [12].

Effluent Concentrations: 2-Phenoxyethanol has been detected in wastewater effluents from the following industries: paint and ink, organics & plastics, photographic, and mechanical products [2].

Sediment/Soil Concentrations:

Atmospheric Concentrations:

Food Survey Values:

Plant Concentrations:

Fish/Seafood Concentrations:

Animal Concentrations:

Milk Concentrations:

Other Environmental Concentrations:

Probable Routes of Human Exposure: Occupational exposure to the ethylene glycol ethers occurs through inhalation of vapor and dermal contact [11]. 2-Phenoxyethanol's use as a solvent for inks, resins and cellulose acetate and its use as a perfume fixative [13] can expose the general population through dermal contact and inhalation of vapor.

2-Phenoxyethanol

Average Daily Intake:

Occupational Exposure: NIOSH (NOES Survey 1981-1983) has statistically estimated that 96,814 workers are potentially exposed to 2-phenoxyethanol in the USA [9]. NIOSH (NOHS Survey 1972-1974) has statistically estimated that 9,558 workers are potentially exposed to 2-phenoxyethanol in the USA [10].

Body Burdens:

REFERENCES

1. Atkinson R; J Inter Chem Kinet 19: 799-828 (1987)
2. Bursey JT, Pellizzari ED; Analysis of Wastewater for Organic Pollutants in Consent Degree Survey. Contract No. 68-03-2867. Athens, GA: USEPA Environ Res Lab p 97 (1982)
3. Dow Chem Co; The Glycol Ethers Handbook. Midland, MI: Dow Chem Co (1981)
4. Eisenreich SJ et al; Environ Sci Technol 15: 30-8 (1981)
5. Hansch, C, Leo AJ; Medchem Project Issue No 26. Claremont CA: Pomona College (1985)
6. Hidaka H et al; J Photochem Photobiol 42: 375-81 (1988)
7. Lucas SV; GC/MS Analysis of Organics in Drinking Water Concentrates and Advanced Waste Treatment Concentrates: Volume 1, USEPA-600/1-84-020A (NTIS PB85-128221) pp 45, 141 (1984)
8. Lyman WJ et al; Handbook of Chemical Property Estimation Methods. Environmental Behavior of Organic Compounds. Washington DC: American Chemical Society pp 4-9, 5-4, 5-10, 7-4, 7-5, 15-15 to 15-32 (1990)
9. NIOSH; National Occupational Exposure Survey (NOES) (1983)
10. NIOSH; National Occupational Hazard Survey (NOHS) (1974)
11. Parmeggiani L; Encyl Occup Health & Safety 3rd ed Geneva, Switzerland: International Labour Office pp 974-5 (1983)
12. Rivera J et al; Chemosphere 14: 395-402 (1985)
13. Sax NI, Lewis RJ Jr; Hawley's Condensed Chemical Dictionary 11th ed NY: Van Nostrand Reinhold Co p 490 (1987)
14. Shackelford WM, Keith LH; Frequency of Organic Compounds Identified in Water. USEPA-600/4-76-062 (1976)
15. Swann RL et al; Res Rev 85: 23 (1983)
16. Valvani SC et al; J Pharm Sci 70: 502-7 (1981)
17. Windholz M et al; The Merck Index 10th ed Rahway, NJ: Merck & Co Inc p.1046 (1983)

N-Phenylformamide

Synonyms:

Structure:

H—C(=O)—N(H)—C₆H₅

CAS Registry Number: 103-70-8

Molecular Formula: C_7H_7NO

Wiswesser Line Notation: VHMR

CHEMICAL AND PHYSICAL PROPERTIES

Boiling Point: 271 °C [11]

Melting Point: 46.6-47.5 °C [11]

Molecular Weight: 121.15

Dissociation Constants:

Log Octanol/Water Partition Coefficient: 1.15 [6]

Water Solubility: $2.86 \times 10^{+4}$ mg/L at 25 °C [11]

Vapor Pressure: 5.49×10^{-4} mm Hg at 25 °C (estimated) [5]

Henry's Law Constant: 8.45×10^{-9} atm-m³/mole at 25 °C (estimated) [8]

ENVIRONMENTAL FATE/EXPOSURE POTENTIAL

Summary: N-Phenylformamide may be released to the environment in wastewater treatment effluents. If released to the atmosphere, vapor-phase N-phenylformamide is expected to degrade rapidly (estimated half-life of 1.4 hr) by reaction with photochemically produced hydroxyl radicals. If

501

released to moist soil, N-phenylformamide may slowly degrade by chemical hydrolysis or undergo extensive leaching. If released to water, hydrolysis may be an important removal mechanism although the reaction may be slow. The hydrolysis half-life in water at 55 °C and pH 5-9 can be estimated to be on the order of 29-146 days. Volatilization, bioconcentration and adsorption to suspended solids and sediments do not appear to be important aquatic fate processes. Insufficient data are available to assess the relative importance of biodegradation of N-phenylformamide in soil or water. In occupational settings, workers involved in the manufacturing of N-phenylformamide may be exposed through inhalation or through dermal contact.

Natural Sources:

Artificial Sources: N-Phenylformamide has been detected, but not quantified, in US wastewater treatment effluent [4].

Terrestrial Fate: When released to moist soil, N-phenylformamide is expected to slowly degrade by chemical hydrolysis (half-life of 29-146 days at 55 °C and pH 5-9 [2]). Leaching should be possible based on a low estimated Koc of 15.5 [7,10]. Based on the low Henry's Law constant, volatilization from moist soils is not expected to be an important fate process. No data are available to assess the relative importance of biodegradation of N-phenylformamide in soil.

Aquatic Fate: If released to water, N-phenylformamide is expected to undergo slow chemical hydrolysis [2]. The hydrolysis half-life in water at 55 °C and pH 5-9 has been estimated to be on the order of 29-146 days [2]. Based on the low Henry's Law constant, volatilization from aquatic environments is not expected to be an important fate process [7]. Direct bioaccumulation in aquatic organisms and adsorption to suspended solids and sediments are not expected to be significant aquatic fate processes. Insufficient data are available to assess the relative importance of biodegradation of N-phenylformamide in water.

Atmospheric Fate: Based upon the estimated vapor pressure, N-phenylformamide is expected to exist primarily in the vapor phase in the ambient atmosphere although some of the chemical may exist in the particulate phase [3]. N-Phenylformamide is degraded rapidly in the ambient atmosphere by vapor-phase reaction with photochemically produced hydroxyl radicals at an estimated half-life rate of about 1.4 hr [1].

Biodegradation:

Abiotic Degradation: The rate constant for the vapor-phase reaction of N-phenylformamide with photochemically produced hydroxyl radicals has been estimated to be 2.67 x 10^{-10} cm^3/molecule-sec at 25 °C which corresponds to an atmospheric half-life of about 1.4 hr at an atmospheric concn of 5 x 10^{+5} hydroxyl radicals per cm^3 [1]. N-phenylformamide is expected to undergo slow chemical hydrolysis [2]. The hydrolysis half-life in water at 55 °C and pH 5-9 has been estimated to be on the order of 29-146 days [2].

Bioconcentration: Based on the measured log Kow, the BCF for N-phenylformamide can be estimated to be 4.4 using a recommended regression-derived equation [7]. This BCF value suggests that N-phenylformamide would not bioconcentrate significantly in aquatic systems.

Soil Adsorption/Mobility: A Koc value of approximately 15.5 can be estimated based on the measured water solubility and a regression-derived equation [7]; this Koc value suggests very high soil mobility [10].

Volatilization from Water/Soil: The Henry's Law constant suggests that N-phenylformamide is essentially nonvolatile from water [7].

Water Concentrations:

Effluent Concentrations: N-Phenylformamide has been detected, but not quantified, in US wastewater treatment effluent [4].

Sediment/Soil Concentrations:

Atmospheric Concentrations:

Food Survey Values:

N-Phenylformamide

Plant Concentrations:

Fish/Seafood Concentrations:

Animal Concentrations:

Milk Concentrations:

Other Environmental Concentrations:

Probable Routes of Human Exposure: N-Phenylformamide was formerly used as a local anesthetic, analgesic and antipyretic [9]; therefore, consumers using the chemical were probably exposed through injection or ingestion. In occupational settings, workers involved in the manufacturing of N-phenylformamide were probably exposed through dermal contact or through inhalation of dust.

Average Daily Intake:

Occupational Exposure:

Body Burdens:

REFERENCES

1. Atkinson R; Int J Chem Kinet 19: 799-828 (1987)
2. Bergstrand B; Acta Pharm Suec 22: 1-16 (1985)
3. Eisenreich SJ et al; Environ Sci Technol 15: 30-8 (1981)
4. Ellis DD et al; Arch Environ Contam Toxicol 11: 373-82 (1982)
5. GEMS; Graphical Exposure Modeling System. PCCHEM. USEPA (1987)
6. Hansch C, Leo AJ; Medchem project Issue No. 26 Claremont, CA Pomona College (1985)
7. Lyman WJ et al; Handbook of Chemical Property Estimation Methods. Environmental Behavior of Organic Compounds. Washington DC: American Chemical Society pp 4-9, 5-4, 5-10, 7-4, 7-5, 15-15 to 15-32 (1990)
8. Meylan WM, Howard PH; Environmental Toxicology and Chemistry 10:1283-1293 (1991)
9. Northcott J; Kirk-Othmer Encycl Chem Tech 3rd ed NY:Wiley 2: 318 (1978)
10. Swann RL et al; Res Rev 85: 17-28 (1983)
11. Windholz M et al; The Merck Index 10th ed Rahway, NJ: Merck & Co Inc p 605 (1983)

n-Propyl Acetate

SUBSTANCE IDENTIFICATION

Synonyms: Acetic acid, n-propyl ester

Structure:

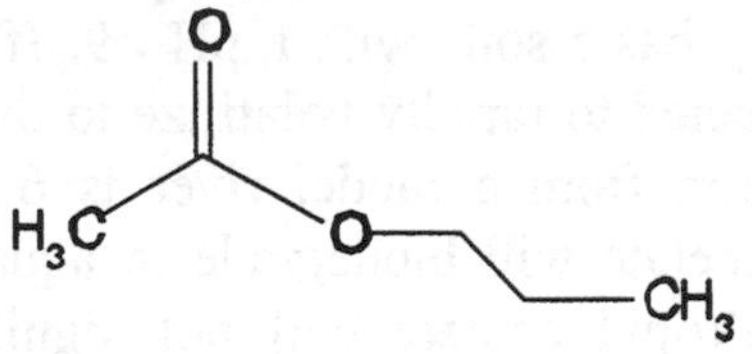

CAS Registry Number: 109-60-4

Molecular Formula: $C_5H_{10}O_2$

Wiswesser Line Notation: 3OV1

CHEMICAL AND PHYSICAL PROPERTIES

Boiling Point: 101.6 °C at 760 mm Hg

Melting Point: -95 °C

Molecular Weight: 102.15

Dissociation Constants:

Log Octanol/Water Partition Coefficient: 1.23 [11]

Water Solubility: 18.9 g/L [29]

Vapor Pressure: 33.7 mm Hg at 25 °C [1]

Henry's Law Constant: 2.18 x 10^{-4} atm m³/mole at 20 °C [14]

ENVIRONMENTAL FATE/EXPOSURE POTENTIAL

Summary: n-Propyl acetate, which is used as a solvent, may be released in fugitive emissions during its manufacture, formulation, or use in commercial products. n-Propyl acetate is also a naturally occurring compound. If released to soil, n-propyl acetate will display very high mobility and it has the potential to leach into ground water. Rapid volatilization is expected to occur from both moist and dry soils. Hydrolysis of n-propyl acetate in soil is not expected to be a significant process except in highly basic soils with a pH >9. If released to water, n-propyl acetate is expected to rapidly volatilize to the atmosphere. The half-life for volatilization from a model river is 6.5 h. Limited data suggest that n-propyl acetate will biodegrade in aquatic systems under aerobic conditions. n-Propyl acetate will not significantly adsorb to sediment and suspended organic matter, nor will it bioconcentrate in fish and aquatic organisms. Hydrolysis of n-propyl acetate in aquatic systems is not expected to be a significant process except under basic conditions of pH >9. In the atmosphere, n-propyl acetate is expected to undergo a relatively slow gas-phase reaction with photochemically produced hydroxyl radicals with experimental half-lives on the order of 4-5 days. n-Propyl acetate may undergo atmospheric removal by wet deposition. The probable routes of exposure to n-propyl acetate are by inhalation and dermal contact during the production and use of this compound. The general public is likely to be exposed to n-propyl acetate by the ingestion of foods in which it is contained.

Natural Sources: n-Propyl acetate may be released to the environment from natural sources. n-Propyl acetate was identified as a volatile component of Concord grape juice [30], Beaufort cheese [7], bananas [21], apples [37] and nectarines [32].

Artificial Sources: In 1989, three companies were listed as producers of n-propyl acetate [28]. The US production of n-propyl acetate amounted to 70.83 million pounds in 1988 [33]. n-Propyl acetate has been used as solvent for cellulose nitrate, chlorinated rubber, and heat-reactive phenolics, but its principal use is as a printing ink solvent [8]. It also finds use as a flavoring agent and in perfumery [27]. n-Propyl acetate may be released to the environment as a fugitive emission during its production, formulation and use.

Terrestrial Fate: If released to soil, calculated soil adsorption coefficients ranging from approximately 19 to 111 [18] (obtained from its water solubility and log octanol water partition coefficient,

506

respectively), suggest that n-propyl acetate will display high to very high mobility in soil [31] and has the potential to leach into ground water. The Henry's Law constant of n-propyl acetate, suggest that this compound will rapidly volatilize from both moist and dry soil. Hydrolysis of n-propyl acetate in soil is not expected to be a significant process except in highly basic soils with a pH >9, as hydrolysis rate constants indicate that this process will be too slow to be environmentally significant under acidic, neutral, and slightly basic conditions [6,8].

Aquatic Fate: If released to water, n-propyl acetate is expected to rapidly volatilize to the atmosphere. Based on a measured Henry's Law constant, the half-life for volatilization from a model river is 6.5 h [18]. The available data indicate that n-propyl acetate will biodegrade in aquatic systems under aerobic conditions [22,25]. From its water solubility and log octanol water partition coefficient, calculated soil adsorption coefficients ranging from 19.4 to 111 [18] and calculated bioconcentration factors ranging 2.5 to 5.1 [18], respectively, indicate that n-propyl acetate will not significantly adsorb to sediment and suspended organic matter, nor will it bioconcentrate in fish and aquatic organisms. Hydrolysis of n-propyl acetate in aquatic systems is not expected to be a significant process except under basic conditions of pH >9, as hydrolysis rate constants indicate that this process will be too slow to be environmentally significant under acidic, neutral, and slightly basic conditions [6,8].

Atmospheric Fate: Experimental rate constants for the gas phase reaction of n-propyl acetate with photochemically produced hydroxyl radicals in the range 2.4-4.1 x 10^{-12} cu-cm/mol-sec [1-6] correspond to an atmospheric half-life ranging 3.39-6.69 days. The relatively high water solubility of n-propyl acetate suggests that this compound may be removed from the atmosphere by wet deposition.

Biodegradation: Pure cultures of <u>Alcaligenes faecalis,</u> isolated from activated sludge, were found to oxidize n-propyl acetate after a short lag period [22]. n-Propyl acetate gave a 5-day BODT of 62% using a settled domestic wastewater seed, and an 80% BODT after 10 days [25]. When the same inocula was added to sea water, the 5-day BODT was 50% and 70% after 10 days [25].

n-Propyl Acetate

Abiotic Degradation: Experimental rate constants for the gas-phase reaction of n-propyl acetate with photochemically produced hydroxyl radicals range from 2.4 - 4.1 x 10^{-12} cm³/mol-sec in the temperature range 23-32 °C [2,3,4,10,13,36], which correspond to an atmospheric half-life ranging from 3.9 to 6.69 days using an average atmospheric hydroxyl radical concentration of 5 x 10^5 mole/cm³ [2]. These values correspond to a removal rate of 1.5% per h for this reaction [17]. The primary products from the reaction of n-propyl acetate with hydroxyl radicals are acetaldehyde and propionaldehyde [12]. A rate constant for the basic hydrolysis of n-propyl acetate in water at 25 °C, 0.087 L/mole-sec [6], corresponds to half-lives of 92 days, 9.2 days, and 22 h at pH 8, 9 and 10, respectively. Hydrolysis under acidic and neutral conditions normally found in the environment is not expected to be significant [8].

Bioconcentration: From the water solubility of n-propyl acetate and its octanol/water partition coefficient, bioconcentration factors of 2.4 and 5.1, respectively, can be calculated by a recommended regression equation [18]. These values indicate that n-propyl acetate will not significantly bioconcentrate in fish and aquatic organisms.

Soil Adsorption/Mobility: From the water solubility of n-propyl acetate and the octanol/water partition coefficient, soil adsorption coefficients of 19.4 and 111, respectively, can be calculated by a recommended regression equation [18]. These values indicate that n-propyl acetate will display high to very high mobility in soil [31].

Volatilization from Water/Soil: Based on a measured Henry's Law constant, the half-life for volatilization of n-propyl acetate from a model river 1 m deep, flowing at 1 m/sec and a wind speed of 3 m/sec is 6.5 h [18]. The Henry's Law constant of n-propyl acetate and its vapor pressure, suggest that volatilization from both moist and dry soil to the atmosphere will be significant fate processes.

Water Concentrations: SURFACE WATER: n-Propyl acetate was detected, but not quantified, in water samples taken from the River Lee, in the U.K., date not given [35].

n-Propyl Acetate

Effluent Concentrations: n-Propyl acetate was detected as gaseous emission from the production of RDX at the Holston Army Ammunition Plant, TN, date not provided, at an emission rate of 1,134 lbs/day [26].

Sediment/Soil Concentrations:

Atmospheric Concentrations: n-Propyl acetate was qualitatively detected in the air of the industrialized Kanawha Valley, WV, in 1977 [9].

Food Survey Values: n-Propyl acetate was identified as a component of Concord grape juice essence [30], and as a volatile flavor component of Beaufort cheese [7], intact, ripening bananas [21] and Kogyoke apples [37]. n-Propyl acetate was identified as a volatile component of the blended fruit of nectarines, and also in a headspace analysis of the intact fruit [32].

Plant Concentrations:

Fish/Seafood Concentrations:

Animal Concentrations:

Milk Concentrations:

Other Environmental Concentrations:

Probable Routes of Human Exposure: The probable routes of occupational exposure to n-propyl acetate are by inhalation and dermal contact during the production and use of this compound. The general public is likely to be exposed to n-propyl acetate by the ingestion of foods [7,21,30,32,37] in which it is contained.

Average Daily Intake:

Occupational Exposure: n-Propyl acetate was found in 1 air sample taken from 11 different auto paint shops in Spain at concentrations of 36.5 mg/m³ [23]. n-Propyl acetate was detected in 3% of the air samples taken from printing industries in Belgium collected beginning in 1983

n-Propyl Acetate

[34]. NIOSH (NOES Survey 1981-83) has statistically estimated that 22,669 workers are potentially exposed to n-propyl acetate in the USA, 88% of which are exposed during the use of trade name compounds in which n-propyl acetate is contained [24].

Body Burdens: n-Propyl acetate was detected in 49.4% of 387 expired air samples obtained from 54 nonsmoking volunteers from Chicago, IL, at a mean concentration of 2.4 ng/L [16]. n-Propyl acetate was described as a normal component of expired air [15]. n-Propyl acetate was detected in 4 out of 10 expired air samples taken from 8 smoking and nonsmoking male volunteers from TX, at concentrations ranging from trace to 2.0 ug/hr, averaging 1.39 ug/hr for the positive samples [5].

REFERENCES

1. Ambrose D et al; J Chem Therm 13: 795-802 (1981)
2. Atkinson R; Chem Rev 85: 69-201 (1985)
3. Atkinson R et al; Adv Photochem 11: 375-488 (1979)
4. Atkinson R; Int J Chem Kinet 19: 799-828 (1987)
5. Conkle JP et al; Arch Environ Health 30: 290-5 (1975)
6. Drossman H et al; Chemosphere 17: 1509-30 (1987)
7. Dumont JP, Adda J; J Agr Food Chem 26: 364-7 (1978)
8. Elam EU; Kirk-Othmer Encycl Chem Tech 3rd Ed. John-Wiley NY 9: 311-3 (1978)
9. Erickson MD, Pellizzari ED; Analysis of Organic Air Pollutants in the Kanawha Valley, WV and the Shenandoah Valley VA. USEPA-903/9-78-007 (1978)
10. Gusten H et al; J Atmos Chem 2: 83-96 (1984)
11. Hansch C, Leo AJ; Medchem project Issue No. 26 Claremont, CA Pomona College (1985)
12. Kerr JA, Stocker DW; J Atmos Chem 4: 263-76 (1986)
13. Kerr JA, Stocker DW; J Atmos 4: 253-62 (1986)
14. Kieckbusch TG, King CJ; J Chromat Sci 17: 273-6 (1979)
15. Krotoszynski B et al; J Chromat Sci 17: 273-6 (1977)
16. Krotoszynski BK et al; J Anal Toxicol 3: 225-34 (1979)
17. Lloyd AC; pp 27-48 in Tropospheric Chemistry of Aldehydes. National Bureau of Standards, Wash DC: NBS-SP-577 (1978)
18. Lyman WJ et al; Handbook of Chemical Property Estimation Methods NY: McGraw-Hill pp 4-1 to 4-33 (1982)
19. Lyman WJ et al; Handbook of Chemical Property Estimation Methods NY: McGraw-Hill pp 5-1 to 5-30 (1982)
20. Lyman WJ et al; Handbook of Chemical Property Estimation Methods NY: McGraw-Hill pp 15-1 to 15-34 (1982)
21. Macku C, Jennings G; J Agric Food Chem 35: 845-8 (1987)

n-Propyl Acetate

22. Marion CV, Malaney GW; J Water Poll Control Fed 35: 1269-84 (1963)
23. Medinilla J, Espigares M; Ann Occup Hyg 32: 509-13 (1988)
24. NIOSH; National Occupational Exposure Survey (NOES) (1989)
25. Price KS et al; J Water Poll Control Fed 46: 63-77 (1974)
26. Ryon MG et al; Database Assessment of the Health and Environmental Effects of Munitions Production Waste Products. Final Report. ORNL-6018 (NTIS DE84-016512) Oak Ridge Natl Labs, Oak Ridge, TN p 217 (1984)
27. Sax NI, Lewis RJSR; Hawley's Condensed Chemical Dictionary 11th ed Van Nostrand Reinhold Co. NY p 496 (1987)
28. SRI International; Directory of Chemical Producers (1989)
29. Stephan H, Stephan T; Solubilities of Inorganic and Organic Compounds in Binary Systems. Stephen H et al. eds, NY, NY 1: 1-79, 1604-43 (1963)
30. Stevens KL et al; J Food Sci 30: 1006-7 (1965)
31. Swann RL et al; Res Rev 85: 17-28 (1983)
32. Takeoka GR et al; J Agric Food Chem 36: 553-60 (1988)
33. USITC; Synthetic Organic Chemicals, United States Production and Sales, 1988: US Interl Trade Comm Washington, DC. USITC Publ #2219 (1989)
34. Veulemans H et al; Ind Hyg Assoc J 48: 671-6 (1987)
35. Waggott A et al; Chem Water Reuse 2: 55-9 (1981)
36. Wallington TJ et al; J Phys Chem 92: 5024-8 (1988)
37. Yajima I et al; Agric Biol Chem 48: 849-5 (1984)

Propyl Ether

SUBSTANCE IDENTIFICATION

Synonyms:

Structure:

CAS Registry Number: 111-43-3

Molecular Formula: $C_6H_{14}O$

Wiswesser Line Notation:

CHEMICAL AND PHYSICAL PROPERTIES

Boiling Point: 89-91 °C

Melting Point: -122 °C

Molecular Weight: 102.17

Dissociation Constants:

Log Octanol/Water Partition Coefficient: 2.03 [7]

Water Solubility: 4,900 mg/L at 25 °C [14]

Vapor Pressure: 62.5 mm Hg at 25 °C [14]

Henry's Law Constant: 3.39 x 10^{-3} atm-m^3/mole [11]

ENVIRONMENTAL FATE/EXPOSURE POTENTIAL

Summary: Propyl ether may be released to the environment as a result of its manufacture and use as an industrial solvent and extractant. If propyl ether is released to soil, it will be subject to volatilization. It will be expected to exhibit moderate mobility in soil and, therefore, it may leach to ground water. It will not be expected to hydrolyze in soil. If propyl ether is released to water, it will not be expected to significantly adsorb to sediment or suspended particulate matter, bioconcentrate in aquatic organisms, hydrolyze, directly photolyze, or photooxidize via reaction with photochemically produced hydroxyl radicals in the water, based upon estimated physical-chemical properties or analogies to other structurally related aliphatic ethers. Isopropyl ether in surface water will be subject to rapid volatilization with estimated half-lives of 3.2 hr and 41 hr for volatilization from a river one meter deep flowing 1 m/sec with a wind velocity of 3 m/sec and a model pond. Propyl ether may be resistent to biodegradation in environmental media based upon screening test data from studies using activated sludge inocula. Many ethers are known to be resistant to biodegradation. If propyl ether is released to the atmosphere, it will be expected to exist almost entirely in the vapor phase based upon its vapor pressure. It will be susceptible to photooxidation via vapor-phase reaction with photochemically produced hydroxyl radicals with an estimated half-life of 30 hours for this process. Direct photolysis will not be an important removal process since aliphatic ethers do not absorb light at wavelengths >290 nm. The most probable route of general population exposure to propyl ether may be via inhalation of contaminated air. Exposure through dermal contact may occur in occupational settings.

Natural Sources:

Artificial Sources: Propyl ether may be released to the environment as a result of its manufacture and use. Propyl ether can be used for the same purposes as diethyl ether [9]. Diethyl ether has many uses including use as an industrial solvent and extractant [8,13].

Terrestrial Fate: If propyl ether is released to soil, it will be subject to volatilization based upon the Henry's Law constant and vapor pressure. It will be expected to exhibit moderate mobility [15] in soil and, therefore, it may leach to ground water, based upon an estimated Koc of 303 [12]. It will not be expected to hydrolyze in soil [4]. Propyl ether may be resistent to biodegradation in environmental media based upon screening test data from studies using activated sludge inocula [10,16]. Many ethers are known to be resistant to biodegradation [1].

Propyl Ether

Aquatic Fate: If propyl ether is released to water, it will not be expected to significantly adsorb to sediment or suspended particulate matter, bioconcentrate in aquatic organisms, hydrolyze [12], directly photolyze [4], or photooxidize via reaction with photochemically produced hydroxyl radicals in the water [2], based upon estimated physical-chemical properties or analogies to other structurally related aliphatic ethers [4,12]. Isopropyl ether in surface water will be subject to rapid volatilization [12]. Using the Henry's Law constant, a half-life for volatilization of isopropyl ether from a river one meter deep flowing 1 m/sec with a wind velocity of 3 m/sec has been estimated to be 3.2 hr at 25 °C [12]. The volatilization half-life from a model pond, which considers the effect of adsorption, has been estimated to be 41 hr [18]. Propyl ether may be resistent to biodegradation in environmental media based upon screening test data from studies using activated sludge inocula [6,17]. Many ethers are known to be resistant to biodegradation [1].

Atmospheric Fate: If propyl ether is released to the atmosphere, it will be expected to exist almost entirely in the vapor phase [5] based upon the vapor pressure. It will be susceptible to photooxidation via vapor-phase reaction with photochemically produced hydroxyl radicals. An atmospheric half-life of 30 hours at an atmospheric concentration of 5 x 10^{+5} hydroxyl radicals per cm^3 has been calculated for this process based upon a measured rate constant [3]. Direct photolysis will not be an important removal process since aliphatic ethers do not absorb light at wavelengths >290 nm [4].

Biodegradation: No data concerning the biodegradation of propyl ether in environmental media were located. An activated sludge aqueous screening study found that the compound was biodegraded quickly after a 13-day lag period with a 63% theoretical biological oxygen demand being measured after 25 days incubation [6]. These screening test results suggest that propyl ether may be resistent to biodegradation in the environment. Another activated sludge screening test study suggested that the compound was degraded only slowly; the study may not have been conducted for a long enough time period as it was indicated that the lag period was >10 days and the test was run for only 14 days [17]. Many ethers are known to be resistant to biodegradation [1].

Propyl Ether

Abiotic Degradation: The rate constant for the vapor phase reaction of propyl ether with photochemically produced hydroxyl radicals has been experimentally measured to be 12.9×10^{-12} cm^3/molecule-sec at 25 °C [3] which corresponds to an atmospheric half-life of 30 hours at an atmospheric concentration of $5 \times 10^{+5}$ hydroxyl radicals per cm^3. Hydrolysis is not expected to be significant under normal environmental conditions (pH 5-9) [12]. Direct photolysis will not be an important removal process since aliphatic ethers do not absorb light at wavelengths >290 nm [4].

Bioconcentration: Based upon the log Kow, a BCF of 21 has been estimated using a recommended regression equation [12]. Based upon this estimated BCF, propyl ether will not be expected to bioconcentrate in aquatic organisms.

Soil Adsorption/Mobility: Based upon the log Kow, a Koc of 303 has been estimated using a recommended regression equation [12]. Based upon this estimated Koc, propyl ether will be expected to exhibit moderate mobility in soil [15]. Propyl ether, therefore, may slowly leach through soil to ground water.

Volatilization from Water/Soil: The half-life for volatilization of propyl ether from a river one meter deep flowing 1 m/sec with a wind velocity of 3 m/sec is estimated to be 3.2 hr at 25 °C [12] based on the Henry's Law constant. The volatilization half-life from a model pond, which considers the effect of adsorption, has been estimated to be 41 hr [18]. Based upon the Henry's Law constant and the vapor pressure, propyl ether will be subject to volatilization from surfaces and near-surface soil.

Water Concentrations:

Effluent Concentrations:

Sediment/Soil Concentrations:

Atmospheric Concentrations:

Food Survey Values:

Propyl Ether

Plant Concentrations:

Fish/Seafood Concentrations:

Animal Concentrations:

Milk Concentrations:

Other Environmental Concentrations:

Probable Routes of Human Exposure: The most probable route of general population exposure to propyl ether is via inhalation of contaminated air. Exposure through dermal contact may occur in occupational settings.

Average Daily Intake:

Occupational Exposure:

Body Burdens:

REFERENCES

1. Alexander M; Biotechnol Bioeng 15: 611-47 (1973)
2. Anbar M, Neta P; Int J Appl Radiation Isotopes 18: 493-523 (1967)
3. Bennett PJ, Kerr JA; J Atmos Chem 10: 29-38 (1990)
4. Calvert JG, Pitts JNJr; Photochemistry. John Wiley & Sons: New York pp 441-2 (1966)
5. Eisenreich SJ et al; Environ Sci Technol 15: 30-8 (1981)
6. Fujiwara Y et al; Yukagaku 33: 111-14 (1984)
7. Hansch C, Leo AJ; Medchem Project Issue No. 26 Claremont, CA: Pomona College (1985)
8. Hawley GG; Condensed Chemical Dictionary 10th ed Van Nostrand Reinhold NY p 435 (1981)
9. Heitmann W et al; Ullman's Encyclopedia of Industrial Chemistry. 5th Comp Rev ed Gerharts W et al Eds. VCH, Weinhein, Fed Rep Germany. p 29 (1987)
10. Heukelekian H, Rand MC; J Water Pollut Control Assoc 29: 1040-53 (1955)
11. Hine J, Mookerjee PK; J Org Chem 40: 292-8 (1975)
12. Lyman WJ et al; Handbook of Chem Property Estimation Methods NY: McGraw-Hill pp 4-9, 5-5, 7-4, 15-16 to 15-29 (1982)
13. Merck; The Merck Index An Encyclopedia of Chemicals, Drugs, and Biologicals 10th ed Rahway, NJ: Merck & Co p 551 (1983)

Propyl Ether

14. Riddick JA et al; Organic Solvents. John Wiley and Sons Inc. NY (1984)
15. Swann RL et al; Res Rev 85: 17-28 (1983)
16. Takemoto S et al; Suishitsu Odaku Kenkyu 4: 80-90 (1981)
17. Urano K, Kato Z; J Haz Materials 13: 147-59 (1986)
18. USEPA; EXAMS II Computer Simulation (1987)

Terpinolene

Synonyms:

Structure:

CAS Registry Number: 586-62-9

Molecular Formula: $C_{10}H_{16}$

Wiswesser Line Notation:

CHEMICAL AND PHYSICAL PROPERTIES

Boiling Point: 185 °C at 360 mm Hg

Melting Point:

Molecular Weight: 136.24

Dissociation Constants:

Log Octanol/Water Partition Coefficient: 4.23 (estimated) [7]

Water Solubility: 1.74 (calculated) [13]

Vapor Pressure: 0.595 (extrapolated) [17]

Henry's Law Constant: 6.14×10^{-2} atm-m^3/mole [11]

Terpinolene

ENVIRONMENTAL FATE/EXPOSURE POTENTIAL

Summary: Terpinolene is released to the environment as a result of its production and emission by plants, especially certain types of trees. It also may be released as a result of its manufacture and/or isolation from plants and subsequent use as a solvent for resins and its use in the manufacture of synthetic resins and flavors. If released to soil, it will be expected to be immobile due to strong adsorption to soil, based upon an estimated Koc. The strong adsorption to soil is expected to greatly limit volatilization from near-surface soil. It will not hydrolyze in soil, but terpinolene may be subject to biodegradation in soil based upon limited data from a study of the treatment of wastewater from kraft wood pulp mills. If released to water, it will not be expected to bioconcentrate in aquatic organisms (based upon an estimated BCF), hydrolyze or directly photolyze. It will be expected to be subject to rapid volatilization based upon an estimated half-life of 3.4 hr for volatilization from a model river one meter deep flowing 1 m/sec with a wind velocity of 3 m/sec. Adsorption to sediment and suspended particulate matter may attenuate the loss of terpinolene from water via volatilization. The half-life for volatilization from a model pond has been estimated to be 41 days. Terpinolene which does not volatilize will be expected to adsorb to sediment and suspended particulate matter, based upon an estimated Koc. No data were located that demonstrate biodegradation of terpinolene in environmental media, laboratory screening tests or biological treatment plants or simulators. However, data from a study of the treatment of wastewater from kraft wood pulp would suggest that the compound may be biodegradable in natural water. If released to the atmosphere, it will be expected to exist almost entirely in the vapor phase based upon its vapor pressure. It will be expected to rapidly degrade in the atmosphere via reactions with hydroxyl radicals ozone and nitrate radicals. The half-lives for vapor phase reactions with photochemically produced hydroxyl radicals and ozone have been calculated to be 1.4 hr and 1.7 to 23 min, respectively, based upon measured rate constants and atmospheric concentrations of $5 \times 10^{+5}$ hydroxyl radicals per cm^3 and $7 \times 10^{+11}$ ozone molecules per cm^3. The half-life for the vapor phase reaction with nitrate radicals in nighttime air has been calculated to be 1.0 min, based upon a measured rate constant and an atmospheric concentration of $2.4 \times 10^{+8}$ nitrate radicals per cm^3. It will not be expected to directly photolyze in the atmosphere. General population exposure to terpinolene may occur through the ingestion of contaminated food, especially fruits such as nectarines and mangoes, and from the inhalation of contaminated air, especially air in some forests. Occupational exposure may occur through the inhalation of contaminated air and dermal contact with solutions that contain the compound.

Natural Sources: Terpinolene is naturally emitted from various plants, especially trees [12]. The compound has been detected in emissions from the following arboreous plants (9 of 17 species studied were positive; detection limit not specified): Scots pine, Siberian pine, Silver fir, Common juniper, Zeravshan juniper, Pencil cedar, Evergreen cypress, Northern white cedar and Chinese arbor vitae; it was also found in emissions from 1 of 5 species of plants studied (Marsh tea) which grow under the canopy of coniferous forests [12]. It has been found in the essential oils of numerous (unspecified) plants and in pine gum terpentines [2].

Artificial Sources: Terpinolene may be released to the environment as a result of its manufacture and/or isolation from plants and subsequent use as a solvent for resins and essential oils, and its use in the manufacture of synthetic resins and flavors [6,10,23]. Terpinolene has been found in the wastewater effluent from pulp and paper mills [6,23] both before and after treatment in aerated lagoons with retention times of 7 days [23].

Terrestrial Fate: If terpinolene is released to soil, it will be expected to be immobile due to strong adsorption to soil, based upon an estimated Koc of 4766 [7,13,19]. Therefore, it will not be expected to leach through soil to ground water. Although the estimated Henry's Law constant suggests that terpinolene may be subject to rapid volatilization from moist near-surface soils, strong adsorption to soil is expected to greatly limit this process. It will not hydrolyze in soil [2], but terpinolene may be subject to biodegradation in soil based upon data from a study of the treatment of wastewater from kraft wood pulp mills [23] which suggest that the compound may be biodegradable.

Aquatic Fate: If terpinolene is released to water, it will not be expected to bioconcentrate in aquatic organisms based upon an estimated BCF of 966 [7,13], hydrolyze [7,13] or directly photolyze [18]. It will be expected to be subject to rapid volatilization based upon an estimated half-life of 3.4 hr for volatilization from a model river one meter deep flowing 1 m/sec with a wind velocity of 3 m/sec [13] which was calculated using the estimated Henry's Law constant. Adsorption to sediment and suspended particulate matter may attenuate the loss of terpinolene from water via volatilization based upon an estimated Koc of 4766 [7,13] and an estimation of the half-life for volatilization from a model pond (half-life, 41 days which considers the effect of adsorption [21]. Terpinolene which does not volatilize will be expected to adsorb to sediment and suspended particulate matter based upon an estimated Koc [7,13]. Although no data were located that demonstrates biodegradation of terpinolene either in environmental media, laboratory screening tests or biological treatment plants or simulators data from a study of the treatment of wastewater from kraft wood pulp mills [23], suggest that the compound may be biodegradable in natural water.

Atmospheric Fate: If terpinolene is released to the atmosphere, it will be expected to exist almost entirely in the vapor phase [8] based upon its vapor pressure. It will be expected to rapidly degrade in the atmosphere via reactions with hydroxyl radicals, ozone, and nitrate radicals [3,4]. The half-life for the vapor phase reaction with photochemically produced hydroxyl radicals has been calculated to be 1.4 hr based upon a measured rate constant [4] and an atmospheric concentration of $5 \times 10^{+5}$ hydroxyl radicals per cm^3. The half-life for the vapor-phase reaction with photochemically produced ozone has been calculated to range between 1.7 and 23 min, based upon two measured rate constants [3] and an atmospheric concentration of $7 \times 10^{+11}$ ozone molecules per cm^3. The half-life for the vapor-phase reaction with nitrate radicals in nighttime air has been calculated to be 1.0 min, based upon a measured rate constant [4] and an atmospheric concentration of $2.4 \times 10^{+8}$ nitrate radicals per cm^3. It will not be expected to directly photolyze in the atmosphere since it will not absorb sunlight [18].

Biodegradation: No data that demonstrate biodegradation of terpinolene either in environmental media, laboratory screening tests or biological treatment plants or simulators were located. Data from a study in 1973 of the treatment of wastewater from both a bleached and an unbleached kraft mill, however, suggest that the compound may be biodegradable in the environment [23]. The compound was detected in wastewater from the bleached and unbleached kraft mills at concn ranging from trace levels to 40 ppb before secondary treatment in aerated lagoons and at concn ranging from not detected to 10 ppb after treatment in the aerated lagoons (volumetric retention time of 7 days; detection limit and trace level concn not defined) [23]. In 3 of the seven sampling periods, the concn of the compound was observed to be decreased after the lagoon treatment period [23]. In 3 of seven sampling periods, however, little or no changes in concn were apparent and in 1 sampling period the concn actually appeared to increase from not detected to 10 ppb [23]. Furthermore, no indication was given of the amounts of compound lost from the wastewater which could be attributed to biological processes versus other processes such as volatilization, although it was acknowledged that at least some of the loss of compound from the wastewaters was due to volatilization [23].

Abiotic Degradation: Terpinolene is highly reactive in a number of photooxidation reactions which occur in the atmosphere and as a result will have a short residence time in the atmosphere [1,2,3,5,6,22]. Terpinolene has been found to be highly reactive in smog chamber studies in which the compound is irradiated in the presence of NOx [1,2,5,22]. In one such study, terpinolene was determined to be 10 times more reactive than β-pinene and 13 times more reactive than isobutene at 28 °C and a terpinolene to NOx ratio of 10 ppb to 7 ppb [22]. The compound is not, however, efficient in the overall production of ozone in these smog chamber studies due in part to its own rapid reaction with ozone [1,2,3,5] and the consumption of large amounts of carbon in the formation of large amounts of aerosols [1,2,5]. In another smog chamber study using a terpinolene/NOx ratio of 8 ppm to 1.3 ppm, 95% of the compound was consumed in one hr and the products observed by infrared spectroscopy were formaldehyde (most abundant product observed), formic acid, carbon monoxide, carbon dioxide, acetaldehyde, peroxyacetyl nitrate, and acetone; these observed products accounted for only 5% of the total carbon contained in the starting amount of terpinolene [2,3,5]. Terpinolene is highly reactive in a number of photooxidation reactions which occur in the atmosphere and as a result will have a short residence time in the atmosphere [3,4]. The rate constant for the vapor-phase reaction of terpinolene with photochemically produced ozone has been measured to be 7.3×10^{-16} cm^3/molecule-sec at 25 °C [3] and 1.0×10^{-14} cm^3/molecule-sec at 22 °C [3] which correspond to atmospheric half-lives of 23 min and 1.7 min, respectively, at an atmospheric concentration of 7 x 10^{+11} ozone molecules per cm^3. The rate constant for the vapor-phase reaction with photochemically produced hydroxyl radicals has been measured to be 2.0 x 10^{-10} cm^3/molecule-sec at 25 °C [4] which corresponds to an atmospheric half-life of 1.9 hr at an atmospheric concentration of 5 x 10^{+5} hydroxyl radicals per cm^3. The overall daylight half-life for these two photooxidation reactions can be calculated to range between 1.7 and 19 min, based upon the measured rate constants and the assumed concn of the oxidants [3,4]. The rate constant for the vapor phase reaction of terpinolene with nitrate radicals present in nighttime air has been measured to be 7.1 x 10^{-11} cm^3/molecule-sec at 25 °C [4] which corresponds to a nighttime atmospheric half-life of 41 sec at an atmospheric concentration of 2.4 x 10^8 nitrate radicals per cm^3. Hydrolysis of terpinolene is not expected to be significant under normal environmental conditions (pH 5-9) [13]. Direct photolysis will not be a removal process in the environment because compounds such as terpinolene which contain only isolated double bonds do not absorb light at wavelengths >290 nm [18].

Bioconcentration: Based upon the estimated log Kow, an estimated BCF of 966 has been calculated using a recommended regression equation [13]. Based upon this estimated BCF, terpinolene may bioconcentrate in aquatic organisms.

Soil Adsorption/Mobility: Based upon the estimated log Kow an estimated Koc of 4766 has been calculated using a recommended regression equation [13]. Based upon the estimated water solubility, an estimated Koc of 3219 has been calculated using a recommended regression equation [13]. Based upon these estimated Koc, terpinolene will be expected to be immobile in soil and will not be expected to leach through soil to ground water [19].

Volatilization from Water/Soil: The half-life for volatilization of terpinolene from a model river one meter deep flowing 1 m/sec with a wind velocity of 3 m/sec is estimated to be 3.4 hr at 25 °C [13] based on an estimated Henry's Law constant of 6.14 x 10^{-2} atm-m^3/mole [11]. The volatilization half-life from a model pond, which considers the effect of adsorption, has been estimated to be 41 days [21]. Based upon its vapor pressure and the Henry's Law constant, terpinolene may volatilize from near-surface soil and other surfaces; adsorption, however, may attenuate this process.

Water Concentrations:

Effluent Concentrations: Terpinolene was qualitatively detected in 1 out of >4000 samples of effluent from 1 of 46 industrial categories (pulp and paper mills) [6]. The compound was detected in 1973 in kraft mill wastewater from both a bleached and an unbleached kraft mill at concn ranging from trace levels to 40 ppb before secondary treatment in aerated lagoons and ranging from not detected to 10 ppb after treatment in the aerated lagoons (volumetric retention time of 7 days; detection limit and trace level concn not specified)[23]. Although no data regarding detection of terpinolene in atmospheric samples were located, the compound may be expected to be found in samples of air from certain types of forests based upon the detected emission of the compound from 9 to 17 species of arboreous plants studied and 1 of 5 plants species that grow under the canopy of coniferous forests (detection limit not specified) [12]. It was qualitatively detected in emissions from the following arboreous plants (detection limit not specified): Scots pine, Siberian pine, Silver fir, Common juniper, Zeravshan vitae; it was also found in emissions from Marsh tea which grows under the canopy of coniferous forests [12]. It was not detected (detection limit not specified) in emissions from the following arboreous plants: Bay-leave willow, Aspen, Balsam poplar, European oak, European birch, Sorb, European larch and European fir; nor was it found in emissions from the following plants which grow under the canopy of coniferous forests: Red billberry shrub, Billberry shrub, Fern and Deciduous moss [12].

Sediment/Soil Concentrations:

Atmospheric Concentrations:

Food Survey Values: Terpinolene was detected in the mixture of volatile components of mangoes grown in Florida at a concn of 2.0 ug/g in fruit stored in a deep freeze at -15 °C for 14 months and at a concn of 1.1 ug/g in fresh fruit [14]. Terpinolene was detected in the mixture of volatile components of 3 of 4 varieties of California nectarines (detection limit not reported) [9]. The concn detected in Sunfree nectarines was 10

Terpinolene

ppb and the concn detected in 2 experimental varieties of nectarines were below the quantitation limit of 10 ppb [9]. The compound was not detected in Flavortop nectarines [9]. In another study, the compound was detected in the mixture of volatile components of unspecified nectarines (detection limit not reported) [20].

Plant Concentrations:

Fish/Seafood Concentrations:

Animal Concentrations:

Milk Concentrations:

Other Environmental Concentrations:

Probable Routes of Human Exposure: General population exposure to terpinolene may occur through the ingestion of contaminated foods, especially fruits such as nectarines [9,20] and mangoes [14], and from the inhalation of contaminated air, especially air in and near certain types of forests [12]. Occupational exposure may occur through the inhalation of contaminated air and dermal contact with solutions that contain the compound.

Average Daily Intake:

Occupational Exposure: NIOSH (NOES Survey 1981-1983) has statistically estimated that 47,312 workers are potentially exposed to terpinolene in the USA [15]. NIOSH (NOHS Survey 1972-1974) has statistically estimated that 49,927 workers are potentially exposed to terpinolene in the USA [16].

Body Burdens:

REFERENCES

1. Altshuller AP; Atmos Environ 17: 2131-65 (1983)
2. Arnst RR, Gay BW; Photochemistry of Some Naturally Emitted Hydrocarbons Res Triangle Park, NC: USEPA-600/3-79-081 p 6 (1979)
3. Atkinson R, Carter WP; Chem Rev 84: 437-70 (1984)

525

Terpinolene

4. Atkinson R et al; Environ Sci Technol 19: 159-63 (1985)
5. Bufalini JJ; Impact of Natural Hydrocarbons on Air Quality Res Triangle Park, NC: USEPA-600/2-80-086 (1980)
6. Bursey JT, Pellizzari ED; Analysis of Industrial Wastewater for Organic Pollutants in Consent Degree Survey. Contract No. 68-03-2867. Athens,GA: USEPA Environ Res Lab p 88 (1982)
7. CLOGP3; PCGEMS Graphical Exposure Modeling System USEPA (1986)
8. Eisenreich SJ et al; Environ Sci Technol 15: 30-8 (1981)
9. Engel KH et al; J Agric Food Chem 36: 549-53 (1988)
10. Hawley GG; Condensed Chemical Dictionary 10th ed Van Nostrand Reinhold NY p 1001 (1981)
11. Hine J, Mookerjee PK; J Org Chem 40: 292-8 (1975)
12. Isidorov VA et al; Atmos Environ 19: 1-8 (1985)
13. Lyman WJ et al; Handbook of Chem Property Estimation Methods NY: McGraw-Hill pp 4-9, 5-5, 7-4, 15-16 to 15-29 (1982)
14. MacLeod AJ, Snyder CH; J Agric Food Chem 36: 137-9 (1988)
15. NIOSH; The National Occupational Exposure Survey (NOES) (1983)
16. NIOSH; The National Occupational Hazard Survey (NOHS) (1974)
17. Perry RH, Green D; Perry's Chemical Handbook 6th ed. NY: McGraw-Hill (1984)
18. Silverstein RM et al; Spectrometric Id Org Cmpd NY: J Wiley & Sons Inc 3rd Ed pp 238-55 (1974)
19. Swann RL et al; Res Rev 85: 17-28 (1983)
20. Takeoka GR et al; J Agric Food Chem 36: 553-60 (1988)
21. USEPA; EXAMS II Computer Simulation (1987)
22. Westberg HH, Rasmussen RA; Chemosphere 163-8 (1972)
23. Wilson D, Hrutfiord B; Pulp Paper Canada 76: 91-3 (1975)

Tetraethylene Glycol

SUBSTANCE IDENTIFICATION

Synonyms: Ethanol, 2,2'-(oxybis(2,1-ethanediyloxy))bis

Structure:

HO$\diagup\diagup$O$\diagup\diagup$O

HO$\diagup\diagup$O

CAS Registry Number: 112-60-7

Molecular Formula: $C_8H_{18}O_5$

Wiswesser Line Notation: Q2O2O2O2Q

CHEMICAL AND PHYSICAL PROPERTIES

Boiling Point: 327.3 °C

Melting Point: -6.2 °C

Molecular Weight: 194.23

Dissociation Constants:

Log Octanol/Water Partition Coefficient: -1.18 estimate [3]

Water Solubility: Miscible [6]

Vapor Pressure: 6.1 x 10^{-7} mm Hg at 25 °C [5]

Henry's Law Constant: 4.91 x 10^{-13} atm-m^3/mole at 25 °C estimate [10]

ENVIRONMENTAL FATE/EXPOSURE POTENTIAL

Summary: Tetraethylene glycol may be released to the environment via effluents at sites where it is produced or used as a solvent for nitrocellulose, lacquers, and coating compositions, and as a plasticizer.

527

Tetraethylene Glycol

Tetraethylene glycol is not expected to undergo hydrolysis or direct photolysis in the environment. The complete miscibility of tetraethylene glycol in water suggests that volatilization, adsorption and bioconcentration are not important fate processes. This is supported by its vapor pressure and the estimated Henry's Law constant, which indicates that volatilization of tetraethylene glycol from natural waters and soil should be extremely slow. A low estimated log BCF suggests tetraethylene glycol should not bioconcentrate among aquatic organisms. A low Koc indicates tetraethylene glycol should not partition from the water column to organic matter contained in sediments and suspended solids, and it should be highly mobile in soil. Aqueous screening test data indicate that biodegradation is likely to be an important removal mechanism of tetraethylene glycol from acclimated aerobic soil and water. In the atmosphere, tetraethylene glycol is expected to exist both in the vapor and particulate phases and reactions with photochemically produced hydroxyl radicals may be important (estimated vapor-phase half-life of 11.5 hr). Physical removal of tetraethylene glycol from air by precipitation and dissolution in clouds may occur. The most probable human exposure would be occupational exposure, which may occur through dermal contact or inhalation at workplaces where it is produced or used.

Natural Sources:

Artificial Sources: Tetraethylene glycol may be released to the environment via effluents at sites where it is produced or used as a solvent for nitrocellulose, lacquers, and coating compositions, and as a plasticizer [8].

Terrestrial Fate: Alcohols and ethers are generally resistant to hydrolysis [11]. They do not absorb UV light in the environmentally significant range (>290 nm) and are commonly used as solvents for obtaining UV spectra [15]. Therefore, tetraethylene glycol should not undergo hydrolysis in moist terrestrial environments, or direct photolysis on sunlit soil surfaces. The estimated Henry's Law constant indicates that volatilization of tetraethylene glycol from moist soil should not be an important fate process [11]. An estimated Koc of 5 [11] indicates tetraethylene glycol should be highly mobile in soil [16]. Aqueous screening test [12,14] data suggest that aerobic biodegradation is likely to be an important removal mechanism of triethylene glycol from acclimated aerobic soil.

Aquatic Fate: Alcohols and ethers are generally resistant to hydrolysis [11]. They do not absorb UV light in the environmentally significant range (>290 nm) and are commonly used as solvents for obtaining UV spectra [15]. Therefore, tetraethylene glycol should not undergo hydrolysis or direct photolysis in aquatic environments. The miscibility of tetraethylene glycol in water suggests that volatilization, adsorption and bioconcentration are not important fate processes. This is supported by the estimated Henry's Law constant, which indicates that volatilization of tetraethylene glycol from natural waters should be extremely slow [11]. An estimated Koc of 5 [11] indicates tetraethylene glycol should not partition from the water column to organic matter contained in sediments and suspended solids; and an estimated bioconcentration factor (log BCF) of -1.13 [11] indicates tetraethylene glycol should not bioconcentrate among aquatic organisms. Aqueous screening test [12,14] data suggest that aerobic biodegradation is likely to be an important removal mechanism of triethylene glycol from acclimated aquatic systems.

Atmospheric Fate: Alcohols and ethers do not absorb UV light in the environmentally significant range (>290 nm) and are commonly used as solvents for obtaining UV spectra [15]. Therefore, tetraethylene glycol should not undergo direct photolysis in the atmosphere. Based on the vapor pressure, tetraethylene glycol is expected to exist both in the vapor and particulate phases in ambient air [7] where vapor-phase reactions with photochemically produced hydroxyl radicals may be important. The rate constant for tetraethylene glycol has been estimated to be 3.38 x 10^{-11} cm^3/molecule-sec at 25 °C, which corresponds to an atmospheric half-life of about 11.5 hr at an atmospheric concentration of 5 x 10^{+5} hydroxyl radicals per cm^3 [2]. The complete miscibility of tetraethylene glycol in water indicates that physical removal from air by precipitation and dissolution in clouds may occur.

Biodegradation: Soil grab sample and river die-away test data pertaining to the biodegradation of tetraethylene glycol in soil and natural waters were not located in the available literature. Yet, a few of aerobic biological screening studies, which utilized settled waste water, sewage, or activated sludge for inocula, indicate that tetraethylene glycol should biodegrade in the environment [1,4,9,12,14]. Tetraethylene glycol biodegraded slowly without acclimation [1,4,9,14]; however, it degraded rapidly with acclimation [12] periods of 45-60 days [14]. Acclimation periods of 45-60 days; 9.58, 71 and 88% BODT of tetraethylene glycol occurred in 5,10,15 and 20 days, respectively, when incubated concentration ranging from 3-10 ppm [14]. A semi-continuous biological treatment simulator showed as 29% loss of tetraethylene glycol in < 24 hr [12].

Abiotic Degradation: Alcohols and ethers are generally resistant to hydrolysis [11]. They do not absorb UV light in the environmentally significant range (>290 nm) and are commonly used as solvents for obtaining UV spectra [15]. Therefore, tetraethylene glycol should not undergo hydrolysis or direct photolysis in the environment. The rate constant for the vapor-phase reaction of tetraethylene glycol with photochemically produced hydroxyl radicals in air has been estimated to be 3.38×10^{-11} cm^3/molecule-sec at 25 °C, which corresponds to an atmospheric half-life of about 11.5 hr at an atmospheric concentration of $5 \times 10^{+5}$ hydroxyl radicals per cm^3 [2].

Bioconcentration: Because tetraethylene glycol is miscible in water, bioconcentration in aquatic systems is not expected to be an important fate process. Based upon the estimated log Kow, a bioconcentration factor (log BCF) of -1.13 for tetraethylene glycol has been calculated using a recommended regression-derived equation [11]. This BCF value also indicates tetraethylene glycol should not bioconcentrate in aquatic organisms.

Soil Adsorption/Mobility: Because tetraethylene glycol is miscible in water, soil adsorption is not expected to be an important fate process. Based on the estimated log Kow, a Koc of 5 for tetraethylene glycol has been calculated using a recommended regression-derived equation [11]. This Koc value indicates tetraethylene glycol will be highly mobile in soil [16], and it should not partition from the water column to organic matter contained in sediments and suspended solids.

Volatilization from Water/Soil: Because tetraethylene glycol is miscible in water, and based upon the estimated Henry's Law constant, the volatilization of tetraethylene glycol from natural bodies of water and moist soils is not expected to be an important fate process [11].

Water Concentrations:

Effluent Concentrations:

Sediment/Soil Concentrations:

Atmospheric Concentrations:

Food Survey Values:

Plant Concentrations:

Fish/Seafood Concentrations:

Animal Concentrations:

Milk Concentrations:

Other Environmental Concentrations:

Probable Routes of Human Exposure: The most probable route of human exposure to tetraethylene glycol is by inhalation and dermal contact.

Average Daily Intake:

Occupational Exposure: The most probable human exposure to tetraethylene glycol would be occupational exposure, which may occur at places where it is produced or used as a solvent, plasticizer and humectant. NIOSH (NOES Survey as of 3/28/89) has estimated that 32,619 workers are potentially exposed to tetraethylene glycol in the USA [13].

Body Burdens:

Tetraethylene Glycol

REFERENCES

1. Alexander M; Biotech Bioeng 15: 611-47 (1973)
2. Atkinson R; Intern J Chem Kin 19: 799-828 (1987)
3. CLOGP; PCGEMS Graphical Exposure Modeling System USEPA (1986)
4. Cox DP; Adv Appl Microbiol 23: 173-93 (1978)
5. Daubert TE, Danner RP; Data Compilation, Tables of Properties of Pure Cmpds, Design Inst for Phys Prop Data, Am Inst for Phys Prop Data, NY, NY (1989)
6. Dow Chemical Co; The Glycol Ethers Handbook. Midland, MI (1981)
7. Eisenreich SJ et al; Environ Sci Technol 15: 30-8 (1981)
8. Hawley GG; Condensed Chemical Dictionary 10th ed Van Nostrand Reinhold NY p 375 (1981)
9. Heukelekian H, Rand MC; J Water Pollut Control Assoc 29: 1040-53 (1955)
10. Hine J, Mookerjee PK; J Org Chem 40: 292-8 (1975)
11. Lyman WJ et al; Handbook of Chemical Property Estimation Methods NY: McGraw-Hill pp 4-9, 5-4, 6-3, 15-16 (1982)
12. Mills EJ Jr, Stack VT Jr; pp 492-517 in Proc 8th Industrial waste Conf Eng Bull Purdue Univ Eng Ext Ser (1954)
13. NIOSH; National Occupational Exposure Survey (NOES) (1989)
14. Price KS et al; J Water Pollut Contr Fed 46: 63-77 (1974)
15. Silverstein RM, Bassler GC; Spectrometric Id Org Cmpd NY: J Wiley & Sons Inc pp 148-69 (1963)
16. Swann RL et al; Res Rev 85: 16-28 (1983)

Tribromomethane

Synonyms:

Structure:

$$H - CHBr_3$$

(structure showing a central carbon bonded to H and three Br atoms)

CAS Registry Number: 75-25-2

Molecular Formula: $CHBr_3$

Wiswesser Line Notation: EYEE

CHEMICAL AND PHYSICAL PROPERTIES

Boiling Point: 149.5 °C at 15 mm Hg

Melting Point: 8.3 °C

Molecular Weight: 252.73

Dissociation Constants:

Log Octanol/Water Partition Coefficient: 2.37 [21] (calc)

Water Solubility: 3,100 mg/L at 25 °C [25]

Vapor Pressure: 5.4 mm Hg at 25 °C [4]

Henry's Law Constant: 5.35 x 10^{-4} atm-m³/mol at 25 °C [34]

ENVIRONMENTAL FATE/EXPOSURE POTENTIAL

Summary: Tribromomethane is released to the environment by industrial activities involving bromine, is formed during the chlorination of water, perhaps by the haloform reaction, and a major source is considered to be marine algae. Release to soil will likely result in leaching to ground water due to the weak adsorption of tribromomethane to soil; volatilization from moist soil surfaces should also occur. Anaerobic biodegradation may occur, but aerobic biodegradation is not expected to be fast. Hydrolysis is expected to be slow. Release of tribromomethane to the atmosphere will result in the reaction of tribromomethane with photochemically generated hydroxyl radicals with an estimated half-life of 325 days. Direct photolysis in the troposphere is not expected to be significant, but may occur in the stratosphere. Human exposure to tribromomethane will most likely occur due to the ingestion of drinking water or air contaminated with tribromomethane.

Natural Sources:

Artificial Sources: Tribromomethane is used as an intermediate in geological assaying, as a solvent for waxes, greases, and oils and as a sedative [24]. It has been hypothesized that tribromomethane may be formed by the haloform reaction, which occurs during the chlorination of water [11]. Tribromomethane is often found in drinking water as a disinfection by-product [29]. Based on monitoring data, mean loadings of tribromomethane in treated wastewater from nonferrous metal manufacturing were 0.017 kg/day and from pulp and paperboard mills were 0.3 kg/day [45]. Marine algae are considered to be a major environmental source of tribromomethane [31].

Terrestrial Fate: The tribromomethane content of secondary wastewater decreased by >10% following percolation through soil treatment basins [7]. The residence time in the basins was 8 hr [7]. Tribromomethane is expected to bind weakly to soil and may, therefore, readily leach. Anaerobic biodegradation may occur, but aerobic biodegradation is not expected to be rapid. Volatilization from moist soil surfaces is expected to be rapid.

Aquatic Fate: The primary fate of tribromomethane in water is expected to be volatilization. Anaerobic biodegradation may also occur but aerobic biodegradation is not expected to be a fast process. Bioconcentration and hydrolysis (half-life = 686 years) are not expected to be an important fate process.

Tribromomethane

Atmospheric Fate: A half-life of 325 days was estimated for the reaction of tribromomethane with photochemically generated hydroxyl radicals in the atmosphere [2]. Direct photolysis in the troposphere is not expected to be a significant degradation process.

Biodegradation: Tribromomethane was incubated with sewage seed at 5 and 10 mg/L for 7 days followed by three weekly subcultures at 25 °C [44]. In the 5 and 10 mg/L cultures, tribromomethane had degraded by 11% and 4%, respectively, at 7 days and by 48% and 35%, respectively, at 28 days [44]. Tribromomethane was not biodegraded in aerobic batch cultures, but >99% of the initial tribromomethane was removed by treatment in a methanogenic (anaerobic conditions) biofilm column (2 day retention time in column) [5]. Under denitrifying conditions, a 60 ug/L solution of tribromomethane degraded to 59, 37, 35, and 2 ug/L after 2, 3, 4, and 6 weeks, respectively [6].

Abiotic Degradation: The rate constant of the hydrolysis of tribromomethane is 3.2×10^{-11} sec^{-1} and the half-life of the reaction is 686 years [33]. A half-life of 325 days was estimated for the reaction of tribromomethane with photochemically generated hydroxyl radicals in the atmosphere [2]. Direct photolysis is not expected to be significant in the troposphere, but by analogy to methyl bromide [39], direct photolysis in the stratosphere may be an important fate process.

Bioconcentration: Using the estimated log octanol/water partition coefficient, a bioconcentration factor of 37.4 was estimated for tribromomethane in fish [32].

Soil Adsorption/Mobility: A Freundlich K value of 1.54 was determined on Keweenaw sandy loam [26]. Using the water solubility, a log soil sorption coefficient of 1.99 was estimated for tribromomethane [32]. Tribromomethane adsorption to soil is expected to be weak from these results and estimations and therefore, tribromomethane is expected to leach.

Volatilization from Water/Soil: The volatilization half-lives of tribromomethane at 25 °C at depths of 6.5 cm and 14.5 cm are 29.3 and 65.4 min, respectively [20]. Using the measured Henry's Law constant, the estimated half-life for the volatilization of tribromomethane from a model river with a wind velocity of 3 m/s, a current velocity of 1 m/s, and a depth of 1 m was 7.2 hr [32].

Water Concentrations: SURFACE WATER: Eugene, OR - 1% samples pos, 1 ug/L [13]. New Jersey - 604 samples, 32.6% pos, 3.7 ppb maximum [36]. New Jersey - 0.6 ug/L mean [38]. New Orleans/Baton Rouge - not detected-0.57 ug/L, 0.25 ug/L mean [38]. Delaware -280 ppb maximum [14]. Niagara River (lower) - not detected-6 ng/L [27]. Lake Ontario - not detected-7 ng/L [27]. Allegheny River (at Pittsburgh, PA) - 501 samples, 1.4% pos, 2 at 0.1-1.0 ug/L and 5 at 1.0-10.0 ug/L [35]. Ohio River (at West View, OH) - 113 samples, 9.7% pos, 10 at 0.1- 1.0 ug/L and 1 at 1.0-10.0 ug/L [35]. Ohio River (at Wheeling, WV) - 539 samples, 0.7% pos, 3 at 0.1-1.0 ug/L, 1 at 1.0-10.0 ug/L [35]. Ohio River (at Parkersburg, WV) - 264 samples, 1.5% pos, 4 at 1.0-10.0 ug/L [35]. Kanawha River (at St. Albans, WV) - 257 samples, 1.5% pos, 4 at 1.0-10.0 ug/L [35]. Kanawha River (at St. Albans, WV) - 257 samples, 31.5% pos, 70 at 0.1-1.0 ug/L, 11 at 1.0-10.0 ug/L [35]. Ohio River (at Huntington, WV) - 530 samples, 17.4% pos, 84 at 0.1-1.0 ug/L, 8 at 1.0-10.0 ug/L [35]. Ohio River (at Portsmouth, OH) - 451 samples, 2.0% pos, 6 at 0.1-1.0 ug/L, 3 at 1.0-10.0 ug/L [35]. Ohio River (at Cincinnati, OH) - 717 samples, 0.3% pos, 2 at 0.1-1.0 ug/L [35]. Ohio River (at Louisville, KY) - 712 samples, 0.4% pos, 3 at 0.1-1.0 ug/L [35]. Ohio River (at Evansville, IN) - 632 samples. 0.9% pos, 5 at 0.1-1.0 ug/L, 1 at 1.0-10.0 ug/L [35]. Tribromomethane was detected but not quantified in surface water samples taken in Narragansett Bay, RI [46]. Arctic seawater - 9.8 ng/L [19]. Arctic ocean surface water - 3-7 ng/L [31]. DRINKING WATER: Various US water treatment plants [18] -not detected-92 ug/L [43]. Unspecified US cities - 12 ug/L mean (of positives), <0.3 ug/L median of all samples [8]. Delaware - 20 ppb in drinking water wells [10]. Tribromomethane was detected but not qualified in drinking water samples taken in Washington, DC [40], Philadelphia, PA [42], unspecified US drinking water [28], Japan [41], England [18]. 35 Water utilities throughout US - median value spring, summer, fall 1988, winter 1989 were 0.33, 0.57, 0.88, and 0.51 ug/L, respectively [29]. Cairo, Egypt - 0.13-5.19 ug/L, 1.56 ug/L avg [16]. GROUND WATER: New Jersey - 1072 samples, 21.9% pos, 34.3 ppb maximum [36]. Delaware - 20 ppb in drinking water wells [10]. Unspecified, randomly selected US sites serving <10,000 persons - 280 samples, 15.7% pos, 2.4 ug/L median (of positives), 54 ug/L maximum

Tribromomethane

[47]. Unspecified, randomly selected US sites serving >10,000 persons - 186 samples, 30.6% pos, 3.8 ug/L median (of positives), 50 ug/L maximum [47]. Unspecified, nonrandomly selected US sites serving <10,000 persons - 321 samples, 27.4% pos, 3.7 ug/L median (of positives), 110 ug/L maximum [47]. Unspecified, nonrandomly selected US sites serving >10,000 persons - 158 samples, 38.0% pos, 5.1 ug/L median (of positives), 68 ug/L maximum [47]. Tribromomethane was detected but not quantified in ground water samples taken in New Jersey [22]. RAIN/SNOW: Oregon (SW of Portland, OR) - Rain samples, not detected-0.50 ng/L, 0.3 ng/L mean [37].

Effluent Concentrations: Tribromomethane residues in treated wastewater effluents from several industries were as follows [45]: Nonferrous metals manufacturing - 49 samples, 6.1% pos, not detected-44 ug/L, 2.1 ug/L mean; Pulp and Paperboard Mills - 18 samples, 5.5% pos, not detected-62 ug/L, 10 ug/L mean [45]. Tribromomethane was detected but not quantified in front and tail brine from bromine industries in El Dorado, AR and Magnolia, AR [15], and in secondary effluents from wastewater treatment plant [17].

Sediment/Soil Concentrations: Tribromomethane was detected but not quantified in sediment samples taken in El Dorado, AR and Magnolia, AR [15].

Atmospheric Concentrations: El Dorado, AR - not detected-27.9 ng/m^3, 8.37 ng/m^3 mean; Lakes Charles, LA - 68.2-734 ng/m^3, 516.9 ng/m^3; Magnolia, AR - not detected-85.8 ng/m^3, 15.5 ng/m^3 [9]. Arctic - 2-46 pptV Aug 1983 - Apr 1984 [3]. Above Atlantic Ocean - 2-28 pptV [12].

Food Survey Values:

Plant Concentrations: Tribromomethane was quantified in several species of algae as follows: Ascophyllum nodosum - 28-520 ng/g, 120 ng/g mean; Fucus vesiculous - 24-62 ng/g, 41 ng/g mean; Gigartina stellata - 3-19 ng/g, 9 ng/g mean [23]. All measurements were made on a dry weight basis [23].

Fish/Seafood Concentrations:

537

Tribromomethane

Animal Concentrations:

Milk Concentrations:

Other Environmental Concentrations:

Probable Routes of Human Exposure:

Average Daily Intake:

Occupational Exposure:

Body Burdens: Tribromomethane was found at concentrations from not detected to 3.4 ng/L and with a mean concentration of 0.6 ng/L in whole blood samples [1]. Not detected in 25 human milk samples [30].

REFERENCES

1. Antoine SR et al; Bull Environ Contam Toxicol 36: 364-71 (1986)
2. Atkinson R; Internat J Chem Kinetics 19: 799-828 (1987)
3. Berg WW et al; Geophys Res Lett 11:429-32 (1984)
4. Boublik T et al; The Vapor Pressure of Pure Substances, Amsterdam, Netherlands, Elsevier Sci Publ (1984)
5. Bouwer EJ, McCarthy PL; Ground Water 22: 433-40 (1984)
6. Bouwer EJ, McCarthy PL; Appl Env Microbiol 45: 1295-9 (1983)
7. Bouwer EJ et al; Water Res 18: 463-72 (1984)
8. Brass HJ et al; pp 393-416 in Drinking Water Qual Enhancement Source Prot (1977)
9. Brodzinsky R, Singh HB; Volatilization Organic Chemicals in the Atmosphere: An Assessment of Available Data. SRI International Menlo Park, CA (1982)
10. Burmaster DE; Environ 24: 6-13, 33-6 (1982)
11. Callahan MA et al; Water-Related Environmental Fate of 129 Priority Pollutants Vol. II USEPA-440/4-79-029B (1979)
12. Class T et al; Chemosphere 15:429-36 (1986)
13. Cole RH et al; J Water Pollut Cont Fed 56: 898-908 (1984)
14. Council on Environmental Quality; Contamination of Ground Water by Toxic Organic Chemicals (1981)
15. DeCarlo VJ; Ann NY Acad Sci 320: 678-81 (1979)
16. El-Dib MA and Ali RK; Bull Environ Contamin Toxicol 48: 378-86 (1992)
17. Ellis DD et al; Arch Environ Contam Toxicol 11: 373-82 (1982)
18. Fielding M et al; Organic Micropollutants in Drinking Water. TR-159 (1981)
19. Fogelqvist E: J Geophys Res 90:9181-93 (1985)
20. Francois CL et al; Travaux Soc Pharm Montpellier 39: 49-58 (1979)

Tribromomethane

21. GEMS; Graphical Exposure Modeling System. CLOG3 (1986)
22. Greenberg M et al; Environ Sci Technol 16: 14-9 (1982)
23. Gschwend PM et al; Science 227: 1033-5 (1985)
24. Hawley GG; Condensed Chem Dictionary 10th ed. NY: Van Nostrand Reinhold p 152 (1981)
25. Horvath AL; Halogenated Hydrocarbons: Solubility-Miscibility with Water, NY: Marcel Dekker (1982)
26. Hutzler NJ et al; Water Resources Res 22: 285-95 (1986)
27. Kaiser KLE et al; J Great Lakes Res 9: 212-23 (1983)
28. Kool HJ et al; Crit Rev Env Control 12: 307-57 (1982)
29. Krasner SW et al; J Amer Water Works Assoc 81:41-53 (1989)
30. Kroneld R and Reunanen M; Bull Environ Contam Toxicol 44:917-23 (1990)
31. Krysell M; Marine Chem 33:187-97 (1991)
32. Lyman WJ et al; Handbook of Chem Property Estimation Methods NY: McGraw-Hill p 5-5 (1982)
33. Mabey W, Mill T; J Phys Chem Ref Data 7: 383-415 (1978)
34. Munz C and PV Roberts; J Am Water Works Assoc 79: 62-9 (1987)
35. Ohio River Valley Water Sanit Comm; Assessment of Water Quality Conditions. Ohio River Mainstream 1980-81 (1982)
36. Page GW; Environ Sci Technol 15: 1475-81 (1981)
37. Pankow JF et al; Environ Sci Technol 18: 310-8 (1984)
38. Pellizzari ED et al; Formulation of a Preliminary Assessment of Halogenated Organic Compounds in Man and Environmental Media USEPA/560-13-79-06 (1979)
39. Robbins DE; Geophys Res Lett 3: 213-6 (1976)
40. Saunders RA et al; Water Res 9: 1143-5 (1975)
41. Shiraishi H et al; Environ Sci Technol 19: 585-9 (1985)
42. Suffet IH et al; pp 375-97 in Identification and Analysis of Organic Pollutants in Water. Keith LH ed. Ann Arbour Science Publ (1976)
43. Symons JM et al; J Amer Water Works Assoc 67: 634-47 (1975)
44. Tabak HH et al; J Water Pollut Cont Fed 53: 1503-18 (1981)
45. USEPA; Treatability Manual Vol 1 Treatability Data USEPA-600/2-81-001a (1981)
46. Wakeham SG et al; Can J Fish Aquatic Sci 40: 304-21 (1983)
47. Westrick JJ et al; J Amer Water Works Assoc 76: 52-9 (1984)

1,2,3-Trichloropropane

SUBSTANCE IDENTIFICATION

Synonyms:

Structure:

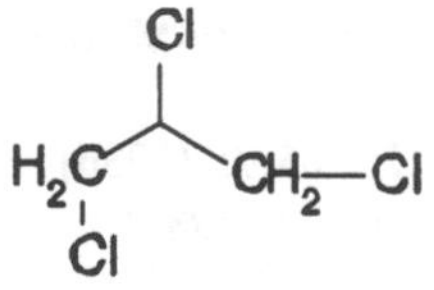

Cl
|
H₂C
|
Cl

CH₂—Cl

CAS Registry Number: 96-18-4

Molecular Formula: $C_3H_5Cl_3$

Wiswesser Line Notation: G1YG1G

CHEMICAL AND PHYSICAL PROPERTIES

Boiling Point: 156.8 °C

Melting Point: -14.7 °C

Molecular Weight: 147.43

Dissociation Constants:

Log Octanol/Water Partition Coefficient: 1.98 [23]

Water Solubility: 1750 mg/L at 25 °C [2]

Vapor Pressure: 3.1 mm Hg at 25 °C [15]

Henry's Law Constant: 3.44×10^{-4} atm-m³/mole at 25 °C, calculated from VP/Wsol

ENVIRONMENTAL FATE/EXPOSURE POTENTIAL

Summary: The release of 1,2,3-trichloropropane to the environment can occur through its manufacture, formulation, and use as an organic solvent. In the atmosphere, reaction with photochemically produced

540

1,2,3-Trichloropropane

hydroxyl radicals should occur with an estimated half-life on the order of fifteen days. The high water solubility of 1,2,3-trichloropropane suggests that rain washout may be an important fate process. If released to soil, 1,2,3-trichloropropane should display high mobility. If released to water, 1,2,3-trichloropropane should not undergo hydrolysis, should not adsorb to sediment and suspended material, and should not bioaccumulate in aquatic organisms. Volatilization from both water and soil to the atmosphere should be an important fate process. The estimated volatilization half-life for a model river is 7 hours . The half-life for volatilization from a model pond can be estimated at 88 hours. Human exposure can occur through the inhalation of vapors during its manufacture, formulation, and use as a solvent.

Natural Sources: 1,2,3-Trichloropropane is an anthropogenic compound, and is not known to exist as a natural product.

Artificial Sources: In 1977, the total USA production of 1,2,3-trichloropropane was approximately 20,000-100,000 pounds. More current production figures could not be found [20]. 1,2,3-Trichloropropane is used in small proportions as a crosslinking agent in the synthesis of polysulfides [8], in the synthesis of hexafluoropropylene [10], as a paint and varnish remover, solvent, and degreasing agent, which may result in environmental release [19]. It is also found as an impurity in nematicides and soil fumigants [1,18].

Terrestrial Fate: If released to soil, the estimated value for Koc, 72 [14], suggests that 1,2,3-trichloropropane would readily leach through the soil [21]. Based on the vapor pressure and the Henry's Law constant, volatilization from either moist or dry soil should be an important fate process.

Aquatic Fate: If released to water, 1,2,3-trichloropropane should not be expected to undergo rapid biodegradation. With a Henry's Law constant of 3.44 x 10^{-4} atm cu-m/mol at 25 °C, volatilization from water should be an important fate process. The estimated half-life for volatilization from a model river 1 m deep, flowing at 1 m/sec, and with a wind velocity of 3 m/sec is about 7 hours [14]. The half-life for volatilization from a model pond can be estimated at 88 hours [22]. 1,2,3-Trichloropropane should not hydrolyze, and adsorption to sediment should not be an important fate process. With an estimated BCF of 9 [14], 1,2,3-trichloropropane should not be expected to bioaccumulate in aquatic organisms.

Atmospheric Fate: If released to the atmosphere, 1,2,3-trichloropropane would be expected to undergo degradation by photochemically produced hydroxyl radicals; the half-life can be estimated at 15.3 days [4]. The water solubility of 1,2,3-trichloropropane, suggests that rain washout may be an important pathway for atmospheric removal. Direct photochemical degradation should not occur.

Biodegradation:

Abiotic Degradation: The experimental hydrolysis rate constants (at 25 °C) for 1,2,3-trichloropropane are 1.8×10^{-6} hr^{-1} under neutral conditions, and 9.9×10^{-4} $M\text{-}hr^{-1}$ under basic conditions [9]. This translates to a half-life of about 44 years in water at pH 7. The half-life for the reaction with photochemically produced hydroxyl radicals in the atmosphere can be estimated at 15.3 days [4].

Bioconcentration: A bioconcentration factor for 1,2,3-trichloropropane can be estimated at 9, suggesting that bioaccumulation in aquatic organisms should not occur [14].

Soil Adsorption/Mobility: A Koc of 72 can be estimated for 1,2,3-trichloropropane, suggesting high mobility in the soil [14,21].

Volatilization from Water/Soil: Experimental half-lives for volatilization of 1,2,3-trichloropropane from fresh water lakes and seawater obtained in the laboratory are 92 and 93 minutes, respectively (theoretical values 106 and 114 minutes, respectively; initial conditions 1 mg/L in a cylinder 106 mm diameter x 148 mm deep) [3]. The vapor pressure for 1,2,3-trichloropropane combined with the Henry's Law constant and low Koc (72) [14], suggest that volatilization from soil should be an important fate process. The half-life for volatilization at 20 °C from a model river 1 m deep, flowing at 1 m/sec, and a wind velocity of 3 m/sec is 6.6 hours [14]. The volatilization half-life from a model pond is 88 hours [22]. The experimental half-life for volatilization from a rapidly stirred solution in the laboratory (initial concentration 1 ppm, 25 °C, average depth - 6.5 cm) was measured to be 56.1 minutes [7].

Water Concentrations: SURFACE WATER: 1979-81, (two summer samples, two winter samples) 1,2,3-trichloropropane was qualitatively found in Narragansett Bay, RI [25]. GROUND WATER: Detected in ground water samples in CA and HI at concentrations ranging from 0.1-5.0 ug/L [1,6]. 1,2,3-Trichloropropane has been found in small-scale and large-scale retrospective studies of Californian and Hawaiian ground water, including wells in Oahu and the Central Valley of California. Typical positives, 0.2 and 2 ppb, respectively [5,6]. In 1983-84, it was found in nine out of nine wells on Oahu, HI, maximum detected concentration was in the range of 300-2800 ng/L [18]. DRINKING WATER: 1,2,3-Trichloropropane has been detected at a concentration of 0.2 ug/L in drinking water taken from the Carrollton Water Plant in New Orleans, LA, 1974 [11]. 1,2,3-Trichloropropane was listed as being qualitatively identified in drinking water [12]. In 1975, it was qualitatively determined in drinking water samples in Cincinnati, OH [13]. Determined in the drinking water of Ames, IA [24].

Effluent Concentrations: 1,2,3-Trichloropropane was qualitatively determined in an advanced wastewater treatment plant in Lake Tahoe, CA, 1974 [13].

Sediment/Soil Concentrations: 1,2,3-Trichloropropane has been qualitatively identified in small-scale and large-scale retrospective studies in Californian and Hawaiian soil samples. [5,6].

Atmospheric Concentrations:

Food Survey Values:

Plant Concentrations:

Fish/Seafood Concentrations:

Animal Concentrations:

1,2,3-Trichloropropane

Milk Concentrations:

Other Environmental Concentrations:

Probable Routes of Human Exposure: The probable route of exposure for 1,2,3-trichloropropane is through the inhalation of vapors during its manufacture, formulation in polymers, and during its use as a solvent and degreasing agent.

Average Daily Intake:

Occupational Exposure: NIOSH (NOHS Survey 1972-1974) has statistically estimated that 490 workers are exposed to 1,2,3,-trichloropropane in the USA [17]. NIOSH (NOES Survey 1981-1983) has statistically estimated that 492 workers are exposed to 1,2,3-trichloropropane in the USA [16].

Body Burdens:

REFERENCES

1. Aharonson N et al; Pure Appl Chem 59: 1419-46 (1987)
2. Albanese V et al; Env Tech Lett 59: 1419-56 (1987)
3. Albanese V et al; Environ Tech Lett 8: 657-88 (1987)
4. Atkinson R; Int J Chem Kinet 19: 799-828 (1987)
5. Cohen SZ et al; pp 256-94 in Schriftenr Ver Wasser, Bodenlufthyg 68, Grundwasserbeeinflussung Plfanzenschutzm (1987)
6. Cohen SZ et al; pp 170-96 in Monitoring Groundwater for Pesticides. ACS Symp Ser, Garner WY et al Ed. Washington DC (1986)
7. Dilling WL; Env Sci Tech 4: 405-9 (1977)
8. Ellerstein SM, Bertuzzi ER; Kirk-Othmer Encycl Chem Tech 3rd Ed. John-Wiley NY 18: 814-31 (1981)
9. Ellington JJ et al; Measurement of Hydrolysis Rate Constants for Evaluation of Hazardous Waste Land Disposal Vol I. USEPA/600/3-86/043 (PB87-140349/GAR) (1987)
10. Gangal SV; Kirk-Othmer Encycl Chem Tech 3rd Ed. John-Wiley NY 11: 24-35 (1981)
11. Keith LH et al; pp 327-73 in Ident Anal Org Pollut Water, Keith LH Ed Ann Arbor Press, Ann Arbor MI (1976)
12. Kool HJ et al; CRC Crit Rev Env Control 12: 307-57 (1982)

1,2,3-Trichloropropane

13. Lucas SV; GC/MS Analysis of Organics in Drinking Water Concentrates and Advanced Waste Treatment Concentrates: Vol 3. USEPA-600/1-84-020 (NTIS PB85-128247) (1984)
14. Lyman WJ et al; Handbook of Chemical Property Estimation Methods NY: McGraw-Hill pp 4-1 to 4-33, 5-1 to 5-30, 15-15 to 15-29 (1982)
15. Mackay D et al; Environ Sci Tech 16: 645-9 (1982)
16. NIOSH; National Occupational Exposure Survey (NOHS) (1984)
17. NIOSH; National Occupational Hazard Survey (NOHS) (1974)
18. Oki DS, Giambelluca TW; Ground Wat 25: 693-702 (1987)
19. Sax NI, Lewis RJSR; Hawley's Condensed Chemical Dictionary 11th ed Van Nostrand Reinhold Co. NY p 1178 (1987)
20. SRC; Health and Environmental Effects Document on Trichloropropanes. Environ Criteria Assess Off. USEPA Cincinnati OH (1987)
21. Swann RL et al; Res Rev 85: 17-28 (1983)
22. USEPA; Exams II Computer Simulation (1987)
23. USEPA; CLOGP-PCGEMS Graphical Exposure Model (1986)
24. USEPA; Health and Environmental Effects Profile for Trichloropropane Isomers (External Review Draft) ECAO-CIN-P010 p 11 (1983)
25. Wakeham SG et al; Can J Fish Aq Sci 40: 304-21 (1983)

Triethylene Glycol

SUBSTANCE IDENTIFICATION

Synonyms: 1,2-Bis(2-hydroxyethoxy)ethane

Structure:

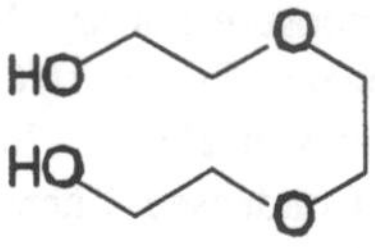

CAS Registry Number: 112-27-6

Molecular Formula: $C_6H_{14}O_4$

Wiswesser Line Notation: Q2O2OQ

CHEMICAL AND PHYSICAL PROPERTIES

Boiling Point: 287.4 °C; 165 °C at 14 mm Hg

Melting Point: -7.2 °C

Molecular Weight: 150.18

Dissociation Constants:

Log Octanol/Water Partition Coefficient: -1.98 estimate [5]

Water Solubility: Miscible [10]

Vapor Pressure: 1.32 x 10^{-3} mm Hg at 25 °C [2]

Henry's Law Constant: 3.16 x 10^{-11} atm-m^3/mole at 25 °C estimate [7]

ENVIRONMENTAL FATE/EXPOSURE POTENTIAL

Summary: Triethylene glycol may be released to the environment via effluents at sites where it is produced or used as a solvent, plasticizer in vinyl, polyester and polyurethane resins, and as a humectant in printing

546

inks, and in the dehydration of natural gas. Triethylene glycol is not expected to undergo hydrolysis or direct photolysis in the environment. The complete miscibility of triethylene glycol in water suggests that volatilization, adsorption and bioconcentration are not important fate processes. This is supported by the estimated Henry's Law constant, which indicates that volatilization of triethylene glycol from natural waters and moist soil should be extremely slow. A low estimated log BCF suggests triethylene glycol should not bioconcentrate among aquatic organisms. A low Koc indicates triethylene glycol should not partition from the water column to organic matter contained in sediments and suspended solids, and it should be highly mobile in soil. River die-away test data demonstrate that biodegradation is likely to be the most important removal mechanism of triethylene glycol from aerobic soil and water. In the atmosphere, triethylene glycol is expected to exist almost entirely in the vapor phase and reactions with photochemically produced hydroxyl radicals should be important (estimated half-life of 11.5 hr). Physical removal of triethylene glycol from air by precipitation and dissolution in clouds may occur; however, its short atmospheric residence time suggests that wet deposition is of limited importance. The most probable human exposure would be occupational exposure, which may occur through dermal contact or inhalation at workplaces where it is produced or used.

Natural Sources:

Artificial Sources: Triethylene glycol may be released to the environment via effluents at sites where it is produced or used as a solvent, plasticizer in vinyl, polyester and polyurethane resins, and as a humectant in printing inks, and in the dehydration of natural gas [6].

Terrestrial Fate: Alcohols and ethers are generally resistant to hydrolysis [8]. They do not absorb UV light in the environmentally significant range (>290 nm) and are commonly used as solvents for obtaining UV spectra [11]. Therefore, triethylene glycol should not undergo hydrolysis in moist terrestrial environments, or direct photolysis on sunlit soil surfaces. The estimated Henry's Law constant indicates that volatilization of triethylene glycol from moist soil should not be an important fate process [8]. An estimated Koc of 2 [8] indicates triethylene glycol should be highly mobile in soil [12]. A series of aerobic river die-away tests, which utilized several differing sources of freshwater [4], suggest that rapid biodegradation is likely to be the most important removal mechanism of triethylene glycol from aerobic soil.

Aquatic Fate: Alcohols and ethers are generally resistant to hydrolysis [8]. They do not absorb UV light in the environmentally significant range (>290 nm) and are commonly used as solvents for obtaining UV spectra [11]. Therefore, triethylene glycol should not undergo hydrolysis or direct photolysis in aquatic environments. The complete miscibility of triethylene glycol in water suggests that volatilization, adsorption and bioconcentration are not important fate processes. This is supported by the estimated Henry's Law constant, which indicates that volatilization of triethylene glycol from natural waters should be extremely slow [8]. An estimated Koc of 2 [8] indicates triethylene glycol should not partition from the water column to organic matter contained in sediments and suspended solids; and an estimated bioconcentration factor (log BCF) of -1.73 [8] indicates triethylene glycol should not bioconcentrate among aquatic organisms. A series of aerobic river die-away tests, which utilized several differing sources of freshwater, suggest that rapid aerobic biodegradation is likely to be the most important removal mechanism of triethylene glycol from aquatic systems [4].

Atmospheric Fate: Alcohols and ethers do not absorb UV light in the environmentally significant range (>290 nm) and are commonly used as solvents for obtaining UV spectra [11]. Therefore, triethylene glycol should not undergo direct photolysis in the atmosphere. Based on the vapor pressure, triethylene glycol is expected to exist almost entirely in the vapor phase in ambient air [3] where vapor phase reactions with photochemically produced hydroxyl radicals may be important. The rate constant for triethylene glycol has been estimated to be 3.38×10^{-11} cm^3/molecule-sec at 25 °C, which corresponds to an atmospheric half-life of about 11.5 hr at an atmospheric concentration of $5 \times 10^{+5}$ hydroxyl radicals per cm^3 [1]. The complete miscibility of triethylene glycol in water indicates that physical removal from air by precipitation and dissolution in clouds may occur; however, its short atmospheric residence time suggests that wet deposition is of limited importance.

Biodegradation: Soil grab sample data pertaining to the biodegradation of triethylene glycol in soil were not located in the available literature.

Triethylene Glycol

However, a series of aerobic river die-away tests, which utilized several differing sources of freshwater, have demonstrated that triethylene glycol should biodegrade rapidly in the environment [4]. At 20 °C, the breakdown of 10 mg/L triethylene glycol was complete within 7-11 days [4].

Abiotic Degradation: Alcohols and ethers are generally resistant to hydrolysis [8]. They do not absorb UV light in the environmentally significant range (>290 nm) and are commonly used as solvents for obtaining UV spectra [11]. Therefore, triethylene glycol should not undergo hydrolysis or direct photolysis in the environment. The rate constant for the vapor-phase reaction of triethylene glycol with photochemically produced hydroxyl radicals in air has been estimated to be 3.38×10^{-11} cm^3/molecule-sec at 25 °C, which corresponds to an atmospheric half-life of about 11.5 hr at an atmospheric concentration of $5 \times 10^{+5}$ hydroxyl radicals per cm^3 [1].

Bioconcentration: Because triethylene glycol is miscible in water, bioconcentration in aquatic systems is not expected to be an important fate process. Based upon the estimated log Kow, a bioconcentration factor (log BCF) of -1.73 for triethylene glycol has been calculated using a recommended regression-derived equation [8]. This BCF value also indicates triethylene glycol should not bioconcentrate in aquatic organisms.

Soil Adsorption/Mobility: Because triethylene glycol is miscible in water, soil adsorption is not expected to be an important fate process. Based on the estimated log Kow, a Koc of 2 for triethylene glycol has been calculated using a recommended regression-derived equation [8]. This Koc value indicates triethylene glycol will be highly mobile in soil [12], and it should not partition from the water column to organic matter contained in sediments and suspended solids.

Volatilization from Water/Soil: Because triethylene glycol is miscible in water, and based upon the estimated Henry's Law constant, the volatilization of triethylene glycol from natural bodies of water and moist soils is not expected to be an important fate process [8].

Water Concentrations:

Effluent Concentrations:

Sediment/Soil Concentrations:

Atmospheric Concentrations:

Food Survey Values:

Plant Concentrations:

Fish/Seafood Concentrations:

Animal Concentrations:

Milk Concentrations:

Other Environmental Concentrations:

Probable Routes of Human Exposure: The most probable route of human exposure to triethylene glycol is by inhalation and dermal contact.

Average Daily Intake:

Occupational Exposure: The most probable human exposure to triethylene glycol would be occupational exposure, which may occur at places where it is produced or used as a solvent, plasticizer and humectant. NIOSH (NOES Survey as of 3/28/89) has estimated that 129,144 workers are potentially exposed to triethylene glycol in the USA [9].

Body Burdens:

REFERENCES

1. Atkinson R; Intern J Chem Kin 19: 799-828 (1987)
2. Daubert TE, Danner RP; Data Compilation, Tables of Properties of Pure Cmpds, Design Inst for Phys Prop Data, Am Inst for Phys Prop Data, NY, NY (1989)
3. Eisenreich SJ et al; Environ Sci Technol 15: 30-8 (1981)
4. Evans WH, David EJ; Water Res 8: 97-100 (1974)

5. Hansch C, Leo AJ; Medchem Project Issue No 26. Claremont CA: Pomona College (1985)
6. Hawley GG; Condensed Chemical Dictionary 10th ed Van Nostrand Reinhold NY p 375 (1981)
7. Hine J, Mookerjee PK; J Org Chem 40: 292-8 (1975)
8. Lyman WJ et al; Handbook of Chemical Property Estimation Methods NY: McGraw-Hill pp 4-9, 5-4, 6-3, 15-16 (1982)
9. NIOSH; National Occupational Exposure Survey (NOES) (1989)
10. Riddick JA et al; Organic Solvents NY:John Wiley & Sons Inc (1984)
11. Silverstein RM, Bassler GC; Spectrometric Id Org Cmpd NY: J Wiley & Sons Inc pp 148-69 (1963)
12. Swann RL et al; Res Rev 85: 16-28 (1983)

Trifluoromethane

SUBSTANCE IDENTIFICATION

Synonyms: Freon 23

Structure:

$$\text{H} - \text{C}(\text{F})(\text{F})(\text{F})$$

CAS Registry Number: 75-46-7

Molecular Formula: CHF_3

Wiswesser Line Notation:

CHEMICAL AND PHYSICAL PROPERTIES

Boiling Point: -84.4 °C [17]

Melting Point: -160 °C [17]

Molecular Weight: 70.02

Dissociation Constants:

Log Octanol/Water Partition Coefficient: 0.64 [7]

Water Solubility: 733 mg/L at 25 °C [9]

Vapor Pressure: 30,000 mm Hg at 20 °C [3]

Henry's Law Constant: 0.095 atm-m³/mole at 20-25 °C (estimated using the vapor pressure and water solubility)

Trifluoromethane

ENVIRONMENTAL FATE/EXPOSURE POTENTIAL

Summary: Trifluoromethane may be released to the environment as emissions resulting from waste incineration or during its use as a refrigerant. It may be released to the soil from the disposal of refrigeration units containing this compound. If released to the soil, trifluoromethane should rapidly volatilize from soil surfaces or leach through soil possibly into ground water. If released to water, volatilization would be the dominant fate process based on a half-life of approximately 2.45 hours from a model river. If released to the atmosphere, all trifluoromethane is expected to exist in the vapor phase. In the troposphere, trifluoromethane reacts slowly with photochemically generated hydroxyl radicals (half-life of 9036 days). This relatively slow half-life in the lower atmosphere suggests that some trifluoromethane may gradually diffuse into the stratosphere (half-life of 20 years). In the stratosphere, trifluoromethane is expected to slowly photolyze and contribute to the catalytic removal of stratospheric ozone. However, no data are available indicating trifluoromethane contributes to stratospheric ozone depletion. From its source of emissions, long distance atmospheric transport is expected to take place due to the stability of the chemical. The most probable route of human exposure to trifluoromethane by the general population is inhalation.

Natural Sources:

Artificial Sources: The rate constant for the vapor-phase reaction of trifluoromethane with photochemically produced hydroxyl radicals has been experimentally measured to be 1.8×10^{-15} cm^3/molecule-sec at 25 °C [2] which corresponds to an atmospheric half-life of about 9036 days at an atmospheric concn of $5 \times 10^{+5}$ hydroxyl radicals per cm^3 [2]. Haloalkanes are essentially inert to reaction with ozone molecules [1]; therefore, trifluoromethane is not expected to react with ozone molecules.

Bioconcentration: Based on the measured log Kow, the BCF for trifluoromethane can be estimated to be approximately 1.8 using a recommended regression derived equation [11]. This BCF value suggests that trifluoromethane would not bioconcentrate in aquatic systems.

Trifluoromethane

Soil Adsorption/Mobility: A Koc value of approximately 116 can be estimated for trifluoromethane based on the measured water solubility and a regression-derived equation [11]; this Koc value suggests high soil mobility [15].

Volatilization from Water/Soil: Based on the measured water solubility and the vapor pressure of greater than 760 mm Hg at 20 °C [3], a Henry's Law constant can be estimated to be approximately 0.095 atm cu-m/mole at 25 °C. This Henry's Law constant indicates that trifluoromethane will rapidly volatilize from water [11]. Based on this value, the volatilization half-life of trifluoromethane from a model river 1 m deep flowing 1 m/sec with a wind velocity of 3 m/sec has been estimated to be approximately 2.45 hr [11]. The volatilization half-life of trifluoromethane from a model environmental pond can be estimated to be approximately 29 hr [16]. The estimated vapor pressure suggests that trifluoromethane should volatilize from dry surfaces.

Water Concentrations: GROUND WATER: Trifluoromethane has been identified, but not quantified, in ground water from a 408-well sample in New Jersey [6]. Trifluoromethane has been detected in 29 of 949 ground water samples in New Jersey with a highest concn of 3.5 ppb [14]. SURFACE WATER: Trifluoromethane has been detected in 33 of 431 surface water samples in New Jersey with a highest concn of 2178.2 ppb [14].

Effluent Concentrations: Trifluoromethane has been detected in stack emissions effluent as a result of waste incineration [10].

Sediment/Soil Concentrations:

Atmospheric Concentrations:

Food Survey Values:

Plant Concentrations:

Fish/Seafood Concentrations:

Animal Concentrations:

Trifluoromethane

Milk Concentrations:

Other Environmental Concentrations:

Probable Routes of Human Exposure: The most probable route of human exposure to trifluoromethane by the general population is inhalation. In occupational settings, it is expected that exposure occurs by inhalation of contaminated air and dermal contact with this compound.

Average Daily Intake:

Occupational Exposure: NIOSH (NOHS Survey 1972-1974) has statistically estimated that 17,721 workers are potentially exposed to trifluoromethane in the USA [12]. NIOSH (NOES Survey 1981-1983) has statistically estimated that 528 workers are potentially exposed to trifluoromethane in the USA [13].

Body Burdens:

REFERENCES

1. Atkinson R, Carter WPL; Chem Rev 84: 437-70 (1984)
2. Atkinson R; Chem Rev 85: 69-201 (1985)
3. Boublik T et al; The Vapor Pressures of Pure Substances; Elsevier Sci Publ; Amsterdam, Netherlands Vol. 17 (1984)
4. Dilling WL; pp 154-97 in Environmental Risk Analysis for Chemicals Conway RA ed NY: Van Nostrand Reinhold Co (1982)
5. Eisenreich SJ et al; Environ Sci Technol 15: 30-8 (1981)
6. Greenberg M et al; Environ Sci Technol 16: 14-9 (1982)
7. Hansch C, Leo AJ; Medchem Project Issue No 26 Claremont, CA: Pomona College (1985)
8. Hawley GG; Condensed Chemical Dictionary 10th ed NY: Van Nostrand Reinhold p 471 (1981)
9. Hine J, Mookerjee PK; J Org Chem 40: 292-8 (1975)
10. Junk GA, Ford CS; Chemosphere 9: 187-230 (1980)
11. Lyman WJ et al; Handbook of Chemical Property Estimation Methods. Environmental Behavior of Organic Compounds. Washington DC: American Chemical Society pp 4-9, 5-4, 5-10, 7-4, 7-5, 15-15 to 15-32 (1990)
12. NIOSH; National Occupational Hazard Survey (NOHS) (1974)
13. NIOSH; National Occupational Exposure Survey (NOES) (1983)
14. Page GW; Environ Sci Technol 15: 1475-81 (1981)
15. Swann RL et al; Res Rev 85: 17-28 (1983)

Trifluoromethane

16. USEPA; EXAMS II Computer Simulation (1987)
17. Windholz M et al; The Merck Index 10th ed Rahway, NJ: Merck & Co Inc pp 597-598 (1983)

Cumulative Index of Synonyms

Cumulative Index by CAS Registry Number

Cumulative Index by Chemical Formula

Milton Keynes UK
Ingram Content Group UK Ltd.
UKHW021931071024
449327UK00022B/1766